Library of
Davidson College

THE COMMONWEALTH of SCIENCE

Archibald Liversidge, FRS (1846–1927)

THE COMMONWEALTH of SCIENCE

ANZAAS and the Scientific Enterprise in Australasia

1888–1988

Edited by Roy MacLeod

OXFORD
UNIVERSITY PRESS

Melbourne
Oxford Auckland New York

OXFORD UNIVERSITY PRESS AUSTRALIA

Oxford   New York   Toronto
Delhi   Bombay   Calcutta   Madras   Karachi
Petaling Jaya   Singapore   Hong Kong   Tokyo
Nairobi   Dar es Salaam   Cape Town
Melbourne   Auckland

and associated companies in
Berlin   Ibadan

Oxford is a trade mark of Oxford University Press

© This collection Roy MacLeod 1988
Contributors retain © in respect of their own contributions.
First published 1988

This book is copyright. Apart from any fair
dealing for the purposes of private study,
research, criticism or review as permitted under
the Copyright Act, no part may be reproduced,
stored in a retrieval system, or transmitted, in
any form or by any means, electronic, mechanical,
photocopying, recording, or otherwise without
prior written permission. Inquiries to be made to
Oxford University Press.

National Library of Australia
Cataloguing-in-Publication data:

The Commonwealth of science: ANZAAS and the scientific
enterprise in Australasia, 1888–1988.
  Bibliography.
  Includes index.
  ISBN 0 19 554683 0.

  1. ANZAAS—History. 2. Science—Australia—History.
  3. Science—New Zealand—History. I. MacLeod, Roy M.
509'.9

Typeset by Graphicraft Typesetters Limited
Printed by Impact Printing, Melbourne
Published by Oxford University Press,
253 Normanby Road, South Melbourne, Australia

**Caelum non animum mutant,
qui trans mare currunt.**

Horace, *Epistles*, Book I.xi.27,
quoted by
Rolf Boldrewood (T. A. Browne),
'Heralds of Australian Literature',
Section I, AAAS, Hobart, 1892

CONTENTS

FOREWORD
*His Excellency, The Governor General,
Patron-in-Chief, ANZAAS*   ix

PREFACE   xi
LIST OF ILLUSTRATIONS   xv
INTRODUCTION   1

PART I LAUNCHING THE ENTERPRISE   17

1. Organizing Science under the Southern Cross   19
   Roy MacLeod

2. From Imperial to National Science   40
   Roy MacLeod

3. The Impulse of Science in Public Affairs, 1945–1986   73
   James Davenport

PART II CHARTING THE SCIENCES   97

4. The Life Sciences: Collections to Conservation   99
   Linden Gillbank

5. The Earth Sciences: Searching for Geological Order   130
   David Branagan and Thomas Vallance

6. The Physical Sciences: String, Sealing Wax and Self-Sufficiency   147
   R. W. Home

7. Chemists at ANZAAS: Cabbages or Kings?   166
   Ian D. Rae

8. Australasian Anthropology and ANZAAS: 'Strictly Scientific and Critical'   196
   D. J. Mulvaney

9. Education, Social Science and the 'Common Weal'   222
   Alison Turtle

PART III SERVING SOCIETY 247

10. Protracted Reconciliation: Society and the Environment 249
    J. M. Powell

11. Developing Nature's Treasures: Agriculture and Mining in Australasia 272
    Bruce Davidson

12. Professional Hygienists and the Health of the Nation 292
    John Powles

13. Social Responsibility of Science: The Social Mirror of Science 308
    Ron Johnston

14. The Political Economy of Technology 326
    Ted Wheelwright and Greg Crough

15. Technology, Employment and Post-Industrial Society 343
    Sol Encel

APPENDICES 359

1. Officers of the Australasian Association for the Advancement of Science (afterwards ANZAAS) 361
2. Sections of the AAAS (ANZAAS) 365
3. Rules and Constitutional Changes 369
4. Local Secretaries and Divisional Representatives of the AAAS (ANZAAS) 373
5. Attendance at AAAS (ANZAAS) Congresses 377
6. AAAS/ANZAAS Medallists and Lecturers 379

A GUIDE TO SOURCES 383
NOTES ON CONTRIBUTORS 404
INDEX 408

In 1988 ANZAAS celebrates one hundred remarkable years of existence, years dedicated to ensuring that the great achievements in the fields of the sciences and social sciences are communicated to us all. That work of comunication will not only help all age groups in the community better to understand the world we live in; it should also inspire young people to embark on careers in one of these fields.

ANZAAS was founded at a most exciting time in the history of science. Between 1888 and 1898 radioactivity, the electron and X-rays were discovered, followed in the next decade by the theory of relativity, genetics and the beginning of quantum theory. But the history of science, and its applications, has not been only the story of discoveries which have improved the quality of life; there has always been an uneasy tension between the promises which the sciences hold out of a brighter future for mankind and the possibilities of the abuse of science in times of war.

Thanks to the activities of ANZAAS, our communities have been given the opportunity to consider the past century's dramatic and far reaching advances in science and technology in the context of their implications for the societies we inhabit. Increasingly over the years, ANZAAS has expanded its interests to encompass all the social sciences. It encourages cross-disciplinary debate on key topics of interest to us all, as well as addressing important ethical issues which may arise in the course of these debates.

In 1985 ANZAAS reaffirmed its commitment to bring the sciences to the widest audience possible, and to encourage discussion and commentary on matters of importance within our region, the Pacific basin. "The Commonwealth of Science" embraces the ethos and philosophy of ANZAAS. It thus provides a fitting testimony not only to its past century of scholarly research but to its future as a dynamic association committed to the advancement of both science and society.

PREFACE

In May 1988 the fifty-eighth Congress of the Australian and New Zealand Association for the Advancement of Science will open in the Great Hall of the University of Sydney. A century ago, in the same venerable surroundings, a huge audience heard the governor, Lord Carrington, open the proceedings and the government astronomer of New South Wales, H. C. Russell, deliver the Association's inaugural address. A century in the history of science, in a part of the world whose European settlement, so deeply implicated in scientific exploration, has encompassed just twice that span, is something to celebrate. This volume forms part of that celebration. It commemorates those who, a century ago and since, have contributed to the advancement of knowledge and its application to the material well-being of our people. For such a well-known fixture of Australian cultural life, the history of ANZAAS (as the Association became) is surprisingly little studied. This book is not, however, an official history of the Association itself. It does not pretend to be an exhaustive history of science in Australasia, much less a complete account of the scientific and other disciplines that ANZAAS has represented or encouraged. Instead, at a time when bicentennials threaten to swamp mere centennials, and when triumphal sets of volumes overshadow mere monographs, this book merely explores some of the directions and limitations of our past efforts to organize and present science, and outlines the disadvantages and contradictions under which the sciences still labour in Australia. Among scientists and social scientists, an ANZAAS congress typically signals an occasion for reporting and reassessment. In this sense, our book constitutes a report to the Sydney Congress of 1988.

The Association's story, told through its reports, and through the wide range of activities it has sponsored, unfolds at many levels. At one, it is the history of a small community of men and women on the fringes of Empire, who appropriated a British culture of science and directed it to colonial and national purposes. At another, it is an account of tensions between the constituencies of our Commonwealth—between state and federal representatives, between the 'practical man' and the academic; the pastoralist and the 'city theorist'; the partisans of research and the advocates of application. At a third level, it is a story of organizational dynamics, the work of factions and committees, and the influence of recommendations on governments. At a deeper level still, it is a history of certain attitudes and mentalities which, reflected by the press and in public discussion, have in different ways influenced the role accorded to the sciences in Australia today. Institutions look to the future by reconstructing their past. Today, our history speaks to the Association's continuing present, and to its task of making science more widely understood, more generously and wisely supported, more directly beneficial to the nation, and more accountable to the public interest.

It is commonly observed that major public institutions are rarely eulogized between their jubilees, and rarely criticized during them. ANZAAS enjoyed a jubilee history at its seventy-fifth anniversary in 1962, with a volume edited for the thirty-sixth Sydney Congress by A. P. Elkin, professor of anthropology at the University of Sydney, trustee of the Australian Museum, and honorary general secretary of ANZAAS between 1926 and 1947. To Elkin, the Association reflected the early desire for federation and co-operation, and the later necessity of interdisciplinary 'integration'. The message of *A Goodly Heritage* was straightforward and directed squarely at the community of science. Its Introduction—'Then and Now'—spoke glowingly of progress, of 'Amazing Scientific Achievement', with 'Research Furrows—Deepened and Narrowed'. A. B. Walkom's suitably entitled 'Short History of the Association' was perhaps adequate recognition of 'then', in an Association preoccupied with 'now'. Over the last quarter century, however, historians have had frequent cause to look to the Association's past, both for information and interpretation. Today there remains much in its ossean archives that demands our attention, and it is to these that our enterprise has turned.

Consistent with the Association's origins comes the inspiration for the present volume, which follows the path charted by a sesquicentenary tribute to the British Association, *The Parliament of Science*, seven years ago. As in that context the discourse of parliamentary self-government supplied a suitable metaphor, so in ours, the discourse of co-operation and the ideal of 'commonwealth' seem apt. In his *Commonwealth of Oceana*, written three centuries ago to counsel and warn his contemporaries, James Harrington drew a prophetic picture of an island nation, founded on principles of natural knowledge, governed by reason and learning, serving the health and liberties of a prudent people. When the Australasian Association for the Advancement of Science began, the concept of federation was still in its adolescence; 'Australasia' might, as Charles Dilke observed, include Australia and Tasmania, but not necessarily New Zealand; Western Australia was connected to the eastern colonies only by sealanes and telegraph wires; and Christchurch was easier to reach from Sydney than was Adelaide. In the 1880s the rhetoric of scientific co-operation offered both motive and means to achieve federation, and, with the dawn of the new century, it offered also a vision of a future in which equity, efficiency, reason, and stability—all the virtues of science—would underwrite the conventions of the political 'commonwealth'. If, in the language of seventeenth century England, liberty consisted in the 'Empire of Reason', so prosperity would derive from the 'Commonwealth of Science'. Given the close identification today of science with the Commonwealth and, until very recently, with the British Commonwealth as well, our title conveys an idea of association that is both symbolic and real.

There is another way, however, in which the language of 'commonwealth' is appropriate. This volume is a pluralist tribute, drawing upon the collective work of scholars from several institutions, disciplines and perspectives. As such, it makes a virtue of necessity: where to find a unifying logic in a field so fragmented? Despite the pioneering efforts of a few, the history of science in Australia and New Zealand is still in its infancy, and a general history is premature. What we have is at best a partial record, emphasizing certain individuals and organizations. Ideally, this book will be read in conjunction with accounts of the CSIRO, the munitions factories, and the developing fields of fundamental research that are now appearing in the scholarly literature. We have rather simply outlined trends and suggested patterns; we have summarized what is known, and voiced concern about what is not. Through the 'passages' of the Association, we have sought bearings of wider significance for Australian science and society.

Two further qualifications of geography and of boundaries must be made. First, in its application to New Zealand, our language of Commonwealth is deliberately ambitious. For reasons both of history and sentiment, New Zealand holds an important place in our account. During the colonial period, New Zealand scientists were pivotal to the progress of intercolonial co-operation, and it is both significant and appropriate that a successful congress was held as recently as 1986 in Palmerston North. New Zealanders, however, will doubtless find in this book less than they wish (and far less than they know) about New Zealand science, either in connection with ANZAAS or in its own right. For this we can but plead guilty; when our attempts to find a New Zealand contributor failed, we endeavoured to 'make do'. This will not satisfy our colleagues across the Tasman, nor does it, us. May we hope that others will take up the brush where our colours ran out. Second, the approach favoured by our authors—as, perhaps, by the Association over much of its history—is one that concentrates on science rather than technology. In an age where the distinction between science and technology has become ambiguous, this bias must be considered in historical context, where (as our concluding chapters suggest) its implications for Australia become clear.

This project could not have been undertaken without the assistance of the Australian Research Grants Scheme, to which members of the larger 'Commonwealth of Science' are always greatly indebted. We thank also the Reserve Bank of Australia, for a timely grant-in-aid. At the outset, the editor had the generous assistance of an informal steering committee, comprising Mr James Davenport (then chairman of council of ANZAAS), Ms Ann Moyal (then editor of *Search*), and Associate Professor Diana Temple (then secretary of ANZAAS). I remain grateful to all three for their constancy and guidance, and to Mr Davenport, in particular, for his close knowledge of the changing ways of scientific organizations in Australia over three decades. To him, as founder and first editor of *Search*, the community owes a special debt.

It is a pleasure finally to record thanks to those who have contributed to this endeavour and who have borne with fortitude and courtesy the intrusive pen of a Sydney editor. We thank those scholars who initially responded to the call—including Dr Louise Crossley, Dr Michael Hoare, Mr Humphrey McQueen, Ms Ann Moyal, Professor Leonie Sandercock, and Professor George Seddon—but for whom illness or other commitments eventually took precedence. We gratefully acknowledge those many librarians and archivists in Britain and in all the capital cities of Australia and New Zealand who responded to our survey of AAAS/ANZAAS records, local accounts, photographs, and memorabilia. We are particularly indebted to Dr David Morley, of the British Association for the Advancement of Science, London, and Mr Peter Lever-Naylor, former executive officer of ANZAAS; to Dr Rupert Best of Adelaide, Professor Boris Schedvin of Melbourne, and Associate Professor David Branagan, who kindly read part of the text; and to Mr Ken Smith, archivist of the University of Sydney, and the staff of the Fisher Library, the Mitchell Library and the State Library of New South Wales. Special thanks for help at a critical stage must go to Ms Joan Radford of Melbourne; and to Ms Kimberley Webber of Sydney, for help throughout the last two years of research and writing. We would also thank Ms Jean Buckley-Moran and Dr L. W. Weickhardt for their great efforts on our behalf; and those who contributed to the history of ANZAAS workshop at the Melbourne Congress in 1985. In collecting materials for individual authors, great service was rendered by Ms Shirley Saunders, Mrs Marilyn Orr, Mr Kurt Oppliger, Mr Andrew Spearritt and Ms Jan Wilson. Our authors may well extend their appreciation to others; on their behalf, I would give special bouquets to Ms Christa Ludlow, Ms Melanie Oppenheimer, and Ms Ruth Bennett. Dr Christine Eslick prepared the index. We remain in perpetuity grateful for the foresight and patience of Ms Louise Sweetland and Ms Ev Beissbarth of Oxford University Press.

Roy MacLeod
Sydney,
Australia Day, 1987

LIST OF ILLUSTRATIONS

Frontispiece: Archibald Liversidge, FRS (1846–1927)

CHAPTER 1

Plate 1   'The Pathfinder' (*Daily Telegraph*, 12 January 1911).
Plate 2   'A Chapter On the British Association' (*Bulletin*, 2 October 1886).
Plate 3   Letter from Archibald Liversidge to the president and council of the Linnean Society of New South Wales, announcing the first meeting of the AAAS, 12 March 1888 (Mitchell Library).
Plate 4   'The Founding Fathers'.
Plate 5   'The Australasian Society of Mutual Bores' (*Bulletin*, 8 September 1888).
Plate 6   Officers of the AAAS at the Melbourne meeting, 1890 (*Liversidge Papers*, University of Sydney Archives).
Plate 7   The president's address in the Great Hall, University of Sydney (*Sydney Mail*, 15 January 1898).
Plate 8   Presentation portrait of members of the AAAS Council, 1898 (Mitchell Library).
Plate 9   AAAS members at Adelaide in 1907 (*Liversidge Papers*, University of Sydney Archives).
Plate 10   Collage of sectional representatives at the AAAS Brisbane, 1909. (*The Queenslander*, 23 January 1909; *Liversidge Papers*, University of Sydney Archives).
Plate 11   The AAAS Meeting, Brisbane, 1909, cover of train brochure (*Liversidge Papers*, University of Sydney Archives).
Plate 12   'Members of the General Committee of the Congress, Adelaide, 1911' (*Daily Telegraph*, 10 January 1911; *Liversidge Papers*, University of Sydney Archives).
Plate 13   1923 Congress, Victoria University College, Wellington (Courtesy of Peter Lever-Naylor).
Plate 14   *Left to right*: Mr David Carment, Sir Hubert Murray and Sir Edgeworth David (*Sun*, 17 August 1932).
Plate 15   'Will science lend a hand?' (Stan Cross, in *Smith's Weekly*, 15 September 1923).

CHAPTER 2

Plate 16   Mueller Memorial Medal, first awarded in 1904.
Plate 17   ANZAAS Medal. Designed by Andor Meszaros, it depicts the progress of knowledge from the pyramids of Egypt

to the radiotelescope: 'only the moon and horizon have remained constant' (*Australian Journal of Science* 28, (September 1965), 98.

Plate 18 'Looking for a Leak at the C.S.I.R.' (*Bulletin*, 13 October 1948).

Plate 19 'The Double Helix' (*Search* 2 (1971), 214).

CHAPTER 5

Plate 20 Harnessing the five-in-hand waggonette during the Glacial Committee expedition by Edgeworth David and Walter Howchin to Crown Point, Central Australia, July 1921 (*Edgeworth David Papers*, University of Sydney Archives).

CHAPTER 9

Plate 21 'A Melbourne University Professor', Professor Henry Laurie (*Melbourne Punch*, 24 July 1890).

Plate 22 'Professor Francis Anderson, M.A.' (*Arts Journal*, 4 (3), University of Sydney 1921).

Plate 23 Department of Public Instruction (NSW). *Report upon the Physical Condition of Children Attending Public Schools in NSW* (Sydney: William Applegate Gullick, (G.P.), 1908).

CHAPTER 10

Plate 24 The 'Gigantic Inheritance' of Australia (redrawn from E. J. Stuart, *A Land of Opportunities*, London, 1923).

Plate 25 Griffith Taylor as interfering marriage broker (*Daily Telegraph*, 25 June 1923).

Plate 26 Taylor's preferred sequence of settlement (adapted from his map in the *Sydney Morning Herald*, 28 February 1925).

CHAPTER 15

Plate 27 'The Scientist Seen as Villain' (*Search*, 1 October 1970).

INTRODUCTION

I

'The Australia that we know was born to the scientific age.' Thus Ernest Scott introduced the history of Australian science to the jubilee congress of ANZAAS, meeting at Canberra in 1939. For the University of Melbourne's professor of history, the 'future of Australia was bound up with the progress of science to a greater degree than with any other of the instrumentalities of civilisations at work within our borders'.[1] Today, the life of science is everywhere obvious. No one surrounded by the ubiquity of Cook and Banks and the eponymy of Botany Bay can escape the significance of science in our colonial heritage; no listener to the ABC's Science Show can fail to appreciate the excitement of recent research; no viewer of Quantum can escape the sense that scientific discovery and innovation, clearly part of our past, also have a vital role in our future. There is around us such a rich intensity of science-based allusion, extending even to our paper currency, with its mindful symbols of knowledge and invention, that Australia viewed by visitors could be easily construed as nothing less than a metropolitan 'scientific culture'.

It is for this reason all the more ironic that the history of science in Australia should be so little cultivated; that few history departments in our nineteen universities consider the history of Australian science in their regular course offerings; and that even fewer of our university science courses discuss the history of those sciences in which Australians have won distinction. General histories may mention Macfarlane Burnet or W. H. Bragg, but we usually look in vain for Robert Ellery or H. C. Russell, David Orme Masson or Alfred Mica Smith. This neglect has not, of course, gone unnoticed. The Australian Bicentennial Authority has a programme to encourage the introduction of 'Australian' science in tertiary courses, and much may come of this. It seems clear that, as F. B. Smith has observed, 'Australian studies which omit our scientific heritage will only reproduce the narrow and conventional.'[2] But the fact remains and is significant in a wider context. This neglect of the history of science parallels a neglect of the sciences themselves as 'objects' of popular and cultural interest. Despite their importance in our everyday lives, scientists and science have been posted to the periphery of our history.

This condition is not unique to Australia (or, for that matter, New Zealand).[3] In part, it merely reflects the presentation of those images Australian historians have wished to convey. Twenty years ago R. W. Connell attacked the 'recurring clichés' of egalitarianism, nationalism and the rights of labour, which for generations have animated and characterized Australian social history.[4] Still they persist, and what people are taught to respect gradually acquires the distinction of a virtue. In their great contests with Nature, Australians are typically seen to value the personal and physical qualities of explorers above the intellectual achievements of research scientists (and to set leaders of manufacturing industry lower still).[5] Nor do the images of Australian heroes—whether of diggers, explorers, or tinkerers and adaptors—sit easily with the elitist, internationalist and managerial image expected of modern scientists.[6] There is a sense in which even practical science and technology are viewed as eccentric, the province of stereotyped boffins in white laboratory coats or solitary inventors in tin sheds or on lonely beaches.[7]

R. M. Crawford, in referring to 'the thin life of the spirit' in Australia, called up an image of Australian 'types of mind' struggling to maturity.[8] Without acceding to such abstractions, or even admitting their necessity, it is clear that few historical stereotypes—from convicts to radical nationalists—have rested comfortably with the historical reality of Australasian science. Urban histories pay little attention to the part 'gentlemen naturalists' or university professors played in reflecting the urban culture of what Manning Clark has called the 'colonial bourgeoisie'. More broadly and deeply, natural history and geology form no part of the Australian legend; there is little rhetoric, in the scientific literature, of mateship and the bush. The literary graces of Jules François Archibald, 'Banjo' Paterson, and John Lang, which quickened the enthusiasm of Vance Palmer, Arthur Phillips, and Russel Ward,[9] turned attention well away from the Muses of colonial chemistry and astronomy, who languished in oblivion. And when, in the 1960s and 1970s, Peter Coleman, Humphrey McQueen and David Walker reassessed the myths of Australian culture, they generally failed to reach the scientific frontier.[10] Despite the successes of Geoffrey Blainey and Michael Cannon in capturing the essence of colonial enterprise,[11] and the gift of *Voss* to its spiritual defence,[12] we still lack an overall sense of science in the context of Australian society.

In a general sense, this perhaps reflects a historical diffidence towards the role and influence of intellectuals in Australian society. Forty years ago, Louis Esson told us that 'Australia is an empty country. We produce wool and cricketers and factory butter and legislative counsellors, but we do not produce ideas.'[13] He was, of course, wrong. But the image persists. Our social history still under-estimates the contribution of learned societies, churches, universities, and liberal movements in general.[14] Their historical interpretation is handicapped by the needless neglect of urban Australian middle-class culture[15] and delayed by the omission of scientists and educators from the discussion of leading Australian writers and philosophers.[16] Where it may be relevant to science, even cultural criticism has too frequently taken ahistorical directions, apparently preferring to emphasize those 'universals' that unite Australian creative genius with the rest of human experience, rather than examining the particular parameters of Australian life.[17]

This is not to say that the history of Australian science has been altogether forgotten: fortunately, its practitioners have seen to that. Several retrospective accounts of colonial science predate Federation, and memorial tributes have regularly appeared, like coral islands of recoverable time, ever since. These, however, have typically reflected the filial pieties of professional colleagues. It was not until the mid-1960s that these island outcrops were linked into a chain of interpretative studies.[18] Then a small number of highly productive, pioneer scholars surveyed the historical geography of scientific organization in Australasia and presented for the first time a coherent picture of colonial scientific enterprise. In the 1960s Bernard Smith, Geoffrey Blainey, and Kathleen Fitzpatrick gave science a new cultural embassy in Australian history.[19] In the 1970s Michael Hoare, Ann Moyal, and Geoffrey Serle,[20] for over two decades in virtually a peerage of their own, produced literary maps and gave others compass directions. Since the mid-1970s, a growing number of other scholars—notably Rod Home, Boris Schedvin, and Lindsay Farrall—have begun to reveal the organization and culture of Australian science in ways that illuminate its professional, political and economic dimensions. A mature tradition of scholarly criticism is now replacing the deuteronomic traditions of colonial history. Such journals as the *Historical Records of Australian Science* (succeeding the *Records of the Australian Academy of Science*, established in 1966), and *Prometheus* (established in 1983) are offering new avenues to publication. Synoptic obituaries in specialist journals are giving way to sustained biographical reassessments. Older models of historical discourse, with progressivist, 'whiggish' overtones, are being replaced by more sensitive accounts of colonial science and its functional and methodological significance to the interests of settler capitalism.[21] Social historians, from George Nadel to Michael Roe and Graeme Davison,[22] are placing colonial science in the context of colonial cultural expectations, while at least one cultural historian[23] has seen in the affairs of science the same factions and patterns of patronage that have dominated the business of colonial culture and learning generally, both before Federation and since.

II

It is not inconsistent with this appreciation to suggest that the history of science in Australia, if it now has its many Parkmans, has not yet found its Carlyle or Macaulay. Today the exploration of Australian science forms part of our rediscovery of Australia's cultural and technological heritage. That rediscovery is taking its place alongside developments in historical demography, women's studies, the history of material and popular culture, and the study of the professions. We await works of synthesis, in which science and technology are given their proper place in Australasian social and economic history, and in their imperial, national, and international dimensions. We await, in short, a new Hancock, who will convey and interpret for us the scientific enterprise, from colony to Commonwealth in the making.

It was once profitable to view Australian science as in many ways reflecting the characteristics of provincial British science, with 'marginal' people at the periphery paying fealty to the 'czars' of metropolitan culture—whether in London's South Kensington and Burlington House or in the centres of excellence of the United States and Europe today.[24] Now, however, we are learning much about the social purposes of science in 'colonial' and national culture. The landscape of colonial science reveals the strong leadership of a few men, frequently dependent on the goodwill and patronage of government; a commitment to empirical utility, as against abstract science; and a close relationship between academic and government science. Then and since, science served Australia as a guide to 'moral improvement' and social organization; as a social elevator for ambitious young men and women; and as a social adhesive for the interests of artisans and managers. Nowhere do we see this more fully illustrated than in the colonial newspaper press; nowhere more spectacularly than in the massive intercolonial exhibitions, which celebrated in cathedrals of iron and glass the ideal of unity between master and man, capitalist and worker, as sanctified by science.[25] These exhibition buildings, where they survive, remind us that, long before science, as agent, could contribute materially to Australian life, science, as rhetoric, served vital social functions, underscoring the virtues of hard work, perseverance, and the search for truth in a land where established values met an uneasy reception and where Nature itself seemed upside down. Science was valued as a repository of idealism and elevating vision and as a cultural antidote to memories of convictism, to legacies of sectarianism, and the deepening excesses of materialism in a land otherwise governed by the pursuit of profit.

Any attempt to place this history of science within a more general historiography must, however, deal with at least two interpretive difficulties. The first requires us to determine precisely the ways in which, beneath the rhetoric of colonial enthusiasts, the sciences actually manifested themselves. There are two aspects to this problem. First, Australian science, as an aspect of Anglo-Celtic enterprise, has been held to be (both here and at 'Home') empirically derivative, intellectually dependent—rarely, if ever, 'high science', and only by courtesy 'high culture'.[26] The 'science' that appeared in the columns of the colonial press, and even the proceedings of the learned societies of the colonies, was frequently 'technological' and 'applied'; what was ordained in the name of colonial science was 'good practice', using modern methods. There was little sense that science should be supported as a cultural overhead (indeed at every turn there were arguments to the contrary), nor that Australians could quickly become creative scientifically. To become so was in effect (and sometimes in fact) to leave Australian culture altogether and enter that larger British or overseas culture— which it was, perhaps, not part of Australian history to retrieve. By this reading, indigenous creativity in science had almost by definition no place in early Australian history. This is clearly an unsupportable proposition, but one difficult to erase.

In its second aspect, the interpretation of Australian science presents a more intricate difficulty. While it is doubtless correct to regard early colonial science from the 1820s to the 1890s as an aspect of the 'common context' of British metropolitan culture, by the 1880s there were sufficiently strong voices—from Matthew Arnold to A. J. Balfour—

distinguishing 'scientific' culture from general culture, both here and in Britain. It may, therefore, become historically necessary to consider Australian science to some extent—however much one may deplore the fact—independently of the 'creative spirit' that actuated colonial endeavour in music, art, literature and architecture. In effect, scientific culture in Australia by the early 1880s had acquired a distinct and popular image. This comprised two important elements. First, its language and practice implied a rationality, a methodological imperative, that transcended traditional knowledge to become a way of seeing, behaving and believing—in this sense, a higher culture. Second, colonial science appeared to validate its claims upon things of this world by the prospect of securing materially useful outcomes from research and teaching. Morality and material well-being became fused and, in the argument 'for' science, interchangeably deployed. The history of science fits uneasily within Australian cultural history because it reflects claims that, in their own context, transcend the limitations of colonial culture. It will remain for scholars to determine whether science in Australia has always transcended Australia, and whether the search for closer association with a reworked version of an Australian legend will ever prove persuasive.

There is one further problem of interpretation. Since at least the 1830s the practice of Australian science and technology—as distinct from its imagery—has been so closely identified with some of the most spectacular developments in Australian economic progress that its neglect by our standard textbooks and courses suggests an attitude more significant than mere ignorance or indifference. Paradoxically, the enemy of science in our textbooks may be not philistinism but familiarity. It is possible to argue that science, like sun and sand, is among the great Australian 'givens'. Its history is like an over-familiar, if uninformed, view of outback landscape: proverbially unproblematic or, less kindly, deeply uninteresting. Science is, therefore, not to be valued less, but less to be praised, criticized or even scrutinized—unless for some reason, it fails to 'work'. In this sense, the history of science can, as yet, have no well-defined place in the Australian historical canon. This will not happen until its achievements are defined, its sources and traditions are debated, its motives and methods questioned, its uses and abuses explained, justified or rejected. A proper history of Australian science (and technology) requires an attitude of rational criticism, not benign acceptance. Without a questioning of historical preferences, we will fail to understand why it became popular to believe that, as knowledge is universal, it matters little if it is 'invented here'.[27]

If this is so, it is perhaps too early to expect general interpretations of the sciences to figure prominently in general historical writing. If Stuart Macintyre's otherwise excellent *Oxford History of Australia, Volume 4 (1901–42)*[28] is an indicator of things to come, we will not find that 1988's bicentennial histories redress the balance. For the present, we must fall back upon the descriptive models and the more specific, if internal, interpretative accounts available in the literature. For Ernest Scott, looking to nineteenth century sources, it was convenient to portray Australian science as pursuing an orderly transition: from an enterprise of discoverers to an age of gentlemen collectors, who placed Australian distances and diversities on the metropolitan map; thence to a period of 'organization', characterized by the unfolding of learned societies and publications, finally culminating in a period of professionalization, during which science became fully ensconced in universities and government agencies, and in the life of the land.[29] The job was, therefore, complete, and the challenges over. Scott was not the first, nor the last, to use the evolutionary metaphor that blended the familial with the biological.[30] The model is pleasingly progressive, without being controversially precise. Geoffrey Serle spoke of our national and intellectual development as occurring in stages from childhood to adolescence, from infancy to maturity.[31] The history of Australian science similarly keeps in step with the political life of the colonies and 'marches to nationhood'. But the difficulty with such models, made famous by George Basalla's influential depiction of the 'march' from colonial dependence to national assertiveness,[32] lies in their omission of conflict and tensions. They leave unasked questions of motive and intent, scoring differences only in tempo and key. The interpretation of Australian science will require a finer set of concepts and categories if we are to see the harmonies and disharmonies that remain of enduring importance.

III

In the search for a historical metre, we confront certain measures, notably of number, distance, isolation and environment, that shaped the perceptions of transplanted Britons in Australia.[33] The apparently large membership—perhaps 5000 in 1900—of the colonial learned societies gives a poor guide to the number of what Nathan Reingold has called active scientific 'cultivators',[34] who can hardly have exceeded 100 before the turn of the century. Even in the 1980s, there are only 40 000 scientists and 50 000 engineers in Australia, of which fewer than 14 000 are in higher education or university research.[35] As late as the Second World War the tyranny of distance was a matter of miles as well as men. The twins of distance are isolation and delay. Colonial states of mind wrestled with the mixed blessings of separateness. Ambivalent loyalties to London (or Edinburgh) and to federalism informed the contradictions of imperial science. Even in the United States—closer-knit internally by the railway and to Europe by frequent and rapid steamships—where political independence was achieved much earlier, an attitude of intellectual defensiveness persisted for generations.[36] Certainly the work of some scientists in Australia, as in the United States, was retarded in infancy, held in a kind of intellectual chrysalis. Accepting isolation brought with it the dangers of what Donald Fleming has called the 'psychology of abdication'[37]—the belief that because it could not be done here, and was done better overseas, let it be done for us, and we will import the final product. Much of colonial science consisted at first of sending new materials and specimens to Britain. Only from the 1850s, with the establishment of university departments and the expansion of government observatories and museums, came substantial recognition of colonial research. Imperial and colonial factors then operated to alter Australians' perspective of themselves and of their metropolitan obligations.[38] From this followed a belated sense of independence in thought, if not precisely leadership in practice. To 'know oneself' was on the way to becoming free, if not (to paraphrase James McAuley) actually independent.

Nowhere is this better illustrated than in the record of astronomy, meteorology, and natural history, in which European Australians slowly acquired the right of eminent domain over the analysis and interpretation of this ancient continent.[39] By the 1870s, when the intercolonial movement began, Australian science had already drawn strength from international interest in the uniqueness of its flora and fauna, the astronomy of its southern skies, the meteorology and mineralogy of its arid landforms, and the confidence of knowing that Australians were best placed to contribute to the store of world knowledge in these fields. The disadvantages of distance were offset by the rewards of discovery, as the global distribution and migration of species prompted new interest in the temporality of origins and the status of transitional organisms. In ways not fully understood, the Australian environment focused the creative imagination of Australian scientists. Possibly even more interesting than the fact were its consequences: Australian science could flourish better, and in new directions, if separated from metropolitan influences, and governed by a spirit that, in A. D. Hope's famous phrase, 'escapes the learned doubt'.[40] Gradually, as the chapters in this volume repeatedly observe, Australians recognized and made choices—drawing upon indigenous materials and wresting significance from the commonplace, or working on problems defined internationally, forming ranks in the army of science, if not always in its advance guard.

This path to 'self-knowledge' was to prove uneven. Dependence upon the metropolis remained a defining characteristic of Australian science. Despite signs of increasing colonial self-reliance, assistance with scientific expeditions, from the 'tropics to the Pole', continued to come from London; British universities provided the postgraduate research training Australians needed and could not obtain locally. With British help, Australians could produce science of world-class interest. From their apparent intellectual vassalage, Australians drew a special strength. Not surprisingly, attempts to claim independence were few and unpopular. Colonial nationalists might dispute the intellectual ownership of certain disciplines to which Australians had legitimate title. But 'scientific nationalism' was not a by-product of Federation. The rapturous welcome given the British Association for the Advancement of Science on its month-long progression through Australia in 1914[41] seemed to epitomize the lingering contradictions of Australian scientific enterprise.

IV

If these factors—among others—supply some context to the understanding of Australian science, their influence is nowhere better illustrated than in the history of the Australasian Association for the Advancement of Science (AAAS). From its origins in the 1880s, the AAAS, to all intents and appearances, crystallized a conception of a scientific commonwealth a generation before Federation, its political equivalent, became reality. The Association proclaimed the value of an academic parliament long before a national academy could be realized; serving the needs of the periphery, it underlined imperial loyalties. 'Before Australia could create an intellectual culture that was not merely imitative', as R. M. Crawford has written, 'there had to be a blending in equal terms of its European heritage and colonial experience'.[42] This the AAAS aptly supplied. The Association, or ANZAAS as it became in 1930, conveyed and sustained images of Australian science, as serving both colonial and imperial functions; where those images came into conflict with the changing priorities of federalism, the Association witnessed their resolution. Travelling with great ceremony from capital to capital, completing a transit of the country every six or seven years, the Association presented a forum for reflection on the state of different specialisms and their contribution to colonial life. ANZAAS, in its origins and quiddities, became a mirror to our history, and in its minutes, reports, and public reception, we find reminders of the ambitions of the architects of Australasian science and of the monuments they sought to build. Through ANZAAS, we also pay tribute to the principles foreshadowed by its architect and founder, Professor Archibald Liversidge, FRS (1846–1927), who, for twenty of his thirty-six years in this country, served as the Association's general secretary.

Looking back over its first century, the history of ANZAAS may be conveniently considered in six overlapping phases. The first formative period, between 1870 and 1888, saw leading figures of colonial science renewing attempts to share experience and circulate information, especially on transcolonial issues affecting astronomy, meteorology, mineralogy, geography, and geology. These men sensed the need for an intercolonial programme, a system of free trade in ideas, of intellectual co-operation that did not compromise colonial loyalties to the city-states of Sydney or Melbourne, Brisbane or Christchurch. Following its inauguration in 1888, we may speak of a 'heroic phase' between 1888 and 1913, where a stronger impulse, mounted by Liversidge and his associates, gave a systematic direction to science. This phase witnessed the flowering of the practical idealism for which the Association became known. Professionally, it confirmed the importance of specialized knowledge for colonial purposes, and the role of government in its continuing support. Domestically, it witnessed the beginning of enquiries and surveys on a national basis, the establishment of a host cycle of congresses and the institution of a federal network, linking the energies of colonial nationalism to the spirit of the imperial federation, and to the new social imperialism of the British Association.[43] During this period, science became an agency of union both within the new

Commonwealth and the wider Empire. These years saw the Association at its most xenophilic, identifying itself with the parent British Association, and conscious of its place among sister Associations in the United States, France and India. As the experiences of wartime demonstrated, these institutional loyalties were to represent far more than mere symbols of imperial obligation.

Following the outbreak of war in 1914, the congress proposed for the following year was postponed—in the event, to 1921—a lapse of eight years. When the officers resumed their places, they had in many ways to begin again—with a younger generation of men and women who nonetheless preserved a remarkable continuity in the Association's affairs. With the 1920s came a third phase of regrouping and reformation—and, in particular, support for Australian-sponsored enterprise in exploration, from New Guinea to Antarctica. This period also saw major steps taken toward the federal sponsorship of scientific research. Through its role in the establishment of the Australian National Research Council (ANRC) in 1921, the Association became identified with the management of Australian science. Through the 1930s, the Association remained the principal scientific organization in Australia, and its reputation was secure. In professional terms, it was steadily overtaken by new bodies, some of which it had spawned, but over which it had little influence; the thin life of academic science in a country that had only five universities, and fewer than 800 scientists, was hard pressed to keep an association like ANZAAS alive. Survive it did however, and by the end of the 1930s it was being encouraged to move beyond its Sections and congresses and to interpret the social responsibilities of science for a wider public.

From the mid-1930s, some council members wanted ANZAAS to become Australia's public apologist for science. Repeated attempts to reassert a more public role for the Association, represented by the launching of the *Australian Journal of Science* in 1938, failed to achieve this, as growing distances separated the representatives of science based in the universities and the Council for Scientific and Industrial Research (CSIR), from the wider demands of the Australian economy. The social relations of science in Australia were to remain largely neglected for years to come. With the end of the war, a fourth phase opened, when ANZAAS, after another interruption of seven years, had (like the British Association) to plant its flag anew. During the 1950s and early 1960s, ANZAAS (again like its British counterpart), continued to reflect academic values, and served as a recognized pillar of the scientific establishment, made the more conservative by the spirit of suburbanism that pervaded the lucky country.[44] In 1955, however, the new Australian Academy of Science took over many of the functions that ANZAAS, especially through its relations with the ANRC, had encouraged for over a quarter of a century. For its part, the Association, labouring under continuing financial and organizational difficulties, and appearing to the world as 'neither adequately federal nor wholly central',[45] sought consolation in its congresses.

Reforming voices, however, asked for more. At the Canberra Congress in 1954, the Association's council established a Future of

ANZAAS Committee to consider the way ahead. Recommendations the following year set out a fresh plan of regional meetings, a restructured council, reduced in size, and public lectures between congresses.[46] The *Australian Journal of Science* began to appear monthly in an effort to reach a general audience. Again, however, the Association's idealism failed to keep pace with changing circumstances. In 1958 Professor M. L. E. Oliphant observed that the Adelaide Congress that year had been highly successful, as judged by the number and quality of specialist papers, but that it had neglected the 'relation of science with the society in which it is carried on'. He issued a plea that the Association be made a 'better instrument for advancing the understanding of science by the people', by taking up controversial questions confronting scientists, social scientists and humanists alike.[47] For the next decade, scarcely a congress passed without some re-examination of the purpose of ANZAAS, the presentation of science (especially through radio and the press),[48] and the role of the numerous Sections, which had bloomed like a hundred flowers.[49]

By the late 1960s, the Association, to which nearly 'all practising scientists in Australia (as well as New Zealand) belonged',[50] entered a fifth phase, one of readjustment. Reforms in its procedures were prompted by concern that over-specialization in Australia, as elsewhere, had damaged the unity of the scientific community. Little attention had been given to the environmental consequences of technology, the political direction of western science, and the capacity of science and technology to bring about desirable social and economic change. There was need in Australia for some organization to confront these issues, in free and open discussion; in responding to the challenges, the Association found a new calling.

In 1969 a committee led by the president of ANZAAS, Sir John Crawford, recommended structural changes to make the Association more efficient and a new set of objectives to make it more open.[51] With the Crawford Report came a sixth phase, a period continuing today, that has seen the beginnings of *Search* and several attempts to make the programmes of ANZAAS more useful to science and relevant to society. In 1972 the Sydney Congress inaugurated twelve major symposia, taking up themes of national importance. At the well-attended Perth Congress in 1973, *Nature* observed that 'extensive open discussion and lobbying with politicians and their staffs were features never previously observed at an Australian Science Conference'.[52] These were stirring times.

Over the last fifteen years, ANZAAS has steadily opened its windows wider, embracing topical issues in technology policy, ecological conservation, nuclear energy and arms control, Aboriginal rights, and population and resource management. The Sydney Chamber of Commerce once glowingly described ANZAAS as the 'mother body of all scientific disciplines in Australia'.[53] Indeed, the Association continued to act in a federal capacity for the encouragement of new subjects (of which, perhaps, women's studies became among the most popular). The reconstruction of the Melbourne Congress of 1985 and the Townsville Congress of 1987 away from Sections and towards themes, conveyed the message that ANZAAS can also serve as a public pleasure ground and as a means of better understanding our regional environment, images that many find attractive.

As the 1980s draw to a close, the Association has altered much to meet its critics. Today ANZAAS has become a travelling emporium in which individuals, ideas, and issues jostle for public attention and press coverage. The tyrannies of time, distance and expense seem, if anything, as severe in the age of aircraft as in the age of steam, and like the British Association, ANZAAS is perhaps unjustly hampered by its own boffin image of apparent inaccessibility.[54] Nonetheless, like the British Association, and to some extent the American Association, ANZAAS still occupies a niche no other organization can fill. When, for a brief interval, it captures the spotlight, it can still provide headlines that people will read. In this confidence, the Association proceeds towards its centenary. By directing public attention to issues where science and technology bear on the well-being of society, the Association has a role to play in helping Australia and New Zealand maintain their places among the innovative and responsible countries of the world.

V

ANZAAS has spread its umbrella over virtually all areas of academic knowledge. This book focuses on only a fraction of the Association's vast contemporary range and, in its emphasis on the natural sciences and relative neglect of the social sciences, cannot avoid doing the Association less than justice. Moreover, a history written through the pages of AAAS/ANZAAS proceedings cannot be, in itself, a complete history of Australasian science. Work now underway—including major histories of the Commonwealth Scientific and Industrial Research Organization (CSIRO), the munitions industries, and the separate disciplines[55]— promises to complement this particular perspective. Science serves Australia today as a cultural expectation, as an economic resource, and as an instrument of defence and foreign policy. In each respect, its history requires detailed study. Nevertheless, the Association's 'biography' does speak to that larger story, and not least by illuminating factors that continue to characterize the life of organized science. In particular, we see in its history the persistence of essential tensions: between the advancement of specialized disciplines and the encouragement of interdisciplinary research; between the pursuit of basic science and the need for linkages between research and productive industry; between the conduct of corporate science and technology, and the imperatives of government regulation; between the values of professional reward and the pressures for social accountability; and between the encouragement of local initiative and the use of foreign capital to stimulate leading industries and overseas markets. Above all, we see through the pages of ANZAAS revealing forms of intellectual dependence on overseas models and precedents, and difficulties that confront the social relations of knowledge in Australian society. These tensions are more pressing now, but they are not new, and the history of the Association affords us a vantage point from which to perceive them more clearly.

Looking over the fifty-eight congresses and their reports and papers, one sees today the steady unfolding of attitudes that have shaped and conditioned intellectual expectations in this country. Like many other Australian organizations, ANZAAS has searched for political and cultural means of rewarding the fugitive gifts of individual excellence without sacrificing the interests of a literate democracy; for ways of securing federal consensus in its proceedings, while preserving the vigour of state and city enterprise. Its successes and failures in the past may make us wince with self-recognition and see more clearly the challenges Australians face in bringing science and technology to bear on the future. As such, this volume is an anticipation, not perhaps of things soon to come, but of possibilities yet to unfold. In interpretative terms, given the nature of a collective exercise, we cannot claim to have pressed back the frontier uniformly or to a great extent, but we can point to the linking of outstations in what must become a spirit of continuing enquiry.

The book is divided into three parts. The first, in two broad chapters, outlines the circumstances that gave rise and early direction to the AAAS as the principal Australasian forum for the collective public expression of scientific enterprise. These chapters draw special attention to the context within which the organization of Australian science responded to meet the demands of scientists themselves. James Davenport, in chapter 3, considers in particular those circumstances which shaped the development of Australian science after the Second World War. He describes the Association, overtaken by the pressure of events, becoming only one of several elements expressing Australian attitudes towards our 'new industrial order'.

In the second part, we divide 'the circle of the sciences' and consider six principal constituencies of the 'commonwealth of science'. In the professional regions familiar to Linden Gillbank, David Branagan, Tom Vallance, R. W. Home, Ian Rae, D. J. Mulvaney, and Alison Turtle, each uses the lens afforded by AAAS/ANZAAS to magnify or clarify traditions that became central to Australian (and to a lesser extent New Zealand) scholarship. The simplest possible framework binds them, but their discussion reveals similar patterns emerging within the several sciences. Each case illuminates the achievements of small professional communities, highly dependent on overseas stimulus and the resources of colonial, state and federal governments; and the influence of strong personalities, whose lengthened shadows persist well after the histories end.

The second part lies in prolepsis to the third, in which we turn from the Sections 'serving the sciences' to the roles of science in 'serving society'. Chapters by J. M. Powell, Bruce Davidson, and John Powles help make the transition, through their accounts of changing Australian attitudes towards conservation, resources, and health. In three concluding chapters, Ron Johnson, Sol Encel, Ted Wheelwright and Greg Crough offer perspectives on the factors that today condition the social responsibilities of scientists, the encouragement of research and development, and the prospects of our using science and technology to achieve Bacon's vision of the 'benefit and use of life'.

As this is written, the ABC's new television series Last Chance for the Lucky Country bemoans the prospect of Australia and New Zealand becoming in science and technology, uncompetitive third world economies. What future course ANZAAS can take in resolving the challenges of our time is by no means agreed. But that it will have clear instructions, and a fair wind, will surely be the wish of all those who greet its centenary return to Sydney in 1988.

NOTES

1. E. Scott, 'The History of Australian Science', *Aust. J. Sci*, 1 (1939), 116.
2. F. B. Smith, 'Australian Studies: A Comment', *Australian Cultural History*, 5 (1986), 114.
3. An important step towards re-establishing the history of science in New Zealand was taken with M. Hoare and L. G. Bell (eds), 'In Search of New Zealand's Scientific Heritage', *Roy. Soc. of NZ Bulletin*, 21 (1983); see also Sir Charles Fleming's 'Science, Settlers and Scholars', *Bulletin of the Roy. Soc. of NZ*, 25 (1987), 1–353.
4. cf., R. W. Connell, 'Images of Australia', *Quadrant*, XII (1968), 9–19, and in particular his views of the critical influence of W. K. Hancock, *Australia* (London: Benn, 1930, 1945).
5. As representative of a large popular literature, cf., Terry Gwynn-Jones, *Heroic Australian Air Stories* (Sydney: Rigby, 1981); Ward McNally, *The Man on the Twenty-Dollar Note: Sir Charles Kingsford-Smith* (Sydney: A. H. and A. W. Reed, 1976); David Parer and Elizabeth Parer-Cook, *Douglas Mawson: The Survivor* (Melbourne: Alella Books, 1983).
6. cf., Frances Wheelhouse, *Digging Stick to Rotary Hoe: Men and Machines in Rural Australia* (Sydney: Cassell, 1966). In this context, T. H. Farrer, *A Settlement Amply Supplied: Food Technology in Nineteenth-Century Australia* (Melbourne: Melbourne University Press, 1980) sets a remarkable, undeservedly neglected, precedent for the new historical 'archaeology' of Australian technology.
7. cf., Robert Ingpen, *Australian Inventions and Innovations* (Sydney: Rigby, 1982); Elena Grainger, *Hargrave and Son* (St Lucia: University of Queensland Press, 1978). This boffin image persists even in official literature; cf., Australian Academy of Science, *From Stump-Jump Plough to Interscan* (Canberra: Australian Academy of Science, 1977).
8. R. M. Crawford, *An Australian Perspective* (Melbourne: Melbourne University Press, 1960), 76, 261.
9. cf., Vance Palmer, *The Legend of the Nineties* (Melbourne: Melbourne University Press, 1954); Russel Ward, *The Australian Legend* (Melbourne: Oxford University Press, 1958, 1966). An interesting attempt to redefine the 'legend' was foreshadowed by G. C. Bolton and D. E. Hutchinson, 'Major Themes in Australian History', *Museum of Australia*, Conference on Australian History (24–25 April 1982), 10–12.
10. Peter Coleman, *Australian Civilisation* (Melbourne: F. W. Cheshire, 1962); Humphrey McQueen, *The New Britannia* (Ringwood: Penguin, 1970); David R. Walker, *Dream and Disillusion: A Search for Australian Cultural Identity* (Canberra: Australian National University Press, 1976).
11. cf., G. Blainey, *Mines in the Spinifex* (Sydney: Angus & Robertson, 1960);

but also his far more influential, *The Tyranny of Distance* (Melbourne: Sun Books, 1966).
12. Patrick White, *Voss* (New York: Viking Press, 1957).
13. Louis Esson, *The Time is Not Yet Ripe* (Sydney: Currency Press, 1973), Act 1.
14. Walker, op. cit., note 10, 204.
15. This observation was made as long ago as 1963 by J. M. Ward, in A. L. McLeod (ed.), *The Pattern of Australian Culture* (New York: Cornell University Press, 1963). The place of 'science' in the middle classes merited more than the disappointing response by Sir Samuel Wadham in the same volume.
16. Consider the absence of science from John Docker's *Australian Cultural Elites* (Sydney: Angus & Robertson, 1974), and H. McQueen, 'Australian Cultural Elites', *Arena*, 36 (1974), 40–9.
17. H. McQueen, 'Images of Society in Australian Criticism', *Arena*, 31 (1973), 44–51.
18. cf., M. Hoare, Science and Scientific Associations in Eastern Australia, 1870–1890 (unpublished PhD thesis, Australian National University, 1974); 'Light on Our Past: Australian Science in Retrospect', *Search*, 6 (1975), 285–90; 'The History of Australian Science: Prospect and Retrospect', *Newsletter of the Australasian Association for History and Philosophy of Science*, 5 (1974), 21–36; for his many subsequent publications, see the *Guide to Sources*.
19. Bernard Smith, *European Vision and the South Pacific: 1768–1850* (Oxford: Clarendon Press, 1960); G. Blainey, op. cit., note 11; and Kathleen Fitzpatrick, *Sir John Franklin in Tasmania, 1837–1843* (Melbourne: Melbourne University Press, 1949).
20. See Moyal's *Scientists in Nineteenth Century Australia, A Documentary History* (Sydney: Cassell, 1976), recently revised and extended as *A Bright and Savage Land* (Sydney: Collins, 1986). An indication of her pioneering work appears in the *Guide to Sources*. cf., Geoffrey Serle, *From Deserts the Prophets Come: The Creative Spirit in Australia, 1788–1972* (Melbourne: Heinemann, 1973).
21. cf., Donald Denoon, *Settler Capitalism: the Dynamics of Dependent Development in the Southern Hemisphere* (Oxford: Clarendon Press, 1983); Ian Inkster, 'Scientific Enterprise and the Colonial "Model": Observations on Australian Experience in a Historical Context', *Social Studies of Science*, 15 (1985), 677–704.
22. George Nadel, *Australia's Colonial Culture: Ideas, Men and Institutions in Mid-Nineteenth Century Eastern Australia* (Cambridge, Mass: Harvard University Press, 1957); Michael Roe, *Quest for Authority in Eastern Australia, 1835–1851* (Melbourne: Melbourne University Press, 1965); Graeme Davison, *The Rise and Fall of Marvellous Melbourne* (Melbourne: Melbourne University Press, 1978).
23. Deborah Campbell, Culture and the Colonial City: A Study in Ideas, Attitudes and Institutions: Sydney, 1870–1890 (unpublished PhD thesis, University of New South Wales, 1982).
24. R. MacLeod, 'On Visiting the "Moving Metropolis": Reflections on the Architecture of Imperial Science', *Hist. Records of Aust. Science*, 5 (1982), 1–16, reprinted in N. Reingold and M. Rothenberg (eds), *Scientific Colonialism: A Cross-Cultural Comparison* (Washington, DC: Smithsonian Institution Press, 1987), 217–50.
25. G. Davison, 'Exhibitions', *Australian Cultural History*, 2 (1982–3), 5–21.
26. Donald Fleming, 'Science in Australia, Canada and the United States:

Some Comparative Remarks', *Proc. 10th Int. Congress History of Science*, 1 (1962), 179–96.
27. This view must surely underscore the message of Donald Horne's superbly ironic *The Lucky Country* (Sydney: Angus & Robertson, 1978).
28. S. Macintyre, *The Oxford History of Australia, Vol. 4, 1901–1942: The Succeeding Age* (Melbourne: Oxford University Press, 1987). It will interest succeeding generations that scarcely any mention of science, and hardly any of technology, crosses these pages.
29. Scott, op. cit., note 1.
30. cf., Dixon R. Fox, 'Civilization in Transit', *American Hist. Rev.*, 32 (1926–7), 753–68.
31. G. Serle, op. cit., note 20, especially chapter 4.
32. G. Basalla, 'The Spread of "Western Science"', *Science*, 156 (1967), 611–22.
33. And, it may be observed, in Canada. See Vittorio de Vecchi, Science and Government in Nineteenth Century Canada (unpublished PhD thesis, University of Toronto, 1978).
34. N. Reingold, 'Definitions and Speculations: The Professionalization of Science in America in the Nineteenth Century', in A. Oleson and S. C. Brown (eds), *The Pursuit of Knowledge in the Early American Republic* (Baltimore: Johns Hopkins, 1976), 33–69.
35. See J. Gani et al., *The Condition of Science in Australian Universities* (Canberra: ANU Press, 1962), 53. For recent calculations, see M. R. Rice, 'Australian Science and Technology Personnel', *Studies for the Australian Science and Technology Indicators Report* (Canberra: Dept of Science, June 1986).
36. cf., Carl Synder, 'America's Inferior Position in the Scientific World', *North American Review*, DXLII (January 1902), 59–72; Bruce Sinclair, 'Americans Abroad: Science and Cultural Nationalism in the Early Nineteenth Century' in N. Reingold (ed.), *The Sciences in the American Context: New Perspectives* (Washington: Smithsonian Institution Press, 1979), 35–54.
37. Fleming, op. cit., note 26.
38. cf. Inkster, op. cit., note 21.
39. cf., K. G. Dugan, 'The Zoological Exploration of the Australian Region and Its Impact on Biological Theory' in N. Reingold and M. Rothenberg (eds), op. cit., note 24.
40. A. D. Hope, 'Australia', in *Selected Poems* (Sydney: Angus & Robertson, 1975).
41. *The Scientific Australian*, XX (September, 1914), 3. Details are in *ANZAAS Archives* (Mitchell Library), MSS 908/21, Minutes, 1914; cf., Sir Ernest Scott, *Official History of Australia during the War of 1914–18* (Sydney: Angus & Robertson, 1936), vol. XI, 31–5; cf., Rosaleen Love, 'The Science Show of 1914: The British Association meets in Australia', *This Australia*, 4 (1) (1984–5), 12–16; Peter Robertson, 'Coming of Age: The British Association in Australia, 1914', *Australian Physicist*, 17 (2) (1980), 23–7.
42. Crawford, op. cit., note 8, 146.
43. cf., Michael Worboys, 'The British Association and Empire: Science and Social Imperialism, 1880–1940', in R. MacLeod and P. Collins (eds), *The Parliament of Science* (London: Science Reviews Ltd, 1981), 170–87.
44. Serle, op. cit., note 20, has graphically described the limitations of Australian intellectual life in this period from which science rarely escaped.
45. 'Report by the Chairman of the Committee of Review', *Search*, I (1970), 4.

46. *Aust. J. Sci.*, 17 (June 1955), 188–91.
47. M. L. E. Oliphant, 'The Future of ANZAAS', *Proc. RACI*, 25 (1958), 433–5.
48. 'Bringing Science to the Public', *Aust. J. Sci.*, 24 (1961), 27–30.
49. See appendix 2.
50. Gani, op. cit., note 35, 17.
51. Sir John Crawford, 'Report by the Chairman of the Committee of Review', *Search*, 1 (1970), 3–5.
52. Quoted in Ann Moyal, 'ANZAAS and the Public Communication', *Search*, 5 (November–December 1974), 589.
53. *ANZAAS Archives* (University of Sydney), Minutes, ANZAAS Council, Item 36, 1–2 March 1971.
54. *The Times Higher Education Supplement* (5 September 1986), 6.
55. Boris Schedvin, *Shaping Science and Industry: A History of Australia's Council for Scientific and Industrial Research, 1926–49* (Sydney: George Allen & Unwin, 1987); A. Ross, The Arming of Australia: The Politics and Administration of Australia's Self Containment Strategy for Munitions Supply, 1901–1945 (unpublished PhD thesis, University of New South Wales, 1986); R. Home (ed.), *Australian Science in the Making: Bicentennial Essays* (Sydney: Cambridge University Press with the Australian Academy of Science, 1988).

PART I

LAUNCHING THE ENTERPRISE

1

Organizing Science under the Southern Cross

Roy MacLeod

'Our colonies are rapidly coming abreast of us', announced the *Saturday Review* of London on 3 November 1888. 'There is a freshness and a breadth about the work of Australian science which, alas!, we rarely find now in that of the old country'.[1] Coming at a time of articles on the depression of trade and the German menace, it seemed good news was welcome, and with this generous flourish England welcomed the new Australasian Association for the Advancement of Science. Three months before, over 850 people, representing the 6 Australian colonies and New Zealand, had gathered in August assembly in the magnificent Great Hall of the University of Sydney, to hear the government astronomer of New South Wales bear inaugural greetings to the largest and most successful gathering of scientists ever seen in Australasia. Thirty-seven years after its American sister,[2] and a decade after its French cousin,[3] the new AAAS was the fifth Association of its kind in the world.[4] Modelled on the British Association for the Advancement of Science, formed in 1831, its objects were, similarly, to give a 'strong impulse and a more systematic direction to scientific enquiry; to promote the intercourse of those who cultivate science in different parts of the Australasian colonies and in other countries; and to obtain a more general attention to the objects of science, and a removal of any disadvantages of a public kind which may impede its progress'.[5]

What was heralded in its own time as a triumph of the intercolonial principle, and seen later as a dress rehearsal for political federation, revealed a great deal about the men and motives of colonial science. The intercolonial spirit represented by the AAAS did not, of course, spring full-blown in 1888.[6] The principles of 'association' were well established in Britain and, subsequently, by the British in India.[7] In Australasia the Tasmanian Society of Natural History had tried to promote intercolonial communication through its journal as early as the 1840s.[8]

Subsequently, Sir William Denison, first as patron of the Royal Society of Van Diemen's Land, and subsequently as president of the Philosophical Society of New South Wales, attempted to unify the activities of the few men of science in the eastern colonies. However, as new colonies were formed—Victoria in 1852, South Australia in 1834, and Queensland in 1859—each established its own learned society.[9] The Adelaide Philosophical Society was set up in 1853; the Philosophical Society of Victoria and the prophetically entitled Victorian Institute for the Advancement of Science followed in 1854; and the Queensland Philosophical Society, in 1859. For reasons both political and cultural, however, each looked to London more than to one another. United as Englishmen, they were disunited as Australians.

In 1857 the Philosophical Institute of Victoria appointed a committee to consider the expedience of organizing 'a system of combined action amongst all the Scientific Societies throughout the colony', but the effort faltered.[10] Difficulties of travel within Australia, the real tyranny of distance, and the insulating conditions of colonial life restricted prospects for scientific co-operation. Across the Tasman, the New Zealand Institute, created in 1867, began to link the several sister societies of that colony. It was not, however, until the 1870s that the 'impulse to associate' quickened the cultural leadership of New South Wales and Victoria. There, over the next two decades, a number of factors converged, contributing to the events of 1888. This chapter will trace this convergence, in the context of those individuals and events that proved influential to this critical stage in the organization of Australasian science. The second chapter will consider the features that characterized the Association's first decades, along with the developments that shaped its reawakening and development in the aftermath of one war and in the shadow of another. In these years Australian science met challenges that were shaping the destinies of larger nations and drawing Australia and New Zealand inexorably in their wake.

SCIENCE IN COLONIAL CULTURE

A visitor to Australia or New Zealand in 1870 could see that colonial science already enjoyed a long and respectable history. Since the 1820s, and especially since the 1840s, vice-regal interest had endorsed science as a measure of advancing civilization and as testament to stability and progress.[11] This appeal to the improving values of natural knowledge, was embodied in philosophical societies, colonial museums, and botanical gardens,[12] and eventually recognized in the appointment of the first professors at the Universities of Sydney and Melbourne.[13] The expansion of colonial wealth and the coming of representative government in the 1850s fuelled rising cultural ambitions, especially among the urban bourgeoisie. The Royal Society of Victoria acquired its title in 1859, followed by the Royal Society of New South Wales in 1866, and eventually societies in the other colonies followed suit.[14] New Zealand could boast scientific societies in each of its four main centres. These styled themselves in name and function either after the premier scientific societies of London and Edinburgh, or after those of the provincial cities. Like the assemblies of provincial England, they were led by men who held positions in government or the universities and depended for their financial survival upon the enlightened voluntarism of the small, generally urban, middle class. For over twenty years, in some cases for much longer, they had advertised the moral virtues and practical benefits of enquiry into the natural history and resources of a 'land half-known'. They had met the requirement of 'gentlemen collectors', and the expectations of the colonial professional classes. By the 1860s, however, the social expectations of colonial science were changing.

In particular Sydney and Melbourne, Adelaide and Brisbane were welcoming, in Michael Cannon's splendid phrase, 'science coming as a friend'.[15] Steam engines, photography, hot-air ballooning and the novelties of zoology, telegraphy and electricity were an increasing part of everyday life—and not only in the capital cities. Leading goldfield towns and rural centres of Victoria and New South Wales established their own scientific and literary societies, field clubs and mechanics institutes, free of the monopoly of the colonial metropolis.[16] In the colonial press, science and technology, at both rhetorical and practical levels, were fashionable and wise. The *Sydney Mail*, for example, had regular scientific and engineering sections, and weekly accounts of mining records and natural history. The *Australian Town and Country Journal* offered regular features under columns headed Science Notes, The Cultivator, Mines and Mining, Facts for Farmers, and Science-Invention. The *Sydney Morning Herald* had a Wednesday column on Science and Arts, and the fortnightly *Illustrated Sydney News* devoted columns to Science Gossip and Scientific Jottings. If the medium of the press was the most frequent messenger of science across intercolonial frontiers, intercolonial and international exhibitions were the most spectacular.[17] In presenting their treasures to the outside world, the colonies vied enthusiastically with each other at the Exposition Universelle in Paris in 1855, at London in 1862, and at Paris again in 1867.[18] In 1866 Melbourne attracted enormous prestige by holding the first Australian Intercolonial Exhibition. In retaliation, the Agricultural Society of New South Wales seized the centenary of Cook's first landing in Australia to stage in 1870 Sydney's first Metropolitan Intercolonial Exhibition, at a grand new purpose-built hall in Prince Alfred Park.[19] This was to New South Wales, the *Sydney Morning Herald* proclaimed, 'what the Great Exhibition of 1851 was to England', celebrating the mastery of nature by the application of science and technology.

Against this increasing popular enthusiasm for the practical, the Royal Societies of New South Wales and Victoria presented a picture of civilized stasis.[20] Their falling numbers mirrored growing impatience with their imperfections: the difficulty of attending monthly meetings, the frequently narrow technical nature of their proceedings, and the perennial delays members suffered in seeing their papers, once presented, finally in print. It was far quicker, as one president admitted, to send work for publication to London than to wait for its publication in Sydney.[21]

By the early 1870s, these circumstances militated for change. Some years ago, Ken Inglis proposed the year 1872 as beginning a new period in Australian history.[22] That year the imperial telegraph was established, finally connecting London and the colonies. Thanks to the cable, A. P. Martin wrote, Australia had become 'as much a part of the empire as Yorkshire'.[23] This development fortuitously coincided with the arrival in Sydney of Archibald Liversidge, aged 26, an Associate of the Royal School of Mines and former student of the Royal College of Chemistry, who was to play a central role in the history of scientific institutions in Australia.[24]

Liversidge, the youngest son of a prosperous London wheelwright, was an undergraduate in Michael Foster's physiology laboratory in Cambridge when he received a call to fill the readership in geology and assistantship in chemistry at the University of Sydney, left vacant by the death of his fellow Londoner, A. M. Thomson.[25] Well served by a half century of museum curators, collectors, explorers and surveyors, the colony of New South Wales had, however, few practising geologists. By far the most senior was the Reverend W. B. Clarke (FRS, 1876).[26] With Clarke, Thomson had been a trustee of the Australian Museum and had helped in the work of the senior scientific body in Australia, the Royal Society of New South Wales. Liversidge inherited many of his predecessors responsibilities. Within a month of arriving in Sydney, he had become a member of the Royal Society of New South Wales and six months later was on its council. Within two years, he had become the Society's honorary secretary and editor of its *Journal and Proceedings*. A shy, retiring bachelor, he devoted his entire energies to its subject from his rooms at the Union Club in Bligh Street, where he lived until 1889. He moved easily among those leading men of Sydney's colonial bourgeoisie—including Charles Rolleston, CMG, Sir Alfred Stephen, and C. K. Mackellar, MLC—who shared his zeal for improvement. With them, his interests spanned metropolitan sanitation and lighting, astronomical observation, geological exploration, and the application of chemistry to the needs of miners and farmers. As the 'second scientist' (after Professor John Smith)[27] at the University, and one of the few in the colony, he pushed through improvements in science teaching, won promotion, and bent himself to the business of 'organizing science' in Australia for the next thirty-five years.

Organizing science was by no means a new experience for Liversidge, who had learned the ways of amateur societies in workingman's London, and the rites of gentlemanly clubs in collegiate Cambridge. He knew the purposes and politics of scientific gatherings—indeed, he had already helped begin one scientific society (the Natural Science Club) in Cambridge, and two others in London (the Chemical and the Geological Societies) he tactfully joined shortly after leaving England. From his personal experience he wished to see installed in Australia a programme of academic and practical scientific instruction, suited for the world of work. Liversidge had no illusions about the difficulty of his task. As in England, science in Sydney was popular, but it did not pay; scientific training was at a discount. As the *Sydney Morning Herald* observed: 'the opportunities for making a living out of science are not numerous here, and those who have to make a living out of it are tolerably hard worked, and have not much strength and leisure for prolonged and profitless research'.[28] Surely not 'all the energies of Young Australia could be absorbed in the pioneering and settling of the continent?' asked the *Australian Town and Country Journal*.[29] Yes, *The Athenaeum* replied, a university man could be had for £20 to £50 a year, but the 'time of a good bullock-driver, fencer, shearer or manager' was worth ten times as much.[30] Science, like Art, had the treble burden of 'educating taste, outliving neglect, securing patronage, and being modest therewithal'.[31] By 1874, Liversidge had achieved some of his university ambitions: that year he was promoted to a personal chair of geology and mineralogy. But beyond the University's horizon, he could contemplate a larger 'visible college' through which science might make an impact upon colonial life.

THE INTERCOLONIAL IMPULSE

What became crystallized in Liversidge's idea of scientific co-operation was in itself not original, nor was its application new. He arrived at a time when intercolonial co-operation had become respectable. Others before him knew that the same sky shone over the whole complex continent; that the weather in one part had relevance to the weather in another; and that if rock formations differed dramatically from one colony to the next, their identification and explanation was required throughout.

From the late 1860s, the New South Wales government astronomer, H. C. Russell, had pressed for a co-operative study of the Australian climate, based on information from all six colonies. In Victoria, Robert Ellery, the government astronomer (one of the two FRSs in the colony), initiated in 1871 an Australian Eclipse Expedition—what Michael Hoare has called the 'first real attempt at formal, intercolonial scientific co-operation on any scale'.[32] Assisted, as was customary, by the Royal Society of London, this invited the co-operation of astronomers and meteorologists of New South Wales, Queensland, Victoria and South Australia. In 1874 Ellery presented the Royal Society of Victoria with a plan to use the newly extended telegraph to communicate astronomical and meteorological forecasts from Adelaide and Hobart to Sydney, Melbourne and Brisbane. That year, the Transit of Venus expedition gave the three government astronomers—Ellery in Victoria, Russell in Sydney, and Charles Todd in South Australia—further experience of intercolonial, as well as imperial, co-operation.[33] Also in 1874 the *Challenger* expedition visited Sydney and prompted colonial scientists to take a comparative perspective. In 1875 the Royal Society of Victoria and the newly formed Linnean Society of New South Wales co-operated in an expedition to New Guinea,[34] and the Royal Societies of New South Wales and Victoria considered joint work in marine biology.

With increasing frequency, this experience revealed the frustrations of intercolonial co-operation on an intermittent basis. Their experience mirrored that of the colonial museums, whose relations were particularly influenced by imperial, rather than intercolonial, considerations. To secure a system of regular contact would require a federation of scientific interests that united 'public' and 'private science', universities and museums, mints and mining departments. Happily, these were already linked by the interlocking directorates provided by the Royal Societies. The fact that scientific societies could achieve such co-operative advantages had been ably demonstrated in New Zealand since 1867, when the federal New Zealand Institute (now the Royal Society of New Zealand) was created 'to encourage the spread of scientific knowledge throughout the country'. This linked the 178 members of 8 'sister academies' under the management–editorship of James (later Sir James) Hector.[35] The New Zealand example was duly admired[36]—not least by Liversidge, who visited Hector in Wellington in 1876 and welcomed his coming to Sydney in 1878–9.

For colonial Englishmen, the easiest model of co-operation to follow was afforded by the peripatetic achievements of the British Association. In September 1875 Sydney's (shortlived) weekly *The Athenaeum* drew attention to the British Association's recent annual meeting and lamented that what scientific news it conveyed would not reach southern shores for six months. To its editor (probably H. W. H. Stephen), the stimulus of the British Association, both to civic culture and to British science, was sorely needed here. An Australian Association was required. Such subjects as colonial coal reserves, theories of gold deposition, geological formations, and indigenous flora and fauna were 'never studied as they should be' in Australia, by Australians; nor, his editorial continued, 'must the essentially practical be omitted ... The whole of the Australian colonies are fast becoming manufacturing and producing centres, and manufacturers and producers nowadays must of necessity set about working scientifically or they may find themselves nowhere in the race.' In the editor's appropriate phrase, to 'acclimatize' such a gathering here was both wanted and necessary, giving 'practical men' and 'mere theorizers and dilettantes' a 'new field to work in'. With 'railways and steamers ... to transport visitors it is surely somewhat singular', he continued, 'that Australia as a whole, or the larger colonies individually, have done ... so little in this direction to advance science, art, and industry'.[37]

Looking abroad, the American Association for the Advancement of Science, formed in 1847, had achieved much towards improving interstate scientific communication between learned societies and inventors across the new republic and Canada. Similarly, the Association Française pour l'Avancement de la Science (motto: *par la science, pour la patrie*), created in 1872 in the aftermath of military defeat, had already served the reforming interests of French scientists. Both models were evident enough in the Australian context. When Liversidge arrived in Sydney, he found reports of the American and British Associations. In 1876 he joined the British Association from Sydney. But he was not the only colonial scientist familiar with its methods. In May that year, in what would be his last anniversary address to the Royal Society of New South Wales, W. B. Clarke proposed that the Society should stimulate its membership to greater efforts by inaugurating Sections similar to those of the British Association and the Royal Society of London. Eventually, eight were formed—Astronomy, Physics, Chemistry and Mineralogy, Geology and Palaeontology, Biology (including Zoology and Botany), Microscopy, Geography and Ethnology, Literature and Fine Arts (including Architecture); and Medical, Sanitary and Social Science. In his adaptive manner, Liversidge supported Clarke's proposed 'Section principle' and during the next three years worked with Russell and Carl Leibius, senior assayer at the Sydney Mint, and joint secretary of the Society, to add to the Society's programme a series of medals, prizes and, most important, the exchange of scientific journals. Soon the Society was known to more than twenty-five societies overseas, and through the efforts of Liversidge and his circle, the Society itself became 'one of the most effectual agencies for making [New South Wales] favourably known abroad'.[38] Liversidge's programme actively employed the 'Exhibition movement', which repeatedly took colonial achievement corporately on show, both in Australia and overseas.[39] In 1876 Victoria and New South Wales exhibited together at the American Centennial Exhibition in Philadelphia. The American Association for the Advancement of Science met that year in Buffalo and celebrated both Anglo-American good will and solidarity among English-speaking men of science. T. H. Huxley and other English scientists were there. Liversidge attended neither meeting but, with C. S. Wilkinson, the New South Wales government geologist, sent mineral exhibits to the Australian pavilion organized by Victoria in Philadelphia and closely followed the exhibition and congress proceedings.[40]

The following year saw further examples of intercolonial cooperation. By 1877, when the telegraph linked all the colonial capitals, Russell and Todd had further developed their 'intercolonial strategy' of weather telegraphy, which by 1879 included Ellery in Victoria and Hector in New Zealand.[41] By 1881, when these four met to press Queensland to join in collecting standardized data, they had become effectively an intercolonial movement in themselves. Their sentiments were shared by a small but enthusiastic Antarctic lobby, notably in Victoria, Tasmania and New Zealand, and by those who held scientific interests in Papua New Guinea.[42] The political and scientific prospects of the island to the north and the continent to the south were recognized by those who saw exploration and discovery as a measure of Australia's sovereignty. By 1878 colonial science thus spoke to colonial imperialism: through science, Australia's influence could reach from the tropics to the Pole.

By the late 1870s, Liversidge surveyed a scene in which all the ingredients of intercolonial co-operation were present. The prospects for colonial improvement were buoyant. Fresh mineral discoveries and the expanding wool trade pushed colonial incomes to record levels, and with this came a spirit of optimism. To advance science and to improve the material condition of colonial life were goals within reach, and the learned societies, invigorated and given a fresh sense of purpose, were the ideal means. In an age of colonial nationalism, Liversidge was not an Australian 'nationalist'. The evidence of later years suggests he remained an Imperial federalist. He joined the Royal Colonial Institute in London in 1879 and regularly sent his papers to its Library. He did not exclude the possibility that imperial co-operation required prior colonial federation. That science could 'foster federation' was not in the foreground of his intentions. Instead, what comes through his papers is a simple longing for an enlarged sense of community.

Following Gerard Krefft's infamous end in 1876 and W. B. Clarke's death in 1878, Liversidge was one of the half-dozen publishing scientists in the colony. The loneliness of his bachelor existence, scarcely relieved by teaching the mere twenty students enrolled at the University, caused him to seek a solution to the tyranny of isolation. This required more than the fellowship of science in New South Wales alone could supply. The larger number of those who 'cultivated' science in all six colonies and New Zealand was not great: perhaps twenty to thirty men had become familiar with the joys of sharing information, exchanging periodicals, and pooling specimens for overseas exhibitions. But they could command more attention if united, particularly if they could count on popular commitment and political support. There was evidence that if the initiative were taken, support would come. In 1831, when the British Association was founded, the population of the United Kingdom was 23 million, and there were about 40 metropolitan and provincial learned societies. In 1875 Australia's combined population was less than 2.5 million, but there were 33 scientific societies, with nearly 3000 members from many walks of life. What they needed was a catalyst. That Liversidge provided.

In 1878 Liversidge took overseas leave, principally to attend the International Congress of Geologists to be held in conjunction with the International Exhibition in Paris that year, and to visit British scientists and friends. The colonial government took advantage of his presence in Paris to appoint him a commissioner for New South Wales at the exhibition, and, as a trustee of the Australian Museum, he was asked to obtain books and specimens for its collections. By a special request from the chancellor of the University, he was also asked to report on the state of British and European technical education. On 26 May 1879 Liversidge returned from Europe with three compelling commitments: to bring about a closer association between men of science in Australia; to extend the range of scientific and technical education provided in the colony; and to establish at Sydney University a school of applied chemistry and mineralogy to equal the best he had seen in Britain. What he had seen and later reported formed the basis of a programme of great importance for the future of education in the colony. While it is only the first of these commitments that concerns us here, it formed but part of a research and educational strategy that embraced all levels of education, all disciplines of science, and both public and private institutions. Again, the Royal Society offered a mechanism for his institutional innovation, at once more accessible and more elastic than the University.

In March 1879, just before leaving England, he wrote to his friend Leibius, a fellow former student of the Royal College of Chemistry,[43] insisting that 'one of the objects of the Sydney Society is to serve as a central institution for the exchange of scientific publications between institutions in Australia and those in foreign countries'.[44] His idea of a 'Sydney-centred' endeavour was the key to co-operation. Scarcely nine days after his returning ship berthed in Darling Harbour, Liversidge reported in general terms on the Geological Congress in Paris to the Royal Society's monthly meeting in Bridge Street. The congress had perhaps not attracted in England the attention it deserved; it had conflicted with the British Association meeting (in Dublin that year), and, he thought, 'many persons were doubtless surfeited with scientific picnics'.[45] Yet the value of having such congresses was clear enough and could be copied with profit, and not limited to geologists. Liversidge fell back upon the exhibition idea the Americans had urged in 1876 and the French in 1878. In 1879 Sydney was set to outdo Melbourne in holding what would be its first International Exhibition. It seemed reasonable to use this occasion to combine the exhibition with a scientific congress. 'I hardly like to propose', Liversidge told his Sydney friends, no doubt wondering what fresh proposals would flow from their dynamic honorary secretary, 'that a Geological Congress should be held because the number who could attend would be such a small one.' However, he added modestly, there was no reason why the Society 'might not with advantage, join with the other scientific societies [of Australia] to hold some special meetings, at which papers could be read and discussed, after the model of the British Association'.[46]

The idea, by now widely canvassed, was plagued with practical difficulties. Despite support from Russell and other friends in the Royal Society, it refused to catch on. By the winter of 1879, Liversidge found his University, his new Technological Museum,[47] and his work for technical education were claiming his full attention. The following year his Report on European Technical Education was published.[48] Perhaps not entirely by coincidence—as he was already a member or fellow of thirteen learned societies at home and in the colony—Liversidge was nominated to the Fellowship of the Royal Society of London. His election in 1882 made him the sole FRS in New South Wales, and (with Ellery and von Mueller in Melbourne) one of only three in Australia. His proposal for a wider association was, however, left to mature, as for the moment, Sydney's energies were directed to the magnificent International Exhibition that arose in the Botanical Gardens, as all the world came to see the Paris of the Pacific.[49]

HOPES DEFERRED

Another five years would elapse before the idea of association came again to prominence. In the meantime, wealth generated by the economic boom of the 1880s continued to widen cultural horizons; the *Bulletin* blazed forth the spirit of a new Australia, and overseas visitors spoke of the vitality of the colonies.[50] In June 1883 the railway brought Melbourne to Sydney; in January 1887, Adelaide to Melbourne, and in January 1888, Sydney to Brisbane, confronting squarely the contradictions in what the Victorian statesman Sir James Patterson called the 'barbarism of borderism'.[51] Successive intercolonial conferences heard premiers debate the merits of free trade and protection, and the imperatives of intercolonial defence. Gradually, colonial federation became a leading issue. The question was not whether, but when, scientists would join the campaign.

Inconveniently, local difficulties arose to preoccupy the Royal Societies of New South Wales and Victoria. Despite the best efforts of Liversidge and Ellery, the Societies had failed to generate a vigorous spirit of 'research'. Of the 500 members in the Royal Society of New South Wales in 1886, for example, only 36 had ever contributed papers to its sessions, and most of these papers were the work of 7 or 8 men.[52] Liversidge himself had given thirty-three papers, making him easily the most productive scientist in the Society. In Victoria, field naturalist clubs were critical of the aloofness and bias of Melbourne's Royal Society.[53] The economic boom had passed by, but the Royal Society of Victoria still suffered in genteel poverty, dependent, like its cousin in New South Wales, upon government for the publication of its annual (and duly delayed) proceedings. The Royal Society of New South Wales had, thanks to Liversidge, new premises in Elizabeth Street, but little new enthusiasm to excite its members.

In reaction, learned discontent bred diversification, which in turn prompted fresh demands for association. In April 1883 E. M. La Meslée and a group of Sydney gentlemen, dissatisfied with the Royal Society's Geographical Section, set out to form a federal Geographical Society of Australasia, for the 'information and benefit of the people of Australasia'.[54] At the same time a proposal to create an Australian branch of the Royal Geographical Society of Great Britain was defeated. As La Meslée put it, before them lay 'a national work', not a colonial one: 'Geography is a science that cannot wait, as our very future depends upon the more or less perfect acquaintance which is gained of the natural resources of the country.'[55] His colleagues looked towards an intercolonial organization, with colonial branches, to serve their purpose. In June an inaugural meeting of the Geographical Society of Australasia was held in Sydney. W. J. Stephens (a founder of the Linnean Society of New South Wales in 1874 and the Zoological Society in 1879, and soon to be Liversidge's colleague in the chair of natural history at Sydney), gave the principal address on the annexation and exploration of Papua and New Guinea. The geographers laid unequivocal claim to a political role and soon acquired a royal title.[56] They found their role within a learned context already adapting to the idea of association.

In the meantime, it fell to that well-adapted animal, the native platypus, to do more than any human being to prompt the federation of Australian science. In late 1883 W. H. Caldwell, of Caius College, Cambridge, came to Australia. Caldwell was the first holder of the Balfour Studentship, created earlier that year to honour Liversidge's contemporary, the distinguished embryologist F. M. Balfour, who had died in 1882. As part of what became Cambridge's programme of 'imperial Darwinism', Caldwell came to Australia to investigate the reproductive mechanism of *Ceratodus* and the monotremes. In August, thanks to the help of Aboriginal trackers, he discovered the oviparous nature of *Ornithorhynchus*. News of his discovery was relayed excitedly from the Burnett River in Queensland, to Liversidge in Sydney, thence by wire to the Biology Section of the British Association, then meeting in Montreal.[57] The announcement that this famous puzzle had at last been solved—and in Australia—had a dramatic impact. For H. N. Moseley, the distinguished physiologist, who had visited Sydney with the *Challenger*, 'no more important telegram in a scientific sense had ever passed through the submarine cables'.[58] All talk was of Australia, and there followed proposals that the British Association hold a future congress in the antipodes.

The British Association welcomed the idea. Since 1884 the British Association had sought to use colonial visits to consolidate its waning influence over 'imperial science', and to maintain its *raison d'être* in keeping with the contemporary expansion of British influence overseas.[59] In Canada federal and provincial governments had proved hospitable, and earlier meetings of the American Association in 1856 and 1882 had given publicity to the progress of science in British North America.[60] Against this background, James Service, the pro-federalist Premier of Victoria, seized the initiative, and wired the British Association an invitation. Alas, the project foundered. *Melbourne Punch* satirized the naivity of British scientists in wishing to 'discover' Australian 'natives',[61] and for practical domestic reasons the proposal died. Distance and expense were at issue. The Canadians had spent $14,000 on 500 British visitors, but there was, as yet, no Anglo-Australian experience, short of convict transportation, that could meet the expenses of so many visitors to Sydney, who wished to be accommodated, moreover, in rather better style.

In September 1884 Liversidge wrote to the *Sydney Morning Herald*, outlining the possibility of a British Association trip, and describing arrangements necessary for steamship and railway travel. Outlining the difficulties, Liversidge suggested, not a meeting of the British Association, but instead a 'federation or union of the members of the various scientific societies in Australia, Tasmania and New Zealand'; to hold its first congress during the centennial celebrations proposed for Sydney in 1888. 'I am sure that such an Association—must come sooner or later if we are to hold our own', he wrote, '[it] would not only do a great deal for the advancement of science in other Colonies, but would also favour their progress in other ways.'[62]

Once again, however, the idea of an association of societies was not immediately pursued. Liversidge's energies were diverted to the new curriculum at the University of Sydney and to the needs of practical chemistry. With the death of Professor John Smith in 1885, and before Richard Threlfall arrived from Cambridge to take up the physics chair in 1886, Liversidge was, with W. J. Stephens, one of only two representatives of academic science in New South Wales. With Stephens he held similar views about 'federal' bodies, but distinguished different meanings of federalism. These meanings mirrored the two prevailing senses in which the phrase was used. In November 1884, for example, the Imperial Federation League was formed in England, with the object of federating 'the Empire for Britain's better security'. In Australia, however, premiers' conferences dwelt on the merits of a 'federated Australia'. Whether imperial defence would be better served by one government, or six, was one issue; whether a 'federated science' would help Australia 'hold its own' remained another.[63]

It was this bifocal character of the federation debate that put a question mark over the particular direction imperial science should take. On St Andrew's Day 1885, T. H. Huxley delivered his presidential address to the Royal Society of London. Huxley, who was soon bitterly to oppose Gladstone's policy of Home Rule, outlined the prospects of a new 'liberal imperialism' of science, within a union of English-speaking scientists: 'Whatever may be the practicability of federation for more or fewer of the rapidly growing English-speaking peoples of the globe, some sort of scientific federation should surely be possible.'[64] 'Nothing is baser than scientific Chauvinism', he added, 'but blood is thicker than water.'[65] Although details of his scheme were unclear, Huxley's suggested federal body was apparently to be directed (or at least co-ordinated) from London.

It is not certain whether Liversidge shared this view in 1879, but by the mid-1880s he had come to hold a different position. Huxley's proposal held all the internal contradictions of imperial federation for the equal representation of British and colonial interests. Reflecting the imperial traditions of the Royal School of Mines, strengthened by personal loyalties both to London and Cambridge, Liversidge was committed to the idea of an 'imperial science' within a 'Greater Britain'. Yet he was also committed to the integrity of an Australian assembly, a commonwealth of science. In 1885, after a decade on the council of the Royal Society of New South Wales, Liversidge was elected its president. He was an FRS, still not yet 40; behind him lay several important achievements: a new faculty of science and science curriculum at the University; a new Technological Museum, with the prospect of a miniature South Kensington in Sydney; and a Board of Technical Education to promote instruction for artisans and managers alike. Now was the moment to press again for a plan of intercolonial co-operation in science. This would at a stroke strengthen the Royal Society's cultural leadership in Australia, establish the Australian colonies securely in the firmament of English-speaking science, and create an organization that would shine like a jewel in the crown of his presidency.

VISIONS REALIZED

In his president's address to the Royal Society of New South Wales in 1886,[66] Liversidge reviewed his proposals of 1879 and 1884, and recast them in a new light. He urged the collaboration of the 'recognised scientific societies' (there were 38) of Australia and New Zealand, comprising on paper between 2000 and 3000 members. The scheme he proposed was virtually identical to that of the British Association: providing for a general committee (or council), with delegates from the different colonial societies; a local committee to prepare congresses; and Sections for individual subjects. Some day the British Association itself might pay imperial court to colonial endeavour; in the meantime, the Australian Association would speak for the colonies. This would not only have an immediately beneficial effect, but would also permanently raise the high-water mark of thought in all the colonies, especially in connection with scientific matters: 'It would tend to stimulate all classes, and disseminate a taste for all branches of knowledge'. In parenthesis, Sydney would become in practice the headquarters of Australian science.

With the backing of his council on 30 June 1886, Liversidge wrote in July to 33 other Australian and New Zealand societies, requesting representatives (one for each 150 members) to meet at a constitutional convention to be held in Sydney on 10 November that year. A Sydney meeting in 1888 would 'take stock' of 'all scientific matters more particularly concerned with Australasia'. The Royal Society, largest and oldest, would be its host; the University of Sydney, its venue.

In August 1886 Liversidge requested government assistance, 'as is done elsewhere', towards 1888. His plan was almost defeated, however, when the Premier, Sir Patrick Jennings, contemplating his government's indecisive arrangements for the colony's centennial, took up instead Liversidge's letter of 1884, and through his agent-general in London, peremptorily invited the entire British Association. The British Association agreed to consider the question at its Birmingham meeting in August, provided the costs of the British visitors could be met. In Sydney on 8 September, however, Jennings met a two-day barrage of criticism led by Sir Henry Parkes in the Legislative Assembly, which forced him to retreat to a more limited visit of between forty and fifty 'eminent scientific persons', costing perhaps £6,000.

On 2 September 1886, *Nature* in London reported that the Premier had included the British Association in his programme for 1888.[67] However, scarcely a week later—ironically, on the same day as his censure by Parkes—the Association's council, meeting at Birmingham, fixed on Bath for its 1888 congress. There were rumours of English protest at what some considered the colony's high-handed insistence— 'in somewhat dictatorial terms', as *Nature* put it—on contributing towards the expenses of so few 'of the most eminent representatives of British science'. But in Sydney, the idea struck even more violent criticism. The *Sydney Morning Herald* thought the Premier's initiative 'rash and premature': what the colony needed were lectures on the general principles of science, for the general public, and encouragement to a 'closer examination of [Australia as] a scientific continent'.[68] Neither could be achieved by a visit from the British Association. 'The British Association would come and go', observed Sydney's *Daily Telegraph*, 'leaving no better organisation than there is at present among colonial scientists'.[69] The *Sydney Morning Herald* and the *Telegraph* in Sydney, and the *Argus* and *Age* in Melbourne supported Liversidge's plan to bring together Australia's own scattered forces. A visit from the British Association would be 'more an opportunity for Australian hospitality than Australian science'.[70]

Instead, the centennial, as Thomas Garrett (the embittered member for Camden) put it to parliament, should be a 'purely Australasian matter'. If this were not done, the 'spirit of flunkyism will be invoked, and some royal prince might attend at the expense of ourselves, or some learned scientist, and the whole interest of the thing would be frittered away...'. When distinguished men visited the colony, the result had 'not been at all satisfactory. They have spent a short time here, then gone home and written books which have grossly misrepresented us'.[71]

Almost the last word was left to Sir Henry Parkes: 'The invitation to the British Association, I imagine, is now abandoned', he told parliament, 'and I trust that the invitations to celebrators in Europe and America are also at an end? Let the government instead', he argued, 'submit some scheme which will be acceptable to all classes of the community.'[72]

In the event, the British Association did not formally reject the proposal, which was merely withdrawn by the agent-general of New South Wales after the Birmingham meeting. The door was left ajar.[73] No outcome could have better suited Liversidge or, indeed, his friends in Victoria. Just the year before Robert T. Litton, editor of the short-lived *Australasian Scientific Magazine*, had remarked with surprise that, given the number of men of learning in Australia, 'there is so little bond of union among them', outside that provided by scientific journals.[74] Earlier that year, Litton had launched the Geological Society of Australasia in Melbourne, which would by 1890 have a membership of over 100, and representatives in New South Wales and Queensland.[75] This moved in a general way towards the same idea. Meanwhile W. C. Kernot, the University of Melbourne's professor of engineering, offered Liversidge warm support, assuring him that Melbourne did not in the least grudge Sydney 'the honour of inaugurating' the congress. Following the preliminary meeting, Kernot, with typical candour, announced: 'no one had any shadow of misgivings to the thing being good, and the sooner it was done the better.'[76] The Royal Society of Queensland also sent its support.[77]

The turnout on 10 November 1886 illustrated precisely the difficulties of intercolonial communication over the previous twenty years. Twenty-seven delegates had been nominated by eighteen of the thirty-three organizations Liversidge approached, but of these, only sixteen came to the meeting at the Royal Society's premises in Elizabeth Street. Seven represented New South Wales, including four officers of the Royal Society (Russell, Liversidge, Wilkinson, and the physician C. K. Mackellar) and two of the Linnean Society of New South Wales (W. J. Stephens and former Newington biology master and editor, J. J. Fletcher). Wilkinson (aged 43) and Liversidge (40) were the youngest men present. The Zoological Society of New South Wales was represented by Liversidge's friends Dr Arthur Holroyd and the ageing medical naturalist and first secretary of the Australian Museum, George Bennett. The Royal Geographical Society of Australasia (NSW) was represented by the retired soldier, Sir Edward Strickland, KCB.

From Victoria, representing the Victorian Engineers Association came K. L. Murray; the Royal Society of Victoria sent Kernot, its president; Litton represented the Geological and Historical Societies of Australasia (Vic.); and W. J. Conder and W. H. Nash, the Victorian Institute of Surveyors. From Queensland came the engineer J. P. Thompson, for the Geographical Society of Australasia (Qld), and Henry Tryon, for the Royal Society of Queensland. Representing New Zealand were S. Herbert Cox, FCS, FGS (formerly one of Hector's geologists in the Victorian survey and later instructor in geology at Sydney Technical College) representing the Nelson Philosophical Society and the Philosophical Institute of Canterbury. The newly formed New South Wales Institution of Surveyors nominated no one, and the delegates of the Institution of Architects of New South Wales and the Engineering Association of New South Wales failed to arrive. No delegates came from the Field Naturalists Club or the Zoological Acclimatization Society of Victoria. Neither the Geographical Society of Australia (Vic.), the Institute of Architects (Vic.) nor the Microscopical Society of Victoria nominated representatives. Only three of the ten New Zealand societies were represented (from five, no reply was received), and no replies came from the three societies in South Australia. The Royal Society of Tasmania appointed two delegates, but neither appeared.[78]

It would have been difficult to regard this group as representative of Australian science, but it was certainly catholic in its diversity. It was in spirit not unlike that inaugural gathering of 'cultivators of science' at the Yorkshire Philosophical Society a half century before.[79] In the British case, they might have been gentlemen; in the Australian, they were both gentlemen and 'players'. These sixteen adopted that day the rules and procedures of the British Association, with slight modifications, and resolved to hold elections for official positions sixteen months afterwards, in March 1888. The first congress was set for September 1888. In six months, Liversidge had at last crystallized scientific, political, and popular feeling, and swept to a victory over nine years in the making.

In July 1888 a report sent to *Nature* outlined the foregoing sequence of events. It regretted, quixotically, that the British Association could not 'see their way to visit Australia during the Centennial year'.[80] It appeared that the colonial association was to be independent, *malgré lui*. It was an association 'thoroughly Australian in character'. But whatever his belief in intercolonial enterprise, Liversidge remained sensitive to his imperial relations. As he wrote to London, the new body was to be an 'Australian offshoot of the British Association'.[81] His use of 'Australian', not 'Australasian', may have been significant. But at the time, the metaphor of family, of organic union which Huxley preferred, Liversidge willingly adopted.

If the thought was father to the deed, however, the deed had to wait. On Christmas Day 1886 Liversidge again sailed for study leave in

Britain. He travelled via Japan and the United States to refresh his contacts and prepare plans for a new chemistry laboratory. His reception over the next fifteen months, culminating in the award of an honorary MA degree at Cambridge, reinforced his position as a leading representative of Australian and 'imperial' science. During his long absence, however, some thought he had given up his interest in the new association, and in the meantime, others rose to organize science, offering alternative models to Liversidge's plan. Perhaps the Imperial Institute, whose doors opened that year in London in celebration of Queen Victoria's Jubilee, could provide the much-discussed imperial focus, uniting colonial scientific societies throughout the Empire with headquarters in London.[82] A third form of federation—an association with a British body regionally represented—had been tried by the British Medical Association. In 1887, however, following a suggestion from the South Australian Branch of the British Medical Association, the first Intercolonial Medical Congress was held—coinciding with the fiftieth year of the colony and the Adelaide International Exhibition. The reception this received endorsed the idea of linked congresses, with Australasian leadership, and confirmed the popularity of the British Association model.[83]

On 7 March 1888, only five days after his return to Sydney from England, Liversidge showed his determination not to follow the 'British federal' or British-sponsored Australasian regional models. Like its new medical cousin, the Australasian Association for the Advancement of Science (AAAS) would link, not supplant, the colonial societies of Australia and New Zealand, and bring them into a common cause. That cause would be the advancement of science and the diffusion of its message—tasks beyond the ability of any one society. Federation for Liversidge was thus a means, not an end in itself.

That day, six men joined him in Elizabeth Street to elect the Association's first officers. Of the seven, the only native-born Australian was the distinguished astronomer, H. C. Russell, one of the first graduates of Sydney, who in 1886 had also become the first Sydney graduate to be elected an FRS. As a popular speaker, past president of the Royal Society (in 1877 and 1885), fellow of the Senate of the University, vice-president of the Board of Technical Education, and a pioneer of 'intercolonialism', he was easily elected president. Liversidge and Bennett were elected honorary secretaries, and Strickland, honorary treasurer. In this way, the Royal Society closed ranks with the Zoological and Geographical Societies. Between April and August, meeting first fortnightly and then weekly, Liversidge, with Russell's help, led this group in organizing the Association. Recalling the British Association's initial practice, the Premier and other leading government and university officers were invited to be vice-presidents, as were the presidents of the Royal Societies of the different colonies.

By 12 March 1888 Liversidge and Bennett had written to all thirty-three societies, reporting the recommendations of the preliminary meeting. A week later, he requested membership lists from all the different societies, so as to circularize each member—2000 altogether.

To this flurry of correspondence he added individual requests; thus, to Hector, whom he invited to be a vice-president:

I hope you will be able to make our first meeting in September next a success by coming over and giving a paper or papers; if you could give an account of the Geological and Geographical exploration of New Zealand since settlement, it would be much appreciated and no one could do it as well as yourself.
I think we shall have a very successful meeting. Will you also kindly draw attention to the meeting through the New Zealand Institute?[84]

Thus, with customary polish, and not a little arm-twisting, Liversidge won his way.

In May it was decided to make the meeting a week earlier, to accommodate the University of Melbourne, so beginning an enduring tradition of the 'conference week'. In the meantime, support grew: from 86 new members in April to 212 in May, and to 501 by early August. Liversidge and Bennett recruited local secretaries for Sections, and chose Section presidents to represent the five eastern colonies and New Zealand. Steamship and railway companies were asked for concessions, and the colonial government was asked for a pound-for-pound grant in aid of expenses. When the new council met again in the Royal Society's rooms on 27 August, at the start of the congress, it was appropriate that Liversidge be thanked by acclamation for 'initiating the movement which had led to the formation of the AAAS'.[85] Colonial science had become 'federated' at last.[86]

NOTES

1. *Saturday Review*, 66 (3 November 1888), 519–520.
2. cf., Sally G. Kohlstedt, 'Savants and Professionals: The American Association for the Advancement of Science, 1848–1860' in A. Oleson and S. C. Brown (eds), *The Pursuit of Knowledge in the Early American Republic* (Baltimore and London: Johns Hopkins University Press, 1976), 299–325; cf., Sally G. Kohlstedt, *The Formation of the American Scientific Community: The American Association for the Advancement of Science, 1848–1860* (Urbana: University of Illinois Press, 1976).
3. cf., Robert Fox, 'The *Savant* Confronts his Peers: Scientific Societies in France, 1815–1914', in Robert Fox and George Weisz (eds), *The Organisation of Science and Technology in France, 1808–1914* (Cambridge: Cambridge University Press, 1980).
4. Fifth, that is, within the western European model. The source of the 'Association idea' has been traced to Switzerland in 1815. (cf., G. V. H. Degen, 'Die Grundungsgeschichte der Gesellschaft deutscher Naturforscher und Ärzte', *Naturwissenschaft*, 11 (1955), 12); and to the better-known German Association founded in the 1820s; cf., Charles Babbage, 'Account of the Great Congress of Philosophers at Berlin', *Edinburgh Journal of Science*, 60 (1829), 225–34; J. F. W. Johnston, 'Meeting of the Cultivators of Natural Science and Medicine at Hamburg', *Edinburgh Journal of Science*, 4 (1830–3), 244; H. Duerner and H. Schipperger, *Wege der*

Naturforschung, 1822–1972 (Berlin: Springer Verlag, 1972). Following the German example, Associations were formed in Italy (1839), Scandinavia (1839), Hungary (1841), Russia (1863), Poland (1869), and Czechoslovakia (1880); cf. R. von Gizycki, 'The Associations for the Advancement of Science: An International Comparative Study', *Zeitschrift für Soziologie*, 8 (1979), 28–49. These, however, were impermanent assemblies; more permanent ones along the 'Anglo-Saxon' model, begun in Britain, were later founded in the United States (1847), South Africa (1902), Spain (1908), India (1914), and Japan (1925). Since the last war, several similar associations have come into existence in Latin America; cf., E. Diaz *et al.*, *La Ciencia Periférica* (Caracas: Monte Avila Editores, 1983); H. Vessuri, 'The Social Study of Science in Latin America', *Social Studies of Science*, 17 (3) (1987), 519–54. In 1986 an African Association was also founded, with its general office in Nairobi.

5. cf., President's Address, *Report of the British Association for the Advancement of Science*, 1 (York, 1831), and *Report of the Australasian Association for the Advancement of Science*, 1 (Sydney, 1888).
6. For introductory background, see M. Hoare, 'The Intercolonial Science Movement in Australasia, 1870–1890', *Records of the Aust. Academy of Science*, 3 (2) (1976), 7–28; and M. Hoare, 'Learned Societies in Australia: The Foundation Years in Victoria, 1850–1860', *Records of the Aust. Academy of Science*, 1 (2) (1969), 7–29; to which I am indebted for some of the information (but not for the interpretation) that follows.
7. For a discussion of the 'principle of association', in both its conceptual and organisational dimensions, see R. MacLeod, 'On the Advancement of Science' in R. MacLeod and P. Collins (eds), *The Parliament of Science* (London: Science Reviews Ltd, 1981), 38.
8. M. Hoare, '"All Things Queer and Opposite": Scientific Societies in Tasmania in the 1840's', *Isis*, 60 (2) (1969), 198–209.
9. M. Hoare, Science and Scientific Associations in Eastern Australia, 1820–1890 (unpublished PhD thesis, Australian National University, 1974).
10. *Trans. Phil. Inst. Vic.*, II (1857), xxxix, xlii, xlvi; II (1858), xxvi. I am indebted to Mr Colin Finney for this reference.
11. Ann Mozley Moyal (ed.), *Scientists in Nineteenth Century Australia: A Documentary History* (Sydney: Cassell, 1976); Ann Moyal, *A Bright and Savage Land: Science in Colonial Australia* (Sydney: William Collins, 1986); D. F. Branagan, 'Words, Actions, People, 150 Years of Scientific Bodies in Australia', *Proc. Roy. Soc. of NSW*, 103 (1971), 123–41.
12. M. Hoare, 'Botany and Society in Eastern Australia' in D. J. and S. G. M. Carr (eds), *People and Plants in Australia* (Sydney: Academic Press, 1981), 183–219.
13. Ronald Strahan *et al.*, *Rare and Curious Specimens* (Sydney: The Australian Museum, 1979); D. F. Branagan (ed.), *Rocks—Fossils—Profs: Geological Sciences in the University of Sydney, 1866–1973* (Sydney: Science Press, 1973).
14. The Royal Society of South Australia received the royal assent in 1880; that of Queensland, in 1884; and of Western Australia, in 1914.
15. Michael Cannon, *Australia in the Victorian Age. Vol. 3 Life in the Cities* (Melbourne: Nelson, 1975), chapter 8.
16. Graeme Davison, *The Rise and Fall of Marvellous Melbourne* (Melbourne: Melbourne University Press, 1978); G. Serle, *The Golden Age: A History of the Colony of Victoria, 1851–1861* (Melbourne: Melbourne University Press, 1963).

17. G. Davison, 'Exhibitions', *Australian Cultural History*, no. 2 (1982–3), 5–21.
18. John Allwood, *The Great Exhibitions* (London: Studio Vista, 1977).
19. John Wade (ed.), *The Sydney International Exhibition, 1879* (Sydney: Museum of Applied Arts and Sciences, 1979).
20. Hoare, op. cit., note 6; cf., Royal Society of NSW, *A Century of Scientific Progress* (Sydney: Roy. Soc. of NSW, 1966), chapter 1.
21. H. C. Russell, 'Presidential Address', *Journ. and Proc. Roy. Soc. of NSW*, 11 (1877), 1–20.
22. K. Inglis, 'The Special Connection', in A. F. Madden and W. H. Morris-Jones (eds), *Australia and Britain: Studies in a Changing Relationship* (Sydney: Sydney University Press, 1984), 38.
23. A. P. Martin, *Australia and the Empire* (Edinburgh: David Douglas, 1889), 63; cited in note 22.
24. D. P. Mellor, 'Liversidge, Archibald (1846–1927)', *Australian Dictionary of Biography*, 5 (1974), 93–4; R. M. MacLeod, *Science Under the Southern Cross: Archibald Liversidge, FRS (1846–1927)* (in preparation).
25. cf., D. F. Branagan and Graham Holland (eds), *Ever Reaping Something New: A Science Centenary* (Sydney: University of Sydney, Science Centenary Committee, 1985).
26. cf., Elena Grainger, *The Remarkable Reverend Clarke* (Melbourne: Oxford University Press, 1982).
27. M. Hoare and J. Radford, 'Smith, John (1821–1885)', *Australian Dictionary of Biography*, 6 (1976), 148–50.
28. *Sydney Morning Herald* (10 July 1886).
29. 'Bringing Science to the Aid of Industry and Enterprise', *Australian Town and Country Journal* (21 March 1874).
30. *The Athenaeum* (13 November 1875); the *Victorian Review* saw the 'coming Australian' as loving field sports, disinclined to recognise authority, and disliking 'mental effort—muscle over mind—all, in short that science was against'. James Hogan, 'The Coming Australian', *Victorian Review*, 3 (November 1880), 103.
31. *Illustrated Australian News* (1 January 1892).
32. Hoare, 'Intercolonial Science Movement', op. cit., note 6, 9.
33. 'Reports of Observations of the Transit of Venus, 8–9 December 1874, made at Stations in NSW', H. C. Russell, *Memoirs of Royal Astonomical Society*, 47 (1882–3), 49–88.
34. Sir William Macleay, 'Address', *Linnean Society of NSW Proc.*, 1 (1877), 95–6. Cited in Mozley Moyal, op. cit., note 11, 197; cf., D. F. Branagan, 'The *Challenger* Expedition and Australian Science', *Proc. Roy. Soc. of Edinburgh*, 73 (10) (1971–2), 85–95.
35. Ian Dick, 'The History of Scientific Endeavour in NZ', *NZ Science Review*, 9 (September 1951), 139–43; C. A. Fleming, 'The Royal Society of New Zealand: A Century of Scientific Endeavour', *Journ. Roy. Soc of NZ*, 2 (1968), 99–114; M. Hoare, 'The Relationship between Government and Science in Australia and New Zealand', *Journ. Roy. Soc. of NZ*, 6 (3) (1976), 381–94.
36. M. Hoare, *Beyond the 'Filial Piety': Science History in New Zealand: A Critical Review of the State of the Art*, 2nd Cook Lecture (Melbourne: Hawthorn Press, 1977); and M. Hoare, *Reform in New Zealand Science, 1880–1926*, 3rd Cook Lecture, (Melbourne: Hawthorn Press, 1977).
37. *The Athenaeum* (11 September 1875), 122–3.
38. W. B. Clarke, 'Anniversary Address', *Proc. Roy. Soc. of NSW*, X (1876), 1–34; *Sydney Morning Herald* (20 May 1876); *Minutes*, General Monthly

Meeting of the Royal Society of New South Wales, 2 May 1877.
39. Davison, op. cit., note 17.
40. Leonard Huxley, *Life and Letters of T. H. Huxley*, 2 vols (London: Macmillan, 1900), vol. 1, 464–7. See Liversidge's preparatory contributions to *Mines and Mineral Statistics of NSW* (Sydney: Philadelphia International Exhibition, 1876), 94–103; 110–15; 115–16.
41. cf., *Minutes of Proceedings of the International Meteorology Conference* (Sydney, November 1879), *Papers of the Legislative Assembly of New South Wales*, 1880, vol. V, 1230–6.
42. E. M. Webster, *The Moon Man: A Biography of Nicolai Miklouho-Maclay* (Melbourne: Melbourne University Press, 1984).
43. Charles (Carl) Adolph Leibius (1833–1893), born in Germany, studied in London, and arrived in Sydney in 1859, where he became assayer to the Mint. He joined the Philosophical Society (later, the Royal Society) of New South Wales in that same year, and remained active in its affairs until his death. He was joint secretary of the Royal Society of New South Wales with Liversidge from 1875 to 1885.
44. *Liversidge Papers* (University of Sydney Archives), Liversidge to Leibius, Royal Society of New South Wales, 24 March 1879.
45. Archibald Liversidge, 'The International Congress of Geologists at Paris, 1878', *Journ. Roy. Soc. of NSW*, 13 (1879), 35–42.
46. ibid.
47. cf., S. Murray-Smith, A History of Technical Education in Australia (unpublished PhD thesis, University of Melbourne, 1966).
48. 'Report upon Certain Museums for Technology, Science and Art, and also upon Scientific, Professional and Technical Instruction, and Systems and Evening Classes in Great Britain and on the Continent of Europe' (13 July 1880), *Legislative Assembly NSW* (Session 1879–1880), vol. 3, 787–1059.
49. Wade, op. cit., note 19.
50. cf., J. A. Froude, *Oceana: or England and her Colonies*, new edn (London: Longmans Green, 1886); Kaye Harman (ed.), *Australia Brought to Book: Responses to Australia by Visiting Writers, 1836–1939* (Sydney: Boobook Publications, 1985).
51. E. Scott, 'History of Australia' in G. H. Knibbs (ed.), *Federal Handbook for the 84th Congress of the British Association in Australia* (Canberra: Commonwealth Government Printer, 1914), 1–5.
52. Archibald Liversidge, 'Presidential Address', *Journ. Roy. Soc. of NSW*, 20 (1886), 1–41.
53. I owe this observation to Colin Finney, who is currently completing a book on the history of natural history in nineteenth century Australia.
54. *Geographical Society of Australasia, NSW and Victorian Branches* (Sydney: Government Printer, 1885), vol. 1, viii–ix; cf., E. M. La Meslée, *The New Australia* (Paris: E. Plon, 1883; translated and edited by Russel Ward, London: Heinemann, 1973).
55. ibid.
56. Unfortunately, with the exception of its South Australian branch, the Geographical Society of Australasia proved an unwieldy organization, and ebbed away by the end of the century. Hoare, 'The Intercolonial Movement', op. cit., note 6, 21.
57. cf., W. H. Caldwell, 'The Embryology of Monotremata and Marsupialia', *Phil. Trans. Roy. Soc.*, 178 (1887), 463–85, esp. 464; for Caldwell, see *Nature*, 148 (8 November 1941), 557–9.
58. 'Monotremes oviparous, ovum meroblastic' read the decisive lines; cf.,

H. N. Moseley, 'On the Ova of Monotremes', *Report of the BAAS*, 54 (Montreal, 1884), 777.
59. R. MacLeod, 'Introduction' in MacLeod and Collins (eds), op. cit., note 7.
60. cf., Rankine Dawson (ed.), *Sir William Dawson: Fifty Years of Work in Canada* (London: Ballantyne Hauson and Co., 1901), 204, 216–17.
61. *Melbourne Punch* (25 September 1884).
62. *Sydney Morning Herald* (17 September 1884).
63. The issues were repeatedly canvassed in Britain and Australia. For a contemporary contribution, see Henry Parkes, 'Our Growing Australian Empire', *Nineteenth Century*, 15 (1884), 138–49.
64. T. H. Huxley, 'President's Address', *Proc. Roy. Soc.*, 39 (1885), 278–301.
65. ibid.
66. Liversidge, op. cit., note 52.
67. *Nature*, 34 (2 September 1886), 434.
68. *Sydney Morning Herald* (9 September 1886).
69. *Daily Telegraph* (3 September 1886).
70. *Argus* (27 August 1888).
71. *NSW Parliamentary Debates, 1885–86*, 22 (9 September 1886), col. 4704.
72. ibid., col. 4702.
73. And reported by *Nature* a year later; cf., *Nature*, 36 (25 August 1887), 398.
74. Robert J. Litton, 'Ourselves', *Australasian Scientific Magazine*, 1 (August 1885), i.
75. cf., D. Branagan and T. Vallance, 'The Geological Society of Australasia (1885–1905)', *J. Geol. Soc. Aust.*, 14 (2) (1967), 349–50.
76. *Sydney Morning Herald* (11 November 1886); *Liversidge Papers* (University of Sydney Archives), Kernot to Liversidge, 19 July 1886).
77. *Royal Society of Queensland Minute Book*, 1 (30 August 1886) as cited in Hoare, 'The Intercolonial Science Movement', op. cit., note 6, 24.
78. Some possible reasons for this are suggested by Hoare, 'The Intercolonial Science Movement', op. cit., note 6.
79. Jack Morrell and Arnold Thackray, *Gentlemen of Science: Early Years of the British Association for the Advancement of Science* (Oxford: Oxford University Press, 1981).
80. *Nature*, 38 (6 September 1888), 437–8.
81. *Kew Papers*, 'NSW and Victorian Letters, 1865–1900', 173, Liversidge to Thiselton-Dyer, 30 September 1888.
82. H. Mortimer-Franklin, *The Unit of Imperial Federation* (London: Swann Sonnenschein, 1887).
83. *Trans. Intercolonial Medical Congress of Australasia* (Adelaide, 1888).
84. *National Museum of New Zealand*, Box 1888, f. 242. Liversidge to Hector, 21 April 1888.
85. *ANZAAS Archives* (Mitchell Library) MSS. 908/1, f. 125, General Committee Minutes, 27 August 1888.
86. cf., R. MacLeod, 'On Visiting the "Moving Metropolis": Reflections on the Architecture of Imperial Science', *Historical Records of Australian Science* 5 (1982), 1–16, reprinted in N. Reingold and M. Rothenberg (eds), *Scientific Colonialism: A Cross-Cultural Comparison* (Washington: Smithsonian Institution Press, 1987), 217–50.

2

From Imperial to National Science

Roy MacLeod

From the tentative beginnings outlined in chapter 1, the Association was to become one of the greatest factors in the spread of knowledge throughout the colonies. By 1938 it was the principal national forum for the presentation of Australian and New Zealand science. This chapter will trace the ideas and events that characterized the congress history of the Association's first decades, and the policies that governed it through the early years of Federation, culminating in the visit of the British Association in 1914, and will follow ANZAAS through the difficult years that ended with the Association's jubilee in 1939. In this half-century, the Association, in alliance with the leadership of Australia's other scientific organizations, faced the challenge of giving stimulus to science in an independent Commonwealth that remained in many ways dependent on the ties of empire.

THE FIRST CYCLE

The first congress, 27 August–8 September 1888, was a celebrated event for the colony, the Royal Society, the University of Sydney, and for Australian science. Warm greetings followed from London.[1] H. C. Russell's presidential address set a high standard of historical argument and public representation, uniting the heterogeneous interests of the different disciplines and factions. The Association was not to be the hobby of a few individuals 'or of one colony', but of 'scientific men and lovers of science'.[2] Its purpose was to 'work up the facts known in every branch of Australian science',[3] and to advance the 'culture of science' in what he called its 'comprehensive sense'. The future Association would surely see science grapple with material 'questions in chemistry, physics and geology; in mining, mineralogy and engineering; in meteorology, water conservation and irrigation, and every other subject that may promote our national advancement'. His final appeal was to echo through the years: 'Science stands or falls as a whole; if we limit it to certain purposes or persons it ceases to be science.' 'This Association', he concluded, 'stands as a protest against the shortsighted and utilitarian policy of those who would cultivate only what they characteristically call the bread and butter sciences.'[4] Certainly, an ill-considered criticism of science as an enemy of general education (moved in a vote of thanks to Russell by the University's chancellor, Sir William Manning) brought a pointed protest from Sir James Hector, which immediately unified the assembly.[5] The proposal to advance equally all the sciences conveyed a Huxleyan vision of material application, based on the appreciation of fundamental principles. The Association was not to be a visionary body, but one led by 'practical idealists', with a programme of useful reforms.

The Congress progressed majestically through a week of ceremonies, *conversaziones*, and excursions. Of the 857 participants, the largest number (560) were, predictably, from New South Wales; Victoria sent 130 representatives; South Australia, 59; Queensland, 46; and Tasmania, 18. Over 40 came from New Zealand, including Sir James Hector from Wellington, and J. A. Pond, S. Percy Smith, and Professor A. P. W. Thomas from Auckland. Hector spoke on the recent Tarawera eruption and played a typically generous role in the Congress's proceedings. R. M. Johnston, the government statistician from Tasmania, expounded upon the novelty of Australian geological and fossil evidence, and Dr E. C. Stirling of Adelaide announced the discovery of the marsupial mole. C. S. Griffith held it was the duty of Australians to explore Antarctica, 'to secure to this colony, universal attention, and the approbation of the entire civilized globe'. Liversidge contributed to the Chemistry Section, giving two papers plus a description of his new chemical laboratories at the University of Sydney, and his colleagues Russell, Threlfall, and Pollock made their mark in Physics. The Hon. John Forrest, speaking to Section E (Geography), surveyed the geographical determinants—the conservation of water, the survey of minerals—which demanded unified efforts, and underlined the value of federation, in which 'colonial history and colonial enterprise' were inextricably connected.[6] The general spirit was summed by the young professor of physics at Adelaide, W. H. Bragg, who wrote excitedly to his fiancée: 'I think this Association is going to do us a lot of good, especially such as, like me, are willing to work, but don't quite know where to begin. Contact with other and more experienced workers will start us off on the right track.'[7]

The new council, with some of the original delegates present, represented the powers of combined interest. In four days in September they met to put all in place. Led by Liversidge, the council included E. H. Rennie, graduate of both Sydney and London, and now professor of chemistry at Adelaide, George Bennett, J. C. Cox, and W. C. Kernot, and a dozen new friends from other societies and government departments. To secure a close federation, vice-presidents were to be limited to the presidents of the colonial Royal Societies. Local committees were to be formed as branch offices, effectively and at a stroke uniting the eastern colonies. Colleagues from the medical world (notably Professor T. P. Anderson Stuart of Sydney and Dr Alan Campbell of Adelaide) proposed holding the intercolonial medical congresses at the same place and time. At Hector's suggestion, the inevitable followed: Sydney (and its Royal Society) was to be the corporate headquarters.

Following the British Association model, thirteen Research Committees of Investigation were formed. Forrest put forward the question of Antarctic exploration; Ellery proposed the 'state of meteorology'; R. L. Jack, the geological record; and Professor J. G. Black of Otago, the state of chemistry, with reference to gold and silver. W. H. Haswell (Stephen's successor at Sydney) suggested the endowment of a biological station; Baldwin Spencer, a bibliography of Australian biology and a committee to consider the protection of native birds and mammals; James Wilson (professor of anatomy at Sydney) the 'construction and hygiene requirements of places of amusement in Sydney'. Liversidge proposed no fewer than six topics, including a census of Australian minerals, and a bibliography on Aboriginal Australia and Polynesia. The topics received a warm response. 'The great need at present', argued J. Steel Robertson in the *Centennial Magazine,* is for 'more extended original research'.[8] Yet the AAAS must also demonstrate its practical worth. After all, the typical learned society had, in the words of the *Australasian,* 'too often condescended, in its dearth of valuable papers, ... to the most trivial and ignorant discussions on rabbit killing or other topics of the day'. The Association would necessarily be an 'institution of a different sort', one that 'represents an important step in the intellectual development of the colonies'. The 'writers who appear before it will be kept up to the mark by the certainty of vigilant and competent criticism'. The AAAS, above all, 'directly challenges public regard'.[9]

Born at the end of the 'long boom' in the economic development of the eastern colonies, the Association was from the start implicated in public affairs and in questions of the day. For example, on the recommendation of Henry Hayter, government statistician of Victoria, the council unanimously agreed to establish a committee to enquire into the conditions of labour, with special reference to strikes, and to make suggestions for their remedy.[10] But its profile was predominantly 'theoretical', and Sections declined to discuss 'matters with which money making is immediately connected in an industrial spirit'.[11] The need to fashion a mutually sustaining relationship between scientific enquiry and material interest posed a difficulty for the AAAS in 1888 that would not quickly disappear. Nor would more practical difficulties of organization and finance. In the Association's management, Liversidge bore the greatest burden. Strickland's death in 1889 made it necessary for him to assume the duties of treasurer, as well as secretary. He was obliged to wait nearly a year before the New South Wales government granted £500 to assist the publication of the first congress proceedings,[12] and few of the Association's financial uncertainties were resolved before the second congress, which opened in Melbourne in January 1890. Happily, Melbourne witnessed the largest attendance (over 1000) in the Association's first fifty years. *Table Talk*[13] welcomed the 'Festive Scientists' to what the *Australasian*[14] called their 'scientific gaieties'. So it continued, through orchestral music and *conversaziones* at the Town Hall, ending with visits to local sites of scientific interest. Baldwin Spencer's *Handbook* was a *tour de force*, and no visitor remained uninformed of the Miocene lava he (or his lady associate) might find under Collins Street, or the two extinct volcanoes within 20 miles of the GPO. 'Science makes "impossible" a thing of the past', said Sir William Clarke, Governor of Victoria, as he welcomed 800 of the delegates to a viceregal entertainment, featuring refreshments under three marquees, perfect weather and military brass bands playing what the *Argus* identified as 'something very like God save the Queen'. [15]

Indeed, what the British Association had done for the spirit of Empire, so the AAAS would do for the colonies. In a long-winded presidential address, the venerable Baron von Mueller (to whom Ellery graciously deferred) pledged the imperial loyalty of the Association: 'while we "science notaries" do not engage in political discussion', he proposed, we might still 'foster ... through our bonds the "union of the Empire"'.[16] The *Argus* considered the Association still an 'experiment'—it remained to be seen 'how far its proceedings would stimulate professional students of science, and enlist the sympathy of the general public'.[17] In the event, twenty-five members were present at the general committee to thank Baldwin Spencer for his organization of the proceedings. The vast, seventy-six page Minerals Census, orchestrated by Liversidge in 1888, was received (and, when published, dominated the congress report). Liversidge himself spoke on hot spring waters, and T. W. Edgeworth David (making his second AAAS appearance) on the New South Wales coalfields. On the recommendation of Section B (Chemisty and Mineralogy) a special committee was set up under Rennie of Adelaide 'to formulate a scheme whereby the assistance of the governments of the various colonies may be enlisted in procuring material for special investigations'.[18]

Six research committees reported; eight were renewed and six others were established to the satisfaction of the council. The Association began to make friends in high places: Richard Threlfall (professor of physics at Sydney) read a paper to Section A on the 'Present State of Electrical Knowledge', which was seen by Sir Samuel Griffith, Premier of Queensland (and member of the Constitutional Committee of the 1891 Federation Convention) and by Bernard Wise, MLA (New South Wales). Both were AAAS members in 1890, and both later sought powers for the federal parliament to control postal, telegraph, and telephone services throughout Australia.[19] From the press, however, there was some criticism of the Sections, of which several were noticeably more thinly attended than the social events.[20] Looking ahead, membership was a key issue. The British Association rules, as first adopted, restricted membership to those who 'belonged to Literary and Philosophical Societies publishing Translations or Journals in the British

Empire'. Almost immediately this was found too limiting, and the council agreed that anyone could join for £1 per year (or £10 for a lifetime). This greater pluralism brought in train its own problems. Professor Ralph Tate of Adelaide (like Liversidge, a former Royal School of Mines man), who led the delegation of seventy from South Australia to Melbourne, urged that sectional committees be more selective in their choice of papers. A wider constituency was not necessarily more 'scientific'. Section F (Economic and Social Science and Statistics) was, for example, criticized for not limiting itself to 'rational investigations'.[21] In other respects, too, Liversidge's wider perspective was not universally accepted. When Section I (Literature and Fine Arts) attracted only three papers, there were murmurs of protest and suggestions that it be discontinued. Possibly, as Edward Morris (professor of modern languages and literature at Melbourne) put it, 'men of letters did not wish to be dragged at the chariot wheels of science'.[22] In the event, David Orme Masson (professor of chemistry at Melbourne) condemned it, suggesting with William Sutherland (a prominent physicist)[23] that a new educational section be created in its place.[24]

For those wishing to keep New Zealand within the 'federation', the third congress venue was an inspired choice. In 1891 the Association went to Christchurch. 'This will not be a voyage and journey for recreation', von Mueller assured his·minister when seeking leave; yet how could it be otherwise?[25] The New Zealand Institute had previously invited the Association to Auckland 'as the most populous city in the colony', but it was the nomination of Liversidge's close friend Captain F. W. Hutton that won the laurels for Christchurch. There, welcomed by the Otago Institute, 235 New Zealanders joined the gathering, with Sir James Hector delivering the presidential address (and editing the proceedings). With this third Congress, the spirit of Australian 'federation' was made superbly manifest. Science had provided 'the first truly effective step towards Federation which has yet been achieved, and I trust that all our workers will continue to be imbued with this spirit'.[26] Hector could see in the Association an extension of the 'New Zealand principle' to the whole of the colonies.[27] He gave personal tribute to the trans-Tasman fellowship begun with Liversidge's first trip to New Zealand fifteen years before:

Politicians should take this well to heart. Let them continue to aid all efforts that will tend to bring scientific accumulations in these colonies into a common store, so that each may discover for what purpose it has been best adapted by Nature; and [by which] each may prosper to the full extent of its natural advantages.[28]

The *Hobart Mercury* agreed: 'in these Unions are splendid evidence of the true spirit of Federation. Science can do for itself ... without much fuss or delay, what politicians are unable to accomplish with infinite talk and delay.'[29] For Hector and his generation, institutional co-operation was central to colonial life. 'Every step in [New Zealand's] reclamation from a wide [sic] state of nature', he told the 500 delegates, 'has depended upon the application of scientific knowledge.'[30] And those who had travelled free by steamship, courtesy of the New Zealand government, or on subsidized rail passes, could happily concur.

In Christchurch, Liversidge stayed at Canterbury College with Hutton, who proposed a recommendations committee (again along British Association lines) to consider the business arising from the Sections. Liversidge agreed, and new rules were drafted, to be approved the following year. The intercolonial network gave young and established scholars access through the Sections, where Masson brilliantly explained the gaseous theory of solution, and W. A. Haswell outlined possible mechanisms of evolution. The press gave space to a paper by Samuel Clemens of Tasmania on the population question, which was to fuel debate on immigration policy in years to come.[31] Not all Liversidge's co-operative impulses had so far fulfilled expectations. Six of the committees of investigation set up in 1888 were deemed to have lapsed in 1891. Four new ones now emerged, however, on Sanitation, the Movement of New Zealand Glaciers, the Chemical Composition of Australasian Mineral Waters, and the Chemical Standards required by New Adulteration Legislation. While Section E (Geography) would not endorse a recommendation that the AAAS memorialize colonial grants for funds to mount an Antarctic expedition, it nevertheless endorsed the desirability of Antarctic exploration—to 'improve the assets of our several Chambers of commerce', as Commander Pasco, RN, Section president put it. This set a precedent for the support of Antarctic research over the next four decades.[32]

From Melbourne onwards, as the Association gathered impetus, the council considered several important constitutional changes. At the outset, the rules vested supreme control in the hands of a general council, comprising delegates from the various colonial scientific societies, one for every hundred members. (What defined an acceptable society was decided on a case basis). This assembly was to act through a general committee, a wider parliament consisting of the council, the officers of the Sections, and all those who read papers; the congresses were to be organized by local committees. By the meeting in Melbourne, however, it was clear that the 'ten commandments' borrowed from the British Association were not sufficiently flexible. For example, the role of the thirty-eight Australasian societies Liversidge had initially approached remained unclear.[33] Certainly Liversidge and the council depended upon their membership, their goodwill, and their premises at congress times, but they had also to distance themselves from strictly local issues. To many, the Royal Societies of New South Wales, Queensland, and Victoria still conveyed an image of bourgeois elitism and social cliques; while other societies were identified with local, particular, and not always compatible political or intellectual interests. Moreover, even if twenty-eight of the societies that agreed to join the enterprise actually sent delegates, the membership of the remaining bodies could not be neglected by an Association that ostensibly spoke to all Australasia. It became important for the Association both to embrace and transcend local societies: to make a federation of science, not a federation of societies.

A committee appointed at Melbourne to consider constitutional revisions reported at Christchurch in 1891, where a new set of forty-four rules was adopted, to be ratified at the Hobart Congress in 1892. These altered the government of the Association in several significant ways (see appendix 3). In particular, they had the twin effect of freeing the Association from the possibility of control by a combination of colonial societies and removing any danger of its acting in competition with them. In practice, the Association (in a strict sense, Liversidge) continued to consult the societies, but it gave notice that, 'as a matter of general policy it did not recognise an extension to other societies, lest claims for special representation should become too numerous'.[34] Societies continued to send delegates to council, although in 1913 the council decided to restrict this right to those that undertook and published research.[35]

In 1892 the 'travelling palladium' made its fourth stop—in Hobart, then celebrating the fiftieth anniversary of the inauguration of the pioneering *Tasmanian Journal* (the forerunner of the *Journal of the Royal Society of Tasmania*). For the first time, an AAAS congress was led by one vice-regal representative (the Governor of Tasmania, Sir Robert Hamilton) and attended by a second (Queensland's Sir Henry Norman). A gathering of 600 (a quarter of them women) heard 128 papers some of which inevitably were distinguished—so it was said—more by their catholicity than by their content. Only about half of these were eventually published, and historians may still ponder what the Archbishop of Hobart had to say about 'Solar Phenomena and their Effects', or Archdeacon Hales about 'the Science of the Unseen'. There were moments of excitement, however, when the press caught wind of an argument between the Anglican and Congregationalist ministers, about which should give the official 'Science Sermon'. In the end, the Congregationalists had the morning service, the Anglicans, the evening, and in this ecumenical spirit the scientific sessions went ahead unruffled. Geographers were told much of Tasman maps, and in the midst of Australia's worst depression of the nineteenth century, Section F (Economics) heard the Association's first woman speaker, Mrs Alex Morton (wife to the Secretary of the Royal Society of Tasmania) tell of the shearer's strike in Queensland, and of the 'unprincipled agitators' of the trade union movement. The 'socialistic addresses by K. Marx, the German Socialist', she observed, were 'calculated to deepen the feelings of unthinking readers against the present condition of things'.[36]

If feelings were deepened, minds were touched by Robert Giffen's lecture on the economics of empire and by H. H. Hayter on the 'urbanization' of Australia. Full advantage of the Sections was taken by W. H. Bragg, making his debut as president of Section A, and by W. M. Hamlet, government analyst of New South Wales, who reported on the conditions of drudgery confronting colonial chemists. Preoccupied with assaying, agriculture, sanitation, and 'criminal investigations incidental to our rapidly growing centres of population, Australian chemists', said Hamlet, 'inevitably occupy places in the rear guard of the advancing army of science.' Perhaps Australians need not remain 'at the outposts skirmishing on the frontier of the knowable. Instead they could use their knowledge toward the well being of society, and so indirectly help the advancement of science.'[37]

For Section I (Literature and Fine Arts) its last congress appearance proved unexpectedly popular. Professor Morris thanked the 'men of science who rule this Association', but warned them in parting that a 'world in which science reigns supreme, where she exercises complete control over education, complete mastery over platform, chair and pulpit, would run the risk of being a dull world'.[38] The Section[39] concluded with a fine lecture on 'Heralds of Australian Literature' by Rolf Boldrewood (T. A. Brown). 'Given the ordinary environments of civilization', he said, 'any British community will exhibit the same personal differentiation of type, and evolve the same average of intellectual development.' The Association seemed fated to reflect its British ancestry, with slight adaptations to its novel environment. An unsuccessful attempt was made to induce T. H. Huxley to come and throw light on this question. Three months later *Nature* printed a lengthy report,[40] and the Association seemed, in British eyes, secure.

Liversidge and his circle had little certainty of anything but the need for hard work. After two sea voyages in succession, there were fears that the Association might be deserted by those with purses too short for long journeys, or by some whose 'devotion to science' as the *South Australian Register* put it, 'could not stand the strain of seasickness'.[41] In September 1893 the Association travelled hopefully to Adelaide where university rooms were commandeered, together with the Public Library and Institute. There took place its fifth and smallest meeting to date (and the smallest in its history). Despite much touting for custom, and the strenuous efforts of Rennie and Bragg, only ninety visitors came, and only seventy papers were read.[42] The timing of the meeting conflicted with university terms, thus losing academic Victorians (Kernot was among the few to make it) and many from the other colonies. Adelaide indeed seemed, as Bragg's experience testified, an isolated place.

Nevertheless, the meeting coincided with the discovery of ancient fossils and, hearing reports by Sir Charles Todd on the meteorological network now linking Australia, received an enthusiastic press reception. Difficulties were noticed, too: almost certainly, most participants came for the social events; many opposed a special status for 'lady associates'; and the absence of discussion after papers left many dissatisfied. Adelaide, however, saw the first grant (£80 to study glacial phenomena in Hallett's Cove) made from the research fund, and the South Australian government contributed £500 towards the congress's publication. Taking into account the poor showing and the fact that so few scientists appeared able to travel, the council decided to abandon the idea of an annual congress and to postpone the next to January 1895.

When the Association travelled for the first time to Brisbane, meeting in the Brisbane Boys Grammar School, the colony was in the grip of deep depression. Sir Samuel Griffith, Chief Justice, suggested that more must be known statistically of its effects. E. M. Shelton, of the Queensland Department of Agriculture, criticized schools for cultivating 'the universal distaste of colonial youth for the better sort of industrial pursuits', and for 'fostering the spectacle of an entire nation holding its breath over the outcome of a game of cricket'.[43] But most speakers took up less sensitive subjects. Liversidge excelled himself, presenting nine scientific papers in three different Sections. Russell and Kernot loyally spoke, as did von Mueller, aged 70, at what would be his last congress. For the first time the Congress issued a series of public recommendations, requesting the extension of legislation to require the notification of infectious diseases and the establishment of a nature reserve at Mount Belleden-Ker.

Welcoming the Congress to Brisbane, R. L. Jack used his presidential address to the Royal Society of Queensland to acknowledge the 'federal council of the scientific workers of these colonies'.[44] But as Liversidge knew, federation, even in science, was not without its critics. What would be the Association's relationship with individual colonial societies which met far more frequently and ostensibly did so much more? Would the dangers of 'over-differentiation', in Jack's words, be remedied by union through the AAAS? Were the problems of the country sufficiently national in scope to justify allegiance to a central body, based in Sydney? And what would be the particular role of the federal union in studying those fields of Australian science 'distinctly her own'?[45]

These were among the questions confronting the council as it closed its first 'cycle of the sciences', with its seventh congress, the second in Sydney. In 1895 Liversidge was nominated to be president in 1897, but as he was to be overseas, the council decided to postpone the congress until 1898. Liversidge opposed the motion, fearing an intermission of three years would cost the Association its momentum, if not its life. Faced by a *fait accompli*, however, he acquiesced. For the two years of his absence, AAAS affairs simmered quietly. As time went by, tributes attached to him, that 'pleasant, frank-mannered gentleman of middle age', as the New Zealand press described him.[46] Von Mueller had spoken of Liversidge's 'genius and circumspect assiduity'; to Hector at Christchurch, he was the 'founder'; while to Edgeworth David he was simply the 'father of Australian science'.[47]

The Sydney Congress and the presidency of Liversidge witnessed the apotheosis of the Association. Virtually daily coverage by all Sydney metropolitan newspapers, beginning as early as March 1897, ensured it a rapturous reception, as the 'fairy tales of science' were unfolded in the University's lecture rooms and laboratories. The opening ceremony in the Great Hall saw the Premier, the Chief Justice, the Speaker of the Legislative Assembly, the President of the Legislative Council, and the heads of government departments gathered to pay homage to the symbols of scientific co-operation. The scientists expected to attend were, promised the *Sydney Morning Herald*, 'not men of local reputation merely, but [men who] have ... done work that is known wherever the English language is spoken, and beyond it'.[48] Over 200 papers on the agenda testified to the vitality of Australian science. For the proseworthy Hamlet, the 'band of science workers' had produced notable results. To judge by the amount of science recorded, his Sydney *Handbook* of 1888, as he put it, was not only out of print, but also 'out of date'.[49]

To the press, the Association's 'federal constitution' had been the 'keynote of its success'.[50] Comparisons with the mother country, and the British Association, received full play: 'the colonial organisation is less comprehensive ... but as conscientious ... and as thorough'—as the *Herald* added reflectively—as the body that was no longer its parent, but now 'its elder brother'. Chemistry and physics were now unashamedly to be treated by reference to local Australasian phenomena, and ethnology and anthropology studied 'by men who have risked disease and violence in their efforts to obtain a thorough knowledge of the subject'. Overall, the Association had 'exploited the present, and looked to the future' of science in Australasia.[51]

Ralph Tate introduced Liversidge in the Great Hall as the 'father of this great scientific movement in Australasia'.[52] In his presidential address Liversidge confirmed the *Herald*'s view and spoke to the advancement of chemistry and the participation of Australia in international science. He would be heard again on these subjects, but this was the Everest of his AAAS career.[53] It was a glorious moment for the Association as well, with six volumes published, over thirty committees reporting, its first research grants awarded, and several recommendations accepted. In particular, the New Zealand and Queensland governments had agreed to create reserves for native flora and fauna, and Queensland had agreed to introduce the compulsory notification of infectious diseases. The council contemplated a progressive future. In the agrarian metaphor of colonial life, the *Daily Telegraph* assured its readers, the 'land had been cleared, the crop sown, and now the pioneers could reflect and consider better methods of doing both'.[54]

FEATURES OF 'FEDERATED SCIENCE'

For the next fifteen years the Association continued to form, albeit intermittently, a focus of Australasian science. In the *Sydney Morning Herald*'s phrase, the congress 'epitomised in a few days the intellectual life of Australasia'.[55] Five years later, the *Herald* renewed this ambiguous compliment: 'It says a good deal of the enthusiasm and determination of its members that this ... Congress has firmly established itself in the face of many difficulties that do not beset ... the British Association.'[56]

The early meetings had established a highly visible image for the congress, as hundreds of be-ticketed and be-tagged participants invaded the leading hotels and finest houses, and monopolized the university buildings, public schools, and town halls of Australia and New Zealand eight times before Federation. The Association's leadership, drawing upon a population of less than a hundred government and academic scientists, was inevitably small, and unavoidably left to those whose creature it was. Despite its rhetoric of openness, it was in fact a fairly closed circle. At the Sydney meeting in November 1886, it was resolved that the president be elected from the colony in which the congress was held. But, as in the British Association, presidents were creatures of the moment; the running of the organization fell upon Sydney. After Strickland's death in 1888, Russell continued as honorary general treasurer until 1904; Liversidge continued as permanent honorary secretary until his retirement from Sydney and return 'home' in 1907. Even then he continued as the Association's representative at British Association meetings in England for the next two decades. When he retired to England,[57] he was succeeded by his protégé, J. H. Maiden (first curator of the Technological Museum and later director of Sydney's Botanic Gardens), who served as general secretary until 1922. There were to be only nine other holders of this office to 1988, six of them in the last twenty years (see appendix 1). Russell's successor, David Carment, served as honorary general treasurer from 1904 until 1934.

Not only in administration did the industry of a few carry the interests of the many. In the sixteen congresses between 1888 and 1923, a group of 67 men and women delivered 5 or more papers, totalling 604; about half of those speaking in 1923 had been members in 1888.[58] Of the fourteen Committees of Investigation appointed at the first congress, Liversidge was responsible for five; but fewer than twenty other men— including Charles Todd, Robert Ellery, H. C. Russell, James Hector, Frederick Hutton, Ralph Tate, W. B. Spencer, Hubert Murray, T. W. Edgeworth David and Walter Howchin—joined the other committees appointed in the first few years. Eventually, these would act like their British Association counterparts, as Baconian 'merchants of light', calling upon many different specialists. By 1900 over twenty of their reports had been published, on subjects ranging from the endowment of a biological station to the improvement of town architecture and sanitation; from the prevention of wheat rust to the fertilization of figs; from the protection of native animals to the encouragement of psychometry; from the prediction of tides to the improvement of museums. For the first three decades, however, they were engineered by a handful of men. It was doubtless necessary for a small number to give a degree of continuity to the Association's affairs. Unavoidably, the continuity had its costs, as well as its benefits.

Russell spoke in 1888 of the Association's promise to bring 'to the front many men ... now scattered through the country who have ability and genius ... whose daily work is of another kind', but who have the 'spare time to take up a limited subject and fill up the whole detail'.[59] This sentiment was echoed by von Mueller in 1890: 'Ours is a kind of scientific federation full of soul. Every one can help'.[60] But against this principle, there seemed to be two important practices. First, for such 'amateurs'—in the Australian context, almost literally hewers of wood and drawers of water—there was in fact little room in the formal congress proceedings. It was a middle-class and at least 'semi-professional' organization from the beginning. Bravely, some congresses sponsored 'working men's lectures', but these feudal reminders of Tyndall, Huxley and Jermyn Street, twenty years and 12,000 miles away, were not a success. Second, the Association did not hesitate to cultivate vice-regal patronage. The Governor of New South Wales was quite forthcoming in 1888, and subsequently Sir William Clarke, Governor of Victoria, and the Earl of Onslow, Governor of New Zealand, made approving remarks that no mere university chancellor could oppose. The Governor of Queensland himself attended the Hobart meeting of 1892 to invite delegates to Brisbane in 1895. Despite his own protestations that he was 'not a man of science', Sir Robert Hamilton, Governor of Tasmania, was elected president at Hobart in 1892, and civic and colonial dignitaries were evident on the platform and in the *conversaziones*, culminating in Sydney in 1898.

This appeal to both the colonial establishment and the ostensibly 'classless' democracy worked well for the Association, certainly at a social level. It would also be wrong to under-estimate the importance of the social events, especially the excursions. In 1898, Liversidge claimed that these entertainments should not be looked upon as 'mere pleasure-making functions; they perform a real part of the work of this Association'.[61]

Yet how far could the Association reach the Australian public? There were three principal limitations. First, the AAAS was an urban body. Some thought that, like the British Association, the AAAS would in time alight in provincial towns such as Bathurst or Ballarat, thus (as in Britain) stimulating local endeavour. This was, alas, not to be, although Liversidge had it in mind as late as 1909.[62] Second, there was a bias towards the eastern 'capital' cities, and no strong inclination to include Perth. There was, to be sure, as much vying by potential hosts as one would find among intercolonial exhibitions. But the council—strongly influenced, one suspects, by Liversidge and the New Zealanders at first—decided its cycle well in advance and with an eye to securing large numbers. It was the fear of losing members, even of dissolving altogether, that seems to have impelled the choice of Christchurch and Hobart before Adelaide: places less costly to reach by sea, and with a larger expected following once there. In 1907 council voted against going to Brisbane in 1909, fearful, it would seem, both of that city's summer heat and small attendances—until the Mayor of Brisbane and the Royal Society of Queensland protested at this 'slight' to their state's pride.[63] In the event, Liversidge's preferences for a Tasmanian summer were overruled, but in return he secured Sydney for the thirteenth meeting in 1911.

Among a people gathered for the most part in five or six major cities, it is difficult to know how far 'capitalization' in Sydney restricted the Association's impact, or the diffusion of its message. The economic crash—beginning in 1892 and continuing through the decade—possibly retarded general public interest, although the period saw increasing attention to technical and commercial education. The economic environment almost certainly limited the Association's influence on the pastoral, agricultural and mining communities—prompting H. C. L. Anderson, director of agriculture in New South Wales, to refer to the pressures of the 'city theorist' and 'clerical agriculturist' 'who live on the farmer, not on the soil'.[64] But a third factor—the character of 'congress science' and its presentation—limited its popular appeal even within the cities. Inevitably, few discoveries new to science were announced. Instead, as the *Argus* predicted in 1890, the Sections cultivated detailed papers, often of an extremely technical nature.[65] Jan Todd and Ian Inkster have calculated that for every 'general' paper given between 1892 and 1895, there were two 'local' papers, and for every one 'theoretical' paper, there were ten 'empirical' ones.[66] This pattern applied throughout the early decades. Press reports repeatedly questioned whether audiences at Sections much exceeded those waiting their turn to speak. The presidential addresses, both general and of Sections, were opportunities to rise above specialization and present a wide picture. At the end of the first cycle, Liversidge admitted that the 'opening meeting of the Session is almost the only opportunity, and ... certainly the most fitting occasion to review what we have accomplished and to consider what we might do for the advancement of science'.[67] But even a 'statesman of science' of the experience and stature of Hector felt obliged to apologize, as one who had never enjoyed 'opportunities for gaining experience as a teacher and public speaker, so to balance his words as to avoid offending the specialists with crude and imperfect statements, and, at the same turn, escape wearying the general audience'.[68] Under the circumstances, it was unfair to expect, as the *Sydney Morning Herald* warned its readers, 'those authoritative summings up of the progress of knowledge which one so eagerly looked for in England'.[69] Not even the British Association was free from the besetting sin of specialization, of course, but, like its British parent, the AAAS seemed unable to find, let alone accept, a solution.

The Association's scope of activity can in some sense be gauged by its financial accounts. Receipts from congresses rarely averaged more than £800 and varied widely, especially in the early years. Thus Melbourne earned £2,081 in 1890, and Adelaide, £426 in 1893. Melbourne had attracted 1162 'members and ladies' in 1890 with a 20 shillings subscription. This was reduced to 5 shillings in 1893 but raised again to 10 shillings in 1900. Balances (after publications, excursions, and other costs) returned to headquarters were small, varying from a 'peak' of £111 in 1890 to a 'trough' in 1902, when the 395 participants at Hobart generated only £83 for Sydney.[70] Government contributions averaged £500, although Victoria gave £1,000 in 1890; rarely did government grants cover costs.[71] Accrued funds for special purposes, including the research grants Liversidge wished to award, grew very slowly. What became the research fund was, in fact, the entire capital accumulated by the Association. Under Liversidge, Russell and their successors, this sum reached only £3,000 by 1910.[72]

If the organization's financial history suggests how contrained the leadership might be, it does not necessarily mirror the Association's overall impact. The circulation of congress proceedings reached a large number—perhaps 2000 at best—for the most popular congresses. And if the Association conveyed its oral message to no more than a few hundred, it nevertheless enjoyed enormous newspaper coverage at home and overseas, and in consequence, perhaps an unexpected influence. From the outset, the colonial press was sympathetic (except to Section I, Literature and Fine Arts) and in turn was warmly acknowledged. As in Britain, the occurrence of such an event in the depths of the 'silly season' was an editor's godsend. Too much ink had dried since the 'Mudfog Papers' of Dickens's day; there were few scurrilous attacks on the Association, and even the *Bulletin* limited its cartoons to mild, almost perfunctory, parodies. By the 1880s the advantages of science were, after all, there to be seen, or dismissed, according to taste, but the congresses, so clearly 'sound' events, transcended caricature. *Sydney* and *Melbourne Punch* made half-hearted attempts at visual puns, but their jokes were stale. Whatever humour there was at the congresses arose inadvertently, or from that gift for self-parody that only the most serious possess.

Perhaps the Association, bound in solid volumes, suffered from a lack of humour in its public persona. Moral earnestness dominated its formal proceedings; classical music, its *conversaziones*; and formal dress parties—in the grounds of government houses and universities—its assemblies. Editors greeted the congresses with solemn columns, often doing far more than the Association itself to define its sober purpose. Thus the *Argus* sympathized with the lonely savant: when our scientific men 'are forced to pursue their solitary labours in separate cities', with few opportunities of meeting, they were apt to become 'monotonous and one-sided'. Congresses would stimulate the 'flagging, scientific worker—broaden his vision, and give science a higher place in public esteem'.[73] This required not frivolity but hard work. According to the *Hobart Mercury*, the public languished for want of general instruction; the Association through science rescued men and women by offering spiritual salvation against preoccupation with material things, while also (through the encouragement, among other things, of Antarctic research) 'serving a valuable commercial enterprise'.[74] There was something for everyone—except, perhaps, the majority who resisted the claims of science or education entirely. For those who gave papers—the purpose was simply to meet and to reap more publicity in a week than most could hope otherwise to achieve in years of colonial publication—and, of course, to tour, to explore parts of the country many had never seen, with excursions that more than compensated the railways and steamship companies for their generous travel concessions.

At the height of the Sydney Congress in 1898, these achievements and imperfections were well known. By the early 1900s, several centrifugal forces were fragmenting the body scientific. The coming of Australian Federation in 1901 strengthened state departments of agriculture, and in some cases geological surveys, but gave little impulse to scientific research on a federal basis. The 'loss' of New Zealand distanced the islands; New Zealanders would have only one further meeting (Dunedin in 1904) during the next quarter century. The *Sydney Morning Herald* mourned that the 'general reader' was being increasingly isolated from the scientific professions.[75] Finally, within the next 'cycle' of meetings, from 1898 to 1911, the Association's leadership began to change. Leibius and Bennett died in 1893 and Tate in 1901; Hutton retired in 1904 and died the following year; Liversidge retired in 1907; Russell died the same year, and Hector, Howitt, and Ellery, in 1908. Bragg left for England in 1909, and Kernot died the same year. Adelaide and the Congress of 1907 'captured in glorious photographs of excursions to E. C. Stirling's "at home" at Mount Lofty and to the Long Gully National Park' was possibly the last occasion on which the 'old guard' met. Only 335 members attended, but the occasion was memorable. It was Liversidge's last congress, but he was everywhere in evidence, with Alfred Howitt, the distinguished explorer and anthropologist, in the chair.[76] As before, Liversidge explained the accounts, sited the next congress meetings, oversaw the work of the recommendations and sectional committees and the award of the new Mueller medal (see appendix 6), and sponsored the creation of a new class of associates, for 'ladies and matriculated students', at 10 shillings a year. But the Association was now in different hands, and looked to different leadership.

In the event the leadership passed to men of similar mind, who sailed the federated flagship well into the new century. These included George Knibbs, who later moved to wider responsibilities under the Commonwealth government; Baldwin Spencer and Orme Masson of Melbourne; Charles Hedley, A. B. Walkom (both of the Australian Museum); and Henry Chapman, R. H. Cambage, Horatio Carslaw, and John Madsen of Sydney. Edgeworth David became president in 1904, and David's former student E. C. Andrews succeeded Maiden in 1922. All would seek greater ways of preserving an association very similar to that founded three decades before.

Their task was not made easier by the country's slow recovery from the 1890s, followed by the distractions of the Boer War. In Sydney Liversidge noted the decline in membership of the Royal Society of New South Wales, from 457 to 368, its lowest since 1885.[77] The other Royal Societies suffered similarly. By 1914 those of Queensland and South Australia had only 100 members, Tasmania, 200, and Victoria, only 94.[78] Liversidge attributed this decline to the splintering effect of new societies (including the AAAS), to the 'suburban spread' of Sydney and Melbourne, and to the neglect of social evenings, which had figured so prominently in the 1870s.[79] The growing population of the country and its capitals should have offered greater scope.[80] Yet, from 1900 to 1913, congress attendances swung between only 335 and 820, and did not exceed 1000 again until 1926. Still, this was not discreditable, as comparative attendance records of fraternal associations overseas suggest. The Melbourne Congress of 1900 summoned forth 137 papers and 693 participants; whereas the Bradford meeting of the British Association that year had 295 papers and 1915 participants—only about 50 per cent more papers, and 60 per cent more people from a population six times more numerous. The American Association, meeting in 1900 in New York, had only 241 papers and 434 participants. In 1911 the AAAS meeting at Sydney actually improved its proportional participation record against the British Association at Portsmouth. By then, however, the American Association, meeting in Washington DC, had soared to 860 papers and a total membership of over 8000.

Inevitably, in this and many other respects, American science surged ahead in the 1900s, capitalizing on the new state universities, the growing wealth of the private institutions and the endowments of major corporations. The early years of Federation, by contrast, did little to bolster the faith of the 'practical idealists' in Australia,[81] whose hard work had kept the Association alive. In his last presidential address to the Royal Society of New South Wales in 1901, Liversidge proposed a 'National Australian Academy' of a 'federal character', based in the federal capital, with membership based on merit: to bring together 'the best intellect of all the states for more systematic consideration and discussion of matters than is possible at the meetings of the Association'.[82] But these appeals for Australian science met a disappointing public response. In 1902 Edgeworth David told the University of Sydney's jubilee congregation that science suffered under a disability from which Europe and America were exempt: 'there is ... little or no scientific opinion in the people of Australia'.[83] Sir Samuel Griffith agreed: 'The great defect in Australian life is the want of apprehension of the value of knowledge in itself.'[84] By 1904, 1104 degrees had been awarded in arts at Sydney, but only 47 in science, and no science doctorates at all.[85] The *Sydney Morning Herald* sympathized with the plight of the science professor: 'the author of an ideal republic of learning would hardly take Sydney as the fabric of his dream'.[86] Universities were dominated by elementary teaching, with their few staff typically holding huge classes and encompassing vast areas of knowledge.[87] Research was hardly expected of them. Of the 291 papers published by University of Sydney staff by 1905, over 100 had been published by Liversidge alone. During his presidency in 1898, Liversidge endeavoured to raise support for an Australasian *Nature* (which he entitled the *Australian Journal of Science*) by sending 7000 circulars to 'landowners, government officers of a certain rank, barristers and solicitors and all professional men'.[88] *Nature* applauded him,[89] but he failed to find sufficient support.[90]

Possibly the greater affluence of Australians and the burgeoning production of wheat, wool, and minerals,[91] undermined the very purposes for which Liversidge and Russell had struggled. In 1898 Beatrice Webb visited Sydney and recorded her conversations with professors who were depressed by the 'utter indifference of well-to-do Australians for learning of any kind'.[92] 'The real work of manhood in Australia,' sighed the *Sydney Morning Herald*, was mainly 'directed towards a practical end'.[93] Even so, the continuing exodus of engineers from a country where manufacturing accounted for less than 20 per cent of GDP[94] was discouraging,[95] and Bragg's return to England dismayed those who saw 'a tendency for our best scientific workers to drift away'.[96] Chemistry, perhaps the leading industrial science, had, as yet, no systematic organization in Australia. None of the four universities, as Charles Fawsitt (Liversidge's successor) observed, had anyone working specifically in organic chemistry.[97] When, after 1912, state governments increased endowments to the universities, new chairs were created in botany and organic chemistry at the University of Sydney and the situation improved. Sydney's population of 638 000 supported a university with 19 science posts; Manchester, with a population of 716 000, had 23. There were still, however, only two major science faculties in Australia, as against at least ten in the United Kingdom. And if science chairs grew slowly in number, fewer students chose science subjects in their arts degrees.[98] Although technical education—especially in agriculture, trades and commerce—expanded outside the universities dramatically between 1910 and 1914, this scarcely lent support to the advancement of research and may, indeed, have been at its expense.

To the Association, reaching its twenty-first birthday in 1909, these considerations offered a special challenge. Its place in Australian society (and, to some extent, New Zealand's) was secure; its purposes, as the *Sydney Morning Herald* put it, were to give Australasian scientists 'that personal contact with master minds that is so essential to original thought', and to 'popularise science'. The first of these it had achieved; the second would continue to require more 'missionary spirit'.[99] Welcoming the third Sydney Congress, the *Herald* suggested that Sections substitute the discussion of ideas for the reading of papers, and pleaded for more popular expositions. To this prompting, the Association showed little response. Moreover, and possibly for this reason, it could not act as a significant lobby at the federal level. Its one spectacular public success between 1900 and 1914 arose from its appeal to new Australian nationalism, with the sponsorship of Mawson's Antarctic expedition in 1912–13. If the Association was a 'parliament' of science, it did not yet speak to the Commonwealth.

It was in many ways symbolic of the resilient imperial sensibilities of its leadership that the Association's crowning achievement came with the visit of the British Association in 1914. Liversidge had never let the idea die, and after his return to England maintained diplomatic liaison with the British Association's general secretary in London. For its part, the British Association, with its mandate of 'social imperialism' in science,[100] had pursued a progression around the Empire. Travelling to South Africa in 1905, it had played a part in bringing together divided provinces and won the new Union academic respectability in England. In 1897 and 1909 it had visited Toronto and Winnipeg and tied Canadians to the Union Jack. The British Association was willing to visit Australia if funds could be found.

For the new generation, such a visit was of great intellectual and political importance. Led by Orme Masson, president-elect in 1909, a deputation waited on Alfred Deakin, Prime Minister, seeking £10,000 to cover costs. Deakin, a leading advocate of 'imperial efficiency', supported the measure; and Andrew Fisher, succeeding him in 1910, confirmed the visit's importance as an 'additional step towards imperial unity . . . one likely to be of great value to the Commonwealth'.[101] The press heralded the scheme as a means of making Australia 'as producer' better known.[102] Even Liversidge planned to return for the occasion.[103] The meeting was approved in London amid the festivities of George V's coronation and scheduled, with what would become tragic irony, for August 1914.

Thus arose the eighty-fourth meeting of the British Association, and its first and last in Australia. Orme Masson was chairman of the organizing committee. David Rivett, his brilliant young student, recent Rhodes scholar and Deakin's son-in-law, became the organizing secretary and established a reputation for administrative skill that rocketed him to national prominence. The Commonwealth granted £15,000 for the passage of 150 official representatives, 'including selected Dominion and foreign scientists'.[104] More than 300 British, American, European, and other foreign and Dominion scientists sailed to Australia to enjoy 'bone-shaking' railway journeys between the state capitals.[105] Nearly 5000 Australians enrolled, producing a record attendance for both the British and Australian Associations.[106] Newspaper stories followed the Association everywhere.[107] Special handbooks were prepared for the occasion, and virtually all Australia's scientific institutions played host.[108] In Rivett's plan, 'broad discussions on large problems' were made the feature of sectional work, and joint meetings between Sections were held. That between Sections A and B in Melbourne was particularly memorable, as Ernest Rutherford described the use of the Wilson cloud chamber and the brilliant work of H. G. J. Moseley on the X-ray spectra of the elements. For the first time senior British scientists, including Oliver Lodge, W. J. Pope, A. S. Eddington, W. J. Sollas, William Bateson, and Henry Tizard gained a sense of the Australian continent. For their hosts, it was a time to praise science in its imperial connection.[109]

Australians unquestionably gained much from the congress's recommendations. These included proposals to establish a solar observatory in Australia and to investigate further the chemistry of plant products, the physiology of marsupial reproduction, the biology of coastal islands and the Barrier Reef, and the problems of Australasian stratigraphy.[110] On these matters, British words carried twice the weight of Australian. But the attachment to the British Association ran deeper, and Australians welcomed Britain's recognition of their country's maturity. Only today does it seem paradoxical that it fell to an Englishman, Professor Frederick Trouton, to suggest that the imperial visit marked the 'scientific coming of age of Australia'.[111] The diffidence with which this tribute was accepted demonstrated loyalties, symbolized by Moseley's death on Gallipoli, that would long survive the First World War.

SCIENCE FOR THE COMMONWEALTH

There comes a time in the development of a young country when systematic and scientific organisation must take the place of rough and ready, happy-go-lucky methods of exploiting national resources.

With these words Francis Anderson greeted Englishmen visiting Australia in 1914.[112] But it would take a war a world away and time to bring about this change—and even then the change would be only partial, uncertain, and incomplete. The Great War saw Australian and New Zealand scientists generally mobilized for the imperial effort. Over a hundred chemists, including some of Orme Masson's students, went to work in British munitions factories, and scores of Australian and New Zealand geologists and mining engineers served on the Western Front. The adventurous Sydney geologist (Lieutenant-Colonel) Edgeworth David, and his colleague in physics (Captain) James Pollock, distinguished themselves in the field, as did Major Douglas Mawson in technical intelligence; Archibald Liversidge came from retirement to work with Richard Threlfall, then removed to the Admiralty; while Orme Masson, remaining in Melbourne, worked to improve Australia's capacity for waging 'scientific war'.

During the war, with its president (Edgeworth David) and most of its members on active duty at home or overseas, the AAAS adjourned its congresses, first to January 1915, then to January 1917, and finally for the duration. In the interval, the now venerable J. H. Maiden in Sydney and Baldwin Spencer (president elect and acting president) in Melbourne kept the Association officially alive. In mid-1915 Britain took steps towards the formation of an Advisory Council for Scientific and Industrial Research, and encouraged other countries in the Empire to follow suit. In 1916 this led to the formation of the Department of Scientific and Industrial Research (DSIR).[113] Early in the same year, moving with British example, Australia's Prime Minister, W. M. Hughes supported proposals to create in Australia an Institute of Science and Industry, similarly inspired 'to apply to the pastoral, agricultural, mining and manufacturing industries the resources of science in such a way as to more effectively develop our great heritage'. In 1920 this Institute was founded under the directorship of Orme Masson,[114] but almost immediately ran foul of state–federal rivalries and conflicting political purposes. When S. M. Bruce succeeded W. M. Hughes as Prime Minister in 1923, advice was sought from Britain; and in June 1926, a decade after the DSIR was created in London, the Institute was reorganized as the Council for Scientific and Industrial Research (CSIR) and at last given an endowment of £500,000 to encourage research, for industrial application mainly in agriculture, for routine testing, and the training of research scientists.[115] Even then, the national effort was scarcely adequate to national requirements, and David Rivett used his presidential address in 1924 to criticize both government and the AAAS for failing to remind the Australian public of the 'paramount importance of scientific investigations on organised national lines'.[116]

In these years, while the organization of research was debated, the AAAS recast its future. The long wartime intermission of eight years in its activities—a generation had passed since what Maiden called the 'scientifically remote period of January 1913'—in fact gave the Association a new lease of life. 'The war had startled us out of sleepy ways and narrow grooves', Maiden wrote to Edgeworth David; and in the economic circumstances of the day, 'we must not be surprised if criticisms are being levelled at institutions such as ours.'[117] With memories of high explosives and poison gas still fresh, the progressive image of science had not escaped unscathed. In the wake of war, some called for science to become more socially responsible; others, for scientists to assume political control of science; and still others, for scientists to remove themselves wholly from the market place, in hopes of recapturing a lost, perhaps mythical, 'purity'.[118] All voices urged scientific institutions to speak to a wider public. Even the British Association had emerged from war to face similar criticism, by members who saw its practices fall short of their expectations. It was necessary, as Maiden said, for both Associations 'to get a move on'.[119]

In 1920, in what proved to be his valedictory report, Maiden produced a blueprint for the future of the AAAS. This plan embraced four points. First, at congresses, the formal reading of papers was to be discouraged; congress reports would be taken as read, as works of reference and record. Second, the Association was to appeal beyond its 'creditable' but small membership, to the 'average citizen', to 'men on the land' and in 'industry', urging them 'to take an active interest in science and what it stands for'. A subscription of 10 shillings (instead of £1) a year would surely attract more men and women 'sympathetic with scientific aims', yet 'absent from our list', Maiden argued. Third, Sections should attempt to discuss ideas, rather than technical points, and consider broad policies as well as detailed research. 'The political aspects of the Australian mandates in the Pacific', he suggested, 'should have their counterpart in the discussions of the scientific questions involved.' Fourth, and most important, the Association should advocate that Australians take 'the most prominent part in the systematic exploration of their wonderful country'.[120]

Over the next twenty years, Maiden's vision was to be endlessly discussed, but only partly realized. In ironic tribute to wartime disruption, the fifteenth (and first postwar) Congress scheduled for Hobart, was held instead in Melbourne because of a maritime stewards strike. January 1921 was a time for new beginnings. Edgeworth David, back from the war, wished to see new initiatives, and, under his leadership the council took action. The now twelve Sections were 'relettered'; the Association's first woman local secretary, Georgina Sweet, joined the ranks of the great and good; E. C. Andrews the geologist and Charles Hedley the biologist were appointed to promote liaison with the newly formed Pan-Pacific Science Association. A cycle of congress venues was arranged with New South Wales, Victoria, and Tasmania agreeing to alternate with the 'more remote states' of South Australia and Queensland. The tyranny of distance still dominated the Association's proceedings. Adelaide sought to hold the Congress of 1924, to coincide with its University's fiftieth jubilee; it was first opposed, until, with beguiling casuistry, Douglas Mawson protested that South Australia was not 'really remote' after all. Perhaps 'Adelaide in midsummer was remote, but if the meeting were held in September it could be very agreeable' he suggested. In any case, 'Hobart was more remote, on account of strikes', and 'Adelaide was not really as remote as Hobart.'[121] In the end Mawson won. Eventually even the existence of Western Australia, after nearly thirty years, was at last formally recognized—though Perth did not actually join the congress cycle until 1926. Finally, 'ANZAC' bridges had to be rebuilt, and the sixteenth Congress was scheduled, by courtesy of the New Zealand Institute, for Wellington in 1923.

Following Maiden's recommendations, the Association vigorously debated its future. The council inaugurated inter-sectional meetings, to 'facilitate discussion on subjects of special importance to the Commonwealth', and Sections recommended federal government support on a number of projects involving major expenditure.[122] For many participants, however, the storm-clouds of war had scarcely parted. The nation had lost heavily. Only in experience had science gained much: and two of the addresses at Hobart/Melbourne were devoted to the wartime experiences of their respective disciplines. Section B (Chemistry) went so far as to urge the federal government to ensure that all future chemical plants be 'erected with a view to ready adaptability to war work in case of need'.[123] In a similar vein Baldwin Spencer's presidential address that year emphasized the need for a closer association of science workers throughout the Empire.

All these activities, prospective and actual, left fundamental questions unresolved throughout that 'mean decade', as R. M. Crawford termed the 1920s. The nation emerged weary from war with a heavy burden of imperial obligation. There were tendencies, most evident in art and literature, to shut away the rest of the world, or to seek an introspective Australian identity.[124] In 1921 James Bryce, the well-travelled former ambassador to the United States, saw only 'material interests [holding] the field of discussion', and these are 'discussed as if they affected only Australia, or Australia only in the present generation'. Nobody, he added, 'looks back to the records of experience for guidance, nobody looks forward to conjecture the results of what is being attempted today. There is little sense of the immense complexity of the problems involved, little knowledge of what is now being tried elsewhere, little desire to acquire such knowledge.'[125]

One development had within it a potential stimulus that Australian science greatly needed. As war in Europe drew to a close, the victorious allies sought ways of continuing the scientific co-operation those years had fostered. In 1918, following discussions in Paris and London, a mission led by the American astronomer George Ellery Hale recommended that an International Research Council be formed. This federation, the forerunner of the present International Council of Scientific Unions, required adhering bodies in each of the sixteen Allied countries. In Britain the obvious agency was the Royal Society and in the United States, the National Research Council/National Academy of Science, both of which had substantially served the Allied war effort. In 1901 Liversidge had proposed for Australia an organization of this kind, resembling the continental Academies and based in the new federal capital, where it would form the nucleus of a cultural establishment, including a national museum and a federal university. But this was typically farsighted. In postwar Australia, there was still no federal body except the AAAS, so, in August 1919, following an official invitation conveyed through the Royal Society of New South Wales, the AAAS was invited to elect the officers of what would become the new Australian National Research Council (ANRC). 'This duty of the General Council of the Association', Maiden wisely observed, 'is of great responsibility as regards the welfare of Australian science.'[126] With its new 'executive' role through the AAAS, the ANRC could represent both 'élite' science and the democratic intellect. A provisional council was formed: R. H. Cambage was elected honorary secretary, and on his return from Europe, Edgeworth David was chosen to be its first president (1919–22).

On the second day of the 1921 Congress, the AAAS council considered the new ANRC. T. H. Laby, FRS—perhaps, after Mawson, Liversidge's most famous assistant at Sydney, and now professor of physics at Melbourne—moved a set of resolutions aimed at making a 'more effective organisation of science in Australia'. Addressing, by implication, the Association's inadequate coverage, he proposed a committee to consider the formation of a society of 'scientific workers', for the purpose of 'advancing science education, promoting research, discussing and publishing the results of scientific investigation, and recognising and encouraging distinguished scientific work'. This new society would consider steps to 'advance the application of science', through industry, government and defence. Laby's proposal—that 'scientific opinion may have more influence in its proper domain'— exposed again the sensitive position of the AAAS. Next day, after what the council minutes called a 'hearty and instructive debate', the Association left the implications of Laby's motion in abeyance and established instead the ANRC.[127] For a decade and more the question of science and its larger social responsibilities would slumber in Australia.

The ANRC was begun with 100 ordinary and 50 associate members, representing each of the 18 scientific disciplines, roughly approximating the 13 sections of AAAS. In 1921 the provisional council of 1919 was confirmed and re-elected, with Edgeworth David serving as chairman of the new executive (from then until his death in 1934); Cambage, honorary secretary, and Professor H. G. Chapman, treasurer. Of the twenty-eight men on the new council, twenty-five were university scientists, and eighteen of them professors—virtually the whole intellectual establishment of the country's five universities. These men inevitably led other developments in the organization of science. Of the twenty-eight, five were also members of the Preliminary Advisory Council of Science and Industry in 1916, and eight were to figure in the reorganization that led to the creation of the CSIR.[128] Curiously, George Julius, consulting engineer, president of the ANRC, 1932–7, and chairman of the CSIR, was not among them; nor was he a member of the AAAS council.

In 1923 the ANRC broke away from its AAAS moorings and thereafter submitted its reports only by courtesy to the Association. The two organizations were, however, destined to remain in close relation, with many of their senior members continuing the 'interlocking directorate' presaged in 1888. In 1926 the council of AAAS joined with the ANRC in pressing for the creation of the CSIR. Orme Masson, as president of the ANRC, the 'only independent scientific advisory body' available to the federal government,[129] was instrumental in reconstituting the Institute of Science and Industry into the CSIR as an independent statutory authority, not subject to the usual constraints of the Public Service Board and 'free from direct political control'.[130] Through the CSIR's state committees, the ANRC sustained regular liaison with the major source of government research funds. For the next thirty years the ANRC encouraged international activities, especially through the International Council of Scientific Unions, the Pan-Indian Ocean Science Association, and the Pacific Science Association. Six AAAS committees, involving international co-operation (including solar physics, seismology, tidal surveys, and physical and chemical constants) passed to its jurisdiction, and these were followed by other projects on Antarctic research, land utilization, soil classification, and marine biology. The ANRC published several journals, including *Oceana* and the *Australian Journal of Science*, begun in 1937, and sponsored over fifty research expeditions in anthropology alone, establishing Australian leadership in this and other fields.[131] The ANRC, however, suffered from inadequate resources (made worse by the notorious defalcations of its treasurer), and from recurring disagreements among its fellowship.

The ANRC encountered most of the representational difficulties—both regional and disciplinary—inherent in ANZAAS. To sustain high standards, it had to reward scientific excellence by restricted entry. Indeed, in 1931, Orme Masson proposed that the ANRC be called the 'Australian Academy of Science'. But there was great dissatisfaction at its lack of representative connection with the other, proud and independent scientific societies. As the AAAS had found, a 'national' perspective was clouded by the contradictions of 'federalism'. David Rivett designed a compromise whereby the ANRC kept its federal character and sought merely to co-ordinate local efforts, but even this proved contentious.

To preserve a federal constituency, not only of representatives, but also of the nation's leading scientists, therefore became a principal aim of both the ANRC and ANZAAS council. At the Auckland Congress in 1937, a new class of 'Fellows' of ANZAAS was created. This new body of 115 scientists, elected on a basis that satisfied regional interests, was constituted as the membership of a 'new' ANRC on 3 December 1937. By yielding, as Elkin put it, 'to the geographical and political divisions of Australia', it attempted to become as 'representative and democratic as possible—though seeking a high standard of membership'. The reassertion of federalism was taken further in 1939, when Fellows were formed into state committees, which in turn elected the executive committee of the ANRC. But the formal position of 'federative science' could not be so easily squared with either the reward of intellectual excellence, the direction of innovative industry, or the efficient management of human resources. The ANRC acted in effect as an Academy, encouraging research and advising the Commonwealth government on scientific matters until 1955. Then, its 'chequered history' no longer 'inspiring confidence in men of science or governments',[132] it was replaced by the Australian Academy of Science.

In retrospect, its fate seems harsh, given the poverty of support for science in the period. But as later chapters show, a similar judgement fell upon the AAAS, with which it had such an 'organic relation'.[133] In 1923, E. C. Andrews had succeeded Maiden as general secretary. An 'intellectual grandchild' of Liversidge, he attempted to restore the status of the Association 'to that which it held prior to the 1914 meeting of the British Association'.[134] Now that the 'pioneer work' had been completed, he argued, the 'best methods of using the Association in the service of the country at large ... needed revision'.[135] Uncoordinated technical papers at specialist Sections could hardly remain the principal activity of an Association that aimed at educating the public. Indeed, the Association had to move from a passive mode to a more active one, sponsoring studies on Australia's flora and fauna, on its dwindling native peoples, on its climate and resources, and on the health, hygiene, and education of its people.

Despite these brave efforts, however, the Association altered little. A disappointingly small congress at Wellington in 1923 heard the new president, G. H. Knibbs, the dynamic and erudite Commonwealth statistician, declaim on 'Science and its Service to Man': 'as a people, we lack a due appreciation of systematic knowledge. Our hope is to see a new spirit born here.'[136] Whence was this inspiration to come? The Association had been greatly impoverished. During its eight-year slumber, no new resources had come in; its budget had shrunk to £1,100, and although its reserve fund remained at its prewar level of £4,000, investments returned only £140 per annum. Hobart 'earned' £200. Funding a permanent home for the Association remained a problem, and the honorary secretary's office was only a cubby hole in the Royal Society's library at a rental of £10 a year.[137] If funds were one nagging issue, another was the Association's name. In 1923, the term 'Australasian' was reported to be disagreeable to the New Zealand government.[138] At the time, the council opposed a change, but change proved inevitable, and finally came in 1930, when titular 'ANZAAS' was born. In fact, the new title made little difference, as New Zealand thereafter played a small and diminishing role in the Association.[139] New Zealand congresses came only at 'sabbatical', seven-year intervals, too infrequent to help New Zealand science, and too brief to bring the two countries more closely together.

Through the 1920s, the internal problems of the Association remained, as congresses marched from Adelaide to Perth, Hobart to Brisbane, and back to Sydney again in 1932. Edgeworth David and Mawson made impressive discoveries, many inventors flourished, and geographers—of whom the most notable was T. Griffith Taylor—kept exploration vividly before the public. With the Depression, the finances of the Association worsened. Its programme of Sections and excursions continued fundamentally unaltered, and its recommendations became more urgent, but now its place on the national stage was shared by new and powerful professional institutes—the Royal Australian Chemical Institute, the Australian Branch of the British Medical Association, and other specialist institutions, some of which had grown from ANZAAS congresses. Where substantial research was required, its work had been taken out of its hands by the ANRC, or taken over by the CSIR. Indeed, by the 1930s the CSIR had acquired a culture of its own, reflecting essential tensions between its pragmatic, mission-oriented consultants' research for the rural, livestock, and forest product industries, and the British-inspired 'ideal of autonomy' espoused by Rivett, for twenty years its chief executive officer.[140] That so many of the CSIR's efforts were directed towards the primary agricultural and pastoral industries and to the eradication of pests reflected objectives of the day, which were influenced in turn by Australia's role in the imperial economic system.[141] That so little research was aimed at the needs of manufacturing, or industrial research and development would become in later decades a source of great concern.

To some of these issues, ANZAAS was alive. Papers appeared regularly on the problems of primary production, economic recession, geographical determination, indigenous flora and fauna, and ways to control the devastating prickly pear.[142] But at the same time, congresses were increasingly relegated to at best, a 'co-ordinating' role, lending a platform to the country's impoverished researchers and technical workers. By 1939 these numbered perhaps 150 in the universities and about 600 in CSIR and the defence munitions establishments, all told perhaps 800 (excluding engineers), in a country whose population had reached 4.5 million people.[143] Outside government laboratories, science had negligible resources. The universities were hard pressed. Potential researchers went overseas, principally to Britain, if they sought higher degrees, and they remained there, especially if they worked in the physical sciences. An independent Australian scientific ethos, signs of which were so visible in the Federation period of the 1880s and 1890s, had all but disappeared in a context that seemed to celebrate intellectual conformism, dependence upon Empire, and an unreflective pragmatism that apparently left the generation of theory (and science-based innovation) to others.

Writing after the war, R. M. Crawford, who brought history prominently to ANZAAS in 1939, spoke of the mid-1930s as a great turning point in the history of Australia. The expansion of the CSIR in support of industrial research, the establishment of the Australian Broadcasting Commission (ABC) and the Commonwealth Literary Fund, and fresh developments in art and music symbolized for him the retreat of the 'Australian legend' and the arrival of a new cultural dawn.[144] Yet only a little of its light filtered in upon the affairs of ANZAAS. By the late 1930s, from its new Sydney premises at Science House in Gloucester Street, ANZAAS (not unlike its British counterpart) was serving a set of largely outdated objectives.

In 1937 the ANZAAS council, approaching its jubilee at Auckland, bowed to the inevitable by amending its 'objects' of 1888. It deleted the now archaic language of Liversidge's day; henceforward, and to our own day, it was to serve simply 'the advancement of knowledge', and to 'promote a spirit of co-operation between scientific workers and scholars and those in sympathy with science and scholarship generally, especially

PLATE 1
'The Pathfinder' (*Daily Telegraph*, 12 January 1911).

PLATE 4
The 'Founding Fathers'

in Australia and New Zealand'.[145] Federation had given way to co-operation and scholarly unionism, in the Commonwealth of Science.

Meeting these objectives was to prove more than a matter of constitutional change. In 1888 the principal difficulty facing Australian scientists was their geographical and intellectual isolation, both from overseas sources of information and stimulus, and from one another. Their sense of isolation served both to strengthen imperial ties with Britain and to retard enterprise on a national basis. By the 1930s the principal difficulties facing Australian scientists were financial and intellectual. In Australia, as elsewhere, there were small, often uneconomic numbers working in many fields. ANZAAS took upon itself the task of confronting the new 'tyrannies' of specialization and the isolation of the disciplines from one another. In the terms of its objects, it succeeded in 'fostering communication between scientists in all disciplines'. This was helpful, but in the context of the 1930s, not enough. Not always did it succeed in fostering communication between scientists and the general public. To many of the issues raised by the Depression, ANZAAS as an organization remained oblivious. One of the few to expose this failure was T. H. Laby, who told the Sydney Congress in 1932 that 'the progress of science has created more social and political problems than mankind has been able to solve. Might it not be to the advantage of mankind', he asked, 'if some of those adept in the methods of the physical sciences turned their attention to the solution of the economic and political problems which their discoveries have created?'[146]

Few, however, rose to answer. For the most part, the pursuit of science remained an activity remote from the general public. Scientists remained objects of eccentric enthusiasm and polite caricature. Overall, the image of science had become faintly negative, and even destructive. Writing on the Melbourne Congress in 1935, S. H. Roberts, later vice-chancellor of the University of Sydney, spoke of the 'scientific odds-and-ends' of the Sections. Hearing nothing more than dire agricultural forecasts from the economists, he told the *Sydney Mail* he wished to escape, 'into a green field and ... a Nature that still looked clean and wholesome'.[147] With all its victories over pests and plagues, with all its triumphs in the primary industries, science had not brought a new spirit into the land and the cities. Instead Australian science was seen as the preserve of a narrow, pragmatic utilitarianism—serving certain established interests, advancing the Australian legend of exploration and the conquest of disease, but distanced from everyday industry and manufacture, and deeply embedded within the conservative political climate of the day. ANZAAS itself remained a public relations office for science—and by now, for the social sciences as well—but it seemed barely conscious of the new, more diverse and dissident cultural and intellectual interests that were growing outside the ranks of government and the universities.[148]

For those promoting national efficiency and progressive imperialism, it was easy to see ANZAAS as serving a stabilizing function, necessary to the official status quo. In this, ANZAAS was quite in step with the British Association and even the American Association for the Advancement of Science, whose congresses between 1932 and 1938 also witnessed repeated, and largely unsuccessful, attempts to reconcile the promotion of research and scientific planning with the objectives of a democratic society. As an organization ANZAAS could do no more than reflect the preoccupations of its leadership, and reform without radical change was the watchword of the leadership of science, here as elsewhere. In Britain and Europe science in these years was on the defensive: implicated in technological unemployment of the Depression and in the cold rationalism of social engineering. By 1934 the British Association had become virtually a public apologist for science, and by so doing had, in the opinion of one observer, saved itself from oblivion.[149] This had been done at a price. In 1936 at the British Association's Blackpool meeting, science came under attack in Britain for being isolated from the community, and for not solving—whether by 'humanism' or good management—the critical problems of modern industrial society. The Division for the Social and International Relations of Science, created in 1938, set the British Association on a new, potentially more politically interested course.[150] But ways for Australian scientists to focus on the uses (and abuses) of science and the wider responsibilities of science to society had to be found outside the traditional avenues of the British Association. The formation of the Australian Association of Scientific Workers, spurred by the British Association meeting in 1939, grew directly from a desire to find such routes.[151] In Australia, science was less implicated in social crisis; and the prevailing pragmatism made for acceptance. But this acceptance only delayed the inevitable. The British Association entered the war, and emerged from it, with a deep deference to the values of pure science, and a commitment to national objectives, cautioned by a deep suspicion of any attempt to impose upon science the apparatus of public scrutiny, or political control. *Sidere mens eadem mutato*: ANZAAS could scarcely avoid a similar destiny.

In January 1939, with war in Europe drawing near, ANZAAS celebrated its jubilee. Appropriately, it met for the first time in the nation's federal capital. It was suitable that a historian be president on such a historic occasion, and Professor Ernest Scott of Melbourne spoke to the 'more enlightened spirit' he saw then diffusing Australia. For Scott, this 'enlightenment' consisted in a more intelligent recognition of what Bergson defined as the essential object of science—to increase our 'influence over things'. To increase knowledge, and its influence, were objects close to the men and women who remembered the difficulties of doing science in the colonial climate of fifty years before. War and Depression had not made the task easier. But the challenge to look beyond research, to the betterment of society, was not part of his programme. To another participant, the past spoke in different ways. Also to Canberra came the best-known historian of the future, H. G. Wells. His topic was 'The Role of the English in the Development of the Human Mind', but as shadows fell across Europe, he spoke more generally of the future of the human race. 'Our historical imaginations', he said, 'are living today in a vastly enlarged system of perspectives, and we know ... that all our working conceptions of behaviour and destiny are provisional and that human nature and everything about it is being carried along upon an irreversible process of change.'[152] Would the world crisis provoke Australia to recognize the 'shape of things to come'?

In 1906 the *Sydney Morning Herald* had reflected on the prospects of the new Commonwealth: 'It is no mere thing for Australians to advise each other that they must take themselves seriously—in their methods of education, their political powers and privileges, their own self-culture, and the necessity for widening their outlook.' 'We cannot afford', the editor continued, 'to remain behind the rest of the world, for it cannot be too often repeated that the period of our isolation has come to an end.'[153] In 1938, three decades later, the historian C. Hartley Grattan surveyed this country from the perspective of a deeply sympathetic American visitor and agreed. Profoundly moved by the country's great potential, he saw a future unfolding for Australia, with its institutions 'self-consciously moving towards a rounded and balanced industrial system' aided by science and technology wisely deployed. But bold steps were needed to release 'the productive capacity of the nation for the benefit of ... the people'.[154] Australia could not advance, or even maintain its position in the world, he argued, if its thinking continued to be dominated by a 'crown colony psychology'. 'What is it', he asked, 'Australians really want?' He could detect no clear response.

The elements for a new Australia lay scattered, uncoordinated and diffused. Whether the pieces would be assembled, and whether science and technology could be applied fully to the service of the nation would require answers that were, in the event, delayed by the coming and passing of another war. Even then, however, the grail of Grattan's quest would remain elusive. Much would depend upon the vision of a new generation of Australians, with a belief in their future, and a determination to build a nation worthy of themselves.[155]

NOTES

1. cf., 'AAAS', *Chemical News*, 58 (1888), 118; *Nature*, 38 (1888), 437–8; 623.
2. H. C. Russell, 'Presidential Address', *Report of the AAAS*, 1 (Sydney, 1888), 12.
3. ibid., 11.
4. ibid., 14.
5. 'Science and the Arts', *Sydney Morning Herald* (29 August 1888).
6. C. A. Fleming, *Science, Settlers and Scholars: The Centennial History of the Royal Society of New Zealand*, Roy. Soc. of NZ Bulletin, 25 (1987), 270; *Report of the AAAS*, 1 (Sydney, 1888), 106–12; 168–83; 303–12; 352–7; 359; 413–15.
7. Quoted in R. W. Home, 'The Problem of Intellectual Isolation in

Scientific Life: W. H. Bragg and the Australian Scientific Community, 1886–1909', *Historical Records of Australian Science*, 6 (1) (1984), 19.
8. J. Steel Robertson, 'Natural Science in Australia', *The Centennial Magazine*, 2 (7) (February 1890), 523–7.
9. *Australasian* (1 September 1888), 490–1.
10. 'Committees of Investigation appointed at the General Committee of the Sydney Meeting, 31 August 1888; No. 1. Conditions of the Labour Committee', *Report of the AAAS*, 1 (Sydney, 1888), xxxii.
11. *Rockhampton Bulletin* (2 June 1888).
12. *ANZAAS Archives* (Mitchell Library), MSS.988/1, f. 152, Publications Committee, 7 September 1888; f. 183, 12 September 1888.
13. *Table Talk* (10 January 1890).
14. *Australasian* (18 January 1890).
15. *Argus* (15 January 1890).
16. Ferdinand von Mueller, 'Inaugural Address', *Report of the AAAS*, 2 (Melbourne, 1890), 25–6.
17. *Argus* (7 January 1890).
18. Minutes of the General Committee, 14 January 1890, *Report of the AAAS*, 2 (Melbourne, 1890).
19. J. A. La Nauze, '"Other Like Services": Physics and the Australian Constitution', *Records of the Aust. Academy of Science*, 1 (3) (1968), 6–44.
20. *Argus* (15 January 1890).
21. ibid.
22. *Australasian Home Reader*, 1 (1892), 4–5.
23. W. A. Osborne, *William Sutherland: A Biography* (Melbourne: Lothian Pub. Co., 1928).
24. *ANZAAS Archives*, op. cit., note 12, f. 188 (7 June 1890).
25. Quoted in Edward Kynaston, *A Man on Edge: The Life of Baron Sir Ferdinand von Mueller* (Ringwood: Allen Lane, 1981), 355.
26. Sir James Hector, 'Presidential Address', *Report of the AAAS*, 3 (Christchurch, 1891), 4.
27. ibid., 21.
28. ibid., 4.
29. *Hobart Mercury* (6 January 1892).
30. ibid.
31. cf., Neville Hicks, '*This Sin and Scandal*': *Australia's Population Debate, 1895–1911* (Canberra: Australian National University Press, 1978).
32. 'Presidential Address, Section E', *Report of the AAAS*, 4 (Hobart, 1892), 130.
33. *ANZAAS Archives*, op. cit., note 12, Minutes, MSS. 988/1, f. 41 (12 April 1888).
34. 'Proceedings of the General Council, Second Meeting, 10 January 1913', *Report of the AAAS*, 14 (Melbourne, 1913), xviii.
35. As the Pharmaceutical Society found to its cost. See 'Pharmacy and the Australasian Association for the Advancement of Science', *Chemist and Druggist of Australasia* (February, 1895), 18.
36. Mrs Alex Morton, 'The Evolution of Hostility Between Capital and Labour', *Report of the AAAS*, 4 (Hobart, 1892), 547–53.
37. W. H. Hamlet, 'Presidential Address, Section B', *Report of the AAAS*, 4 (Hobart, 1892), 49–50.
38. Professor E. E. Morris, 'Presidential Address, Section 1', *Report of the AAAS*, 4 (Hobart, 1892), 173.
39. From its ashes rose the Home Reading Union, which sustained for many years a programme of continuing education.

40. *Nature*, 45 (3 March 1892), 422–7.
41. *South Australian Register* (30 April 1892).
42. cf., the response in *Nature*, 48 (6 July 1893), 229.
43. E. M. Shelton, 'Our Defences Against Low Prices of Farm Products', *Report of the AAAS*, 6 (Brisbane, 1895), 721–8.
44. R. L. Jack, 'Presidential Address', *J. Roy. Soc. Qld* (1895), xiii.
45. *Sydney Morning Herald* (1 October 1897).
46. *Liversidge Papers* (University of Sydney Archives), Box 17, unreferenced newspaper clippings, Christchurch, 1891.
47. *South Australian Register* (5 January 1907).
48. *Sydney Morning Herald* (3 January 1898).
49. W. M. Hamlet, *Handbook to the Sydney Congress* (Sydney, 1898), preface.
50. *Sydney Morning Herald* (1 October 1897).
51. *Sydney Morning Herald* (3 January 1898).
52. *Daily Telegraph* (7 January 1898).
53. Liversidge was elected as the first life member of the Association at Hobart in 1902.
54. *Daily Telegraph* (7 January 1898).
55. *Sydney Morning Herald* (6 January 1898).
56. *Sydney Morning Herald* (9 January 1904).
57. *Sydney Morning Herald* (17 December 1907); *Evening News* (6 December 1907).
58. I. Inkster, 'Scientific Enterprise and the Colonial "Model": Observations on Australian Experience in a Historical Context', *Social Studies of Science*, 15 (1985), 694.
59. H. C. Russell, 'Presidential Address', *Report of the AAAS*, 1 (Sydney, 1888), 11.
60. Ferdinand von Mueller, 'Presidential Address', *Report of the AAAS*, 2 (Melbourne, 1890), 3.
61. Archibald Liversidge, 'Presidential Address', *Report of the AAAS*, 7 (Sydney, 1898), 7.
62. Archibald Liversidge, 'The Australian [sic] Association for the Advancement of Science', *Nature*, 82 (30 December 1909), 264.
63. *Advertiser* (11 January 1907).
64. H. C. L. Anderson, 'Presidential Address, Section G', *Report of the AAAS*, 5 (Adelaide, 1893), 164.
65. *Argus* (7 January 1890).
66. Inkster, op. cit., note 58, 696.
67. Liversidge, op. cit., note 61, 5.
68. Hector, op. cit., note 26, 1.
69. *Sydney Morning Herald* (9 January 1904).
70. 'Balance Sheet for Tasmania–Hobart Session, 1902', *Report of the AAAS*, 9 (Hobart, 1902), xlv.
71. At Brisbane in 1909, for example, 2000 copies of the conference proceedings cost £455, but expenses exceeded £900. Only £66 was remitted to Sydney that year. 'Balance Sheet, Brisbane Meeting', *Report of the AAAS*, 13 (Sydney, 1911), lxii.
72. 'General Balance Sheet', ibid., lxi.
73. *Argus* (7 January 1890).
74. *Hobart Mercury* (6 January 1892).
75. *Sydney Morning Herald* (6 January 1898).
76. See Mary Howitt Walker, *Come Wind, Come Weather: A Biography of Alfred Howitt* (Melbourne: Melbourne University Press, 1971), 121; 272 et passim.

77. The Royal Society's membership continued to fall, reaching 293 in 1933. Today, however, it has risen to around 360.
78. 'Miscellaneous Notes on Australia, People, Activities' in George Knibbs (ed.), *Federal Handbook for the 84th BA Meeting* (Canberra: Commonwealth Government Printer, 1914), 596–7.
79. Archibald Liversidge, 'Presidential Address', *Proc. Roy. Soc. of NSW*, 35 (1901), 5–6.
80. Between 1881 and 1891 Sydney's population grew from 225 000 to 383 000 and by 1900 accounted for over 35 per cent of the state's population. Melbourne had grown to over 500 000 people, and accounted for over half of Victoria's population.
81. cf., *Daily Telegraph* (31 January 1912).
82. Liversidge, op. cit., note 79, 13. See M. J. Lewis, 'The Royal Society of Australia: An Attempt to Establish a National Academy of Science', *Records of the Aust. Academy of Science*, 4 (1) (1978), 51.
83. T. W. Edgeworth David, 'University Science Teaching', *Record of the Jubilee Celebrations of the University of Sydney* (Sydney: Brooks and Co., 1903), 117.
84. ibid., 24.
85. cf., 'Science in Australia', *Sydney Morning Herald* (9 January 1904).
86. *Sydney Morning Herald* (28 April 1905).
87. cf., F. Anderson, 'Education Policy and Development', in G. H. Knibbs, op. cit., note 78, 541–3.
88. *Liversidge Papers*, op. cit., note 46, Box 19, Liversidge to Baldwin Spencer; *Report of the AAAS*, 11 (Adelaide, 1907), xxiv.
89. *Nature*, 73 (18 January 1906), 279.
90. *Liversidge Papers*, op. cit., note 46, Box 19; Angus & Robertson to Liversidge, 18 May 1905. There were several attempts to form a journal of general science in Australia. For many years, however, none would compete with the *Scientific Australian*, a popular quarterly (later monthly) devoted to the 'Arts and Industries', and published in Melbourne from 1895 to 1924.
91. cf., Gerald Lightfoot, 'Manufacturing, Industrial and Commercial Development of Australia' in Knibbs, op. cit., note 78, 463.
92. A. C. Austin (ed.), *The Webb's Australian Diary* (1898) (Melbourne: Sir Isaac Pitman and Sons Ltd, 1965), quoted in Kaye Harman (ed.), *Australia Brought to Book* (Sydney: Boobook Publications, 1985), 148.
93. *Sydney Morning Herald* (2 July 1904).
94. *Australians: An Historical Atlas* (Sydney: Fairfax, Syme and Weldon, 1986), 126.
95. *Sydney Morning Herald* (12 August 1910) and (4 January 1912).
96. *Sydney Morning Herald* (12 January 1909) and (3 July 1909).
97. *Sydney Morning Herald* (26 March 1909).
98. *Daily Telegraph* (31 December 1912).
99. *Sydney Morning Herald* (12 January 1909).
100. cf., M. Worboys, 'The British Association and Empire: Science and Social Imperialism, 1880–1940' in R. M. MacLeod and P. Collins (eds), *The Parliament of Science* (London: Science Reviews Ltd, 1981), 170–87.
101. *Hansard* (House of Representatives), 9 August 1910, 1267; *Daily Telegraph* (12, 13 August 1910).
102. 'Science and Australia', *Sydney Morning Herald* (17 December 1909).
103. *Liversidge Papers*, op. cit., note 46, Box 17, BAAS, MSS. note, 24 June 1910.

104. O. J. R. Howarth, *The British Association for the Advancement of Science; A Retrospect* (London: BAAS, 1931), 134, 137.
105. Rohan Rivett, *David Rivett: Fighter for Australian Science* (Melbourne: Dominion Press, 1972), 55.
106. Howarth, op. cit., note 104, 134–40.
107. cf., *Australian* (1 August 1914); *Bulletin* (27 August 1914); *Daily Telegraph* (22, 27 August 1914).
108. cf., Rosaleen Love, 'The Science Show of 1914: The British Association meets in Australia', *This Australia*, 4 (1) (1984–5), 12–16.
109. See the tribute by Orme Masson's daughter, Marjorie, 'An Army to Make You Wise', *The Lone Hand* (1 August 1914).
110. *Report of the BAAS*, 84 (Sydney, 1914), 45 *et passim*.
111. Cited in Peter Robertson, 'Coming of Age: The British Association in Australia, 1914', *Australian Physicist*, 17 (2) (1980), 24.
112. Francis Anderson in Knibbs, op. cit., note 78, 520.
113. R. M. MacLeod and E. K. Andrews, 'The Origins of the DSIR: Reflections on Ideas and Men', *Public Administration*, 48 (1970), 23–48.
114. D. P. Mellor, *The Role of Science and Industry* (Canberra: Australian War Memorial, 1958), 17.
115. cf., George Currie and John Graham, *The Origins of CSIRO: Science and the Commonwealth Government 1901–1926* (Melbourne: Melbourne University Press, 1966). I am indebted here, and elsewhere for the advice of Professor Boris Schedvin, whose definitive history of CSIRO was not available to us at the time of writing.
116. David Rivett, 'Presidential Address, Section B', *Report of the AAAS*, 17 (Adelaide, 1924), 195.
117. *ANZAAS Archives*, op. cit., note 12, MSS. 1613, 'Second Progress Report on the AAAS', 18 December 1920, 11.
118. R. M. MacLeod and Kay MacLeod, 'The Social Relations of Science and Technology, 1914–1939', in Carlo Cipolla (ed.), *Economic History of Europe* (London: Collins, 1976), vol. 5: *The Twentieth Century*, part 1, 301–35.
119. J. H. Maiden, 'Presidential Address', *Report of the AAAS*, 15 (Melbourne, 1921), 11.
120. ibid., 12–13.
121. 'Proceedings of the General Council', *Report of the AAAS*, 16 (Wellington, 1923), xviii.
122. ibid., xxv–xxx.
123. ibid., xxix.
124. See G. Serle, *From Deserts the Prophets Come: The Creative Spirit in Australia, 1788–1972* (Melbourne: Heinemann, 1972), chapters 6–8.
125. James Bryce, *Modern Democracies*, 2 vols (New York, 1921), vol. 2, 251; quoted in Serle, ibid., 148.
126. J. H. Maiden, 'Second Progress Report on the AAAS, op. cit., note 117, 8.
127. *Report of the AAAS*, 15 (Melbourne, 1921), xxiii.
128. That is, Professors T. H. Laby, T. Lyle, R. D. Watt, D. O Masson, and A. J. Gibson. The latter included Professors R. D. Watt, R. H. Cambage, H. G. Chapman, D. O. Masson, G. H. Knibbs, G. Lightfoot, H. A. Woodruff and A. E. V. Richardson; cf., Currie and Graham, op. cit., note 115.
129. J. Ronayne, *Science in Government* (Melbourne: Edward Arnold, 1984), 364.
130. Mellor, op. cit., note 114, 19.

131. A. P. Elkin, 'The Australian National Research Council', *Aust. Journal of Science*, 16 (1954), 206.
132. As Professor Ronayne reminded us in 1978, there is no formal history of the ARNC, beyond a brief account by A. P. Elkin on which this passage draws heavily: see Elkin, ibid., 203–11. See also Ann Moyal, 'The Australian Academy of Science: The Anatomy of a Scientific Elite', *Search*, 11 (1980), 231–8; 281–8. The ANRC papers, now catalogued in the National Library, Canberra, would repay careful study.
133. A. P. Elkin, 'ANZAAS: A History', *Aust. J. Sci.*, 25 (1962), 3.
134. Andrews, op. cit., note 121, xxiv.
135. ibid., xxiii.
136. G. H. Knibbs, 'Presidential Address', *Report of the AAAS*, 16 (Wellington, 1923), 44, 46.
137. 'Proceedings of the General Council', op. cit., note 121, xxiv.
138. ibid., xxvi.
139. For details, see Fleming, op. cit., note 6, 270–2.
140. cf., Rivett, op. cit., note 105; cf., C. B. Schedvin, 'The Culture of CSIRO', *Australian Cultural History*, 2 (1982/3), 76–89.
141. G. Currie and J. Graham, 'Growth of Scientific Research in Australia: The Council for Scientific and Industrial Research and the Empire Marketing Board', *Records of the Aust. Academy of Science*, 1 (1968), 25–35.
142. ibid.
143. cf., Gani *et al.*, *The Conditions of Science in Australian Universities* (Canberra: Australian National University, 1962), 10–11. I am indebted to Professor Boris Schedvin for more accurate figures.
144. R. M. Crawford, *An Australian Perspective* (Melbourne: Melbourne University Press, 1960), cf., Peter Coleman (ed.), *Australian Civilization* (Melbourne: F. W. Cheshire, 1962), 1.
145. 'Constitution and By-Laws and Rules of Procedure', *Report of ANZAAS*, 23 (Auckland, 1937), xxxiii–xliv.
146. T. H. Laby, 'Science and the Depression', *Report of ANZAAS*, 21 (Sydney, 1932), 432.
147. *Sydney Mail* (30 January 1935).
148. cf., Serle, op. cit., note 124, 154, 160–6.
149. Peter Collins, 'The British Association as Public Apologist for Science, 1919–1946' in MacLeod and Collins (eds), op. cit., note 100, 211–36.
150. J. G. Crowther, *The Social Relations of Science* (Cambridge: Macmillan, 1941), 631.
151. Jean Moran, Scientists in the Political and Public Arena: A Social-Intellectual History of the Australian Association of Scientific Workers, 1939–1949, (unpublished MPhil thesis, Griffith University, 1983).
152. H. G. Wells, 'The Role of the English in the Development of the Human Mind', *Report of ANZAAS*, 24 (Canberra, 1939), 340–9.
153. *Sydney Morning Herald* (7 April 1906).
154. C. Hartley Grattan, 'The Future in Australia', *Australian Quarterly*, 10 (1938), 13, 22, 28.
155. ibid., 29.

3

The Impulse of Science in Public Affairs, 1945–1986

James Davenport

Since the end of the Second World War Australia has participated in the rapid technological changes that have profoundly affected the economic and social conditions of the entire western world. The advent of the 'atomic age' demonstrated the immense destructive power achieved by a government through the carefully orchestrated use of science and scientists. Less dramatic, but of greater significance to economic and social conditions, were the developments, stimulated by the demands of war, in electronics, jet propulsion, computers, materials science, and antibiotics, and carried through to peacetime. Inevitably, 'advanced technology has become the key to both economic and military power',[1] with the corollary that 'organized knowledge as a political force should no longer be underestimated'.[2] As a result, involvement in science by governments in the western world increased markedly after the war. Vannevar Bush's report to the United States President in 1945, *Science the Endless Frontier*, argued that money spent on basic research would automatically result in economic advantage. Expenditure on research in the United States increased significantly in the years following the report, slowed down in the 1970s, and has risen again subsequently. Such increases in science budgets could not continue indefinitely and strategies for the rational deployment of scientific resources developed. Science policy now has a considerable literature, reflecting the difficulties that the development of such strategies has entailed.[3]

In Australia, the image of science in the community has undergone vicissitudes common to all developed countries. Once perceived in economic terms as an 'endless frontier', science is now subject to continuing debate concerning both its social costs (and benefits) and its moral values.[4] Whatever the outcome of that debate, the need to exploit science and technology for our national well-being has become strikingly clear. At the end of the nineteenth century, wool and gold had given Australia the highest per capita income in the world. Our dependence on the export of agricultural and mineral commodities, with little value added from processing, has continued to the present and made us vulnerable to the volatility of commodity prices. The need to counter this vulnerability by the development of export-competitive industries based on sophisticated technology has, until recently, been barely acknowledged by either government or industry. The industrial expansion of the 1950s was initiated by a Labor government on the basis of import replacement using largely imported technology. The small domestic market led to high levels of tariff protection actively promulgated by John McEwen, deputy to Menzies as Prime Minister in the 1950s and 1960s and leader of the Country Party, even though high tariff protection was contrary to the Party's interests. The traditional protectionist philosophies of the pre-Federation colony of Victoria reigned supreme and were supported by an unholy alliance of management and unions. The resulting complacent management and a system of craft unions (institutions inherited from Britain) mitigated against technological innovation and our export of high technology products today is amongst the lowest in the OECD countries.[5] The considerable presence of transnational corporations in Australian manufacturing has also been an impediment to technological innovation. The transnational corporations import their know-how, usually from their parent companies, and they have ignored important Australian innovations.[6] We may ask if we have moved from imperial to national science or merely swopped the British imperium for the imperia of the transnational corporations. If we were indeed the 'lucky country', our luck may well be running out. With 0.35 per cent of the world's population, Australia accounts for 1.6 per cent of the world's scientific literature but only 0.31 per cent of the world's emerging patents.[7] The country fails to exploit an excellent scientific infrastructure for economic advantage.

Ann Moyal has pointed out that Australia, traditionally, has depended upon *institutions* to provide the science necessary for the attainment of national goals, rather than developing *strategies* that focus on economic and social objectives and encourage the necessary science to help achieve those objectives.[8] It is in the nature of the scientific institutions and in the attitudes of the scientific community, as well as in their interactions with political and industrial institutions, that we must seek the answer to why we have failed to take advantage of our excellence in science. In that inquiry the reasons why attempts to put in place a 'strategic' approach to science policy have been slow and, until the late 1970s, ineffectual, should emerge.

J. K. Galbraith has pointed out that 'the educational and scientific estate, with its allies in the larger intellectual community ... like any new political force lacks confidence in its own objectives'.[9] The scientific community in Australia has frequently been confronted with intellectual luddites, both in politics and in the business community,[10] which erodes confidence. On the other hand, the scientific community, with notable exceptions, imbued with an ethos of autonomy and elitism and a naive faith in the 'science push' model of economic growth, have not been effective lobbyists. More money for science has been their cry but the response from Realpolitik demands accountability. ANZAAS, especially since 1970, has attempted to raise the consciousness of the scientific community on the wider social and economic issues. Support for the Association since then, however, has declined, partly owing to the lack of active recruitment on its part but more generally because of many scientists' overwhelming involvement in their specialist concerns and their consequent lack of interest in the broader problems.

The brief overview that follows considers some of the many influences that have shaped and directed the impulse of science in public affairs. The focus inevitably will be on the scientific and educational institutions, including ANZAAS, that have played a significant role in the scientific community. But science is not merely an intellectual activity and its impulse cannot be examined in isolation from economic, political and social considerations.

SCIENCE IN THE AFTERMATH OF WAR

The Second World War demonstrated the need for Australia to become industrially self-sufficient. Our imperial role as a source of raw materials for the factories of Great Britain was seriously undermined during the war. The vulnerability of sea transport and the collapse of Britain's traditional role as defender of Australia, let alone as supplier of manufactured goods needed to prosecute a war, led to major efforts to marshall Australia's scientific and technological manpower in order to establish secondary industries.[11] By 1945 Australia was at last moving from a prewar economy largely dependent on agriculture and mining towards becoming an industrial nation. D. P. Mellor noted that 'at the turn of the century, only about 12 per cent of adults were employed in secondary industry as compared with 30 per cent in primary industry. By 1950, the proportions were almost completely reversed.'[12]

A major role in the industrialization of Australia in the postwar period was played by the Commonwealth Department of Post-war Reconstruction, formed in 1943 by the Labor government. This sponsored a Secondary Industries Commission charged with advising the government on the orderly transition of Australia's manufacturing from war to peace. Many of the government factories devoted to supporting the armed forces during the war were handed over to private industry. A report of the Department of National Development published in 1952 listed the products of 43 000 factories, representing 200 major industries.[13] Nevertheless, it must be emphasized that these developments were motivated by import replacement and largely relied on imported technology to serve a domestic market. Our principal exports were still primary products and subsequent protectionist policies ensured that exports of manufactured goods would be uncompetitive.

In 1946 the government, mindful of the significance of minerals to the Australian economy, established the Bureau of Geology, Geophysics and Mineral Resources. Its surveys of Australia's geology contributed especially to the exploitation of monazite in beach sands, the discovery of oil in many parts of Australia, and the discovery of large deposits of uranium in South Australia, Queensland, and the Northern Territory. A Defence Research and Development Policy Committee was also established. Mellor commented that 'science thus became more closely integrated with the defence of Australia than it had been in the war of 1939–45'.[14] The most ambitious of all the postwar developments of the forties was the Snowy Mountains Hydroelectric Scheme, first mooted in 1914. Confidence in scientific and engineering technology engendered during the war years facilitated the launching of such a vast scheme.

The Council for Scientific and Industrial Research (CSIR, later CSIRO) dominated the Australian scientific scene in the postwar years, as it had done before the war. Its part-time chairman was Sir George Julius, a leading industrialist and engineer, and its chief executive officer was A. C. D. Rivett. The latter advocated 'complete autonomy for the practising scientist' and was inspired in this view 'by believing implicitly, almost blindly, in the strength and moral values of British institutions'.[15] CSIR(O) developed a high reputation, especially in its contributions to Australian agriculture,[16] and its ethos of scientific autonomy was coupled with 'the absolute values of science and the quest for the intellectual mastery of nature', so that CSIR(O) made major contributions to fundamental knowledge in the areas where it considered that existing knowledge was inadequate to solve the practical problems. Among CSIR's initiatives after the war were the provision of sophisticated equipment for the National Standards Laboratory, essential for the development of precision engineering and optical industries, and the formation of the Division of Industrial Chemistry and divisions devoted to the processing of wool. The radar experts assembled during the war formed the nucleus of the Division of Radiophysics and it directed its activities to three major fields: rain and cloud physics, radioastronomy, and the application of radar to air and marine navigation. Australia subsequently became a world leader in these three fields. These immediate postwar developments indicated increasing emphasis on research of benefit to secondary industry. But a review of CSIRO thirty years later considered that the support for secondary industry remained inadequate and commented that the links between the CSIRO and industry were at best tenuous.[17] As well as being the major recipient of government scientific spending, CSIR was the principal source of scientific advice to government, making David Rivett the most influential scientist in the country. R. G. Casey, prominent Liberal and a former minister in charge of CSIRO, wrote in 1972 that 'Julius as Chairman and Rivett as Chief Executive Officer each made important contributions to the building of an institution of a new and unique character, and with a most ambitious range of research interests which have been described as taking place in a social and political climate that was on the whole at that time rather indifferent to scientific research.'[18]

During the period of postwar tension between the Soviet Union and the western nations, indifference turned to antagonism. Rivett, in common with leading scientists throughout the world, believed passionately in the freedom of scientists to pursue their research and to communicate their results internationally with the rider that 'if there is to be secrecy it is better for defence work of this kind to be carried out by special staffs and laboratories not connected with CSIR'.[19] A British scientist who had been a security officer during the war inspired an attack on CSIR in a Sydney weekly suggesting that it was slack on security and allowed scientists who might be Communists to work on projects that might involve national security. This led to attacks on Rivett by members of the Liberal–Country Party opposition in federal parliament,[20] which Oliphant described as 'one of the most shocking episodes in the history of the Commonwealth Parliament'.[21] These attacks were designed to discredit the Labor government, but they also indicated extreme anti-intellectualism and cold war hysteria.[22]

The science and Industry Research Acts (1926–39) gave CSIR no authority to engage in defence work for which strict secrecy was essential. Nevertheless, CSIR had inevitably participated in defence work during the war. After the war the Radiophysics Laboratory, which had been involved in military radar, took up peace-time activities in which secrecy was not essential. The Division of Aeronautics, however, continued with military work and became more deeply involved when the Long Range Weapons Project was established in 1947.[23]

In August of 1948 Prime Minister Chifley, without consulting CSIR, appointed H. C. Coombs, director of Post-War Reconstruction and W. E. Dunk, chairman of the Public Service Board, to report on CSIR. In December 1948 the Labor government, in response to the report, passed an Act empowering it to transfer any or all of the activities of CSIR irrespective of its connection with defence to the control of various government departments under the Public Service Board. The Opposition attack upon CSIR in the House in September–October left the Prime Minister with little option in this matter.

There was an immediate and powerful response from the scientific community, which culminated in a long and reasoned letter from the Australian National Research Council (ANRC) to J. J. Dedman, minister in charge of CSIR, setting out the damage that the proposed changes could do to the quality of research in CSIR.[24] Dunk, to his credit, impressed by the arguments against the Act by both CSIR and the scientific community generally, suggested significant changes to Dedman. While the Act of 1948 remained on the statute book another Act, the Science and Industry Act 1949, was passed in April of that year, and the Council for Scientific and Industrial Research became the Commonwealth Scientific and Industrial Research Organization. Under this Act, CSIRO retained scientific direction of its affairs with a chairman supported by a small executive drawn from its own ranks, and an advisory council drawn widely from the community. Conditions of work, promotion, and salaries remained directly under the control of the scientific executive, but the Public Service Board was responsible for the loyalty of staff. Under the Act of December 1948, the Division of Aeronautics was transferred to the Department of Supply.

The attack by the Opposition on CSIR was part of a wider hysterical campaign against scientists, especially members of the Association of Scientific Workers, some of whom were connected with left-wing political movements. The frequently unjust villification of some individuals led to both the loss of some quite brilliant scientists to other countries and the discouragement of others who remained. Jean Moran has argued that this 'deep-seated xenophobia with a strong anti-intellectual thrust ... effectively put an end to the public articulation of social responsibility in science for a generation of scientists in Australia'.[25]

ANZAAS, in abeyance during the war years, held its first postwar congress in Adelaide in 1946. Until the Second World War, and for some years thereafter, ANZAAS was undoubtedly the premier scientific society in the country. It was closely associated with the Australian National Research Council, which comprised the fellows of ANZAAS. Most leading Australian and New Zealand scientists contributed to its administration in various roles and at various times. To be president of the congress or president of a Section was a high honour (and indeed it still is) and most scientists from industry, government or academia took advantage of the congresses to present specialist papers and, less frequently, to address themselves to the wider issues of concern to the government and the community. The congresses were held at approximately eighteen month intervals and were mounted in rotation around the six capitals of Australia, and about once every ten years in New Zealand. During the week of the congress, the proceedings received wide media coverage, especially by the press at the congress venue, but also in the major dailies in Melbourne and Sydney. It was invariably referred to as 'The Science Congress' and there was no other comparable national meeting.

Very little research was undertaken in the Australian universities owing to shortage of staff and lack of equipment.[26] Before the war no Australian university offered a doctorate. The University of Melbourne offered the degree in 1945 and the other universities soon followed suit. Nonetheless many talented students still went overseas, especially to Britain, for postgraduate training and many did not return. CSIR had its own scheme of overseas studentships in order to recruit staff of the appropriate calibre.

Physicists, chemists, and engineers were represented by professional institutes, which were branches of their British parents. Their main concern was with professional standards and the interests of scientists and engineers in industry, as well as those in universities and government, but they did not provide meetings for the reporting of research. There were virtually no societies representing the scientific disciplines and no national journals for the publication of research until 1947 when the ANRC, in conjuction with CSIR, established as *Australian Journal of Scientific Research*.[27] By default, the Sections of ANZAAS had become the venue for the communication of research. The Association thus operated like a federation of specialist societies, which detracted from its principal objectives: 'the advancement of knowledge and to promote a spirit of co-operation between scientific workers and scholars and those in sympathy with science and scholarship generally'. Within the Sections scientists only communicated, by and large, with each other. The *Australian Journal of Science,* published by the ANRC, included research communications, but these dealt largely with Australian biology and geology, as did also the proceedings of the state Royal Societies, with the notable exception of that of New South Wales in which the physical sciences were represented.

Science education received a boost after the war owing to the operation of the Commonwealth Reconstruction Training Scheme, which was responsible for the rehabilitation of ex-service men and women, providing them with professional training in universities and technical colleges. As well, the Labor government established a Universities Commission in 1942, which selected students for courses in science and engineering. These were exempted from war service and, for the first time, were paid a small stipend. These initiatives resulted in an injection of seven million pounds into universities and colleges by the end of the 1950s. The amount of money available for research in universities, however, was still pitifully small compared to CSIR. The additional funds dried up as the reconstruction scheme completed its task.

The objectives of science education and the nature of the institutions charged with effecting it were, at this time, the subject of debate. Fears were expressed that Australia might be blindly following the pattern of British universities which, in contrast to institutes of technology in Europe and the United States, had failed to produce sufficient graduates oriented towards careers in industry. Initiatives to transmute the Sydney Technical College into an Institute of Advanced Technology resulted in high expectations that the British models were no longer to prevail. The Sydney Technical College, established in 1885, enjoyed a very high reputation. Its diplomates were widely sought after in industry and its diploma in chemistry was recognized by the Royal Institute of Chemistry in Great Britain, the only diploma in the British Commonwealth outside Great Britain to be so recognized. In 1946 R. J. Heffron, Minister for Education in New South Wales, consulted Carl T. Compton, president of the Massachusetts Institute of Technology, and shortly afterwards established a developmental council for a New South Wales Institute of Technology. Courses were held in the Sydney Technical College pending the development of a new site at Kensington. An Act incorporating the new institute was passed by the New South Wales government in 1949 but 'at the last moment, parliament acting on suggestions from Britain, changed that name to the NSW University of Technology'.[28] The erosion of the original concept, based on advanced institutes of technology in Europe and the United States, had begun. In subsequent years, the addition of faculties such as arts and medicine brought about another change of name. The institution became the University of New South Wales, which was similar to the other state universities. Subsequently, institutes of technology were established in

Sydney and in the other major cities in Australia, but they functioned as Colleges of Advanced Education and were subject to state government control, a far cry from the original concept.[29]

The reduced status of the Sydney Technical College, when many of the best staff left to join the new university at Kensington, deprived New South Wales of high quality training at an intermediate level necessary for flourishing industries. As originally conceived, the students at the Institute of Technology were to have work experience in industry, and close ties with industry were to be maintained at the research level.[30] In 1952 hopes were high that the New South Wales University of Technology might be able to 'recover for the British Commonwealth some of that initiative for founding new industries on the basis of more recent scientific knowledge, which has been lost to the United States and Germany'.[31] Despite this hyperbole, nothing approaching such high expectations was achieved nationally. As late as 1986, it was reported that 'the number of PhD mathematicians, scientists and engineers graduating from our universities is small. Of that small number, the number choosing research careers in industry is minute.'[32]

SCIENCE IN THE MENZIES ERA

R. G. Menzies was elected Prime Minister in 1949 and dominated the political scene in Australia until he retired in 1966. His attitudes, pro-British and elitist, had an important influence on the development of both the Australian Academy of Science (AAS) and the Australian National University (ANU). His perception of science and its national role was a dominant factor in delaying the development of a strategic science policy. These perceptions were revealed in 1958, in an oration to the Australasian Medical Congress.[33] In the previous year, Russia had launched Sputnik 1 and 2 and in 1953 had exploded its first hydrogen bomb; it seemed that Russia may have been ahead of the west in advanced technology. Menzies considered science to be essential for the progress of mankind: 'without more and more scientists and more and more scientific research, the material future of our people cannot be assured and enlarged'. Nevertheless, he did have nostalgic reservations: 'in those far-off ridiculous days when the higher education was classical and humane, when men stored up in their minds a love of knowledge of letters and philosophy and history, our fathers had not reached this strange state of worship for the machine and obedience to its demands'. These reservations about technology were countered, however, by his essentially scholarly and liberal attitudes: 'let us have more scientists, and more humanists. Let the scientists be touched and informed by the humanities. Let the humanists be touched and informed by science, so that they may not be lost in abstractions derived from outdated knowledge or circumstances.' His real concern was the fear that Russia may be using science for economic and military advantage more effectively than the democracies, and that this Russian advantage was

PLATE 5
'The Australasian Society of Mutual Bores' (*Bulletin*, 8 September 1888).

PLATE 6
Officers of the AAAS at the Melbourne meeting, 1890 (*Liversidge Papers*, University of Sydney Archives).

PLATE 10
Collage of sectional representatives at the AAAS Brisbane, 1909. (*The Queenslander*, 23 January 1909; *Liversidge Papers*, University of Sydney Archives).

the result of complete direction of 'both the programme of research and the activities of the research workers'. In this he saw a 'danger that we may be urged to adopt the Russian method in order to achieve the Russian result'. Ironically he was making a plea for scientific freedom that had been so virulently attacked by some of his political colleagues when they were in opposition ten years earlier, though he had distanced himself from that attack at the time. This rejection of scientific planning and direction, perceived as a socialist approach, was to influence the conservative government under Menzies and his successors until the early seventies when they lost government. Rejection of formal science policy arrangements was encapsulated in Prime Minister Gorton's notorious comment in 1968, that committees set up to determine science policy were only groups of individuals 'pushing the barrow' for their own disciplines.

Moves to establish a postgraduate university, similar to Princeton in the United States, began before the Menzies government came to office. In 1946, mindful of the inadequacies of the state-funded universities and concerned at the loss of science graduates overseas, the Labor Prime Minister, J.B. Chifley, and H.C. Coombs, Director of Post-War Reconstruction, consulted M.L. Oliphant in London. Oliphant, an expatriate Australian, was professor of physics at Birmingham and a leading scientist in the atomic bomb project. Chifley 'understood Australia's need to upgrade all its tertiary education facilities' and 'realised that modern industry depended increasingly on the output of university graduates'.[34] Their meeting led to the formation of the Australian National University in Canberra. The original concept of a Princeton model was to be considerably modified. In 1950 Oliphant returned as director of the School of Physical Sciences in the new university and thereafter played a dominant role in Australian scientific affairs.

The proposal to establish the Australian National University immediately raised debate and controversy within the academic communities of the established universities. Would a national university drain federal funds from the other universities? Would the proposed national university reverse the brain drain? Was it desirable to divorce undergraduate teaching from research in any university?[35] The establishment of an advanced research institute would have circumvented some of this debate, but the need to provide postgraduate training required a university. The Act of 1946 establishing the University provided for amalgamation with the existing Canberra University College, as Canberra at this time was too small to justify the maintenance of a separate undergraduate institution. This was strongly opposed by Oliphant, who favoured a new national, elitist institution. Menzies, as leader of the Opposition, deplored the name and suggested that it should be called Canberra University and went on to say, 'if institutions of this kind are to be real, *and not merely scientific shops*, they must develop some esprit de corps, their own character and colour, and make their own mark upon those who pass through them' (emphasis added).[36] Nevertheless, when Prime Minister, Menzies strongly supported the University. In light of his traditional, and especially anglophile, attitudes, he would have preferred the new university to be an antipodean Oxbridge[37] rather than an antipodean Princeton, and D. B. Copland, the first vice-chancellor, also advocated 'a balanced structure on the pattern of the long established universities of older civilisations'.[38] When the Canberra University College finally merged with the University in 1960, Oliphant was bitterly disappointed at the abandoning of the original concept of a research university. The tensions resulting from this 'shot gun marriage' still persist.[39] The elitist tradition and performance of the Australian National University have never been critically examined and the state universities had to wait almost ten years for the Murray Commission to persuade the government of the necessity for their adequate support.

Mindful that postwar reconstruction funds for universities had ceased, Menzies in 1950 asked a committee to report on the financial situation of the universities, and following its recommendations a State Grants (Universities) Act was passed in November 1951, which provided one million pounds a year for the next three years. By 1956, however, it was apparent that these funds were hopelessly inadequate. Menzies then invited Sir Keith Murray, chairman of the University Grants Committee in Great Britain, to head an inquiry into the future of Australian universities. The Murray Committee, reporting in 1957, recommended a massive programme of new buildings, increased triennial grants for recurrent expenditure, and the establishment of a permanent University Grants Committee.[40] In a remarkable gesture reflecting the seriousness of the plight of the universities, Menzies persuaded the Labor Opposition to vote with the Government in implementing the Murray Committee's recommendations. Thus started a long period of expansion of university education.

The continuing saga of the problems of tertiary education continued with the publication of the Martin Report in 1965.[41] This resulted in increased funding for tertiary institutions and the establishment of new Institutes and Colleges of Advanced Education, different from, but complementary to, the universities. The traditional roles of the Commonwealth and the state governments in the funding of tertiary education were reversed.

In 1951, on the occasion of the Commonwealth Jubilee, a seminar entitled Science in Australia was conducted by the Australian National University 'to examine the past and present of scientific effort in Australia in relation to the practical and intellectual needs of the country and the growth of science elsewhere'.[42] Oliphant was principal organizer of the seminar and Rivett took the chair. In light of the objectives, some aspects of the organization of the seminar were disquieting. The number of participants was small and the proceedings were held in camera.[43] It was essentially a meeting of the élite in Australian science. The press were excluded, so any influence on public opinion or on the politicians was minimal. The attack on science in the parliament in 1948 may well have made them wary of public exposure. A final session was devoted to discussion of science policy and senior representatives of industry were invited to participate, but they were not admitted to the discussions until the final section on science and industry. One industry leader made a plaintive protest 'that many had not been able to hear the discussions that had taken place earlier in the week. It was, therefore, a little difficult to contribute adequately and in a constructive way.'[44] The organizers of the symposium were clearly not inclined to concern themselves with the 'dark satanic mills'.

It was in this élite and politically naive atmosphere that the proposal to establish an Australian Academy of Science was made. The leading advocates of the Academy were two Fellows of the Royal Society—Oliphant and D. F. Martyn. Both were strongly supported by Rivett. The proposed Academy was intended to perform many of the functions of the Australian National Research Council (ANRC) and the resulting problem of duplication exercised the minds of the participants at the symposium. Oliphant's view was that 'he believed the Australian National Research Council was no longer the effective representative of the country's science—that Australia needed a national body such as the Royal Society of London to stimulate and give coherence to its scientific life'.[45] Oliphant's biographers go on to comment that 'no one else in an allegedly egalitarian country such as Australia, especially after eight years of Labor government, had been prepared to stand up publicly and sponsor such an elitist idea'. Others at the symposium were more kindly disposed towards the ANRC and I. W. Wark, who as chief of the Division of Industrial Chemistry of CSIRO had close ties with industry, asked, 'cannot we establish confidence in our own ANRC so that its advice will be sought and taken as is that of the Royal Society in UK or the National Academy of Sciences in USA?' He went on to point out that, 'at the present time, the Commonwealth Government has given to a group of three men—able and eminent men, but none of them a scientist, and all with a similar background of economics and finance—the responsibility of advising it on how to assist the Universities. Since much of the proposed assistance must be for scientific research, I regard it as disappointing that the group has not consulted such a body as the ANRC.'[46] His answer probably lay with Oliphant's suggested Academy. 'It raised us', he said later, 'from level of the ANRC to another level, where they [the political leaders] felt that they could talk to us.'[47]

After considerable consultation during the next two years between the ANRC and eleven Fellows of the Royal Society resident in Australia, the ANRC agreed to abolish itself, thus clearing the way for the establishment of the Academy. By the time the Academy was founded in 1953, a conservative government under Menzies found no difficulty in supporting an elitist body, especially one based on the prestigious Royal Society of London.[48] Menzies not only guaranteed its financial future but was also influential in obtaining a royal charter for the new Academy.

The Academy assumed the major functions of the ANRC, but it lacked a number of its favourable characteristics. Although the ANRC mainly represented academic research interests, it was in close contact with large numbers of scientists in industry and government by having balanced state representation and by meeting during ANZAAS congresses. The Academy did not make provision for the election of Fellows for achievements in fields other than that of pure research. Martyn summed up the Academy as 'we are whom we elect', and the ANRC asked, pointedly, 'does all wisdom reside in eminent research workers?'[49] Although the Academy provided for regional advisory committees in its charter and by-laws, such committees were never appointed. Within a year of the founding of the academy, H. C. Webster, professor of physics at the University of Queensland, was to write an acerbic letter to the *Australian Journal of Science* pointing out that of the fifty-seven Fellows of the Academy only one was resident in Queensland. He concluded that 'it is not fit to become the national scientific body of Australia, nor is it fit to receive moneys from the national exchequer. It should be treated as what it actually is, a parochial scientific club.'[50] Even by 1969, there were only 3 Queensland Fellows in a total of 135. Webster's attitude to the Academy was quite widespread and has persisted to the present time, even within the fellowship itself. In 1969, there were thirty-seven Fellows from the ACT, only two less than for the whole of New South Wales. And although this may well reflect the excellence of scientific research in the Australian National University and the divisions of the CSIRO in Canberra, another well-known phenomenon prevails—not only 'we are whom we elect' but also 'we elect whom we know'.

The ANRC had represented not only the natural sciences, but also the social sciences. The Academy confined its membership to the former. In this it was following the Royal Society and had the strong support of Menzies, who would have seen the social sciences as a hot bed of trendy, if not down-right socialist, ideas. At its dissolution, the ANRC spawned a Social Sciences Research Council, which eventually became the Academy of Social Sciences.

Of the two options open to the scientific community at that time—to reform the ANRC or to establish a new academy—the former may well have been in the best interests of the country and the scientific community. That is not to say that excellence in research should not be rewarded with dignity and prestige. A body such as the Academy is best suited to represent Australia in international scientific councils. It was in the high hopes of the founders that the Academy would be a powerful influence on government, representing the aspirations of the community of science and helping to shape the economic future of Australia, that it was ineffective.

New Zealand provides us with a contrasting history in the evolution of its Royal Society.[51] The New Zealand Institute, established by an Act of Parliament in 1867, incorporated regional scientific societies. In 1933 it became the Royal Society of New Zealand. It retains close links with the regional societies and also with specialist scientific societies through its Member Bodies Committee. At the same time it elects a fellowship, thus recognizing excellence. This structure has been described as a 'unique contribution to scientific organization, a situation achieved in few other countries'.[52]

The foundation of the Academy in 1953 and the concomitant abolition of the ANRC inevitably eroded the status of ANZAAS. As well, since the 1950s a large number of specialist societies have been founded, and their meetings have become the preferred venues for the communication of original research. Thus much of the support previously given to ANZAAS dwindled. In 1954 a review committee consisting of the Executive Committee and the vice-presidents was appointed 'to consider the implications of the winding up of the ANRC upon the position of the Association',[53] but the review committee failed to address itself to these basic issues. Its main recommendations were that the functions of ANZAAS should be:

1. Publication of the *Australian Journal of Science* [which it had taken over from the ANRC];
2. Holding of conferences;
3. Distribution of small research grants;
4. Organization of regional groups when members in a region consider it warranted.

It did report, however, that there was general opposition to specialized papers at meetings, but that there was a role for review papers intended to bring scientists up-to-date in fields outside their own, and for symposia, especially those involving two or more sections. It even suggested that 'occasionally symposia should deal with purely Australian problems'! Only at the end of the report, almost as an afterthought, was it conceded that 'there should be a major effort to take science to the public'.

The review committee's report was hardly an exciting blueprint for the future. It was not surprising that Oliphant in his presidential address to the Adelaide Congress in 1958 stated bluntly that 'the concept of ANZAAS as an organ of the relation of science with the society in which it is carried on has virtually disappeared'.[54] Oliphant also succinctly summed up another shortcoming of ANZAAS: 'between congresses ANZAAS virtually dies'. Media coverage of the congresses was always extensive, but between congresses nothing newsworthy arose from the Association. The *Australian Journal of Science*, now the organ of ANZAAS, had never been effective as a bridge between the scientific community and the media. As well, the media had no science correspondents who could distil news stories from scientific publications. Under the ANRC, the journal had been a two monthly publication. The council of ANZAAS, concerned at the growing cost of the publication of congress proceedings, decided in 1958 that the journal should appear monthly and that the increased space should carry the Section presidential addresses and lists of Section papers by title. The presidential addresses, some of which were highly specialized, appeared for many months after the congress. This engendered a sense of *déjà vu*, and denied space to more timely articles on issues facing the scientific community and the nation at large. Moreover, as the journal was produced entirely by voluntary effort, it was not surprising that the news and views section of the journal reported little more than elections to the Royal Society and the Australian Academy and appointments to the staffs of Australian universities. The Australian media found little here that was worth reporting. A new, grey, cover design, introduced in 1962, seemed to characterize its contents as well.

Oliphant in his presidential address to ANZAAS in 1958 also stated with foresight that, 'here in Australia we have not yet begun to create instruments of policy which take care of the basic science and technology on which our long-term future depends', and went on to warn that 'the men of science who create new concepts, the technologists who make them practically useful and the economists and sociologists who must fit them into the social structure—all must play an important and increasing part in the formulation of policy, if disaster is to be avoided'. However, he made no concrete proposals on how this was to be achieved.

At the Sydney Jubilee Congress of ANZAAS in 1962, a symposium was held on federal aid for scientific research in Australian universities. The discussion was based on a *Current Affairs Bulletin* of the University of Sydney on science policy especially produced for the congress. It advocated a scientific advisory council as well as a national science foundation, essentially a funding body.[55] A resolution from the symposium, transmitted by ANZAAS to the Prime Minister, recommended the establishment of the foundation but omitted the scientific advisory council. More money for science was acceptable but direction of research activity by a government appointed council was rejected. This may have been reasonable in the context of university research, but there was an apparent failure to realize that more money for university research could only be considered in the light of other government commitments to research, and who was to advise government on such priorities?

In 1963, the first meeting of the ministers responsible for science in the twenty OECD countries took place. Only four of these countries had special ministers for science. Australia, in common with most of the other countries, was represented by the Minister for Education. At this meeting, a system for reviewing science policy in member countries was established, but Australia was to wait ten years and the election of a Labor government before it took advantage of the OECD review. In the interim, the Academy, supported by the chairman of CSIRO, Sir Frederick White, and the Australian Atomic Energy Commission, continued to lobby for the establishment of science policy machinery. Further elements were set in place with the establishment of the Australian Research Grants Scheme in 1965, the passing of the Industrial Research and Development Grants Act in 1967, and the addition of science to the federal education portfolio in the same year. In March 1965, just before he retired, Menzies made a statement to the House of Representatives predicting the establishment of an advisory body to co-ordinate government expenditure on research (policy for science), though no action resulted. The president of ANZAAS, R. N. Robertson, in commenting on the Prime Minister's statement, noted that 'such bodies will not work if they consist only of scientists, who, by nature, are optimistic and forward-looking with faith in what science can do. For some this leads to an almost blind assumption that the solution to all our problems lies in more research and that enough money for scientific research is just a bit more than they have.'[56] This was a polite version of Gorton's 'pushing the barrow'. Robertson, who later became president of the Academy, recognized that national economic and social priorities needed to be established before rational decisions could be made on how to divide up the research-money cake (science for policy).

'THE LONGEST RUNNING SHOW'

The events leading from Menzies's prediction of the establishment of a science policy advisory body in 1965 to the establishment of an effective Australian Science and Technology Council (ASTEC) in 1976 has been described as the 'longest running show in Australian science'.[57] In 1968 the Academy established the Science and Industry Forum. The membership of the Forum consisted of leaders in industry, government, academia, and selected a total of seventy-five Fellows. It met in an enclave reminiscent of the Science in Australia symposium that had spawned the Academy. Its first, rather slim, report was on science policy machinery for Australia. Malcolm Fraser, as Minister for Education and Science in 1969, made a major statement on science policy to the Forum in response to this report.[58] He argued that 'we may be wisest to continue our pragmatic evolutionary approach seeking advice from different people as different projects arise'. He did, however, announce the launching of Project SCORE, dealing with the compilation of data on scientific expenditure based on the OECD system decided upon by the ministerial meeting in 1963. This annual exercise has been carried out by the Department of Science in its various forms ever since and is a necessary component of science policy.

In 1972 the Liberal–Country Party government, under Prime Minister McMahon, which had been in office for twenty-three years, fell, and a Labor government, led by E. G. Whitlam, was elected. Just before its defeat, the McMahon coalition government had finally established an Advisory Committee on Science and Technology, reporting to the Prime Minister through the Minister for Education and Science. The first Advisory Committee consisted of representatives of industry, commerce, and science and was unacceptable to the incoming government, which abolished it. They wanted representatives from consumers, environmentalists, social scientists, and scholars of science policy. It was reasonable to have some representation from the community at large who, after all, would be at the receiving end of science policy. The Labor government produced a Green Paper, then a White Paper, and invited a team of OECD examiners to report on Australia's needs.[59] In an effort to place before the government views of a wide spectrum of the scientific community, ANZAAS formed a Science Policy Commission, which reported in 1974.[60] Eventually in January 1975 the government appointed the Australian Science and Technology Council (ASTEC), which reported to a ministerial committee chaired by the Prime Minister. During 1975, in a number of important decisions involving science, ASTEC was virtually ignored. In November of that year the government was dismissed. Thus the high hopes of the scientific community that a Labor government would be more effective than its predecessors in establishing a science policy machinery were not realized.

A Science Task Force was set up in 1975 to report on 'the conduct and co-ordination of scientific work carried out, financed and/or supported by the government', as part of the Coombs Royal Commission on Australian Government Administration. It was a daunting task entrusted to three academics, three representatives of CSIRO, one political scientist, and one industrial research leader. When its report was published,[61] it was trenchantly criticized by observers experienced in science policy studies basically because the report was 'overwhelmingly a report on policy for science', emphasising scientific autonomy at the expense of using science to achieve national goals.[62] One of its more controversial recommendations was the abolition of the Department of Science, reflecting, perhaps, that Department's inadequacies at that time in formulating policy. In 1975 the incoming Fraser government dismissed most of the Labor appointees to ASTEC and set up a small group to advise on its future. The next year, the government accepted the recommendations of the advisory group, and ASTEC was permanently established, answerable to the Prime Minister and with an adequate secretariat within his Department. Fraser, formerly opposed to formal science policy machinery, had experienced a change of attitude, probably arising from his long experience as Minister for Education and Science and minister in charge of CSIRO. The chairman of the now permanent ASTEC was Sir Geoffrey Badger, a professor of chemistry at Adelaide, who had been president of the Academy of Science and of ANZAAS. ASTEC began its work with a broad-scale review of science and technology in Australia and moved on to a number of specific issues. The show had come to an end. By 1979 Ann Moyal was able to state that 'it is apparent that mammoth changes are taking place in the organisation of Australian science and technology. The impetus has come from the changed climate of public opinion, from political responses, and from the establishment of ASTEC.' There were, however, still gaps in the system—'large questions of concentration, overlap, balance, and flexibility remain and cannot all be resolved by ASTEC'.[63] Moyal identified a lack of co-ordination of the roles of government scientific institutions and of mechanisms of technology assessment as two major shortcomings.

The chequered history of the Department of Science also reflected the confusion during the 'longest running show'. Science had remained part of the Department of Education until the election of the Labor government in 1972, when it became a separate department, but with a minister who was not even in the Cabinet and who was the most junior minister in the government. At the end of the Labor government's term of office, the department became a Department of Science and Consumer Affairs. W. L. Morrison, the first Labor Minister for Science, had pointedly declared that there were no votes in science; perhaps there were some in consumer affairs! In 1978 it became the Department of Science and the Environment under the Fraser government, and with the election of a Labor government again in 1983, it changed its responsibilities to Science and Technology. In 1985 Technology was transferred from Science to the Ministry of Trade, Industry and Commerce, a senior portfolio with Cabinet representation. Science was once again naked and alone and among the two or three most junior ministries. There is little doubt that technology is of the utmost importance to trade, industry, and commerce, but the divorce of science and technology was disquieting in view of their interdependence.

THE WINDS OF CHANGE

The fading perception of science as the 'endless frontier', beginning in the 1960s and gaining momentum during the 1970s, was compounded of many factors, especially concern at the social consequences of technological change, environmental concerns, fear of technology devoted to military ends, and the soaring costs of research and development. The ANZAAS response to the changing circumstances began in the 1960s as a result of initiatives at the local level, though they were more zephyrs than gales. Congress organizers began to introduce into congress programmes symposia that dealt with issues of public interest. A division was formed in New South Wales in 1966 as a response to the social and economic implications of advances in science and technology, and to various anti-science movements, which began with the publication of Rachel Carson's *Silent Spring* in 1962. The 'public articulation of social responsibility in science' was returning. Divisions that had been formed in Western Australia and in Papua New Guinea in 1957, were responses to isolation from the centres of population. The divisions provided the climate and organization for popular lectures and symposia, which were usually well covered by the local media, and which countered, to some extent, the accusation that 'ANZAAS dies between congresses'. The New South Wales division mounted an active programme that included discussion of important issues such as automation[64] and natural resources.[65] These activities of the New South Wales division of ANZAAS in its early days indicated an effective role for the Association in the shaping of public affairs. Finally, in 1970, ANZAAS underwent a major reorganization as a result of the report of a committee of a review chaired by Sir John Crawford.[66] Major results of the reorganization were the formalization of divisions in the constitution, leading to their establishment in the remaining states, provision for a long-term chairman of council, but retaining an honorific president, and the replacement of the *Australian Journal of Science* with a new journal, *Search*. It had a striking new format, and its editorial policy was 'to publish such articles that will inform the scholar about interesting advances in fields outside his own area of competence' and to 'publish articles which deal with the social and economic consequences of advances in science and technology'.[67] *Search* was subtitled 'Science, Technology and Society' to emphasize this latter objective. Slowly *Search* developed into a major, indeed the only, forum for the Australian scientific community, and its contents covered extensively, and in depth, public issues such as science policy, innovation, communication in science, pollution and environmental concerns, as well as the more 'in house' interests of scientists. The new spirit culminated in the Perth Congress in 1973, which was the largest and most successful ever held, involving a remarkable range of participants from both the public and private sectors, including five federal ministers.[68] Pockley described the congress as 'the political blooding of Australian science',[69] and a correspondent for *Nature* observed that 'extensive open discussion and lobbying with politicians and their staffs were features never previously observed at an Australian science conference'.[70] Certainly with the upsurge in interest in Australia's natural resources, especially mining in Western Australia, the theme—Science, Development and the Environment—was timely. It was also the first ANZAAS congress after the election of the Whitlam Labor government, and the political involvement was a reflection of that government's concern with new directions and policy initiatives.

An Australian Academy of Technological Sciences was founded in 1976. It arose from a failure of the Australian Academy of Science to evolve mechanisms by which applied scientists and engineers working in industry could be elected to fellowhip, although this had been achieved by the Royal Society in Britain. The new Academy was more responsive to the ebb and flow of public affairs than had been the Australian Academy of Science. It also opened its annual symposia on important policy issues to a wide cross-section of the scientific and industrial community in contrast to the 'closed shop' approach of the AAAS in its Science and Industry Forum. The prediction by Moyal that 'it is certain that the Government of the day will turn increasingly for advice to the diverse and practically oriented expertise of the Academy of Technological Sciences' is being fulfilled.[71]

The need for major changes in the structure of the Australian economy had been evident since the seventies. Government, faced with alternating 'boom and bust' economic conditions largely predicated upon international commodity prices and the lessening of the importance of many traditional manufacturing industries, seemed concerned only with their success at the next election and did little to face the long-term policies needed to effect the restructuring. The bureaucrats advising the government were usually the products of faculties of law or economics, notorious for their narrow orientation; an understanding of, or sympathy with, a role for science and technology in national affairs was rarely present. A former head of Treasury recently suggested that the Department of Science 'is performing no useful service and should be abolished'![72] Science and technology was but one factor, albeit an important one, in the complex of necessary policy initiatives. In the OECD league we ranked near the bottom in the export of technology-based products as well as in the percentage of our gross national product devoted to research and development. Spending on research and development fell from 1.34 per cent of gross domestic product in 1968–9 to 0.95 per cent in 1983–4. Most of this fall was due to a contraction in research and development performed by the private sector.

Financial support for CSIRO had risen from approximately 0.1 per cent of gross domestic product in the 1950s to almost 0.2 per cent by 1970. Thereafter, though fluctuating, it remained at this level. There was, however, increasing disquiet concerning the thrust of CSIRO research. In 1977 an inquiry into the CSIRO chaired by A. J. Birch of the Australian National University reported to the government. While recognizing the prestige of the organization, the inquiry noted that 'it exists to serve the interests of the community, rather than those of the scientists in it, and must not be allowed to lose touch with community affairs'.[73] Some of the problems of CSIRO stemmed from its age and size. Its age profile, with a system of tenured appointments, made it vulnerable to the complaint that it was too ingrown. It had been brilliantly successful in agricultural research, but its contribution to industrial research was less impressive. The persistent managerial crises resulted from the ideal of scientific autonomy so carefully fostered by Rivett and the impossible task confronting its small executive, drawn largely from within its own ranks, of analysing and understanding every sector of the Australian economy to which it could make a contribution. The Birch inquiry rejected any subdivision of the organization despite its monolithic structure. Separate statutory bodies responsible for defined sectors of the economy could have solved many of the problems without sacrificing scientific excellence, but any such suggestions had always been vigorously opposed by its staff, so proud were they of its national and international status. The inquiry approached this problem by recommending that divisions be grouped into institutes that were related to sectors of the economy or broad discipline areas, and by the strengthening of the outside advisory council, which lacked, however, independent research and secretarial assistance, which it needed to operate effectively. The grouping of divisions into institutes added another administrative layer to the organization without necessarily solving its problems.

Signs that the economic significance of science and technology was finally receiving due political attention began with the election of the Hawke Labor government in 1983. By 1985, the collapse of commodity prices in the international markets gave added momentum to the necessary policies. In the contribution of science and technology to these policy initiatives Senator Button, as Minister for Industry, Trade and Commerce, and Barry Jones, as Minister for Science, played a leading role. The latter was a remarkably well-informed minister and his book *Sleepers, Wake!*[74] dealing with the consequences of technological advance, generated international interest. Initiatives of the Hawke government dedicated to the strengthening of industrial research included a 150 per cent taxation concession for research and development conducted by industry, a National Industry Extension Service, and various mechanisms for the encouragement of new high technology industries. As well, in 1984, OECD was invited to report, for the second time, on Australia's science and technology policies. By 1984–5 there were encouraging signs that both industrial research and development and the formation of high-technology companies was dramatically increasing.

'Future Directions for CSIRO' was also on the agenda in the form of a report from ASTEC, commissioned by the Prime Minister.[75] The report was acted upon in 1986 with a Science and Industry Research Legislation Amendment Bill. The major change to CSIRO was to establish a top-level policy body, the board, separate from management. The new board consisted of part-time members, drawn from outside CSIRO, with scientific and industrial experience, and with the chief executive as the only full-time member. Neville Wran, a former Premier of New South Wales and president of the Labor Party, was appointed chairman, indicating a stronger political presence for the organization. The Board was charged with preparing three to five year plans for research directed at strengthening industrial research and continued assistance to primary industry, especially in increasing added-value by the processing of export commodities. In this task the board was assisted by non-statutory channels of advice, replacing the earlier advisory councils. Again subdivision of CSIRO was rejected, even though separate bodies charged with responsibility towards defined areas of the economy would have simplified policy making, finance, and administration without necessarily sacrificing scientific excellence.

Although a Senate Standing Committee on Science and the Environment had been established in 1976, and a House of Representatives Standing Committee on Environment and Conservation was formed in the same year, it was not until 1986 that a Representatives Standing Committee on Science was proposed by the government. Obviously environmental concerns had been more politically influential than science, even though managing the environment could not be adequately pursued without a sound scientific basis. A government invitation to the Academy of Technological Sciences to form a working party on space science and technology in 1985, the appointment of Neville Wran as chairman of CSIRO in 1986, and the involvement of the two academies, the Institute of Engineers, and the Federation of Scientific and Technological Societies in formulating the government's science and technology budget were further signs that science was assuming a more significant political role that had been the case since the end of the Second World War.

CONCLUSION

In the preface to *Australia: The Daedalus Symposium,* an American expresses concern at 'whether Australia will continue to do well in the fiercely competitive economic world of the future, whether so easygoing a society will be able to accommodate itself to the demands of a new kind of industrial order, and how all this may serve to transform the country, leading it in a wholly new direction'.⁷⁶

The answer to this question is now emerging. Australian governments have slowly moved towards recognizing the role of science and technology in our future and setting in place the apparatus for at least a 'policy for science'. The complementary 'science for policy', predicated upon the existence of clearly defined national goals and recognition of how science can contribute to their achievement, is less well developed, though there have been promising recent advances. Our scientific institutions were slow in achieving a significant role in these initiatives, but they have, in recent years, forged more effective links with government. The scientific community has shed the yoke of the largely British-inspired concept of autonomy and to a much greater extent than heretofore are accepting a role in the market place. In 1986, in introducing the Bill that restructured CSIRO, the Minister for Science heralded the 'new kind of industrial order':

Australia must develop as a nation that makes much greater use of its intellectual resources. If we are to arrest the relative decline in Australia's standard of living, Australia must become a more intellectually active scientific and technological culture. However, simple adoption of technology from overseas at a greater rate, or more effectively, will not create the required environment in Australia for acceptance and development of new technologies. A strong local capability in research and development, directed to the opportunities of the future, not the past, is essential.⁷⁷

Much still remains to be done in the reform of scientific and educational institutions and the results of these initiatives will only become fully operative well into the future. It is to be hoped that the myopic and *ad hoc* approaches that have characterized the period under review will not return with the frequent changes of government that may be expected in the times of economic uncertainty that lie ahead.

NOTES

I am indebted to Ann Moyal for very helpful discussions during the preparation of this chapter, without committing her to the views expressed therein.

I use the term 'science' generally in the sense of organized knowledge of both natural and social phenomena and of technology. Distinctions between these will be made when appropriate.

1. David Dickson, *The New Politics of Science* (New York: Pantheon, 1984), 3.
2. Don K. Price, 'Endless Frontier or Bureaucratic Morass', in Gerald Holton and Robert S. Morison (eds), *Limits of Scientific Inquiry* (New York: W. W. Norton, 1979), 77.
3. J. Ronayne, *Science in Government* (Melbourne: Arnold, 1984). The Bush

Report, which led to the formation of the National Science Foundation in the United States, assumed the linear or 'science push' model of innovation, now seen to be inadequate. This model is discussed in Ronayne, 43 ff.
4. Holton and Morison, op. cit., note 2.
5. Barry Jones (Minister for Science), *Science and Technology Statement, 1984–85* (Canberra: AGPS, 1985), 49.
6. Transnational corporations in Australia rejected co-operation with the CSIRO in developing atomic absorption spectroscopy, which has been described as the most important analytical advance of the twentieth century. Finally, the CSIRO collaborated with two small Melbourne companies who lacked the marketing and service facilities for a world-wide operation. These two companies merged and finally were taken over by an American company, which now manufactures a range of scientific instruments in Australia which are successfully marketed throughout the world. The significant point of the story is that the transnational corporation came to Australia, not for a market, but for Australian scientific expertise.
7. A. J. A. Healy (ed.), *Science and Technology for What Purpose? An Australian Perspective* (Canberra: Australian Academy of Science, 1979), 377.
8. Ann Moyal, 'The Effect of Institutional Evolution on Science Policy', in Healy, ibid., 68–83.
9. J. K. Galbraith, *The New Industrial State* (London: Hamish Hamilton, 1967), 382.
10. Hugh Stretton, 'The Quality of Leading Australians', in Stephen R. Graubard (ed.), *Australia: The Daedalus Symposium* (Sydney: Angus & Robertson, 1985), 197–230.
11. D. P. Mellor, 'The Role of Science in Industry', in P. Hasluck, *Official History of Australia in the War of 1939–45* (Canberra: Australian War Memorial, 1957).
12. ibid., 705
13. Department of National Development, *The Structure and Capacity of Australian Manufacturing Industry* (Canberra: AGPS, 1952).
14. Mellor, op. cit., note 11, 690
15. C. B. Schedvin, 'The Culture of CSIRO', *Australian Cultural History*, 2 (1982–3), 81.
16. Led by the tyranny of infertility and aridity, the CSIRO became a world leader in arid zone research. It may be fitting to ask whether this has led to the exploitation of agricultural land in Australia that costs the community more in ecological damage and government subsidies to marginal farmers than the community receives in return.
17. A. J. Birch (Chairman), *Report of the Independent Inquiry into the Commonwealth Scientific and Industrial Research Organisation* (Canberra: AGPS, 1977).
18. Rohan Rivett, *David Rivett: Fighter for Australian Science* (Melbourne: Dominion Press, 1972), vii.
19. ibid., 199
20. ibid., 1–14.
21. Milton Cockburn and David Ellyard, *Oliphant* (Adelaide: Axiom Books, 1981), 187.
22. One of the attackers, Archie Cameron, a former Country Party leader and Minister deplored the money being spent to establish the Australian National University and boasted that his education had ceased when he was twelve. Cockburn and Ellyard, ibid., 147.
23. The Division shared a building with the Division of Chemical Industry and 'wire mesh screens had to be built down passages and stairways to

separate secrecy-controlled Aeronautical Research of CSIR under Laurence Coombes from its sister Division under Ian Walk'. Rivett, op. cit., note 18, 200.
24. Anon., 'Freedom of CSIR', *Aust. J. Sci.*, 11 (1949), 147–50.
25. Jean Buckley-Moran, 'Australian Scientists and the Cold War', in B. Martin, C. M. Ann Baker, C. Manwell, and Cedric Pugh (eds), *Intellectual Suppression* (Sydney: Angus & Robertson, 1986), 18–19.
26. Institute of Physics, 'The Place of Physics in Australia', Part I, *Aust. J. Sci.*, 10 (1947), 33–8; Part II, ibid., 67–73; Part III, ibid., 97–104. This statement on the parlous state of physics research in Australian universities applied as well to all scientific disciplines.
27. Anon., 'New Milestone in the Publication of Scientific Work in Australia', *Aust. J. Sci.*, 10 (1948), 95.
28. Mellor, op. cit., note 11, 679–80.
29. Ironically, in 1986 the president of the New South Wales Institute of Technology made a plea for its reorganization as a university since Colleges of Advanced Education received 'no Government funding for research'. *Sydney Morning Herald*, Feature (9 September 1986). Such government policy mitigates against the development of tertiary education geared to the development of Australian industry.
30. Anon., 'The NSW University of Technology', *Aust. J. Sci.*, 13 (1950), 3–7.
31. ANU, *Science in Australia* (Melbourne: Chesire, 1952), xiv.
32. J. G. Sekhan, 'We Need a New Type of Doctorate', *Sydney Morning Herald* (August 1986).
33. R. G. Menzies, 'Modern Science and Civilization', *Aust. J. Sci.*, 20 (1958), 251–7.
34. Cockburn and Ellyard, op. cit., note 21, 145.
35. This debate was conducted in the pages of the *Australian Journal of Science*. A public meeting, 'The Place of the National University', was held at the Hobart Congress of ANZAAS in January 1949. *Aust. J. Sci.*, 11 (1949), 195. The issue of teaching and research also provoked a lively debate. ibid., 5, 61, 81–4.
36. F. K. Crowley, *Modern Australia in Documents, 1939–1970*, 2 vols (Melbourne: Wren, 1973), vol. 2, 147.
37. The Menzies government provided money to establish University House, a residential college with a Master and High Table modelled on Oxbridge, as part of the Australian National University. In 1953, when University House had been built, Menzies was appalled by its contemporary architecture. Cockburn and Ellyard, op. cit., note 21, 173. Perhaps he preferred a 'Gothick' style as being appropriate to the Oxbridge model.
38. ANU, op. cit., note 31, ii–iii.
39. Kate Legge, 'Too Much Tension at the National Brains Trust', *National Times* (20 June 1986), 26–7.
40. Anon., 'Report on the Committee on Australian Universities', *Aust. J. Sci.*, 20 (1958), 161.
41. Anon., 'Tertiary Education in Australia', *Aust. J. Sci.*, 28 (1965), 1–10.
42. ANU, op. cit., note 31, v.
43. Anon., 'Science in Australia', *Aust. J. Sci.*, 14 (1951), 1–2.
44. ANU, op. cit., note 31, 176–7.
45. Cockburn and Ellyard, op. cit., note 21, 203.
46. ANU, op. cit., note 31, 39–40.
47. Ann Moyal, 'The Australian Academy of Science: The Anatomy of a Scientific Elite', *Search*, 11 (1980), 233.

48. Menzies, in the 1930s, when a Minister in the Victorian state government, had promoted the foundation of an Australian Academy of Art based on the British Royal Academy to counter the prevailing contemporary art movement. It was a dismal failure. With an Academy of Science, he was on safer ground, as there was no antagonism between contemporary and traditional science.
49. Moyal, op. cit., note 47, 232.
50. H. C. Webster, 'The Case for Queensland', *Aust. J. Sci.*, 17 (1954), 105.
51. C. A. Fleming, 'Science, Settlers and Scholars. Centennial History of the Royal Society of New Zealand', *Roy. Soc. of NZ Bulletin*, 25 (Wellington, 1987).
52. M. E. Hoare, *Bulletin of the Archives Committee of the New Zealand Library Association*, no. 8 (July 1976), 5–8.
53. Anon., 'The Future of ANZAAS', *Aust. J. Sci.*, 17 (1955), 181–91.
54. M. L. E. Oliphant, 'Science and the Survival of Civilization', *Aust. J. Sci.*, 21 (1958), 8–16.
55. Anon., 'Federal Aid for Scientific Research in Australian Universities', *Aust. J. Sci.*, 25 (1962), 201–2.
56. R. N. Robertson, 'Science and Leadership in Democracy', *Aust. J. Sci.*, 28 (1965), 100–4.
57. Peter Pockley, 'Under a Cloud of Committees', *Nature*, 267 (1977), 476.
58. Malcolm Fraser, 'Government Approaches to Science', *Aust. J. Sci.*, 31 (1969), 410–5.
59. OECD, *Examiners' Report: Australia* (Canberra: AGPS, 1974).
60. S. Encel, 'Report of the ANZAAS Science Policy Commission', *Search*, 5 (1974), 560–88.
61. J. R. Philip, 'The Science Task Force Recommendations', *Search*, 7 (1977), 93–6.
62. Jarlath Ronayne, 'The Uneasy Alliance: Science and Politics in Australia', ibid., 85–9; Paul F. Gross, 'Diversity and Adaptability in Science Policy: A Potential Trade-Off between Autonomy and Influence', ibid., 89–92.
63. Moyal, op. cit., note 8.
64. G. W. Ford (ed.), *Automation: Threat or Promise? Impact and Implications for Australia* (Sydney: Law Book Company, 1969).
65. A. J. Sinden (ed.), *The Natural Resources of Australia. Prospects and Problems for Development* (Sydney: Angus & Robertson, 1972).
66. Sir John Crawford, 'Report by the Chairman of the Committee of Review', *Search*, 1 (1970), 2–5.
67. J. B. Davenport, 'The Art of the Possible', ibid., 1–2.
68. Ann Mozley Moyal, 'ANZAAS and the Public Communication of Science', *Search*, 5 (1974), 589–94.
69. Peter Pockley, 'Government-science Links Enter a New Era', *National Times* (20 August 1973), 11.
70. Anon., 'Scientists go Political', *Nature*, 244 (1973), 534.
71. Moyal, op. cit., note 47, 286–7.
72. John Stone, 'The Mixture Largely as Before', *Sydney Morning Herald* (8 May 1985).
73. A. J. Birch, op. cit., note 17.
74. Barry Jones, *Sleepers, Wake!* (Melbourne: Oxford University Press, 1982).
75. ASTEC, *Future Directions for CSIRO. A Report to the Prime Minister* (Canberra: AGPS, 1985).
76. Graubard, op. cit., note 10, viii.
77. Barry Jones, *Hansard (House of Representatives)*, 17 September 1986, 853.

PART II

CHARTING THE SCIENCES

4

The Life Sciences: Collections to Conservation

Linden Gillbank

In Australasia today they are everywhere—men and women weaving their contributions into the fabric of the life sciences. From the Tanami Desert in Central Australia, to the moist forests of East Gippsland, Victoria, and from the Great Barrier Reef to New Zealand's Fiordland, university, government, and company ecologists and taxonomists are determining and describing sites of botanical and zoological significance, documenting the distribution of indigenous species and feral invaders, and studying the complexity of interactions between organisms in their habitats, before those habitats are irretrievably lost.

Other biologists are cossetted in hospitals and research institutes, the Department of Scientific and Industrial Research (DSIR), the Commonwealth Scientific and Industrial Research Organization (CSIRO) and government departments, universities, museums, zoological and botanical gardens, and herbaria. They are developing antivenoms and vaccines, and are unravelling the causes of hypertension, heart disease, and diabetes, and the relationships between diet and health. They are studying the reproductive physiology and phylogenetic relationships of marsupials and monotremes, and are investigating *in vitro* fertilization in livestock and humans. They are breeding better varieties of wheat, lupins, and clover, and are investigating the chemistry of indigenous plants. Some of their work involves co-operation with amateur volunteers, or with private industry.

Such is the ubiquity and diversity of biological work in Australasia in the 1980s, involving scientists specialized in many fields: biophysicists and physical or organic chemists; biochemists, physiologists, anatomists, embryologists, neurologists, and endocrinologists; pathologists, immunologists, haematologists, and epidemiologists; parasitologists, microbiologists, virologists, and bacteriologists; mycologists, nematologists, entomologists, ichthyologists, herpetologists, and ornithologists; microscopists, geneticists, ecologists, and taxonomists. Most have an Australasian university or tertiary college degree; many have overseas postdoctoral experience. Their work is reported in Australasian and foreign journals, and at conferences within Australasia and overseas. They are members of both the Australian or New Zealand and the international scientific community.

THE TRADITION OF COLONIAL BIOLOGY

The kaleidoscope of Australasian biological activity has not always been so brilliant. For much of the nineteenth century, Australia and New Zealand, their inhabitants and visitors, were used by European scientists as sources of biological material and information to bolster or undermine European biological ideas. As Australasia was explored, plant and animal specimens, including important type specimens, were dispatched by both official and unofficial collectors to swell private and public collections in museums and botanical gardens throughout Europe.[1] Such specimens carried their monetary, curiosity, and scientific value across the equator.

The nineteenth century world depended on plants directly and indirectly for all its food, drugs, and clothing, and some of its buildings and industrial processes. Imperial powers sought to control the cultivation of useful plants, with colonial botanical gardens providing crucial testing grounds for the suitability of plants to new climates. The Australasian botanical gardens were part of the Kew-controlled network of British colonial gardens, whose function was to manipulate global botanical resources for the economic benefit of Britain.[2] Australasian government botanists and botanical garden directors and curators were appointed with the approval of the directors of Kew Gardens, and they dutifully sent plants back to that centre of British botanical wisdom.[3] Nourished by the accumulated taxonomic information, colonial floras, including Joseph Hooker's *Handbook of the New Zealand Flora* and Bentham's *Flora Australiensis*, were written at Kew.[4]

Because the British colonial fauna was not accorded the economic potential of its flora, there was no institutionalized zoological equivalent of the highly co-ordinated Kew-centred network. Although zoologists, including Gould and Bennett, wrote books about the Australasian fauna and enhanced European collections with Australasian specimens, there were no nineteenth century published *Faunas*.[5] In contrast to the imperially initiated botanical gardens, Australasian museums and zoological gardens arose in response to local demands. With a preference for exotic rather than local specimens they mimmicked their European counterparts.[6] All three institutions were integral components of nineteenth century Australasian life, providing centres for scientific enquiry, popular education, and recreational pleasure.

THE 1880s: THE DEVELOPMENT OF A PRO-EVOLUTION AUSTRALASIAN-BASED BIOLOGY

In Australasia in the 1880s, biologists were employed in government departments, universities, museums, and zoological and botanical gardens. Unlike their late twentieth century successors, they were practically all male and had come, sometimes minus formal qualifications, from the Northern Hemisphere, generally Britain. Many were 'amateur' in the sense that they earned a living in non-biological capacities, such as army officers, ministers of religion, or medical practitioners. In the late nineteenth century such amateur status did not preclude men, and the rare woman, from gaining expert and respected status in the scientific community. Although biological work was reported to the established colonial scientific societies, by the 1880s, the new Field Naturalists Club of Victoria, the Linnean Society of New South Wales and other smaller natural history and field naturalists societies provided the principal meeting grounds for biologists.[7] Their work was published in these societies' journals, and in Northern Hemisphere journals including *Nature*, *Proceedings of the Zoological Society*, *Philosophical Transactions of the Royal Society*, and *Quarterly Journal of Microscopical Science*. From 1888, the Australasian Association for the Advancement of Science (AAAS) provided the first intercolonial forum for biologists in Australasia.

During the 1880s and early 1890s, Australasian colonial governments, aware of their dependence on the produce of farms and forests, created departments or sections focusing on agriculture and forestry, which included new biological, mainly botanical, positions. Plant pathologists, entomologists, and foresters began to enter the public service. Universities increased their biological, especially zoological, options with the creation of chairs and lectureships in 'biology', which by the turn of the century would replace all Australasian chairs of 'natural science' and 'natural history' (table 4.1). For the first time, science degrees were provided by each of the first four Australasian universities, allowing undergraduates to study three years of biological subjects. Agricultural, horticultural, and pharmaceutical colleges, and Sydney's Technological Museum were established.[8] Crucial to the development of Australasian biology into the twentieth century was the injection of Darwinian evolutionary ideas through the young and often highly trained men who came to fill these new government and university positions.

By the 1880s much had been published about the flora and fauna of Australasia. In the wake of the Darwinian debate, new interest was generated. Could the strange antipodean animals and plants provide answers to the evolutionary questions tantalizing European biologists? Australia and New Zealand provided truly wonderful laboratories—the only land masses of relatively undisturbed vegetation to be explored by Europeans since the development of Linnaeus's classification system and Darwin's ideas about natural selection. What fine fossicking fields for phylogenetic relationships! As the biology papers presented to the first few AAAS congresses show, prodigious, though often unfocused,

TABLE 4.1
Nineteenth century biological chairs and science degree courses at Australasian universities

	Professor of natural history or natural science	Professor of biology	BSc degree course first offered in
1. University of Sydney	1882–90 W. J. Stephens	1890–1914 W. A. Haswell	1883
2. University of Melbourne	1854–99 F. McCoy	1887–1919 W. B. Spencer	1887
3. University of Adelaide	1875–1901 R. Tate	—	1882
4. University of New Zealand			1884
Auckland University College	c. 1881–5 A. P. W. Thomas	c. 1886–post-1900 A. P. W Thomas	
Canterbury College	—	1880–93 F. W. Hutton	
University of Otago	1877–9 F. W. Hutton	1880–97 T. J. Parker	

collecting, cataloguing, and naming of organisms new to science continued into the 1890s. For the first time, however, a pro-evolutionary Australasian-based investigation of the indigenous biota was developing. No longer would Australasia be a mere mining site for European biologists.

No longer was a classical education with an interest in natural history (Stephens) or palaeontological experience (McCoy) adequate for appointment to a biological chair. In the 1880s, into this biologist's paradise, with its unique flora and fauna and its few resident biologists, some of whom still clung to pre-Darwinian views, came young science graduates steeped in 'modern' evolutionary ideas and skilled in 'modern' histological techniques, thoroughly equipped to participate in the exciting and fashionable fields of comparative embryology, anatomy, and physiology.[9] How upsetting was it for such bright young biologists as Baldwin Spencer, Australia's first professor of biology, and his assistant lecturer, Arthur Dendy, to work in a university where the professors of natural science and physiology still refused to accept evolutionary ideas, in a colony where the government botanist continued to doubt that plant species might vary due to natural selection?[10]

The new biologists arrived in lands still liberally endowed with a grand diversity of habitats not yet ruined by the destructive footprints of Europeans, and were welcomed into thriving natural history societies that wedded the interests of the middle class amateur and those of the professional biologist in a productive foray into the local flora and fauna. Certainly Spencer appears to have enjoyed and benefitted from his participation in the Field Naturalists Club of Victoria meetings and excursions.[11] Without such a satisfying initiation into the biology of the local indigenes, would he have plunged so readily into microanatomical, taxonomical, and later zoogeographical studies of the Australian fauna? Would he have so rapidly become such a public advocate for conservation,[12] including the reservation of Wilson's Promontory as a national park?

The 1880s provided Australasia with the essential ingredients—the individuals and institutions, the ideas and techniques—for a renewed exploration from within Australasia of its own flora and fauna, an exploration spurred on by evolutionary and economic incentives. Australasia was still dependent on Europe for its biological ideas and manpower, but it was developing the structures for its eventual biological independence, structures that would grow and proliferate to nurture Australasian biological research through the twentieth century.

AFTER THE 1880s: EVOLUTIONARY AND ECONOMIC INCENTIVES

Spencer's work before the turn of the century was typical of the new post-Darwinian Australasian biology: the search for new species that could be 'missing links' in the linear sequences of Huxley's evolutionary hierarchies, and the survey of the terrestrial fauna, whose distribution could be explained by past land bridges.[13] Addresses to Section D (Biology) around the turn of the century elaborated this theme for various groups of organisms.[14]

In 1890 James T. Wilson, an Edinburgh medical graduate and the University of Sydney's foundation Challis Professor of Anatomy, founded the first Australian school to focus primarily on the Australian vertebrate fauna. Aided by his physiology demonstrator, Charles J. Martin, he initiated research into Australian marsupial and monotreme anatomy, physiology, and embryology, not as mere elaboration of novelty, but with the aim of explaining the place of these peculiar mammals in the evolutionary scheme of things. James P. Hill, initially Haswell's biology demonstrator, joined with Wilson and Martin to shape marsupial and monotreme research into the twentieth century.[15]

By the turn of the century, another group whose focus was unashamedly Australian was well established. Although its origins were provoked by economic rather than evolutionary interests, it too applied modern methods to Australian material. Joseph Henry Maiden, when appointed in 1881 curator of Sydney's new Technological Museum,[16] was neither well qualified nor experienced in any biological field. By the turn of the century, however, he had set the museum on its extensive and productive investigation into economically promising aspects of the Australian flora, including eucalyptus oils,[17] and had himself become the government botanist and director of Sydney's Botanic Gardens. Maiden wrote the botany chapter in the *Handbook of Sydney* for the first AAAS congress, and in 1907 was president of Section D.

Much phytochemical work on Australasian plants had already been carried out, mainly in Europe.[18] Maiden realized that to stimulate local research and industrial development, this information had to be made readily available. This the museum achieved through its collections, research, reports to the local Linnean Society and the AAAS,[19] and books about such useful native plants as wattles, native pines, and eucalypts. Thanks to Maiden, R. T. Baker, and H. G. Smith, an Australian institution was focusing continuing scientific effort on biological problems in Australia, rather than waiting for results from research laboratories on the other side of the equator.

Thus the turn of the century witnessed economic incentives to investigate the indigenous flora and an evolutionary incentive to examine the native fauna: the flora in government instrumentalities and university botany departments, and the fauna in university biology (later zoology) and anatomy departments.

As their reports to prewar AAAS congresses indicate, government botanists investigated plant diseases, including wheat rust, and attempted to breed better varieties of crop plants, including wheat;[20] they investigated pasture and poisonous plants;[21] they studied the timber potential of their diminishing forests, and considered planting exotic softwoods.[22] They read *Flora Australiensis*, noted any recent additions to their colonial or state herbarium collections, worried about the receding indigenous vegetation, and compiled colonial or state floras.[23] In 1907 Maiden pleaded for a new Australian flora to be prepared by Australian botanists and published in Australia. It should, he said, be 'based on the most modern lines of taxonomic research, modified, indeed by our own special knowledge of our plants and their affinities'.[24]

As part of their newfound obligation to investigate the indigenous flora and fauna, Australasian biologists recognized the need for various facilities. As the papers, resolutions, and committee reports of the first fourteen AAAS congresses show, biologists used this intercolonial forum to argue for updated regional floras, for biological field stations, for regional biological surveys, and for the compilation of bibliographies of relevant published work. They discussed the inappropriateness of European terms for Australian vegetation, and of European common names for Australian plants. They also argued for the formation of reserves for the preservation of native flora and fauna, and the reservation of forest areas for the preservation of water quality.[25] The AAAS performed an important role in providing a forum for the discussion of these and other biological issues, where biologists working in widely separated parts of Australasia could express their common concerns, even when these conflicted with the policies of their governments.

One of the AAAS's first research committees illustrates the importance of such efforts. The Protection of Native Birds and Mammals Committee was formed at the first congress following a motion by Professor Baldwin Spencer. During the five years of its existence, it included representatives from all the colonies except Western Australia; professors from the four universities (Spencer, Stephens, Haswell, Tate, and Thomas), men from two museums (Morton from Tasmania and

Tryon from Queensland), and the expert amateur ornithologist A. J. Campbell. It allowed ideas and efforts already initiated in individual colonies to receive wider scientific and public support. The experience of the Field Naturalists Club of Victoria in considering the protection of birds and the reservation of Wilson's Promontory was brought by Professor Spencer, A. J. Campbell, and the Reverend J. J. Halley. The involvement of A. F. Robin and S. Dixon ensured the consideration of the reservation of the western end of Kangaroo Island, an issue already canvassed by the Field Naturalists Section of the Royal Society of South Australia.

The committee's report, presented to the 1893 congress, detailed strategies for the protection of the Australasian fauna, and recommended the proclamation of government-funded reserves, under the control of local honorary trustees, in all Australasian colonies.[26] Some of the areas specified, including Flinders Chase and Kangaroo Island (South Australia), Freycinet Peninsula (Tasmania), Mount Bellenden-Ker (Queensland), and Resolution Island (New Zealand) were set aside as 'nature reserves' or 'national parks' in response to pressure from the AAAS provoked by this committee.

The AAAS reflected the participants and activities of the biological community. The first four presidents of Section D were professors at the four universities: Tate (1888), Thomas (1890), Haswell (1891), and Spencer (1892). Most of the biological community—government, university, and amateur—contributed papers. The first fourteen AAAS congresses reveal the dominant interest in indigenous plants and animals, their naming, their physiology and anatomy, and their distribution, and the gradually declining participation of the amateur biologist, who continued to be served by natural history societies.

By the 1913 congress, the last before the arrival of the British Association for the Advancement of Science and the First World War, biologists were stepping out from Australasian, rather than European, institutions into the field and forest, not merely to collect specimens to enhance European collections and ideas, but to answer questions of their own choosing. While Professors Spencer and Tate, stepped further than most, most Australian biological investigations remained within the arable, moderate-rainfall crescent of the continent. While some biologists, such as Wilson and Hill, would return to Britain to complete their academic careers, others, such as Spencer, Haswell and Hutton, would see no such need. The ties with Europe were beginning to weaken.

BETWEEN THE WARS: ECONOMIC AND ECOLOGICAL IMPERATIVES

Between the wars, both fundamental and applied aspects of Australasian biology were developed. Laboratory and field work developed to reveal increasingly minute details of the inner workings of organisms, and to elaborate the organism's place within its environment. Biology looked toward the arid interior of Australia and aided the pastoralist. It served both national and imperial needs.

As Professor Ewart noted in his 1921 presidential address to Section D of the first postwar congress, the destructiveness of war had stimulated the physical sciences, while stultifying those sciences allied to the productive activities of agriculture and forestry. Despite the relief and enthusiasm with which individuals and institutions were returning to, and enjoying the fruits of, these biology-based activities, Melbourne's foundation professor of botany felt the need to spell out justifications for fundamental biological research.[27] While benefits to be gained from an improved understanding of crop, stock, and human disease were obvious, their dependence on basic research was not so obvious. In the 1920s and 1930s this was a recurring theme in public addresses; biologists repeatedly noted advances in agriculture and medicine that depended on the work of university-trained researchers. Basic research in fields as diverse as systematic entomology, marine zoology, marsupial anatomy, plant physiology, and saltbush ecology, could all be justified in terms of their potentially useful applications.[28] Such arguments also reflected the desire to justify the role of universities in nurturing biological research. Australasian universities responded by establishing applied biology departments—medical, agricultural, and veterinary— rather than new biology, botany, or zoology departments. The University of Adelaide's Waite Agricultural Research Institute, for example, was established in 1924. Applied biology societies and their journals arose to sustain and stimulate such research.

After the war, systematic studies on living and fossil indigenous plants and animals (especially invertebrates) were resumed, including the naming and description of new species.[29] The distribution of the indigenous flora and fauna, as well as their chemistry, anatomy, and physiology, were of interest in the continuing consideration of its origins and evolution. Now, however, theories of continental drift were beginning to provide explanations for past floral and faunal links between Australia, New Zealand, and their neighbours.[30] As an ecological awareness developed, attention was focused on vegetation rather than flora. As suitable techniques were developed, attention was focused on metabolism rather than chemistry, and on cytology and genetics.

While the productive capacity of the soil and native pastures had definitely declined, neither the need for timber and paper, nor the diversity, frequency, and severity of plant and animal diseases, and pests had decreased. The rabbit was still rampant. The Europeans' nutrient-loving crops and cloven-hooved, heavy-grazing stock, the source of their wealth, were continuing to destroy the very environment on which their survival depended.

Applied biologists whose skills were harnessed by governments in the 1890s—plant pathologists and entomologists—were still in demand. They joined geneticists, agronomists, biochemists, and others to address the needs of the pastoral, agricultural, and horticultural industries. Better-yielding and disease-resistant varieties of crop and pasture plants were bred. By the 1930s, half a century after William Farrer had conceived the idea, a variety of wheat, Gabo, was bred to combine high yield and good baking quality with rust and drought resistance.[31] The addition to some Australasian soils of nitrogen, courtesy of subterranean clover, phosphorus as superphosphate, and the trace elements, manganese, copper, zinc, and molybdenum, was found to remove nutrient-deficiency diseases and increase yields of crops and pastures.[32] Vast areas were made available for pastoral and agricultural use. Vaccines were developed against animal diseases including sheep 'black disease' and bovine pleuro-pneumonia. The cause of foot and mouth disease, tomato spotted-wilt, sugar cane mosaic, and banana 'bunchy top', the enigmatic virus, its structure and modus operandi yet to be elucidated, taxed the minds of chemists, microbiologists, and plant pathologists.[33] The physiology of xerophytes and fruit-ripening was being investigated.[34]

The classic exemplar of biological control was effected. By 1920 the introduced prickly pear cactus had affected over 25 million hectares of grazing and arable land. In the mid-1920s the introduction of its natural predator, the cochineal insect *Cactoblastis cactorum*, provided a strikingly cost-effective example of pest control.[35] Could natural predators also control such unwanted immigrants as sheep blowfly, buffalo fly, codlin moth, grasshoppers, blackberry, St John's Wort, and skeleton weed? This particular ecological procedure offered great promise.

The problems confronting biologist and primary producer in the interwar years were in the main not new, but their accumulated severity was. The gradual razing of forest cover, the gradual impoverishment and erosion of the soil, the gradual overgrazing of native pastures, and the gradual spread of pests and diseases all reached intolerable levels. The ugly reality was described in 1923 by the Institute of Science and Industry:

From plant diseases alone the loss has been estimated at £5,000,000 annually. An attempt to estimate the loss from the sheep fly gives as much as £4,000,000 in a bad year. Prickly pear already covers an area in Australia considerably greater than the total area under all forms of cultivation. New South Wales alone has expended £600,000 during the past fifteen years in an attempt to keep back the cattle tick pest. The loss from fruit diseases and pests is estimated at £1,000,000 annually.[36]

While officers of the state Departments of Agriculture responded to agricultural and horticultural crises, they had no opportunity for long-term research. They did investigate plant and animal diseases, including flag smut in wheat and mastitis in dairy cattle, as well as plant and animal nutrition, and the irrigation of vineyards and crops. However, neither they, nor the various agricultural colleges, nor the universities, most of which by 1930 had departments of agriculture, were managing to devote adequate resources to the snowballing problems of primary industry.

A New National Biology: The CSIR and DSIR

For an effective onslaught against problems that showed no respect for state boundaries, nor any readiness to respond to simple controls, national rather than state research efforts were required. In recognition of this, the 1923 Pan-Pacific Science Congress in Sydney resolved 'that the Federal Government should set aside adequate funds for the establishment, equipment and maintenance of a Federal Bureau of Entomology for the necessary research in this connexion'.[37] Not surprisingly, both the Department of Scientific and Industrial Research (DSIR) and the Council for Scientific and Industrial Research (CSIR) were initially primarily agricultural research institutions. The need for concentrated, fundamental, specialized biological research was so great that a 1927 meeting between representatives of CSIR and state Departments of Agriculture welcomed some Commonwealth involvement in agricultural research, constitutionally and traditionally a responsibility of the states.[38]

Political and economic conditions in Britain also fostered the founding of these national research institutions. In 1926, the year the DSIR and CSIR were established, the British government set up the Empire Marketing Board to encourage efficient primary production in the Empire and its marketing in Britain, thereby ensuring Britain's food and fibre supplies. CSIR and DSIR provided convenient institutions through which to channel Board grants for research into factors adversely affecting primary, especially pastoral, production in Australia and New Zealand.[39] In the late 1920s, both dominions received grants for pasture and entomological research. The Board helped finance arid grassland research at Adelaide, a DSIR experimental orchard, an insectary in Canberra, and a CSIR laboratory in Townsville to investigate tropical animal diseases.

The CSIR's and DSIR's early work illustrates the continuing imperial dependence of these purportedly national scientific research institutions. Projects chosen for their prospective benefit to the Mother Country were granted much-needed finance by the Empire Marketing Board. To control entomological research throughout the Empire, the Board had its own entomological laboratory at Farnham Royal, which was popularly known as the 'parasite zoo',[40] where entomologists would research and breed promising predators of known pests, leaving the dominions and colonies to distribute and study them in the field. If the Board had its way, Australia and New Zealand would remain dependent upon Britain for fundamental entomological research. Furthermore, British rather than Australian biologists advised the government on research strategies of new CSIR divisions. In the conclusion to his 1928 report on the creation and organization of an Australian Forest Products Laboratory, A. J. Gibson of the Indian Forest Service noted the value of Australian forests, not to Australia, but to the British Empire.[41]

Between the wars, Australasia still depended to some extent on Northern Hemisphere institutions to supply biologists, research grants, and training. In the 1920s, the Royal Society of London funded ecological work at the University of Adelaide, and a field study of natural hybrids in New Zealand.[42] Biologists came from North American and British institutions to head university departments and CSIR divisions.[43] Australasian biologists were also sent overseas to study or to gain experience in North American and European institutions.[44]

David Rivett, CSIR's first chief executive officer, was not prepared to allow the biological work of CSIR to remain derivative by handing all fundamental problems over to Britain: Australian institutions must research Australian biological problems. Despite Farnham Royal's preference for recruiting Britons who would remain in Britain, Rivett insisted that Australians should be able to gain experience there.[45] Equally he was not prepared to be dependent on Australian university or state agriculture department laboratories for CSIR work: CSIR must have its own laboratories.[46]

There were other glimmers of growing nationalism in Australasian biology in this period: several national professional organizations and their journals were founded;[47] a Commonwealth Forestry Bureau and a Forestry School were established;[48] a Commonwealth Herbarium was being discussed;[49] and a National Museum of Australian Zoology was established.[50] Also, while British-trained scientists continued to fill Australasian biological positions, the interwar years saw a significant shift towards Australasian graduates filling such positions. For example, six Australian biological professors during the 1920s and 1930s had degrees from a single Australian university—Sydney, under Haswell and Wilson.[51] Of the first chiefs of the first four CSIR divisions, three had Australian training or experience.[52] Many Australasian-trained biologists continued to go abroad to gain postgraduate qualifications and experience, but certain Australasian biological departments were becoming influential in the training of future research biologists: Adelaide and Sydney university botany departments, under Professors Osborn and Wood, were certainly so for ecology.[53]

Ecological Emphases

Between the wars, the new science of ecology was very important in Australasia: it provided a new holistic way of looking at the living world, and it stimulated work in other biological sciences. Moreover, it was useful in the study of forests, pastures, weeds, insect pests, plant and animal diseases, and the problems of soil erosion and overgrazing. Insect ecology flourished in the euphoric slipstream of the prickly pear success.

Although it developed much earlier in New Zealand,[54] the young science of plant ecology emerged in Australia after the First World War. It was stimulated by the birth in 1913 of the international *Journal of Ecology*, and visits both before and after the war by overseas botanists, some of whom were attracted by the 1914 British Association for the Advancement of Science meeting in Australia, others by the arid-zone vegetation and its research at Adelaide. After the war, botanists in Australia began to ecologically examine a range of vegetation types from Mount Kosciusko and the Blue Mountains, New South Wales, to the salt marshes near Sydney; from the forests of Tasmania and Victoria, to the saltbush vegetation of inland South Australia.[55]

The more intensively ecologists studied the Australasian vegetation, with its often subtle variations, the more they provoked further systematic work on the flora. A single species in an herbarium collection or a published flora may in its natural habitats be found to include several closely related species or hybrids. Leonard Cockayne and Harry Allan slated the strictly herbarium method of species making. Their extensive taxonomic and ecological studies of the New Zealand flora revealed a surprising number of natural hybrids, something not then adequately considered in Australia.[56] In the 1920s and 1930s, a vigorous Australasian plant ecology was providing new ammunition for the publication of new Australasian floras and comprehensive regional biological surveys.[57] Biochemical, physiological, and morphological studies were required to explain how plants survived where and when they did. In laboratories at Adelaide University and its Waite Institute, metabolic responses of plants to drought conditions were studied. The emerging field of biochemistry had definite links with the developing field of plant ecology.[58]

In the early 1920s, in the absence of any national ecological institution, society or journal, the AAAS's Ecology Committee performed the useful function of collecting information about Australian plant ecological work. The committee, which included university and government botanists, compiled a Bibliography of the Australian Floristic and Ecological Plant-Geography; discussed the suitability of European terms and graphical representations for Australian vegetation; prepared a provisional list of the main vegetation types; and, with the intention of preparing a vegetation map of Australia, wrote to the *Pastoralists' Review* requesting information about the distribution of various types of vegetation, including saltbush, grassland, scrub, open forest, and desert country.[59]

Provoked both by the needs of the pastoral industry and a biological interest in the unknown, the arid inland saltbush areas became the focus of intensive ecological interest in the 1920s and 1930s. Australia's pastoral industry depended on native pasture plants. In the drier areas overgrazing was taking its toll of these plants. Pastoralists wanted to know how many sheep they could graze in 'good' and 'drought' years. Why were some pasture species disappearing? Why were some areas being converted to barren deserts? With the establishment in the mid-1920s of a vegetation reserve and field laboratory in the saltbush area at Koonamore in South Australia, Osborn, Wood, and their colleagues could study processes operating within the arid, nutrient-deficient, overgrazed pastoral ecosystems, and determine the conditions required for regeneration of the valuable drought-tolerant vegetation.[60]

So great was national concern about soil erosion and drift in arid and semi-arid Australia, that in the mid-1930s, CSIR zoologist Francis Ratcliffe used his ecological skills to determine causes and possible remedies for the problem of soil loss in parts of northern South Australia and south-west Queensland. He investigated the effects of drought, wind, sheep, and rabbits on palatable and unpalatable plants of saltbush and mulga scrub regions.[61]

Ecological and related studies of xeromorphic and other types of indigenous vegetation highlighted two things: the inadequacy of the accumulated data on the flora and fauna, and the need to preserve it in the face of increasing habitat destruction. Resolutions passed at ANZAAS congresses in the 1930s reflected the insistent concern of biologists for indigenous flora and fauna preservation and study. Following 1932 Section D resolutions for unifying the administration of state fauna and flora legislation, and for implementing a biological survey of the Australian fauna, a Biological Survey of Australia Committee was appointed at the 1935 ANZAAS Congress.[62] In 1939, the Committee argued for government support for the forthcoming Interstate Faunal Conference and for 'more adequate National Reserves for the preservation of the indigenous flora and fauna'. It recommended a Department of Biological Survey within CSIR, to co-ordinate and initiate the required systematic and ecological work.[63] The idea was not new. In 1921, at the first postwar congress, Spencer had recommended an immediate 'co-ordinated investigation into the land and fresh water fauna and the flora of Australia and Tasmania', involving natural history societies and museums.[64] Even earlier echoes—from the AAAS's pre-Federation Fauna Protection Committee—remained. Now, however, there was recognition of the need for federal involvement in legislation and research, and for a systematic survey of the Australian fauna.

The 1939 ANZAAS resolutions were so extensive, so full of conservational hope ... and then came the war.

WARTIME BIOLOGY

During the war, food and fibre from its antipodean dominions were crucial for Britain's survival; both the DSIR and CSIR mobilized biological research for the war effort. Some war-induced research lapsed after the war, while other research flourished to occupy scientists to the present, including northern Australian pastoral and agricultural research, and phytochemical research on the Australian flora.

The CSIR and DSIR investigated the dietary needs of civilians and soldiers, and the processing of food to enhance its nutritional value and storage life. DSIR's Soil Bureau and Divisions of Agronomy and Plant Diseases joined forces to develop an instant linen flax industry from successful prewar trials.[65]

Wartime necessity was also the mother of biological research into the indigenous flora, both aquatic and terrestrial. With Japanese supplies of agar cut off, the CSIR's Division of fisheries and the DSIR's Botany Division began surveying indigenous algae for commercially exploitable agar.[66] So little was then known of the Australasian distribution of the required red algae. While the DSIR's Botany Division and the CSIR's Division of Plant Industry investigated the feasiblity of growing pharmacologically important exotic plants to supply Australasian citizens and allied troops with much-needed drugs,[67] the indigenous flora was scrutinized for drugs. Australian plants, whose useful chemicals had been discovered last century, were looked at with new interest. Rutin to repair capillary fragility, aesculin to screen out ultraviolet radiation, and scopolamine as an antidote for motion sickness and bomb shock appeared particularly attractive.[68]

In New Zealand and Australia commercial drug extraction ceased soon after the war, but research into the pharmacological potential of the Australian flora continued. During the war, a collaborative phytochemical investigation of the Australian flora was begun by the CSIR with help from the University of Sydney's Pharmacology Department and the University of Melbourne's Physiology Department. After the war, other university departments, keen to resume productive peacetime research, participated in the Australian Phytochemical Survey,[69] and in 1959 *A Phytochemical Register of Australian Plants* was published. Recently the CSIRO has developed a computerized database, PHYT, which includes all post-1940 published papers on the phytochemistry of Australian plants—a productive legacy from the war.

During the war, Drug Houses of Australia (DHA) had commercially extracted tropane alkaloids from indigenous species of *Duboisia*. In the 1950s, however, both the CSIRO's *Duboisia* research and DHA's alkaloid extraction ceased.[70] Australia returned to being an exporter of the dried *Duboisia* leaf and an importer of prepared drugs. The ghost of Joseph Henry Maiden would have shed a tear at the loss of the research and exploitation of this native plant to Northern Hemisphere institutions.

The war provoked institutions to direct their biological research northwards. Partly in response to Australian National Research Council advice for the need to scientifically survey and develop the resources of northern Australia, the CSIR established, within Plant Industry, the Northern Australian Regional Survey Section, which compiled an inventory of northern Australian flora and landforms.[71] The agricultural and pastoral exploitation of northern Australia still requires much ecological and physiological research.[72]

POSTWAR DEVELOPMENTS

Scientific Specialization

By the 1950s, with the growth of biological research in the CSIRO, the DSIR, and state government departments, together with the expansion of universities, their granting of doctoral degrees, and the funding of biological research with Australian Research Grants Committee (ARGC) money, the population of Australasian biologists was sufficient to sustain national specialist societies. An increasing awareness and interest in environmental and conservation issues in the 1960s and 1970s also stimulated indigenous flora and fauna research.

Postwar congresses reflected the diversification and expansion of Australasian biological research and interests. ANZAAS attempted to satisfy the need for increasingly specialized interests by providing more

Sections. After 1960 the three postwar biological sections—Botany, Zoology, and Physiology—were progressively divided into six—Botany, Plant Pathology, Zoology, Biochemistry, Food Science and Nutrition, and Physiology. Microbiology had arisen out of the Medical Science Section, while the prewar Sections of Pharmaceutical Science, Veterinary Science, and Agriculture and Forestry survived intact. The provision of these Sections was, however, not enough. In the 1950s and 1960s many independent specialist biological societies arose, sometimes in tandem with an equivalent ANZAAS Section (table 4.2).[73] Moreover, between 1950 and the mid-1970s, about a dozen Australian and half a dozen New Zealand new specialist biological journals appeared. The majority were published by CSIRO and DSIR, and embraced botany,

TABLE 4.2
Postwar Establishment of Biological Societies and ANZAAS Sections

Society		ANZAAS Section	
Australian Biochemical Society	1955	1954 Physiology and Biochemistry	N
		1969 Biochemistry	17
Malacological Society of Australia	1956		
Australian Mammal Society	1958		
Australian Society of Plant Physiologists	1958		
Australian Society for Microbiology	1959	1954 Medical Science, National Health and Microbiology	I
		1955 Microbiology, Epidemiology and Preventative Medicine	I
		1969 Microbiology and Immunology	14
		1972 Microbiology	14
Ecological Society of Australia	1960		
Australian Society for Limnology	1961		
Australian Society of Herpetologists	1964		
Australian Entomological Society	1965		
Australian Plant Pathology Society	1969	1969 Plant Pathology	18
Australian Society for Reproductive Biology	1969		
Australian Systematic Botany Society	1973		

zoology, marine and freshwater research, agricultural research, and wildlife research.[74] Some, such as *Australian Mammalogy* and the *Australian Journal of Ecology*, were born of independent specialist societies.

While this proliferation of 'extra-ANZAAS' societies demonstrates the failure of ANZAAS to satisfy all the needs of Australasian biologists, some of these societies were born of ANZAAS, and some initially held their meetings in conjunction with ANZAAS. In bearing and nourishing the new societies, ANZAAS was mother as well as failure. One lively offspring was the Australian Mammal Society. Conceived during a field expedition to collect the hairy-nosed wombat during the 1958 Adelaide Congress, the society was officially born the following week at that meeting. A year later, at its inaugural meeting immediately before the Perth Congress, the Australian Mammal Society was officially baptized.[75] Other societies had similar origins. The 1958 Congress also spawned the Australian Society of Plant Physiologists. Plant Physiologists had met previously in conjunction with Section M (Botany).[76] Until the late 1960s this society and the Australian Biochemical Society held their meetings in conjunction with ANZAAS.

ANZAAS continued to serve biologists by nurturing committees to formulate recommendations to governments and by providing a forum for the discussion of issues that span specialist discipline boundaries and are of national importance, such as conservation, biological surveys, and national floras.

National Flora and Fauna Research and Conservation

Postwar ANZAAS papers and symposia continued to emphasize the need for further study and conservation of the indigenous biota, a concern voiced by biologists through the Association throughout its lifetime. Section D's Biological Survey of Australia Committee provoked successive postwar congresses to repeat their prewar pleas for the establishment within the CSIR and then CSIRO of a department to undertake systematic and ecological work throughout Australia and to co-ordinate the survey and conservation of the indigenous flora and fauna. And there was the rabbit, or rather rabbits, millions of them, their numbers swollen by wartime and postwar shortage of manpower and materials for their control, and the lack of successful trials with the myxoma virus. By the mid-1940s, rabbits were posing unbearable agricultural problems.

In 1949, in response to both ANZAAS and the rabbit, CSIRO's Wildlife Survey Section was established to investigate native and introduced mammals and birds of economic importance.[77] Its dual *raison d'être* was pest control and conservation-oriented management of the indigenous fauna, which have remained the guiding principles of the section and subsequent division. Initially, however, Australia's major pest, the rabbit, was the sole focus of the small Wildlife Survey Section under F. N. Ratcliffe.[78]

Throughout the 1950s, biologists pleaded for some expression of the conservation side of the Wildlife Section: successive congresses advised federal and state governments of the urgency of conserving and surveying the indigenous flora and fauna.[79] Given the institutional limitations on biological research, perhaps it is fortunate that so many native animals have been defined as possible pests. Just as Ratcliffe's pioneering work on the flying fox was provoked by the perceived need for its control, so basic information about many Australian birds and animals has accrued from studies to control them. From the 1950s through into the 1970s, the CSIRO's Wildlife Section/Division investigated a wide range of supposed pests of crops and pastures, including waterfowl, eagles, galahs, cockatoos, emus, and kangaroos, resulting in the accumulation of basic data about various biological phenomena, including reproduction of marsupials and waterfowl, and physiological adaptations to arid conditions.[80] Vertebrate ecology was at last taking shape in Australia, and with it a knowledge of the Australian fauna. There was still, however, no systematic Australia-wide survey of the flora or fauna, nor any modern published flora or fauna.

During the past three and a half decades, the *Flora of New Zealand* has been prepared and published in four volumes,[81] while pressure from both ANZAAS and the Australian Academy of Science (AAS) has eventually led to the progressive publication of the multi-volume *Flora of Australia*.

ANZAAS's Flora of Australia Committee, established in 1958,[82] recommended a twenty year, fourteen taxonomist project.[83] In 1962 the Australian Academy of Science made two urgent recommendations to the government—the appointment of an editor and staff to compile a comprehensive flora of Australia, and the establishment of a Museum of Australian Biology, to conduct a national biological survey. In 1968 the Australian Academy of Science proposed a less-centralized organization to oversee both projects.[84]

Systematic botanists had long used ANZAAS to argue for a new flora. Following the 1971 Congress, ANZAAS's Systematic Botany Committee submitted three proposals for less ambitious preliminary publications. The Australian Academy of Science's promptly established Standing Committee for a Flora of Australia, which included university and state taxonomists, decided to initiate one of those proposals—the preparation of an index of Australian plant names. Using Academy funds and the CSIRO's research and library facilities, the preliminaries to a new *Flora of Australia* were at last under way.[85]

There was still no biological survey. In 1972, however, following increasingly widespread interest in and support for conservation issues, the Academy's proposal was considered by the House of Representatives Select Committee on Wildlife Conservation, which subsequently recommended a biological survey that was incorporated into the election platform of each of the major parties in the federal election.[85]

In 1973 the Academy's proposal finally bore fruit. The federal Labor government established the Australian Biological Resources Study (ABRS) to fund taxonomic and ecological work on the Australian biota, in existing state and Commonwealth institutions. ABRS was a funding and co-ordinating body rather than another Commonwealth research institution (e.g. a Museum of Australian Biology), largely because of the very healthy state, by the 1970s, of university research, and the taxonomic work and collections in state herbaria and museums. Increasing interest in biological control, conservation of endangered species, and ecological studies had increased the pressure for taxonomic information from these institutions. Had a Museum of Australian Biology been established as suggested in 1962, its relationships with state institutions would have resembled those existing between the CSIRO and state departments.[87] Under ABRS, the *Flora of Australia*, planned to comprise over sixty volumes, is at last being prepared and published volume by volume, as is the ten volume *Fauna of Australia*.[88]

Over the past century biologists have used the AAAS and ANZAAS to express their concern for the need to protect the flora and fauna in the face of habitat destruction. During that century, national parks have been established for a range of reasons in a range of regions. Earlier parks provided recreation and fresh air for urban dwellers and sanctuary for refugee native species whose survival was perceived as doubtful. Some were established for their scenic value, some to preserve water catchments, and some because the land was considered unproductive. Despite the term 'national' park, they were established by the states not the Commonwealth. In recognition of past *ad hoc* acquisitions of such reserves and the need to preserve viable representative samples of all types of ecosystems, in 1958 the Australian Academy of Science instigated the first national review of national parks.[89] Following important state surveys of existing national parks and reserves, the final report in 1968 recommended 'that adequate representative areas of at least the most important environments and habitats are set aside permanently as national parks and reserves'.[90]

As part of the International Biological Program (IBP), Professor Ray Specht co-ordinated a conservation survey of Australian vegetation, to show how a national system of reserves could be created to conserve the maximum possible number of species. Each state herbarium provided a list of rare and endangered plant species. The results and recommendations were published in the monumental work *Conservation of Major Plant Communities in Australia and Papua New Guinea*, the major recommendation being to launch a concerted programme to establish a national network of well-buffered ecological reserves throughout Australia, containing a diversity of ecosystems, each large enough to ensure the survival of all species within it.[91]

The Australian Academy of Science celebrated the report with a symposium in 1974, which confirmed the need for ecological reserves and made recommendations relating to their establishment.[92] Following further discussion at the 1975 ANZAAS Congress, the states, the CSIRO, and the Department of Environment and Conservation would sponsor an Ecological Survey of Australia that would build upon the Specht Report. Specht has further refined his assessment of major plant communities in Australia, using the database TAXON,[93] and a National Conservation Strategy has been discussed and published.[94]

IN RETROSPECT

The past century has witnessed vast changes in biological endeavour in Australasia. In some senses, however, we have come full circle: in the present there are echoes of the past. Twentieth century floras of Australia and New Zealand are replacing thoe of last century. To handle increasing volumes of biological data, computerized data bases are replacing card systems and published bibliographies. Current pleas for national reserves echo some of Mueller's ideas about 'Floral Commons', which 'should be reserved in every great country for some maintenance of the original vegetation, and therewith for the preservation of animal life concomitant to peculiar plants'.[95] With a better ecological understanding of the biota, however, biologists now argue for reserves to include all types of Australasian ecosystems and thereby preserve the genetic diversity of the indigenous flora and fauna. Imperially inspired herbarium and museum collections have become integral components of national and international conservation-oriented biological projects. International objectives have replaced imperial ones.

Teamwork in biological investigation is not new. However, the nineteenth century groups of amateur and professional biologists that collected biological data on field naturalists' excursions and scientific expeditions have been replaced by multidisciplinary teams of specialists in both field and laboratory. At the turn of the century, comparative anatomical studies of the Australian fauna were aimed at determining their position in the evolutionary scheme of things. Today modern techniques are tuned to answer similar questions; the 'comparative anatomy' of DNA and protein molecules is being used to elaborate phylogenetic relationships of the Australian fauna. Evolution and economics continue to shape Australasian biological endeavour. Integral components of current Australasian biological work—national parks, wildlife research, the Australian Biological Resources Study, *Flora of Australia*, and specialist societies—are all legacies of AAAS and ANZAAS, which continues to provide a forum for the discussion of broad biological issues.

NOTES

Owing to the greater accessibility of Australian than New Zealand records, to the common bias of the AAAS and ANZAAS toward Australian rather than Australasian biological problems, and to editorial pressure, the majority of examples are Australian rather than New Zealand. The research assistance of Dianne Chambers, Ruth Lane, and Christa Ludlow is gratefully acknowledged. Information, not all of which could be included in such a small chapter, was received with thanks from scientists including R. D. Croll, L. T. Evans, M. H. Friedel, W. J. Griffin, H. G. Higgins, R. L. Hughes, H. H. G. McKern, G. B. Sharman, R. L. Specht, A. D. Thomson, and L. J. Webb.

1. Much has been written about the nineteenth century collection of Australasian biological specimens, for example, A. A. Abbie, 'The History of Biology in Australia', *Aust. J. Sci.*, 17 (1954), 1–9; R. T. Baker, 'Biology', *A Century in the Pacific* (Sydney: William H. Beale, c. 1914), 68–72; E. Scott, 'The History of Australian Science', *Report of ANZAAS*, 24 (Canberra, 1939), 4–9; A. Mozley Moyal, *Scientists in Nineteenth Century Australia: A Documentary History* (Melbourne: Cassell Australia, 1976), 10–29; A. Moyal, *A Bright and Savage Land—Scientists in Colonial Australia*, (Sydney: Collins, 1986); C. M. Finney, *To Sail Beyond the Sunset—Natural History in Australia 1699–1829* (Sydney: Rigby, 1984); C. R. Twidale, M. J. Tyler and M. Davies (eds), *Ideas and Endeavours— The Natural Sciences in South Australia* (Adelaide: Royal Society of South Australia, Inc., 1986), see early parts of chapters on botany and zoology; G. Bentham, *Flora Australiensis: A Description of the Plants of the Australian Territory* (London: Lovell Reeve and Co., 1863, reprinted, Amsterdam: A Asher and Co., 1967), vol. 1, 7–18; F. M. Bailey, 'Concise History of Australian Botany', *Proc. Roy. Soc. Qld*, 8 (1891), xvi–xli. L. A. Gilbert, Botanical Investigations of New South Wales 1811–1880 (unpublished PhD thesis, University of New England, 1971); L. A. Gilbert, 'Plants, Politics and Personalities in Nineteenth Century New South Wales', *J. Roy. Aust. Hist. Soc.* 56 (1970), 15–35; J. J. Fletcher, 'On the Rise and Early Progress of our Knowledge of the Australian Fauna', *Report of the AAAS*, 8 (Melbourne, 1890), 69–104; K. G. Dugan, 'The Zoological Exploration of the Australian Region and its Impact on Biological Theory' in N. Reingold and M. Rothenberg (eds), *Scientific Colonialism: A Cross-Cultural Comparison* (Washington: Smithsonian Institute Press, 1987), 79–150; L. Cockayne, *The Vegetation of New Zealand* (Leipzig: Verlag von Wilhelm Engelmann, 1928), 8–12.
2. L. Brockway, *Science and Colonial Expansion: The Role of the British Royal Botanic Gardens* (New York: Academic Press, 1979), 74–6, 100–2.
3. Mozley Moyal, op. cit., note 1, 39–54. Even Mueller, in the late 1850s during his early years as Victoria's government botanist, sent his best specimens for naming to Kew and kept no duplicates. In later years he built up his own extensive herbarium of Australian plants and kept duplicates of specimens that he sent to Kew (Bentham, op. cit., note 1, 11–12.) See also early chapters of R. T. M. Pescott, *The Royal Botanic Gardens Melbourne; A History from 1845 to 1970* (Melbourne: Oxford University Press, 1982); and L. Gilbert, *The Royal Botanic Gardens Sydney; A History 1816–1985* (Melbourne: Oxford University Press, 1986).
4. Mozley Moyal, op. cit., note 1, 172–84. Joseph Hooker had visited New Zealand in 1840, and published several papers on its flora. Soon after succeeding his father as director of the Kew Gardens in 1865, his *Handbook of the New Zealand Flora* (1864–7) was completed. Meanwhile, despite the fact that so many of the necessary plant specimens were supplied by Mueller, the seven volume *Flora Australiensis* (1863–78) was likewise being prepared at Kew rather than in Australia. As well as various books on the flora of Victoria, Mueller prepared his own *Fragmenta Phytographiae Australiae* (1858–82) and *Systematic Census of Australian Plants* (1882), in which the plant names were not always those originating from Kew. See also Bentham op. cit., note 1, 12–18; and J. M. Powell, 'Exiled from the Garden. Von Mueller's Correspondence with Kew, 1871–81', *Victorian Historical Journal*, 48 (1977), 313–20, and 'A Baron under Siege: von Mueller and the Press in the 1870s', *Victorian Historical Journal*, 50 (1979), 18–35.

5. Fletcher, op. cit., note 1, 78. Mozley Moyal, op. cit., note 1, 60–84. Although not as extensive as the Kew-centred network, Richard Owen did establish a widespread zoological network. His collectors in Australia included George Bennett, Gerard Krefft and Ronald Gunn (A. Mozley Moyal, 'Sir Richard Owen and his Influence on Australian Zoological and Palaeontological Science', *Records of the Australian Academy of Science*, 3 (2) (1975), 41–56.) In 1904 F. W. Hutton's *Index Faunae Novae Zealandiae* was published for the Philosophical Institute of Canterbury, by Dulau and Co., London.
6. C. F. I. Jenkins, *The Noah's Ark Syndrome (One Hundred Years of Acclimatization and Zoo Development in Australia)* (Perth: Zoological Gardens Board, Western Australia, 1977); S. K. Kohlstedt, 'Natural Heritage: Securing Australian Materials in 19th Century Museums', *Museums Australia* (December 1984), 15–22.
7. M. E. Hoare, 'The Intercolonial Science Movement in Australasia 1870–1890', *Records of the Australian Academy of Science*, 3 (2) (1976), 12–14, 17–18; J. H. Willis, 'The First Century of the Field Naturalists Club of Victoria', *Victorian Naturalist*, 97 (1980), 93–106. Half the papers presented in the Biology Section of the Melbourne AAAS Meeting in 1890 were presented by members of the Field Naturalists Club of Victoria.
8. The first Australasian agricultural college, Lincoln College, was established in New Zealand in 1880, the Pharmacy College, Melbourne in 1881, Burnley Horticultural College (near Melbourne), in 1891, and Roseworthy Agricultural College, South Australia, in 1883. The Technological, Industrial and Sanitary Museum of New South Wales was established in 1881 (see note 16).
9. D. J. Mulvaney and J. H. Calaby, '*So Much that is New*': *Baldwin Spencer 1860–1929, A Biography* (Melbourne: Melbourne University Press, 1985). The first four chapters and the beginning of chapter 8 describe Spencer's biological training.
10. W. B. Spencer was professor of biology at the University of Melbourne while F. McCoy was still professor of natural science, and G. B. Halford was professor of general anatomy and physiology: von Mueller was then Victoria's government botanist. For the Darwinian debate in Australia see A. Mozley, 'Evolution and the Climate of Opinion in Australia, 1850–76', *Victorian Studies*, 10 (4) (1967), 411–30; Mozley Moyal, op. cit., note 1, 186–99. B. Butcher, The Reception and Impact of Darwinism in Australia, 1860–1914 (PhD thesis, University of Melbourne in progress).
11. Mulvaney and Calaby, op. cit., note 9, 96, 10–2, 148–9, 151–2, 157.
12. 'Committees of Investigation', *Report of the AAAS*, 1 (Sydney, 1888), xxxiv.
13. Fresh from comparative morphological and microanalytical work at Oxford, Spencer focused on such lesser known indigenous animals as earthworms, planarians, leeches, sponges, frogs, and lungfish. His choice of new species to describe and name was determined not merely for their novelty, but according to their evolutionary or zoogeographical interest. The examination of the fauna on King Island in Bass Strait during the Field Naturalists Club of Victoria excursion there in 1887 only months after his arrival in Melbourne, was designed to investigate faunal relationships between King Island, mainland Australia, and Tasmania. (W. B. Spencer, 'The Fauna and Zoological Relationships of Tasmania', *Report of the AAAS*, 4 (Hobart, 1892), 892–124; also Mulvaney and Calaby, op. cit., note 9, 97, 101–2, 149). That sensitized Spencer to looking at the fauna in terms of its past and present distributions, an interest he would develop

further, even before his participation in the Horn Expedition into Central Australia in 1894.
14. R. Tate, 'On the Influence of Physiographic Changes in the Distribution of Life in Australia', *Report of the AAAS*, 1 (Sydney, 1888), 312–25; Spencer, op. cit., note 13; F. W. Hutton, 'On the Origin of the Struthious Birds of Australasia, *Report of the AAAS*, 4 (Hobart, 1892), 365–8; C. Hedley, 'The Faunal Regions of Australia', *Report of the AAAS*, 5 (Adelaide, 1893), 444–6; A. Dendy, 'The Cryptozoic Fauna of Australia', *Report of the AAAS*, 6 (Brisbane, 1895), 99–119; W. B. Benham, 'The Geographical Distribution of Earthworms and the Palaeogeography of the Antarctic Region', *Report of the AAAS*, 9 (Hobart, 1902), 319–43; W. V. Legge, 'The Zoographical Relations of the Ornis of the various sub-regions of the "Australian Region", with the Geographical Distribution of the Principal Genera therein', *Report of the AAAS*, 10 (Dunedin, 1904), 217–85.
15. Mozley Moyal, op. cit., note 1, 223–6. One of Haswell and Hill's students, Theodore T. Flynn (father of film star Errol) in 1911 became the foundation professor of biology at the University of Tasmania where his marsupial research influenced a generation of zoologists who carried their interest in marsupial and monotreme genetics and physiology to other Australian universities into the second half of the twentieth century (e.g. Professors G. B. Sharman and R. L. Hughes).
16. The Museum's initial collection comprised the residue of specimens from the Sydney International Exhibition of 1879. The committee of management first met in January 1880. The first staff member, the acting secretary, was appointed in May 1880 (McKern, pers. comm. 1987).
17. As the new field of phytochemistry grew, methods were developed for the detection and extraction of useful chemicals including oils from eucalypts and other Australian plants. The eucalyptus oil industry, which had been pioneered in the middle of the nineteenth century in Victoria by Bosisto and Mueller, was hungry for refinements, as increased demands for oils of known composition and properties paralleled their discovery. While Maiden, and the botanist R. T. Baker and the chemist H. G. Smith, began as virtual novices, some of their research paralleled European developments. As progress occurred in the taxonomy of the native flora, and in the organic chemistry of their oils and other components, these men contributed to and benefited from such developments. Perhaps because phytochemical work at the museum began when phytochemistry was still a new discipline, it could remain near the cutting edge of that discipline. While essential oil and other phytochemical research continued at the museum for over three-quarters of a century, even before the First World War it was spreading into university chemistry departments, in some of which, phytochemical work on indigenous plants continues today, for example, in the department of chemistry at the University of Western Australia. H. H. G. McKern, 'The Natural Plant Products Industry of Australia', *Proc. Roy. Aust. Chem. Inst.*, 27 (1960), 296–99; H. H. G. McKern, 'Research into the Volatile Oils of the Australian Flora, 1788–1967', *A Century of Scientific Progress: The Centenial Volume of the Royal Society of New South Wales* (Sydney: Royal Society of New South Wales, nd (c. 1968)), 310–20; J. H. Maiden, 'The Chemistry of the Australian Indigenous Vegetation', *Report of the AAAS*, 6 (Brisbane, 1895), 25–57.
18. J. H. Maiden, 'Bibliography of the Chemistry of Indigenous Australian

Vegetable Products', *Report of the AAAS*, 1 (Sydney, 1888), 183–93. It included New Zealand as well as Australian plants.

19. Maiden, Baker and Smith contributed twenty-five papers to AAAS Congresses between 1888 and 1913, e.g. R. T. Baker, 'A census of Victorian Eucalypts and Their Economics', *Report of the AAAS*, 14 (Melbourne, 1913), 294–310; R. T. Baker, 'Contribution to a Discussion on the Economics of the Eucalypts', *Report of the AAAS*, 14 (Melbourne, 1913), 330–6; H. G. Smith, 'Discussion of the Eucalypts and their Products', *Report of the AAAS*, 14 (Melbourne, 1913), 116–25. See also E. H. Rennie, 'The Chemical Exploitation, Past, Present, and Future of Australian Plants', *Report of the AAAS*, 18 (Perth, 1926), 1–35.

20. Report of 'Rust in Wheat Committee', *Report of the AAAS*, 3 (Christchurch, 1891), 547–50; S. Fish, 'The History of Plant Pathology in Australia', *Annual Review of Phytopathology*, 8 (1970), 13–17; N. White, 'A History of Plant Pathology in Australia' in D. J. and S. G. M. Carr (eds), *Plants and Man in Australia* (Sydney: Academic Press, 1981), 42–5; D. McAlpine, 'Rust and Stinking Smut in Wheat', *Report of the AAAS*, 9 (Hobart, 1902), 610; D. McAlpine, 'Smut Experiments in Victoria', *Report of the AAAS*, 12 (Brisbane, 1909), 590; W. Farrer, 'Making and Improvement of Wheats for Australian Conditions', *Report of the AAAS*, 7 (Sydney, 1898), 908; W. S. Campbell, 'An Historical Sketch of William Farrer's Work in Connection with his Improvements for Australian Conditions', *Report of the AAAS*, 13 (Sydney, 1911), 525; C. W. Wrigley, and A. Rathjen, 'Wheat Breeding in Australia', Carr and Carr, ibid., 100–17.

21. F. Turner, 'Fodder Plants and Grasses of Australia', *Report of the AAAS*, 2 (Melbourne, 1890), 586–96; F. Turner, 'The Supposed Poisonous Plants of Western Australia', *Report of the AAAS*, 7 (Sydney, 1898), 910–19; S. Everist, 'The History of Poisonous Plants in Australia', Carr and Carr (eds), op. cit., note 20, 225–31.

22. W. Gill, 'Deforestation in South Australia: Its Causes and Probable Results', *Report of the AAAS*, 5 (Adelaide, 1893), 527–36; W. S. Campbell, 'Forestry in New South Wales', *Report of the AAAS*, 7 (Sydney, 1898), 958–61. H. J. Colbourn, 'Some Aspects of Tasmanian Forestry', *Report of the AAAS*, 9 (Hobart, 1902), 582–93.

23. By 1903, colonial or state floras had been prepared by the government botanists of Victoria (von Mueller), New South Wales (Moore), Queensland (Bailey) and Tasmania (Rodway), and for South Australia by Adelaide University's professor of natural science (Tate). In 1906 Cheeseman's *Manual of the New Zealand Flora* was published.

24. J. H. Maiden, 'A Century of Botanical Endeavour in South Australia', *Report of the AAAS*, 11 (Adelaide, 1907), 161. To facilitate the work of students of the Australian flora, whether inside or outside Australia, Maiden also recommended the publication within Australia of descriptions of all new Australian plants.

25. Report of Australasian Biological Station Committee, *Report of the AAAS*, 2 (Melbourne, 1890), 354–55; 'Section D Resolutions and Discussions', *Report of the AAAS*, 8 (Melbourne, 1900), xxx; 9, (Hobart, 1902), xlix–l; 13 (Sydney, 1911), xlii, lx. Report of the Committee for the Biological and Hydrographical Study of the New Zealand Coast, *Report of the AAAS*, 13 (Sydney, 1911), 362–5; *Report of the AAAS*, 14 (Melbourne, 1913), 336–7; T. W. E. David, 'The Aims and Ideals of Australasian Science', *Report of the AAAS*, 10 (Dunedin, 1904), 4–7, 21–23; F. von Mueller, Inaugural

Address, *Report of the AAAS*, 2 (Melbourne, 1890), 10–11; Maiden, op. cit., note 24, 159–60.

26. Report of Committee appointed to make Recommendations for the Protection of Native Fauna, *Report of the AAAS*, 5 (Adelaide, 1893), 241–2; B. H. Woodward,' National Parks and the Flora and Fauna Reserves in Australasia, *Journal of the W. A. National History Society*, 2 (4) (November, 1907), 13–27.

27. A. J. Ewart, 'Biology During the War and After', *Report of the AAAS*, 15 (Melbourne, 1921), 134–44.

28. T. G. B. Osborn, 'The Factors Influencing the Regeneration of Vegetation in the Arid Parts of Australia, with Special Reference to the Pastoral Industry', *Report of the AAAS*, 18 (Perth, 1926), 823–4; W. C. MacKenzie, 'The Importance of Zoology to Medical Science', *Report of the AAAS*, 19 (Hobart, 1928), 235–42; W. J. Dakin, 'Some Recent Researches in Marzine Zoology and their Bearing upon Current Problems of Today', *Report of ANZAAS*, 22 (Melbourne, 1935), 147–64; J. G. Wood, 'The Plant in Relation to Water', *Report of ANZAAS*, 24 (Canberra, 1939), 281–90.

29. Ewart, op. cit., note 27, 137–41.

30. L. Harrison, 'The Composition and Origins of the Australian Fauna, with Special Reference to the Wegener Hypothesis', *Report of the AAAS*, 18 (Perth, 1926), 332–96; R. H. Cambage, 'Presidential Address', *Report of the AAAS*, 19 (Hobart, 1928), 1–29 (see section on Origin and Development of Portion of the Australian Flora, pp. 7–14); G. E. Nicholls, 'The Composition and Biogeographical Radiations of the Fauna of Western Australia', *Report of ANZAAS*, 21 (Sydney, 1932), 93–138; H. E. Le Grand, 'Specialities, Problems and Localism—The Reception of Continental Drift in Australia 1920–1940', *Earth Sciences History*, 5 (1) (1986), 84–95.

31. S. L. Macindoe, 'History of Breeding for Resistance to Stem Rust of Wheat in Australia', *Report of ANZAAS*, 22 (Melbourne, 1935), 324–5; W. L. Waterhouse, 'Some Aspects of Plant Pathology', *Report of ANZAAS*, 24 (Canberra, 1939), 251–9; Wrigley and Rathjen, op. cit., note 20, 121–3. A good overview of the importance of biological science to Australasian primary industry is provided by Cambage, op. cit., note 30, 4–7.

32. F. C. Cradock and C. A. Hawkins, 'Soils and Fertilizers in New South Wales', *A Century of Scientific Progress: The Centenary Volume of the Royal Society of New South Wales*, note 17, 233–7; G. W. Leeper, 'Australia and Trace Elements' in Carr and Carr (eds), op. cit., note 20, 35–41; C. G. Stephens and K. H. Northcote, 'Soil Science' in Twidale *et al.* (eds), op. cit., note 1, 74.

33. B. T. Dickson, H. D. Wright, H. R. Carne and R. J. Noble, 'Filtrable Viruses', *Report of ANZAAS*, 21 (Sydney, 1932), 437–9; H. R. Wallace, 'Plant Pathology' in Twidale *et al.* (eds), op. cit., note 1, 158–9.

34. J. S. Turner, 'The Development of Plant Physiology in Australia' in Carr and Carr (eds), op. cit., note 20, 13–14, 17.

35. T. H. Johnston, 'The Australian Prickly-Pear Problem', *Report of the AAAS*, 16 (Wellington, 1923), 347–401; W. B. Alexander, *Natural Enemies of Prickly Pear and their Introduction into Australia*, Institute of Science and Industry Bulletin No. 29 (Melbourne, 1925); C. B. Osmond and J. Monro, 'Prickly-Pear' in Carr and Carr (eds), op. cit., note 20, 194–222.

36. G. Lightfoot, *The Co-operative Development of Australia's Natural Resources*, Institute of Science and Industry Pamphlet no. 3 (Melbourne, 1923), 9.
37. 'Resolutions passed by the Congress', G. Lightfoot (ed.), *Proceedings of the Pan-Pacific Congress, Australia, 1923* (Melbourne: Government Printer, nd), vol. I, 36. The suggestion of federal involvement was not new. At the 1893 AAAS meeting a plea had been made for a Federated State Board of Horticulture, which should include plant pathologists and entomologists. (A. Molineux, 'A Plea for an Intercolonial State Board of Horticulture', *Report of the AAAS*, 5 (Adelaide, 1893), 537–9.)
38. L. T. Evans, 'Fifty Years of Plant Research', *Nature*, 261 (1976), 655; F. W. G. White, 'A Personal Account of the Historical Development of CSIRO', *Nature*, 261 (1976), 634; J. D. Atkinson, *DSIR's First Fifty Years* (Wellington: Department of Scientific and Industrial Research, 1976), 23–4; A. D. Thomson, 'Evolution of Plant Research is DSIR: Impetus for the Establishment of Plant Diseases Division at Mt Albert', *Crop Research News*, no. 18 (1976), 15–16.
39. G. Currie and J. Graham, 'Growth of Scientific Research in Australia: The Council for Scientific and Industrial Research and the Empire Marketing Board', *Records of the Australian Academy of Science*, 1 (3) (1968), 25–35; C. B. Schedvin, 'Environment, Economy and Australian Biology, 1890–1939', *Historical Studies*, 21 (82) (1984), 17, 26; C. B. Schedvin, *Shaping Science and Industry: A History of Australia's Council for Scientific and Industrial Research* (Sydney: George Allen & Unwin, 1987).
40. Currie and Graham, op. cit., note 39, 34.
41. A. J. Gibson, *A Forest Products Laboratory for Australia: Justification for its Creation, Outline of its Organization and Rough Estimate of Cost*, CSIR Pamphlet no. 9 (Melbourne, 1928), 24.
42. R. L. Specht, 'Australia', in E. J. Kormondy and J. F. McCormick (eds) *Handbook of Contemporary Developments in World Ecology* (Connecticut: Greenwood Press, 1981), 388; L. B. Moore, 'Harry Howard Barton Allan', *Proc. Linn. Soc. Lond.*, 175 (1963–4), 190.
43. Thorburn Brailsford Robertson, educated at the University of Adelaide, taught at the universities of California and Toronto before accepting the chair of physiology at Adelaide in 1919. Bertram Thomas Dickson, professor of plant pathology at Canada's McGill University, became the first chief of CSIR's Division of Economic Botany in 1928.
44. In 1919 and 1920 Isaac Herbert Boas, while officer in charge of the Forest Products Laboratory in Perth, visited similar research laboratories in North America, England, Europe, and India. Following his appointment in 1928 as the first chief of CSIR's Division of Forest Products, he again visited overseas research institutions. Studentships were provided for two men to study wood preservation and wood chemistry at the Forest Products Laboratory, Madison, USA. (S. Preston, 'Division of Forest Products: A Brief History', *CSIRO Forest Products Newsletter* no. 350 (April/May 1968), 5–6.) A problem common to each of the early CSIR divisions was the shortage in Australia of qualified biologists. Three men were sent to train at Farnham Royal, and two to the Department of Entomology at the Imperial College of Science and Technology, London University. R. J. Tillyard, *The Work of the Division of Economic Entomology for the year 1828–29*, CSIR Pamphlet no. 13 (Melbourne, 1929), 8.
45. R. Rivett, *David Rivett: Fighter for Australian Science* (Blackburn, Victoria: The Dominion Press, 1972), 108–9.

46. White, op. cit., note 38.
47. The Australian Society for Experimental Biology, the Australian Veterinary Association, the Australian Institute of Agricultural Science, the New Zealand Institute of Foresters, and the Institute of Foresters of Australia, and the *Australian Journal of Experimental Biology and Medical Sciences*, *Australian Veterinary Journal*, *Journal of the Australian Institute of Agricultural Science*, *New Zealand Journal of Forestry* and *Australian Forestry* were born in the 1920s and 1930s.
48. A Commonwealth Forestry Bureau and an Australian Forestry School were established in Canberra in the mid-1920s. The prime mover and lead actor was Lane Poole, former conservator of forests for Western Australia and forest advisor to the Commonwealth Government. These federal institutions were to supply much-needed trained foresters, and to organize and co-ordinate research into various problems of silviculture and forest management throughout Australia. A. Meyer, *The Foresters* (Hobart: Institute of Foresters of Australia, Inc., 1985), 10–12; L. T. Carron, *A History of Forestry in Australia* (Canberra: Australian National University Press, 1985), 250–5.
49. Although it would not materialize till decades later, a Commonwealth herbarium was being discussed. While consideration was being given to the position and composition of botanical gardens in Canberra, consideration was also being given to the establishment in the national capital of a herbarium that restricted its interest and expertise to Australian plants. Such a herbarium should publish an *Australian Index* which would be the equivalent of the wider ranging *Kew Index*. L. Rodway, 'The Desirable Commonwealth Herbarium', *Report of the AAAS*, 19 (Hobart, 1928), 626–8; 'Summary of Resolutions', *Report of the AAAS*, 19 (Hobart, 1928), xxv.
50. In 1924 an Act was passed to establish a National Museum of Australian Zoology around Colin MacKenzie's zoological and anatomical collection of Australian animals. It was later renamed the Australian Institute of Anatomy. M. MacCallum, 'MacKenzie, Sir William Colin', *Australian Dictionary of Biography*, 10 (1986), 307.
51. Flynn, Biology, Tasmania, 1911–31; Goddard, Biology, Queensland, 1923–48; Harrison, Zoology, Sydney 1923–8; Hunter, Anatomy, Sydney, 1922–4; T. H. Johnston, Zoology, Adelaide, 1922–51 and S. J. Johnston, Zoology, Sydney, 1918–22. L. Farrall, 'The Biological Professoriate in Australia 1850–1940: Towards a Collective Biography', Prepared for the Conference on Scientific Colonialism 1800–1930: A Cross Cultural Comparison, held at the University of Melbourne, May 1981, 35.
52. T. Brailsford Robertson (Animal Nutrition), R. J. Tillyard (Economic Entomology), and I. H. Boas (Forest Products).
53. T. G. B. Osborn was professor of botany at the University of Adelaide from 1912 to 1927 and then at the University of Sydney from 1928 to 1937. J. G. Wood, a former student of Osborn, was lecturer in charge of Botany from 1928 until his appointment to the botany chair in 1935, which he held until his death in 1959. Eight ecologists who would become leaders in Australian ecological research during the 1950s and 1960s studied in these departments under T. G. B. Osborn and J. G. Wood, the 'father and son' of Australian ecology. Specht, op. cit., note 42, 409. See also E. L. Robertson, 'Botany' in Twidale *et al.* (eds), op. cit., note 1, 116–19, 143–6; D. Branagan and G. Holland (eds), *Ever Reaping Something New—A Science Centenary* (Sydney: University of Sydney, 1985), 167–9.
54. Before the war, several New Zealand botanists published ecological papers,

for example Leonard Cockayne, even before 1910, such as his 'On the Importance of New Zealand as a Field for Botanical Study and Research', *Report of the AAAS*, 10 (Dunedin, 1904), 290–7. New Zealand plant ecological studies were so well developed that before the war Cockayne had prepared *The Vegetation of New Zealand* in the series *Die Vegetation der Erde*, which provided an ecological rather than a strictly floristic interpretation of the New Zealand vegetation (Cockayne, op. cit., note 1, 15–20). For a history of economic ecology in New Zealand see L. Cockayne, 'New Zealand Economic Plant Ecology', *Imperial Botanical Congress, 1924, Report of Proceedings* (Cambridge: Cambridge University Press, 1925), 259–69.

55. T. G. B. Osborn, 'The Biological Factor in the Study of Vegetation with Special Reference to Australian Conditions', *Report of the AAAS*, 19 (Hobart, 1928), 613–26; L. Rodway, 'Some Ecological Features in Tasmania', *Report of the AAAS*, 17 (Adelaide 1924), 730–8; Ecology Committee Report, *Report of the AAAS*, 17 (Adelaide, 1924), 138; Ecology Committee Report, *Report of the AAAS*, 18 (Perth, 1926), 89–90; Specht, op. cit., note 42, 387–8, 395.
56. L. Cockayne and H. H. Allan, 'The Bearing of Ecological Studies in New Zealand on Botanical Taxonomic Conceptions and Procedure', *Journal of Ecology*, 15 (1927), 234; H. H. Allan, 'The Significance of Hybridism in the New Zealand Flora', *Report of ANZAAS*, 20 (Brisbane, 1930), 429–77.
57. Osborn (1928) op. cit., note 55, 613–16.
58. Wood, op. cit., note 28, 287.
59. Appointed to the original ecology committee at the 1913 AAAS meeting were Dr L. Cockayne, J. H. Maiden, R. H. Cambage, Dr Morrison, Professors T. G. B. Osborn and A. J. Ewart, L. Rodway, and Dr C. S. Sutton. After the war, at successive AAAS meetings to which the committee reported (1921–6), A. A. Hammilton, Dr Ethel McLennan, Miss M. Collins (Mrs Sheils), R. T. Patton, Professor A. A. Lawson, and C. A. Gardner were appointed. Despite Cockayne's inclusion, New Zealand work was not included in the bibliography. Ecology Committee Report, *Report of the AAAS*, 16 (Wellington, 1923), 109; 17 (Adelaide, 1924), 138; 18 (Perth, 1926), 89–90; CSIR produced the first national vegetation map and native pasture survey: J. A. Prescott, *The Soils of Australia in Relation to Vegetation and Climate*, CSIR Bulletin no. 52 (Melbourne: CSIR, 1931) and A. McTaggart, *Survey of the Pastures of Australia*, CSIR Bulletin no. 99 (Melbourne: CSIR, 1936).
60. Osborn (1926) op. cit., note 28; Osborn (1928) op. cit., note 55, 622–4; E. Cheel, 'A Review of the Flora of the Arid and Semi-arid Regions of Australia', *Report of ANZAAS*, 23 (Auckland, 1937), 307–37; A. A. Hall, R. L. Specht and C. M. Eardley, 'Koonamore Vegetation Reserve, South Australia, 1926–62', *Aust. J. Bot.*, 12 (1964), 205–64.
61. Ratcliffe, an ex-EMB officer, had already investigated the habits of the purported pest, the flying fox, for CSIR; F. N. Ratcliffe, *Soil Drift in the Arid Pastoral Areas of South Australia*, CSIR Pamphlet no. 64 (Melbourne, 1936); F. N. Ratcliffe, *Further Observations on Soil Erosion and Sand Drift, with Special Reference to South-Western Queensland*, CSIR Pamphlet no. 70 (Melbourne, 1937); F. N. Ratcliffe, *Flying Fox and Drifting Sand: The Adventures of a Biologist in Australia* (Sydney: Angus & Robertson, 1947), 187–332. In New Zealand Cockayne had already (1917–21) been investigating pasture palatability and regeneration, following overgrazing, in semi-arid Central Otago. Cockayne, op. cit., note 1, 21.
62. The Committee comprised Professors W. E. Agar, T. Harvey Johnston, G. E. Nicholls, W. J. Dakin, and E. J. Goddard, Drs J. Pearson, G. A.

Waterhouse, and R. J. Tillyard, and A. H. Chisholm. *Report of ANZAAS*, 22 (Melbourne, 1935), xxxi.

63. Summary of Resolutions passed by the General Council: Section D, *Report of ANZAAS*, 24 (Canberra, 1939), xxx–xxxi.
64. *Report of the AAAS*, 15 (Melbourne, 1921), xxxi.
65. H. R. Marston and M. C. Dawbarn, *Food Composition Tables*, CSIRO Pamphelt no. 107 (Melbourne: CSIRO, 1941); Atkinson, op. cit., note 38, 55–9.
66. E. J. F. Wood, *Agar in Australia*, CSIR Bulletin no. 203 (Melbourne, 1946); Atkinson, op. cit., note 38, 59.
67. The plants included *Atropa bella-donna*, *Digitalis purpurea*, and *Papaver somniferum*, as sources of atropine, digitalis (containing cardiac glycosides), and opium. A. Thomas, 'Early days of Botany Division', *Botany Division [DSIR] Newsletter*, no. 16 (1976), 7–8 and no. 26 (1977), 2; J. R. Price, The Australian Phytochemical Survey, unpublished manuscript.
68. *Eucalyptus macrorhyncha* as a source of rutin, *Bursaria spinosa* as a source of aesculin, and two species of *Duboisia* as sources of important tropane (Atropine-like) alkaloids, hyoscine, also known as scopolamine, and hyoscyamine. McKern (1960) op. cit., note 17, 299–303.
69. L. H. Briggs, 'Some Australasian Plant Products', *Report of ANZAAS*, 28 (Brisbane, 1951), 16–17; Price op. cit., note 67; Anon, 'Phytochemical Survey', *Rural Research*, 26 (1956), 26–30.
70. J. F. T. Grimwade, *A Short History of Drug Houses of Australia Ltd to 1968* (Melbourne: J. F. T. Grimwade, 1974), 60–3; McKern (1960) op. cit., note 17, 300–1. The only Australian research group currently interested in *Duboisia* alkaloids is in the pharmacy department of the University of Queensland. During the past two decades, Dr W. J. Griffin and colleagues have reported their phytochemical and pharmacological findings at ANZAAS meetings; see W. J. Griffin, '*Duboisias* of Australia', *Pharmacy International*, (December 1985), 305–8.
71. Australian National Research Council: Advice to Commonwealth Government, *Report of ANZAAS*, 25 (Adelaide, 1946), xl. Following its name-change in 1950 to Land Research and Regional Survey, it continued to send teams of scientists specialized in a range of disciplines, to conduct thorough surveys in the undeveloped parts of Northern Australia. C. S. Christian, 'The Revolution in Agriculture in Northern Australia', *Aust. J. Sci.*, 22 (1959), 138–47; and R. A. Perry, '*Pasture Lands of the Northern Territory, Australia*', CSIRO Australian Land Research Series no. 5 (Canberra: CSIRO Divison of Land Research and Regional Survey, 1960). Its Alice Springs Field Centre, established in 1953, included ecological and pasture plant introduction research, CSIRO *Central Australian Laboratory: a Centre for Research on Management of Semidesert Rangelands*, (Canberra: CSIRO Division of Land Resources Management, 1981), 3–5, 10. In 1956 the Australian Academy of Science sponsored a symposium in Brisbane entitled Man and Animals in the Tropics'. The resultant recommendations included the establishment of an Institute of Tropical Physiology in north-western Queensland, cattle and sheep breeding for tropical conditions, and the study of plant–soil–atmosphere–water relationships and the nitrogen cycle under arid tropical conditions. (F. Fenner and A. L. G. Rees (eds), *The Australian Academy of Science: The First Twenty-five Years* (Canberra: Australian Academy of Science, 1980), 178.) In 1959 CSIRO'S Division of Tropical Pastures was established from a section within the Division of Plant Industry. In 1960 CSIRO's Northern Cattle Research Laboratory, later named the Tropical Cattle

Research Centre, was established at Rockhampton, A. G. Eyles and D. G. Cameron, *Pasture Research in Northern Australia—Its History, Achievements and Future Emphasis* (St Lucia, Queensland: CSIRO Division of Tropical Crops and Pastures, 1985).

72. The current state of knowledge of the vegetation of northern Australia is revealed in the papers on the functioning of tropical and subtropical plant communities presented to a conference arranged by the botany department of the University of Queensland as part of that university's seventy-fifth anniversary celebrations, and published in H. T. Clifford and R. L. Specht (eds), *Tropical Plant Communities—Their Resilience, Functioning and Management in Northern Australia* (St Lucia, Queensland: Botany Department, University of Queensland, 1986). The plant communities discussed range from rainforests to woodlands and heathlands, from grasslands to wetlands and coastal dune vegetation. Chapter 4 'Plant Science Research in Tropical-Subtropical Australia (North of Latitude 30°S)' details suitable research strategies for further research in these inadequately understood plant communities, possibly as part of the Australian contribution to the international 'Decade of the Tropics'.

73. CSIRO, *Australian Scientific Societies and Professional Associations* (Melbourne: CSIRO, 1971).

74. B. J. Walby, 'Australian Journals of Scientific Research', *Nature*, 261 (1976), 661–2; A. D. Thomson, 'A History of the Publications of Scientific and Agricultural Research in New Zealand', *Crop Research News*, 18 (April, 1976), 22.

75. G. B. Sharman, Pers. comm. 1986.

76. J. S. Turner, op. cit., note 34, 30.

77. In both Australia and New Zealand, problems provoked by introduced animals led to the postwar establishment of national research groups, which have developed ecological studies of the indigenous fauna. Following a survey of introduced mammals of economic importance in New Zealand, a Wildlife Section was established within DSIR in 1948 to further survey introduced mammals. The next year CSIRO followed suit with the establishment of its Wildlife Survey Section. Both sections were initially very small. Both were accorded divisional status by the early 1960s. Both initially focused their attention on the rabbit. A. D. Pritchard, *Ecology Division Research Report 1984* (Lower Hutt, NZ: Ecology Division, Department of Scientific and Industrial Research, 1985), 6; H. J. Frith, 'Wildlife Research' *Nature*, 261 (1976), 637.

78. In 1950 field studies in the Murray Valley, the myxoma virus, spread by the mosquito, was at last successful in rapidly reducing rabbit population numbers. In collaboration with the Australian National University and state departments, the intricacies of this promising example of biological control were explored; the ecological relationships between different strains of the virus, its various vectors, and its host, the rabbit, under a range of environmental conditions. Frith, op. cit., note 77, 637. F. Fenner and F. N. Ratcliffe, *Myxomatosis* (Cambridge: Cambridge University Press, 1965); C. B. Schedvin in the second volume of the history of the CSIRO (forthcoming).

79. e.g. Resolutions from Sections. *Report of ANZAAS*, 29 (Sydney, 1952), 361–2, Australian Fauna and Flora Conservation Committee.

80. Frith, op. cit., note 77, 638. C. S. Christian, and R. A. Perry, 'Arid Land Studies in Australia' in W. G. McGinnies and B. J. Goldman (eds), *Arid Lands in Perspective* (Arizona: University of Arizona Press, 1969), 216–18; A. A. Strom, 'Conservation as a Goal' in J. A. Sinden (ed.), for ANZAAS,

The Natural Resources of Australia (Sydney: Angus & Roberston, 1972), 64–6.

81. Between 1954 and 1960, in a unit of DSIR's Botany Division maintained specifically for the production of a new flora, H. H. Allan, assisted by Lucy Moore, completely revised the second (1925) edition Cheeseman's *Manual of the New Zealand Flora*, to produce the first volume of the new *Flora*, which was published in 1961. Subsequent volumes have been published in 1970, 1980, and 1985. DSIR, 'Taxonomy', *Botany Divsion Triennial Report* (Wellington: New Zealand, 1960) and subsequent Reports.
82. *Report of ANZAAS*, 33 (Adelaide, 1958), 231. Professor J. G. Wood initiated the ANZAAS discussion after Hansjoerg Eichler, the keeper in the State Herbarium of South Australia, in 1975 recommended that his staff be permitted to undertake Australia-wide botanical studies that could be used towards a new *Flora of Australia*.
83. S. T. Blake, 'A New Flora of Australia', *Aust. J. Sci.*, 23 (1960), 173–6.
84. Fenner and Rees, op. cit., note 72, 64–8; 'Proposal to Establish a Biological Survey of Australia': Report by the Fauna and Flora Committee of the Australian Academy of Science, *Aust. J. Sci.*, 31 (1969), 377–82.
85. A. S. George, 'The Background to the *Flora of Australia*', *Flora of Australia*, vol. 1 (Canberra: 1981), 7–9; [Ride, W. D. L.] 'Flora of Australia', *Search*, 7 (1976), 327; Fenner and Rees, op. cit., note 72, 70–1.
86. ibid., 69.
87. W. D. L. Ride, 'Towards a National Biological Survey, The Australian Biological Resources Study', *Search*, 9 (1978), 73–5.
88. From 1973 the Australian Biological Resources Study (ABRS) was under an Interim Council. Following support from ASTEC and the Committee of Heads of the Australian Herbaria, the government formally established the Australian Biological Resources Study within the Department of Science in 1978. Its major objective was the collection and publication of data on the identity and distribution of the Australian flora and fauna. In 1979 the Bureau of Flora and Fauna was established within the Department of Science and the Environment. Through the ABRS, the Bureau coordinates, edits, compiles and publishes several series of books and databases. The *Flora of Australia* describes all the plants known to be native or naturalized in Australia—over 40 000 species—a dramatic increase over the 8125 species included in Bentham's *Flora Australiensis* (op. cit., note 1) which was completed over a century ago. The first volume was launched at the thirteenth International Botanical Congress in 1981 in Sydney. The *Fauna of Australia* describes the biology, taxonomy, evolution, and history of discovery of all known terrestrial and aquatic animals in Australia. The *Zoological Catalogue of Australia* is a computer database and multi-volume directory to the most comprehensive and recent biological and taxonomic information available on Australian animals. The Bureau of Flora and Fauna also publishes the *Australian Flora and Fauna Series*, and develops as databases the *Australian Plant Name Index* and the *Census of Australian Vertebrate Species* (now available though the CSIRONET System).
89. Fenner and Rees, op. cit., note 71, 75–7. For a history of National Parks see W. Goldstein (ed.), *Australia's 100 Years of National Parks* reprinted from *Parks and Wildlife*, 2 (April, 1979), and S. Bardwell, National Parks in Victoria 1866–1966 (unpublished PhD thesis, Monash University, 1974), and 'A History of National Parks in Victoria', *Journal of the Royal Historical Society of Victoria*, 56 (4) (December, 1985), 10–18.

90. *National Parks and Reserves in Australia* (Canberra: Australian Academy of Science, 1968), iii, 16.
91. R. L. Specht, Roe, E. M. and Boughton, V. H., 'Conservation of Major Plant Communities in Australia and Papua New Guinea', *Aust. J. Bot. Suppl. Series*, 7 (1974), 1–667; R. L. Specht, 'The Report and its Recommendations', *A National System of Ecological Reserves in Australia*, Australian Academy of Science Report, no. 19 (Canberra, 1975), 11–21.
92. F. Fenner (ed.), *A National System of Ecological Reserves in Australia* (Canberra: Australian Academy of Science, 1975); F. Fenner, 'A National System of Ecological Reserves in Australia', *Search*, 6 (1975), 108–11.
93. R. L. Specht (ed.), 'Major Plant Communities in Australia: An Objective Assessment', *Aust. J. Bot.* (in press). From 3803 plant communities in 605 ecological surveys plus 907 individual rainforest communities, 378 floristic groups have been recognized.
94. Anon., *A National Conservation Strategy for Australia* (Canberra: Australian Government Publishing Service, 1984). This document details proposals agreed to at a conference in Canberra in June 1983.
95. F. von Mueller, 'Inaugural Address', *Report of the AAAS*, 2 (Melbourne, 1890), 10.

5

The Earth Sciences: Searching for Geological Order

Thomas Vallance and David Branagan

GEOLOGY — A NECESSITY

From unpromising beginnings in a penal outpost at Sydney in 1788, European settlement, in less than a hundred years, spread across the whole Australasian region. The story of this expansion is also a story of growing concern for earth materials, of geological concern. The first settlers had found good building stone and brick clay conveniently situated. Australia's first commercial mineral export, a shipment of coal to Bengal, followed just eleven years later in 1799. As activity diversified with the spread of settlement a need to know the earth continued unabated. The explorer and surveyor Thomas Mitchell put it neatly in 1838, a time when pasture was paramount: 'It is only where trap, or granite, or limestone occur, that the soil is worth possessing, and to this extent every settler is under the necessity of becoming a geologist.'[1]

A century after Governor Phillip's founding venture of 1788, British Crown colonies divided between them the entire extent of Australia, Tasmania and New Zealand. Many had prospered by minerals and in each there was at least a small community whose interest in geology went beyond the purely practical. Indeed by 1888 every one of the colonial governments either had an established geological survey or had recently sponsored investigation by professional geologists. The capital cities, and many larger provincial towns, already possessed museums with geological collections, in some cases curated by appropriately trained staff. Learned societies had emerged in most of the colonies and through their meetings and journals helped spread scientific information. In Sydney, Melbourne, Adelaide, and in New Zealand (notably at Dunedin), universities proclaimed local aspirations to culture. Thus far only the University of Sydney had dignified a teacher with the title professor of geology, but each of the institutions in its own way acknowledged a place for earth sciences. The School of Mines in the

University of Otago, Dunedin, gave a particular focus to that place, one matched in Australia by the schools of mines on the Victorian goldfields. The achievements of a century were impressive, but they were also extraordinarily dispersed, and the efforts of individuals and institutions were still largely uncoordinated.

The century's growth in geological knowledge of the region was no less a product of that dispersal, yet it was an impressive product, if only because of observations in country so recently unknown. Furthermore, it was knowledge derived largely from the work of people trained in Europe, some of them in the period to about 1840 when earth sciences were slowly acquiring (in Europe) characters still regarded as effective. Geological work in the colonies accordingly tended to reflect attitudes and ideas picked up at home: early investigators in Australia brought with them the then-fashionable belief that Europe was the geological model of the world. For them the rocks and strata of Europe, their constituents and arrangement, were sufficient guides to establishing correlation in the remote south land.[2] Coal, for instance, had to occupy in Australia a stratigraphical place equivalent to that of European coal. It was an attractively simple view that held adherents in Australia long after the geologists of Europe had lost faith in so-called universal formations. The age of Australian coal, in fact, became the subject of protracted intercolonial dispute among geologists appealing to different sorts of European authority.[3] By 1888, however, that particular problem had been resolved to the satisfaction of most colonial geologists. The key had come not from Europe but from the work of more independently minded colleagues in India.

COLONIAL CONFIDENCE AND GEOLOGICAL MAPS

Newer recruits to Australasian geology, most of them still trained in European schools, seem to have taken heed of the lesson. With Europe no longer their infallible guide they became readier to trust their own judgement. A once-standard practice of seeking 'higher' opinions from European experts began to wither under local attack. In that regard, the Adelaide palaeontologist Ralph Tate, himself an Australian resident only since 1875, was notably forthright in his warnings of the 'danger' of entrusting diagnostic work to outsiders.[4] Yet difficulty attended this refreshing new sense of confidence. Colonial circumstances determined that scientists, even groups of them, would be widely separated one from another. Intercolonial travel was no common activity for most. Knowledge in a territorial subject like geology therefore continued to grow as distinct gatherings related to particular colonies. Information from further afield had to be gleaned from correspondence or publications, of which the most comprehensive sources were works sponsored by individual colonial governments and related to their territory. R. A. F. Murray's book on Victoria[5] and R. M. Johnston's on Tasmania[6] are representative of this latter genre. Each author strove for some wider perspective but his treatment, perforce, was tied to his patron's domain.

Similar problems attached to those characteristic tokens of geologists' work, geological maps.[7] Production of any but the simplest, most diagrammatic sorts of geological map was beyond the capacity of private bodies in the colonies. Governments had to be involved and they, understandably, were reluctant to trespass. Since the geological survey in Victoria had set an example in the 1850s, other colonies had sponsored geological mapping and issued map sheets covering part or all of their territory. However, the efforts, like the results were uncoordinated. Colours, symbols, even scales, varied as if to thwart matching across colonial boundaries. Yet so many of those boundaries were works of men, not of nature, and increasingly geologists found the non-natural constraints irksome.

James Hector's sketch map of the geology of New Zealand broke new ground. When it appeared, in 1869, that colony still had its separate provincial legislatures. Hector himself had started his geological career in New Zealand as an employee of the province of Otago. He had irritated his provincial masters by removing to Wellington in 1865 to lead geological efforts for the central government; with the map he showed his readiness to override local politics. As it turned out, Hector was anticipating the abolition of provincial governments in New Zealand achieved seven years later. The situation with the colonies of Australia and Tasmania was far different. Certainly, politicians had talked for years about federation, of federation even including New Zealand and other islands in the Pacific Ocean, but the governments in the Australian region were too deeply entrenched even to contemplate their disappearance in any act of union. Australians had no central government, and no Hector, it must be admitted. It is the more impressive, therefore, to note that by 1875 a geological map of Australia and Tasmania in two sheets had been issued by the government of Victoria. Geologists across the country contributed readily to a project they perceived as significant. R. B. Smyth, head of the geological survey in Victoria at the time and compiler of the map, may have been a controversial figure, and no great geologist, but his vision is not in doubt. Within two years a reissue of the map had to be made, and in 1887 a new, larger-scale map, in six sheets, became available, again by courtesy of the colony of Victoria.[8] The welcome extended to these works surely reflected growing interest among Australasian geologists to look beyond their local borders, to establish closer rapport with their neighbours and the geology of their neighbours' colonies.

THE AUSTRALASIAN ASSOCIATION FOR THE ADVANCEMENT OF SCIENCE AND EARTH SCIENCES

The centenary year 1888 arrived, with politicians still talking of federation. New South Wales celebrated the anniversary, almost on its own, and stood aloof from the affairs of the Federal Council established by act of the British parliament in 1885. Colonial science and scientists emerged with far more credit. In 1884 Archibald Liversidge, then professor of chemistry and mineralogy in the University of Sydney, proposed establishment of an association that would bring together the disparate and scattered exponents of science in the Australasian colonies. Not only geologists had been working towards a wider vision; Liversidge's proposal was welcomed by professional and amateur scientists alike. Thanks largely to his enthusiasm, the first congress of the Australasian Association for the Advancement of Science (AAAS) met as a body, more than 800 strong, at Sydney in August 1888. Earth scientists voted with their registrations. Geologists, mining men, and surveyors constituted about one-tenth of the total, and of course, a far greater proportion of members from the professions of science, a fact evident when the congress divided into specialist sections along lines taken long before by the British Association for the Advancement of Science, Liversidge's model for AAAS. Liversidge had gauged aright the spirit of the Australasian scientific community. The meeting ended with a resolution to foregather in 1890 at Melbourne.

From the start the sectional organization of the AAAS and later ANZAAS implied no exclusiveness. It was common for members to be active in various Sections. Indeed the first congress president, H. C. Russell, government astronomer of New South Wales, ranged widely. Geologists might gravitate naturally to their Section (Section C), but they were to be found also with the mathematicians and physicists, the chemists, biologists, geographers, engineers and so forth. The congress presented opportunity for making or enlarging contacts in all branches of science and for learning of relevant studies in other colonies. Members of Section C, most of them from New South Wales and Victoria, heard their president, R. L. Jack, government geologist of Queensland, dilate on salient features of the geology of his colony. That most remarkable of police magistrates, A. W. Howitt, explained to the Section something of his spare-time researches in eastern Victoria on metamorphism, a subject at the time barely considered elsewhere in Australasia. The list of contributors and contributions was impressively diverse. The AAAS congress at Sydney in 1888 established its credentials as the intercolonial forum for science.[9]

COMMITTEES OF INVESTIGATION

The momentum achieved in Sydney was not allowed to fall quite away until revived at Melbourne in 1890. The bulky record of the first congress contains not only addresses and papers but also announcements that various 'committees of investigation', appointed at Sydney, were to report in Melbourne. Fourteen such committees came into being in 1888. They covered the spectrum of congressional interest and, more particularly, focused attention on topics perceived as deserving encouragement or clarification. Not surprisingly, the founding father of

AAAS, Liversidge, urged into being an Australasian Mineral Census Committee to advance knowledge of the subject he had transformed in New South Wales.[10]

With Liversidge as secretary, and representatives (some chemist-mineralogists, some geologist-mineralogists) from eastern mainland Australia and New Zealand, the committee had a considerable collection of data to report in Melbourne. Liversidge co-opted detail from South Australia for Melbourne, but Victorians had to be informed of their representatives' dilatoriness. Having done what it could, the committee ceased to exist in 1890, still lacking census data from Tasmania, Victoria and Western Australia. Two years later, a committee for Tasmania's minerals came into being, a committee convened by the librarian and amateur of science A. J. Taylor and including an even more remarkable amateur, W. F. Petterd, who practically compiled the list published by the Royal Society of Tasmania in 1893. Liversidge's enthusiastic example bore fruit in unusual ways. For instance he was involved from 1891 until 1898 on a committee, under the secretaryship of the New Zealand agricultural chemist George Gray, investigating the composition and properties of the mineral waters of Australasia. Appropriate to his background, Liversidge served as bridge between chemists and geologists in the AAAS but not even he could excite many of his chemical colleagues in Section B (Chemistry and Mineralogy) about minerals. After a few years, the chemists surrendered mineralogy to Section C without a struggle.[11]

GEOPHYSICAL CONSIDERATIONS

Physicists and mathematicians took a more lasting interest in fields relevant to earth sciences. Indeed, through committees prompted by members of Section A (Astronomy, Mathematics, Physics, and Mechanics), the Association effectively became a promoter of geophysical research. A committee to investigate and report on the seismological phenomena of Australasia came into being in 1888, not unexpectedly with a New Zealander as secretary, and he (James Hector) the only geologist in the group. There was no report ready for Melbourne (1890) but the following year at Christchurch came impressive amends in the form of compiled earthquake records for New Zealand and discussion of seismic phenomena there. The group soon became known simply as the Seismological Committee. With changes in personnel that maintained the geological representation among a preponderance of physicists, the committee continued until at least 1921. What seems like a resurrection of seismic interest within the Association in 1939 did not last.[12]

Interest in terrestrial magnetism led in 1898 to a committee to report on a magnetic survey of New Zealand. The Adelaide-born physicist C. C. Farr had convinced Pietro Baracchi, government astronomer of Victoria, of the need for such a survey and he, in turn, arranged for Farr to begin work in Dunedin. Farr thus began his distinguished career in

New Zealand. As secretary, Farr kept the committee going until 1907 when, it seems, increasing academic burdens forced him to stand aside. Meanwhile, however, he had also served on another AAAS committee, established in 1902, to report on progress of studies in terrestrial magnetism throughout Australasia, which had been served initially by his friend Baracchi as secretary.

That committee was still in existence in 1921, but Farr's connections with geologists lasted longer, as shown by his collaboration with Sir Douglas Mawson on problems of terrestrial magnetism in the years after Farr retired as professor in Christchurch in 1935. Magnetism, in fact, also remained a subject of constructive concern to the Association. In 1932 it granted funds towards the cost of a geophysical survey of the meteorite craters at Henbury, central Australia. At the same time, Section A urged the Association to contribute towards purchase of a magnetometer brought to Australia and used here by the pioneering Imperial Geophysical Experimental Survey. After it became clear the instrument would be lost to the country if the Association did not act generously, joint purchase with the Council for Scientific and Industrial Research (CSIR, now CSIRO) was agreed and ANZAAS became a distinctly practical sponsor of geophysical research.[13]

Physicists, mathematicians, and geologists in 1901–2 were also involved in a committee investigating diamond drill bores and measurement of underground temperatures. The subject of gravity, however, seemed to be regarded as more geologists' business. The Committee for the Determination of Gravity in Certain Critical Localities had an exclusively geological membership when set up in 1913. Even after it was renamed the Committee for the Study of Earth Movements by Horizontal Pendulums in 1921 and the physicist E. J. Pigot of the Riverview Observatory (Sydney) was added, the former majority remained, with Pigot in fact balanced by the botanist J. H. Maiden. The secretary at this time, the geologist L. A. Cotton, was himself busy with pendulum work on the Burrinjuck Dam site and the committee, until 1926, brought to attention several studies of wide significance.[14]

GEOLOGICAL CONCERNS

As the Association and its committees built interdisciplinary bridges so also did they provide means for sharing within-discipline information and focusing collective attention on critical issues. The published sources of Australian geology were already extraordinarily dispersed. In the days before the emergence of local societies and government agencies publishing science, newspapers had been used by colonial scientists. By 1888 the old dependence on ephemeral media was not quite extinct. The Australasian Geological Record Committee of AAAS in 1888 set out to establish the current range of sources. Its secretary, the palaeontologist Robert Etheridge, with one of his fellow members, R. L. Jack, had recently prepared a notable bibliography of Australian geology.[15] The record committee sought to continue that effort. In the event came disappointment: reports covering only Queensland and New Zealand were received at Melbourne in 1890. Some remarkably obscure sources add interest to the reports but although the committee was reappointed with a new secretary, it was heard of no more. Perhaps the literature survey then planned by the Geological Society of London and issued from 1894 was thought to remove need for local effort. If so, the 'replacement' was a poor substitute for Australasian geology.

Other formal matters of concern to geologists became subjects for committees. At the Melbourne meeting of 1890, the draughtsman Arthur Everett—the man chiefly responsible for the 1887 geological map of Australia mentioned earlier—read a paper urging 'unification of the colouration of geological charts'. Here was a business of major concern to geologists and it is no surprise to find a committee of heads of colonial surveys, with F. W. Hutton of New Zealand as secretary, appointed to consider unifying colours and symbols for geological maps. Was uniformity even desirable? Opinions presented at the next meeting (Christchurch in 1891) varied so widely from colony to colony there seemed no chance of an agreed recommendation. On that score the representatives behaved more like politicians than scientists.

A more neutral theme became the business from 1892 of the AAAS committee on the photography of geological surveys. As it emerged later, a main purpose was to encourage photography as an adjunct to fieldwork and to organize and catalogue 'geological photographs'. Richard Daintree had set a fine example, first in Victoria and then in Queensland, of how photography could assist survey work. The committee lasted from 1892 to 1904 under the secretaryship variously of the Victorian architect and amateur naturalist J. H. Harvey and the New South Wales geologist J. M. Curran. Its reports were rarely exhaustive, but the notable photographic records of geological sites made at the time, and still extant, bear witness to a useful and productive purpose.[16]

GLACIAL PROBLEMS

At the AAAS Congress in 1888 geologists, at least those from eastern Australia, had reason to think of glacial problems. From India had come the evidence about glaciation in late Palaeozoic time that helped make sense of the coal problem. In 1886 R. D. Oldham, from India, and T. W. Edgeworth David of the New South Wales Geological Survey discovered glacial evidence within the coal-bearing succession of the Hunter River region, a find of great significance in fixing a match with India.[17] Not many decades before, glaciation had been assigned to only Pleistocene time. Now there was increasing evidence glacial conditions had existed at other, earlier, periods of earth history.

The Australasian Glacial Evidence Committee, set up in 1888 with Ralph Tate as secretary, had an exciting problem as its concern. And with it, the excitement lasted. The Association kept its glacial committee, variously constituted of course, until 1946. Edgeworth David followed Tate as secretary, and under his enthusiastic guidance the committee flourished. There can be little doubt this was one of the most active and fruitful of all the Association's committees of investigation. Its canvass was kept broad; its reports cover evidence of glacial action in Australia with ages ranging from Precambrian to Pleistocene. Effort to encourage study of the movement of existing glaciers led to a separate AAAS committee being formed at Christchurch in 1891 for the purpose; sadly, it disappeared without record. However the Geology and Geography Sections of the British Association meeting held in Australia in 1914 combined to recommend systematic recording of glacial phenomena in New Zealand (in conformity with the methods adopted by the Commission Internationale des Glaciers).[18]

Linked to glacial matters was the Cainozoic and Quaternary Climate Committee, which acted through the 1920s. The ice-bound continent of Antarctica, however, proved a longer-lasting attraction. Here was scope for more than glacial studies and from its inception in 1888 the Antarctic Exploration Committee had widely representative membership. By 1911 there were subcommittees in all the colonies under the presidency of the Melbourne chemist D. O. Masson and 'a supreme commander' Douglas Mawson. Therein is a sign of the role that the AAAS and its committee played in support of practical exploration and science.[19]

STRATIGRAPHIC MATTERS

The interest that directed attention to glacial evidences in 1888 had, for the geologists, a particular stratigraphic element. The late Palaeozoic succession, in which Oldham and David found signs of glaciation, continued to present problems. Permo-Carboniferous, the name devised by Etheridge and widely adopted, was recognized as a compromise. In view of the considerable effort expended on these strata from 1888 it is surprising that not until 1911 was an AAAS committee set up to consider questions of classification and nomenclature. W. S. Dun, palaeontologist to the New South Wales Geological Survey, was secretary for some years; his successor, C. A. Sussmilch of the New South Wales Department of Technical Education, long kept affairs moving. The committee indeed only just survived him, disappearing in the 1950s. Problems confronted by the committee clearly were perceived at the outset as peculiarly Australian, but New Zealand representation emerged in 1921, the year in which progress was acknowledged by a change of name to the Carboniferous and Permian Rocks Committee. Additional interest lies in this being the first of the Section C groups to appoint a woman member, Lucy Hoskins (later Lucy Hanrahan) of Western Australia in 1932. And to demonstrate that this was no aberration, she was succeeded (in 1935) by Kathleen Prendergast from the same state.[20]

A clearer New Zealand involvement marked the committee set up in 1935 to consider correlation of the Tertiary rocks of Australia and New Zealand. It continued to function until 1946 and did much to increase rapport between workers on both sides of the Tasman Sea—the name for the water mass between Australia and New Zealand, a name incidentally on maps in no small measure because the AAAS at Christchurch in 1891 resolved to recommend it as an appropriate one for this part of the Pacific Ocean. The Tertiary Committee always had a strong representation of palaeontologists in its membership. They were active here and also on the Committee for Palaeontological Correlation of Australian and New Zealand Formations that existed from 1923 to 1928. In contrast to the longer-lasting challenges of correlation, the more specialized Present State of Knowledge of Australian Palaeontology Committee, set up in 1890, seems never to have produced a report, and it disappeared without sign of action. A committee formed in 1901 to consider nomenclature of geological formations had a matching currency, but its franchise was revived in 1946, when a new Stratigraphical Nomenclature Committee was appointed with M. F. Glaessner of Adelaide as secretary. Work in that line continues but has passed from the patronage of the Association.[21]

LANDFORMS AND TECTONISM

Another enterprise that reached fruition under other auspices was begun in 1902 with the AAAS Committee for Recording Structural Features. Its original title explained one aim as elucidation of the evolution of the Australasian land surface. The double-barrelled theme—subsurface structures and superficial characters—matched the intellectual breadth of the founding secretary T. W. E. David. However, for reasons that remain unclear, by 1913 it was working in parallel with a separate group, the Physiographic Features of Australia Committee, run by E. C. Andrews, but with David a member. When the two committees united in 1921 with representatives of both the Geography and Geology Sections, it was described as a new group for the Investigation of Structural Features and Land Forms in Australia and New Zealand. Joint secretaryship reflected the involvement of the two Sections, but this new committee seems only to have restored David's vision of 1902. The committee survived until 1946, with what seems an increasing bent towards landforms. Were structural geologists and physical geographers beginning to take separate ways? David's death in 1934 certainly removed a leader who had long managed to maintain a unifying role. At any rate, in 1946 the geologists proclaimed a particular objective to prepare a tectonic map. However this venture by the Section C committee was for a tectonic map of Australia only. New Zealand had gone its own way through the local geological survey. The Australian venture passed from ANZAAS to the Geological Society of Australia in the 1950s.[22]

Government patronage earlier overwhelmed the Section C committee on the occurrence of artesian water, a highly relevant theme for Australia acknowledged by the AAAS in 1921. As the secretary, L. K. Ward of

South Australia, reported in 1923, he had by then received nothing from members and hinted that the government-sponsored Inter-State Conference on Artesian Water of 1921 in Adelaide had stolen the committee's thunder.[23]

IGNEOUS AND METAMORPHIC ROCKS

With its committees of investigation, the Association reflected fashions of research and helped foster them. By the early years of the twentieth century an increasing number of specialists had followed the lead set by A. W. Howitt in developing microscopical petrography. Igneous rocks had first gained the detailed attention of petrographers overseas and systems of classification and nomenclature were proliferating. The first moves within the AAAS to promote petrographic work related to the problems raised by this activity. The committee, appointed in 1902 with secretaries from Western Australia, New South Wales, and Tasmania and with Howitt a member, sought to recommend a uniform system for the nomenclature of igneous rocks. It continued with that aim until 1907, but, it must be admitted, achieved no impressive success.

In 1909 attention moved from systematics to a particular group of igneous rocks then attracting attention in Australasia: those marked by especial richness in alkalis. The inaugural secretary, H. I. Jensen, was then investigating tertiary alkaline volcanic associations in eastern Australia. Patrick Marshall, a noted New Zealand petrologist, was a member of this committee and remained associated with it till its demise in 1928. In 1935 igneous rocks again came within the scope of a Section C committee. But this time the aim was to collate and disseminate news of current petrological research rather than records of petrographic character. By means of this committee, workers gained knowledge of current studies, in particular of overseas work published in little-known or inaccessible journals. The cyclostyled summary of Wager and Deer's study of the Skaergaard intrusion in Greenland, prepared for the committee by A. B. Edwards of Melbourne, is but one example of the practical contribution made by this committee.

Formal consideration of research on metamorphic rocks came relatively late on the ANZAAS scene. The committee set up in 1926, with W. R. Browne of Sydney as secretary, was to consider correlation of the metamorphic rocks of Australasia. Consideration of metamorphic materials, as distinct from attempts at crude 'stratigraphical' matching, became a significant aim with reorganization of the committee in 1937 as the Metamorphic Rocks Committee. Much credit for the redirection is due to Germaine Joplin who served as secretary. Further refinements followed. By 1939 we find a Metamorphism and Assimilation Committee, which led in part to the Current Petrological Committee of 1946, the group that brought together the igneous and metamorphic petrologists.[24]

EXCURSIONS

The British Association, begun at York in 1831 and Liversidge's model for the AAAS, itself owed much to a German organization, the Deutscher Naturforscher Versammlung.[25] Excursions, scientific and more broadly cultural, had been a popular part of its programme since the first meeting, at Leipzig in 1822. As the British body had found them equally acceptable to members, it was hardly surprising that the first AAAS programme also provided for excursions. The Sydney meeting of 1888, in fact, offered a range of visits to places of interest, some local, some more distant and requiring excursions up to four days in duration. Although no particular sectional identity was given, the number of geologists among party leaders must have expressed organizers' confidence that earth scientists would respond supportively to field visits. Thereby those from afar could examine critical occurrences under expert guidance and at the same time strengthen personal relations with colleagues in the informality of the field.

The confidence, *vis-à-vis* geologists, was not misplaced. By the 1900 Congress in Melbourne, specific geological excursions were provided; thereafter sectional excursions became a recurring feature of congress programmes. Preparation of excursion handbooks, furthermore, gave welcome opportunity for collecting up-to-date information. In time the activity was extended to the useful volumes of local scientific conspectus issued to all congress members.[26] Through reports of the excursions we see also how advances in transport technology were turned to service. Geological excursions from Melbourne in 1913 travelled in 'motor cars provided by members of the Automobile Club of Victoria'. By the 1960s there were camping excursions using four-wheel-drive vehicles, and even occasional forays by plane.

The AAAS and then ANZAAS served geological science splendidly with its excursions. Yet the past tense seems increasingly necessary. The impetus once given by the Association has been transferred as geologists form other allegiances of more focused, and national, character.[27]

FEDERATION, THE ASSOCIATION, AND GEOLOGY

The Australian colonies became states of a Commonwealth of Australia on 1 January 1901. New Zealand's terms for joining the federation had proved unacceptable. The two nations would go their separate ways: New Zealand with one central government, Australia with both central and regional (state) legislatures. Geological surveys in the new order retained basically their old attachments. Science, indeed, in the early years of the Commonwealth had little national role in Australia and Federation at first imposed no marked change on the AAAS. Not until twelve years after parliament first met (1927) in the new national capital did the Association convene in Canberra. The delayed involvement stands in contrast with the Australian meetings in 1914 of the British Association, held at the invitation of the Commonwealth government and generously underwritten by grants of its funds.

Commonwealth sponsorship of geological effort was for long limited to territories specially ceded to its charge. To have exceeded those bounds would invite accusations of intrusion on states' rights, or so it was argued. Thus when, in 1912, the Australasian Institute of Mining Engineers (later, the Australasian Institute of Mining and Metallurgy) forwarded to the Minister of Home Affairs a proposal that a federal geodetic, topographical, and geological survey be established, the response was inaction. Others, however—notably Edgeworth David and H. C. Richards—in the 1920s took up the cause of a federal geological survey. Still reluctant to act, the Commonwealth in 1927 appointed W. G. Woolnough as geological advisor to the Department of the Interior and by its Petroleum Prospecting Act of 1926 began to promote oil search in the states as well as territories. The 1920s saw also the beginnings of mineral investigations on an Australia-wide basis by the new Commonwealth Council for Scientific and Industrial Research (later CSIRO). That Council indeed would soon sponsor publication of David's geological map of the Commonwealth and its explanatory notes. Not until 1946, however, did the Australian government take the plunge and set up what amounted to a federal geological survey, the Bureau of Mineral Resources, Geology and Geophysics.[28]

With resources far beyond those available to state bodies, the Bureau assumed the leading role in regional geological effort in Australia and became our most prolific publisher of geological maps and reports. Not surprisingly, too, it became a leading employer of earth scientists. Its growth undoubtedly strengthened the Australian geological community, already enlarged through the 1930s as mining companies came to accept a role for geologists in their mineral exploration programmes.[29] With that growth came also thoughts of more specific alliance. Mining engineers long before had formed their own institute as a professional body. There had indeed been a Geological Society of Australasia represented at the AAAS in 1888, but it was a small and scattered group that did not survive the removal overseas of an enthusiastic president.[30] Geologists were still linked to ANZAAS in 1951 when, during the Brisbane Congress that year, they adopted a draft constitution for a new Geological Society of Australia with E. S. Hills of Melbourne its first president. The Society continued to meet during ANZAAS congresses until 1957, when the Association went to New Zealand, whereas the Society—being an Australian body with no trans-Tasman links—met in Sydney. New Zealand geologists had already begun going their way with establishment of the Geological Society of New Zealand in May 1955.[31]

Formation of the two specialist societies on either side of the Tasman Sea, and their growing support by geologists at the expense of allegiance to ANZAAS, may be seen as an unfortunate breaking of old bonds— bonds made and maintained for so long by the Association to the advantage of Australian and New Zealand scientists. Nevertheless, it must be admitted that scientists are now less isolated and the distances that separate them less tyrannical. Their greater mobility may also enable them to find rapport, as needed, with colleagues in other sciences.

But then even while the AAAS and then ANZAAS committees performed their important tasks, it is noticeable how much less interdisciplinary these groups became as the years passed. Attachment to particular Sections may have fostered the confinement, but was it in some measure also a response to burgeoning information within the section sciences? Whatever the reasons, it is noteworthy that many initiatives once fostered by the Association have been taken up by the geological societies. Within the Geological society of Australia, at least, the debt goes further—its very organization of federal secretariat and state divisions is hardly original and the nationwide specialist groups have some identity of purpose with the old committees the of AAAS and ANZAAS.[32] As if to extend the match, the Society also holds regular geological conventions in various centres across the country.

In some cases, for instance the Tectonic Map of Australia and the ordering of stratigraphical nomenclature, the Geological Society of Australia took over tasks once explicitly sponsored by ANZAAS, and with necessary assistance from the Bureau of Mineral Resources has brought them to fruition. In other directions, most notably by its publications, the Society has taken innovative leads. Its journal, published since 1953, has established an international reputation. Now known as the *Australian Journal of Earth Sciences*, it enlarges Australian science at no public expense, unlike the similarly titled journals for other natural sciences published by the CSIRO.[33] In 1976 the Society was a major sponsor of the twenty-fifth International Geological Congress at Sydney, the first such to be held in Australasia. Ten years later it played a similar role for the twelfth International Sedimentological Congress in Canberra.

Up to 1970 more than one-third of the presidents of ANZAAS came from the geological ranks. There has been none since. Likewise the purely geological content in the conferences has declined. This is to be expected as geologists direct much of their energies to specifically geological matters of both theoretical and practical concern to their profession.

The response from those geologists still involved in ANZAAS Section 3 (formerly C) has been to direct their attention to broader social and political issues with a significant earth science content, such as the mining of uranium, the disposal of radioactive waste, national parks, resources and the environment, and urban development. This has brought earth scientists into contact with other scientists and technicians and has been a very fruitful exercise. Plans for forthcoming conferences and the nature of material published in *Search* indicate that this new direction will continue.

Despite the ANZAAS interest in social and political issues in recent years, however, other umbrella organizations, such as the Australia Geoscience Council and ASTEC have arisen to deal largely with issues of science policy. Such matters are dealt with elsewhere in this volume.

CONCLUSION

In the AAAS and later ANZAAS, the earth scientists of Australia and New Zealand first found an effective communal forum. The Association helped them follow nature, and students of nature, across the boundaries of distance and politics. Its working committees gathered information and focused interests in research. Its meetings and excursions increased friendly, constructive rapport between individual scientists.

Geological science in Australia owes the AAAS and ANZAAS a considerable debt. It is a debt not much admitted nowadays when geologists look to their own specialist societies unaware of what went before. These flourishing societies are in significant ways outgrowths of political as well as scientific developments, and the Association in its earlier days kept pretty much to science, at least in the geological line. Ironically, since the defection of so many professional geologists from its ranks in recent years, earth science themes within ANZAAS have become more directed to the environment and conservation, public issues in which geological content is inextricably interwoven with that of other sciences and non-sciences, and there has been a renewed recognition of the unity of science. Geologists should view quite happily the Association's present efforts to introduce science to Australian youth. Their specialist societies, as yet, have made only sporadic effort to foster this worthwhile purpose. May the Association that did so much for geological science be encouraged by geologists to succeed in increasing awareness of the earth among the youth of Australia and New Zealand.[34]

NOTES

The several versions of this manuscript were typed by Anne Cook and Helen Young.

1. T. L. Mitchell, *Three Expeditions into the Interior of Australia*, 2 vols (London: T & W Boone, 1839), II, 321.
2. T. G. Vallance, 'Origins of Australian Geology', *Proc. Linn. Soc. NSW*, 100 (1975), 13–43.
3. T. G. Vallance, 'The Fuss about Coal' in D. J. and S. G. M. Carr (eds), *Plants and Man in Australia* (Sydney: Academic Press, 1981), 136–76.
4. R. Tate, 'Century of Geological Progress', Presidential Address, *Report of the AAAS*, 5 (Adelaide, 1893), 1–100. See also T. G. Vallance, 'Pioneers & Leaders—a Record of Australian Palaeontology in the 19th Century', *Alcheringa*, 2 (1978), 248.
5. R. A. F. Murray, *Victoria Geology and Physical Geography*, (Melbourne: Government Printer, 1887). Murray was president of Section C at the third meeting of the AAAS (Christchurch, 1891), delivering an address on 'The Past and Future of Mining in Victoria', *Report of the AAAS*, 3 (Christchurch, 1891), 119–27.

144   PART II   *Charting the Sciences*

6. R. M. Johnston, *Systematic Account of the Geology of Tasmania*, (Hobart: Government Printer, 1888). Robert McKenzie Johnston was typical of the devoted 'amateurs' who made a substantial contribution to the AAAS and to Australian geology. Despite his official position for more than thirty years as government statistician in Hobart, he published extensively on geology. His paper entitled 'How Far can Australian Geologists Safely Rely upon the Order of Succession of the Characteristic Genera of Fossil Plants of a far Distant Region, in the Determination of the Order and Relationship of Australian Terrestrial Formations?', *Report of the AAAS*, 1 (Sydney, 1888), 302–11, typifies the search for an Australian geological order inherent in the title of this chapter.
7. D. F. Branagan, 'The History of Geological Mapping in Australia' in D. H. Borchardt (ed.), *Some Sources for the History of Australian Science*, Historical Bibliography Monograph No. 12, (Sydney: University of New South Wales, 1984), 33–46.
8. T. A. Darragh, 'The First Geological Maps of the Continent of Australia', *J. Geol. Soc. Aust.*, 24 (1977), 279–305.
9. In his address Jack called for help, in 'unravelling the structure of Queensland' at 'meetings such as this of those who are . . . not incurious in God's handiwork', *Report of the AAAS*, 1 (Sydney, 1888), 196–206. Howitt's paper is entitled 'Notes on the Metamorphic Rocks of the Omeo District, Gippsland,' ibid., 206–22; see also T. G. Vallance, 'Achievement in Isolation: A. W. Howitt, Pioneering Investigator of Metamorphism in Australia', *Earth Sciences History*, 5 (1986), 39–49.
10. Of the original fourteen committees three were specifically geological. In addition to the Mineral Census (No. 7), they were Australasian Glacial Evidence (No. 8) and Australasian Geological Record (No. 13). The Australasian Seismological Committee (No. 10) came originally from Section A. For more discussion of Liversidge and the Mineral Census Committee see T. G. Vallance, 'Sydney Earth and After: Mineralogy of Colonial Australia, 1788–1900', *Proc. Linn. Soc. NSW*, 108 (1986), 149–81, esp. 172–4.
11. This followed a similar move in Britain. Section B of the British Association was entitled 'Chemistry and Mineralogy' from 1835 to 1894. However the position of mineralogy in the AAAS wavered during the early years. Originally Section B was 'Chemistry and Mineralogy', but for the 1895 Brisbane meeting, Section C is shown as 'Geology and Mineralogy', *Report of the AAAS*, 6 (Brisbane, 1895), 58. At the 1902 meeting the name of Section B was changed to 'Chemistry, Metallurgy and Mineralogy', *Report of the AAAS*, 9 (Hobart, 1902), lxi. In 1913, Section B became 'Chemistry' and Section C 'Geology and Mineralogy', *Report of the AAAS*, 14 (Melbourne, 1913), vii.
12. Seismological Research Committee Report, *Report of ANZAAS*, 24 (Canberra, 1939), 367.
13. *Report of ANZAAS*, 21 (Sydney, 1932), xxxi.
14. *Report of the AAAS*, 18 (Perth, 1926), 87.
15. R. Etheridge and R. L. Jack, *Catalogue of Works . . . on the Geology . . . of the Australian Continent and Tasmania* (London: Stanford, 1881).
16. In the first *Report of the AAAS*, 1 (Sydney, 1888), there are photographic reproductions of biological subjects (plates XIX, XX, XXI). The first use of photographs in a geological context is in reference to Palaeozoic glaciation: T. W. Edgeworth David 'Evidence of Glacial Action in Australia', Presidential Address, Section C, *Report of the AAAS*, 6 (Brisbane, 1895), 58–98, and in a paper by G. Officer, L. Balfour and E. G.

Hogg, 'The Glacial Geology of Coimadai', ibid., 323–30. This follows a paper by Officer and Balfour, 'The Glacial Deposits of Bacchus Marsh', ibid., 321–3, in which the authors defend themselves against the accusation by G. Sweet and C. C. Brittlebank of having 'trespassed on [a] prior claim to the working of the district'. The so-called proprietary rights of individuals to certain field areas has been a long-standing problem for geologists and contrasts to some degree with the comments by Jack, op. cit., note 9.
17. See Vallance, op. cit., note 3.
18. *Report of the BAAS*, 84 (Australia, 1914), lxv.
19. The Antarctic Committee grew from seven members in 1888 to twenty-seven in 1911. *Report of the AAAS*, 13 (Sydney, 1911), xlix. In addition, a Macquarie Island Committee was operating in 1913 with twenty-five members. *Report of the AAAS*, 14 (Melbourne, 1913), 72.
20. There were almost no women professionally engaged in geological work before the 1920s, but geology was popular as a course of study for women and many pursued it as a enjoyable hobby. See D. F. Branagan and H. G. Holland, *Ever Reaping Something New* (Sydney: University of Sydney, 1985).
21. *Report of the ANZAAS*, 26 (Perth, 1947), 28. The pursuit of stratigraphic order since the 1950s has resulted in the establishment of various committees such as that to formalize the naming of rock units within the New South Wales Coal Measures, and a central register in Canberra to record formal names and avoid duplication.
22. The Tectonic Map Committee's structure changed in 1947 when its members were specified as the professors of geology of each state university (or their nominees), the directors of each state Geological Survey (or their nominees), and the director of the Commonwealth Bureau of Mineral Resources (or his nominee) with the power to co-opt. G. D. Osborne of the University of Sydney was secretary. *Report of ANZAAS*, 26 (Perth, 1947), 28. The Geological Society in association with the Commonwealth Bureau of Mineral Resources published its first tectonic map of Australia in 1962 at a scale of $1:2\,500\,000$. This is, strictly speaking, only a stratigraphic-structural map. A second tectonic map of Australia (scale $1:5\,000\,000$), truer to title, was published by the Society in 1971. See Branagan, op. cit., note 7, and Branagan and Townley, op. cit., note 28, for further discussion.
23. While organizational matters concerning research on groundwater were dealt with outside ANZAAS, the Association was the forum for many papers on the topic and the source of limited funding for research. The British Association meeting in Australia in 1915 was the scene of a continuing battle between E. F. Pittman, government geologist of New South Wales and J. W. Gregory, formerly professor of geology at Melbourne. See D. F. Branagan and E. Linn, 'J. W. Gregory, Traveller in the Dead Heart', *Historical Records of Australian Science*, 6 (1984), 71–84.
24. Current Petrological Research, *Committee Report of ANZAAS*, 25 (Adelaide, 1946), xlvii.
25. O. J. R. Howarth, *The British Association for the Advancement of Science: A Retrospect 1831–1931* (London: British Association, 1931), 6–12.
26. The excursion handbooks and the conference handbooks have taken various forms and continue to be useful sources of information. The thirty-sixth congress handbook, A. P. Elkin (ed.), *A Goodly Heritage* (Sydney: ANZAAS, 1962) celebrated the seventy-fifth birthday of ANZAAS. It includes an interesting historical review and a seminal paper by G. H. Packham, 'An Outline of the Geology of New South Wales'. H. L. White,

Canberra, *A Nation's Capital* (Canberra: ANZAAS, 1954), prepared for the thirtieth meeting, is an important reference. Prepared for the ANZAAS meeting in Port Moresby in 1970 was R. G. Ward and D. A. M. Lea, *An Atlas of Papua and New Guinea* (University of Papua and New Guinea, Department of Geography, and Collins Longman, 1970).

27. The Geological Society of Australia and the Geological Society of New Zealand have extended the publication of field guides in the last twenty years. The excursion programme for ANZAAS still continues, and, because of the numbers of geologists in the country, it attracts a viable proportion to meetings.

28. The topic of a Federal Geological Survey was certainly discussed at the AAAS meetings. See, for instance, Resolutions, *Report of the AAAS*, 16 (Wellington, 1923), xli, and D. F. Branagan, 'Putting Geology on the Map', *Historical Records of Australian Science*, 5 (1981), 30–57. The growth of applied geology and geophysics through the 1930s is itself another large story. For further details see D. F. Branagan and K. A. Townley, 'The Geological Sciences in Australia—a Brief Historical Review', *Earth Science Reviews*, 12 (1976), 323–46; R. K. Johns, *History and Role of Government Geological Surveys in Australia* (Adelaide: Government Printer 1976); B. W. Butcher, 'Science and the Imperial Vision: The Imperial Geophysical Experimental Survey, 1928–1930', *Historical Records of Australian Science*, 6 (1984), 31–43.

29. See for instance D. Reid, *They Searched: A History of Exploration Within the Western Mining Corporation*, Part 1 (Adelaide: Western Mining Corporation 1981.)

30. D. F. Branagan, 'The Geological Society of Australasia,1885–1907', *J. Geol. Soc. Aust.*, 23 (1976), 169–82.

31. The Geological Society of Australia did not, however, sever its links with ANZAAS at this time. ANZAAS in Melbourne (1967) and Brisbane (1971) incorporated Geological Society activities. *Programme of the ANZAAS Congress*, 43 (Brisbane, 1971), 50, 56.

32. The Geological Society's specialist groups are Sedimentology, Tectonics and Structural Geology, Coal Geology, Solid Earth Geophysics, Earth Sciences History, Geochemistry Mineralogy and Petrology, Economic Geology, Engineering Geology, and Palaeontology.

33. Although new avenues of publication have emerged for geologists since the 1950s, ANZAAS, through the medium of the *Australian Journal of Science* and its successor *Search*, continues to provide a convenient Australasian (but perhaps largely Australian) medium for the rapid publication of short notes on many earth science topics.

34. A card index on the committees of the AAAS and ANZAAS and relevant geological papers was compiled. Assistance in this research by A. R. Norman, M. P. Branagan, and L. F. Branagan is gratefully acknowledged.

6

The Physical Sciences: String, Sealing Wax and Self-Sufficiency

R. W. Home

When the Australasian Association for the Advancement of Science (AAAS) met for the first time in Sydney in August 1888, the sciences of astronomy, mathematics, physics, and mechanics were grouped together in pride of place in Section A. In part this merely reflected a pattern long since established in the British Association for the Advancement of Science, which in this as in much else served as a model for the new organization. The arrangement was also, however, an indication of the pre-eminence of the astronomical observatories among colonial scientific institutions. So, too, was the appointment of H. C. Russell, director of the Sydney Observatory, as president of the whole congress. The observatories were to provide much of the leadership of Section A and of the Association for many years thereafter.

It is not known what proportion of the 820 people who registered for the Sydney Congress attended the meetings of Section A, because the list of registrants does not indicate sectional affiliation. In all probability, however, the numbers were relatively small. Papers on mathematical topics have never attracted large 'lay' audiences, and physics, too, had by this time become the province of the specialist. Certainly the meeting was dominated by the professionals. R. L. J. Ellery, director of the Melbourne Observatory, was president of the Section and in his presidential address offered a comprehensive survey of 'The present position of astronomical knowledge'. Russell reviewed the history of astronomical observing in New South Wales before the establishment of the new Sydney Observatory in 1858. Richard Threlfall, professor of physics at the University of Sydney, was the Section's secretary and responsible for arranging the programme. He also read three papers, one describing his splendid, recently completed laboratory (a guided tour of which was another feature of the programme), the other two reporting some early achievements of the ambitious programme of electrical research on which he had embarked as soon as he arrived in Sydney. William Sutherland from Melbourne presented two sophisticated papers on molecular theory that would have severely taxed the understanding of any lay members of the audience. Sutherland, though never regularly employed as a physicist, was already a highly respected contributor to kinetic and molecular theory, and he published regularly in the leading international journals.[1] The programme was rounded out by an inconsequential discussion of rock magnetism prompted by an inquiry from England, and a paper by N. A. Graydon, a surveyor from Melbourne, who proposed an aetherial mechanism to explain the variation of barometric pressure and wind. This last was the only paper presented by someone who was not an expert in the field, and it was evidently something of an embarrassment, being the only one presented to Section A that year that was not published in the congress proceedings.

The practising scientists who attended the 1888 Congress were fully conscious of its significance to their work. As Russell noted in his presidential address, 'want of numbers' had long been the bane of colonial scientific activity. This, he said, had been the great weakness of Australia's earliest scientific society, the Philosophical Society of Australasia, when it was formed in Sydney in 1821, and:

unfortunately in many branches of science the same difficulty exists today, and makes it always difficult, and in some sciences impossible, to keep alive societies for their promotion only. Even in those societies which include a number of subjects ... it is often difficult to find enough original work, and perhaps I ought to say enthusiasm, to keep a healthy amount of vitality at its meetings. I am sure there are many here present ... who have often wished, as I have, for something ... which would bring together a greater number of workers in each branch of science, so that ideas might be interchanged, and a little healthy emulation aroused.[2]

Though Russell did not say so, the problem was particularly acute in the physico-mathematical sciences. Here, in contrast to the more popular field sciences, even in Melbourne and Sydney the numbers engaged were miniscule, while elsewhere one found, at most, isolated individual workers. Where they existed, the observatories reigned supreme. Ellery and Russell were the only representatives of these sciences among the élite group of Fellows of the Royal Society of London resident in Australia at this time. They were joined by Charles Todd, director of the Adelaide Observatory, in 1889. All three men were active in local scientific affairs. Ellery was president of the Royal Society of Victoria for nineteen years (1866–85) and contributed regularly to its proceedings. Russell contributed no fewer than sixty-nine papers to the *Journal and Proceedings* of the Royal Society of New South Wales and served as the Society's president on three separate occasions. Todd held various offices in the Royal Society of South Australia, including the presidency, and also contributed to its proceedings.

The work of the three observatories (as of a fourth established in Perth in 1896) was by no means confined to astronomical investigation. On the contrary, they performed several functions of immediate importance to the daily life of their respective colonies. They were responsible for providing a local time service, for calibrating navigational, surveying and other instruments, and for maintaining such standards of weights and measures as were held in the colonies. They provided base-point determinations for the colonial survey departments, and they co-ordinated chains of meteorological observing stations scattered across the landscape and linked to the observatories by the ever-expanding colonial telegraph networks. Astronomical research for its own sake loomed largest at Melbourne, where the 48-inch reflector known as the Great Melbourne Telescope, the largest telescope in the world at the time it was installed in 1868, was the major item of equipment. In addition, the Melbourne Observatory maintained a long-running programme (which still continues) of recording the various elements of the Earth's magnetic field. At its peak, the Observatory's staff numbered a round dozen. Neither the Sydney nor the Adelaide Observatory was equipped or staffed on anything like the same scale, but both maintained significant observing programmes in addition to their day-to-day 'service' functions.

The commitment to systematic meteorological work had given rise to some of the earliest examples of regular intercolonial scientific exchanges. In Australia as elsewhere, meteorology had been transformed in the 1850s and 1860s, as the advent of the telegraph had at last made it feasible for meteorologists, previously largely confined to studying data recorded at single observing stations, to collect and analyse data recorded simultaneously at a widely distributed set of stations. Todd had been a pioneer of telegraphy in Australia and was well placed to exploit for meteorological purposes the South Australian telegraph network over which he presided. With the completion of telegraph links to the other colonies, still larger possibilities had opened up, and in the 1870s the three observatory directors had initiated a system of Australian weather telegraphy that eventually had embraced all the Australasian colonies and enabled Russell, and later Ellery as well, to begin publishing daily synoptic weather charts covering the whole south-eastern sector of the Australian continent.

Inevitably, these developments had led to a need for closer consultation. In 1879 Russell had seized the occasion of the International Exhibition in Sydney to convene an Australian Meteorological Conference, which was attended by Ellery, Todd, and New Zealand's James Hector as well as Russell himself. Two years later, Ellery had hosted a second such conference during Melbourne's great International Exhibition. A third, to be held in Melbourne in September 1888, would be attended by representatives of every Australian colony and New Zealand, and would finally establish a system of daily exchange of weather telegrams covering the entire region.[3]

At this period, mathematics and physics everywhere remained sciences confined almost exclusively to the universities. In the Australasian region, however, such universities as existed were tiny institutions entirely committed to undergraduate teaching and without research ambitions of any kind. In each case, the first group of professors appointed had included a professor of mathematics, almost inevitably a Cambridge graduate—John Shand, a graduate of the University of Aberdeen appointed to the chair in Otago, was the only exception—and usually of considerable distinction. All the mathematics professors were expected to teach some 'natural philosophy' as well, that is, some of the more mathematical aspects of what we should today call physics (for example, dynamics, hydrodynamics, Newtonian planetary theory, and geometrical optics). Lectures on such topics would normally be accompanied by experimental demonstrations performed by the lecturer, but the students were themselves seldom expected to perform laboratory exercises. (Horace Lamb in Adelaide seems to have been alone in requiring his upper-level students to do practical work in the laboratory.[4]) At the University of Sydney, another foundation professor, John Smith, had been responsible for 'chemistry and experimental physics', under which rubric he had offered, every second year, a course of first-year lectures, decorated with experimental demonstrations, on traditionally non-mathematical and experimentally based branches of physics, such as heat, light, sound, electricity, and magnetism. A. W. Bickerton at Christchurch and F. D. Brown at Auckland were, like Smith, professors of chemistry who were expected also to teach some experimental physics. Such research as had been done in these infant establishments was regarded as a professorial hobby rather than as part of the normal function of a university. Among the early mathematics and physics professors, only Lamb had indulged himself in this way, his ten years in Adelaide (1875–85) having been marked by the publication of an impressive series of mathematical papers and also the first edition of his famous treatise on hydrodynamics.[5]

The 1880s had seen a great expansion in the commitment of the universities in Sydney and Melbourne to science. Separate science degrees had been created and several new science professors appointed, including in each case a new professor of physics (or 'natural philosophy', as it was styled at Melbourne).[6] Melbourne had moved first, appointing a local candidate, H. M. Andrew, to its new physics chair in 1881 and a Cambridge graduate, E. F. J. Love, in 1887 as his assistant lecturer. Andrew, however, had died unexpectedly en route to England on leave early in 1888, and he had not yet been replaced at the time of the initial AAAS meeting a few months later. In Sydney, Smith's death in 1885 had enabled the university to rearrange its teaching duties in this area and to appoint a new professor, Threlfall, to take responsibility for all physics teaching in the university. Threlfall had arrived in Sydney in mid-1886 and been joined a year later by an assistant lecturer, J. F. Adair.[7] Meanwhile in Adelaide, Lamb had resigned and been replaced early in 1886 as professor of mathematics and physics by the young W. H. Bragg. By 1888 Bragg, too, had acquired an assistant lecturer, R. W. Chapman.[8] In New Zealand, expansion had occurred only at the University of Otago, where the chair of mathematics and natural philosophy had been divided into two separate chairs in 1886. Shand had retained the chair of natural philosophy, and a new professor of mathematics, F. B. Gibbons, had been appointed.

To the young W. H. Bragg, attending from Adelaide, the 1888 Congress provided an opportunity not to be missed to meet senior colleagues and exchange ideas with them. Enthused by the occasion, he abandoned his earlier plan to return to Adelaide before the end of the conference: 'I find it such a great advantage to meet these men and for a time to live in the atmosphere they create, that I do not think it right to miss the chances given me.'[9] The meeting came, indeed, as a revelation to Bragg as it must have done to others, opening his mind to the possibility of doing original work in physics even in this distant quarter of the globe. At subsequent meetings Bragg took his first tentative steps on the wider scientific stage with presentations of his own. Others, too, including (from the early 1890s) a trickle of research students from the larger departments, profited similarly from the new Association.

The pattern set for Section A at the initial meeting, reflecting as it inevitably did certain general features of the physico-mathematical sciences in Australia at the time, was generally maintained at subsequent congresses up to the First World War. Meetings continued to be dominated by senior staff of the observatories and university physics departments. (The senior mathematics professors, T. T. Gurney from Sydney and E. J. Nanson from Melbourne, were conspicuous by their absence, although at most meetings there were some mathematical papers on the programme.) The number of speakers was generally small and the better-known people were called upon, year after year, for contributions. Multiple presentations by some speakers were occasionally found necessary in order to fill out a programme. With the exception of a few surveyors, almost no professionally employed scientists contributed to the Section who were not attached to one of the major observatories or one of the universities. Schoolteaching aside, there simply were no jobs in the colonies for physicists or mathematicians, outside these institutions. Moreover—the contributions of William Sutherland always excepted—papers presented by other than professionally employed scientists were almost never published, their titles only being recorded. In other words, though oral presentations by non-specialists were tolerated, especially in the smaller centres such as Brisbane and Hobart where it was otherwise hard to fill the programme, in these fields even the local scientific literature was in effect already closed to the amateur.

This is not to say that the other papers presented were narrowly focused professional research reports. Many papers, and especially the Section's presidential addresses, provided broad surveys of particular fields. The best of these, such as Threlfall's 1890 presentation on electromagnetism[10] or Bragg's in 1904 on radioactivity,[11] brought their audiences in contact with the leading edge of current thinking. Other contributions grew out of the activities of the various specialist committees of investigation that to some degree maintained the spirit of intercolonial scientific co-operation between congresses. For example, the first congress saw the establishment of a long-lived seismological committee, which found particular support in geologically active New Zealand and regularly gave rise to papers presented at the sectional meetings. A committee on the tides of Australia, formed at the 1890 Congress in Melbourne and co-ordinated by the enthusiastic Adelaide harbourmaster Captain Alexander Inglis, likewise generated several papers. By contrast, the meteorological committee set up in 1888 seems never to have met, presumably because the co-ordinating role envisaged for it was rendered redundant by the success of the Intercolonial Meteorological Conference that convened in Melbourne within days of the Sydney meeting. Later, forming a specialist committee became a well recognized means of bringing pressure to bear on governments to support favourite projects. Thus the seventh Congress, held at Sydney in 1898, saw the formation of a committee 'to take such steps as are necessary' to persuade the New Zealand government—successfully, as it turned out—to undertake a magnetic survey of its territories. Similarly, at the Brisbane Congress in 1909, Geoffrey Duffield orchestrated the creation of a committee to press, less successfully, for the establishment of a 'solar physics observatory' in Australia.[12]

In the early years, few of the papers presented to Section A reported the results of laboratory-based research. Initially, only Threlfall in Sydney had a physical laboratory worthy of the name. By the early 1890s, Lyle in Melbourne also had a functioning laboratory, and it was from here, shortly afterwards, that the first research students came to present reports on their work,[13] a practice that was later to become a feature of the AAAS meetings. By the first years of the new century, Bragg, too, in Adelaide, had a reasonably satisfactory laboratory in which to work.

In the 1890s both Threlfall and Lyle established significant research programmes in the field of electricity and magnetism, which led to a steady flow of publications in the leading British journals as well as to reports at AAAS congresses.[14] This was the glamour area of physical research at the time, but was supplanted soon afterwards by the exciting new discoveries of radioactivity and X-rays, and the new insights of relativity and quantum theory. All three Australian physics professors were quick to repeat Röntgen's amazing X-ray experiment, and Threlfall even sent a useful paper on the subject to England for publication in the *Philosophical Magazine*.[15] No one in Australia, however, became seriously engaged in research in these new fields until Bragg embarked in 1904 on a series of remarkable investigations of the new ionizing radiations that led in quick succession to his election as a Fellow of the Royal Society of London, his return to England to take up the chair of physics at Leeds, and eventually, in 1915, to the Nobel Prize for physics, which he shared with his Adelaide-educated son.[16] It likewise provided the basis of his address as president of the Congress at Brisbane in January 1909, which also served as his farewell to Australian science.[17]

While Bragg was still in Australia, his success inspired J. A. Pollock, who had succeeded Threlfall (also returned to England) in the Sydney chair in 1899, to commence some research of his own on ionizing radiation.[18] The young Sydney-educated physicist T. H. Laby brought a programme of research in the same field with him from Cambridge's Cavendish Laboratory to Wellington, where he took up the chair of physics in 1909, and this became the basics of his presidential address to

Section A in 1911.[19] But the great bulk of work being done and papers presented continued to deal with more traditional subjects. Seismological and meteorological themes and instrumental developments associated with them, geodesy, terrestrial magnetism, and the tides all remained popular. Astronomical reports, as such, featured less frequently. Electricity and electrochemistry predominated among reports of laboratory work. Among the mathematicians, Alexander McAulay from the University of Tasmania was a regular contributor, as was the new professor at Sydney, H. S. Carslaw, and both took a turn as president of the Section before the First World War. A small group of schoolteachers, of whom the most prolific was E. G. Hogg of Christchurch, helped maintain a modest flow of mathematical papers.

CHANGING PATTERNS

The new century witnessed a marked decline, however, in the fortunes and standing of the colonial observatories. Ellery, Russell, and Todd all retired at about this time, and only at Melbourne, where Ellery was replaced by Pietro Baracchi in 1895, was a vigorous programme of work maintained by their successors. Even at Melbourne the continued vigour did not last. The economic crisis of the 1890s had already led to a general reduction in staffing levels and these had not been restored once the crisis had passed. The formation in 1908 of the Commonwealth Bureau of Meteorology was a grievous blow, since it saw the observatories stripped of the meteorological responsibilities that had previously constituted one of their chief public functions. The advent of a worldwide telegraph system had already greatly reduced the importance of local observatories for navigational purposes in fixing longitudes, and within a few years, with the rise of regular radio communications, local time services based on the local observatory were to be supplanted by a single, national service supplied from Melbourne. Much of the rationale for maintaining separate observatories thus disappeared during these years. At the same time, encroaching city lights made their locations more and more unsuitable for the purely astronomical functions that remained to them. Furthermore, the Perth, Melbourne, and Sydney observatories all found themselves locked into a massive and dreary international programme of photographic mapping of the skies at a time when, elsewhere, the leading edge of the discipline was racing ahead into other, much more exciting fields of activity. Australian astronomy went into a sudden, steep decline.[20] The founding in 1925 of a Commonwealth Solar Observatory at Mount Stromlo was the first sign that this might eventually be arrested, but even so the subject remained in the doldrums until after the Second World War, when Mount Stromlo struck out in new directions under a new and vigorous director, Richard Woolley.[21]

AAAS congresses ceased for the duration of the First World War. So, too, more or less, did Australian and New Zealand activity in the physico-mathematical sciences. Many of the nation's physicists—

especially, of course, the younger ones—went to war. No special value was placed on their talents, however, and as a result some, like Thomas Parnell, lecturer and later professor of physics at the recently founded University of Queensland, simply enlisted in the artillery. Pollock from Sydney was one of the few whose training in physics was put to use; he served on the Western Front under Edgeworth David in the famous Australian Mining Corps, becoming an expert on the listening devices needed to protect his fellow miners from enemy counteraction.[22]

At home, Bragg's erstwhile collaborator in Adelaide, J. P. V. Madsen, now assistant professor in electrical engineering at the University of Sydney, became director of the Army's Engineer Officers Training School. T. R. Lyle, newly retired from his Melbourne chair, became a leading member of the Commonwealth government's Advisory Council on Science and Industry, and scientific consultant to the Royal Australian Navy. Others, such as his successor, T. H. Laby, recruited by the University of Melbourne from Wellington to fill its vacant chair, worked long hours to keep the university teaching programmes going, often chafing as they did so at their inability to contribute more directly to the war effort.[23] Those such as Laby who remained behind could not have managed, however, without the help of a number of young women physics graduates, temporarily employed by the universities as demonstrators to hold the fort for the men who had gone away to war. Expanded student numbers after the war enabled some of these women to stay on when the men returned. They were not encouraged, however, to undertake advanced work, and all remained in junior positions in their departments until close to retiring age.

The return of peace saw a resumption of the regular cycle of AAAS congresses. The *de facto* downgrading of the observatories saw Section A now dominated by university people. A remarkable feature of the first postwar congress, held in Melbourne in January 1921, was an extensive discussion of relativity theory, the first occasion on which this topic had appeared on the conference programme. The new-found interest had been sparked in Australasia as elsewhere by A. S. Eddington's widely publicized announcement late in 1919 that observations taken during an eclipse of the sun had confirmed Einstein's prediction of the gravitational deflection of light rays. Part of the Australian interest derived from the knowledge that several major expeditions were being planned to remote regions of the Australian outback to repeat the observations during another total solar eclipse that was to be visible there in September 1922.[24] In 1921 and subsequently, however, Australian discussions of relativity were dominated by the mathematicians.

The 1921 meeting also witnessed the demise of several AAAS committees, some of them of very long standing, that had been promoted by Section A. These included the Solar Physics, Seismological, Terrestrial Magnetism, Tidal Survey, Physical and Chemical Constants, and Longitude committees. Their fate reflected the rise of a second national scientific organization, the Australian National Research Council (ANRC), a body set up explicitly to manage Australia's formal international scientific linkages. The AAAS committees were allowed to lapse because 'their methods of working involve[d] international cooperation'; accordingly, it was felt, their work should now become the responsibility of the ANRC. (The particular interest of the New Zealanders, who were, of course, not involved in the ANRC, in the work of some of these committees, was conveniently overlooked.)

The published record of the meetings of Section A later in the 1920s reveals the rise of several new Australian institutions in the physicomathematical sciences. Among the first were the geomagnetic observatory established by a private American research foundation, the Carnegie Institution of Washington, at Watheroo in Western Australia, in 1919, and the Research Division of the Commonwealth Bureau of Meteorology, set up soon afterwards in Melbourne. Each in turn was headed by Edward Kidson, who later in the decade was to become director of New Zealand's Meteorological Service.[25] In the late 1920s the first papers appeared from the newly founded Commonwealth Observatory, and from researchers attached to the Radio Research Board newly set up under the aegis of the Council for Scientific and Industrial Research (CSIR, which was, itself, a creation of the 1920s). The radio work burgeoned during the 1930s, evolving as it did so into the study of the physics of the ionosphere and, in the process, linking up with the work being done at Watheroo and with some of that being done at Mount Stromlo.[26] Several of those involved, and the young D. F. Martyn in particular, acquired significant international reputations. In the late 1930s, New Zealand, too, set up a Radio Research Board to foster work in this field. Less visible at AAAS congresses, but of some significance as employers of physics graduates, were the Munitions Research Laboratories established by the Department of Defence at Maribyrnong, Victoria, which served throughout the inter-war period as the *de facto* national centre for metrological standards; the Research Laboratories of the Postmaster General's Department, also in Melbourne, where work was done on the more practical aspects of radio and telephone communications; and the Commonwealth Radium Laboratory (later the Commonwealth X-Ray and Radium Laboratory) established in 1929 in the grounds of the University of Melbourne to maintain a radium service for Australian hospitals and to develop dosimetric and safety standards for radiotherapeutic practice.[27] In Sydney, a number of physicists found employment with the 50 per cent Commonwealth-owned company Amalgamated Wireless (Australasia) Ltd, which established its own research laboratories in the early 1930s to undertake developmental radio research.

Yet all these institutions remained small and they employed few physicists. The majority of those engaged in the discipline in Australia and New Zealand continued to be attached to the universities. Here, too, departments remained small, but most were bigger than they had been before the First World War, and the numbers of students taking out honours and master's degrees rose slowly but steadily as wider opportunities for finding employment as physicists emerged. Although the overwhelming emphasis in university departments continued to be on basic undergraduate teaching, in some, and above all under Laby at

Melbourne, research became a normal feature of departmental life. The effect can be seen in the reports of the AAAS (or, from 1932, ANZAAS) congresses where, from one congress to the next, more and more of the papers presented to Section A take the form of laboratory-based experimental research reports.

No Australian university offered the PhD degree, however, until the Second World War, so few research students continued on for more than a year or two after completing their undergraduate degrees. The most talented were siphoned off to Britain for their advanced training, funded by Empire-wide scholarship schemes such as the 1851 Exhibition awards and the Rhodes Scholarships, or by 'travelling scholarships' endowed at their home universities. (It was a proud boast of Laby, for example, that no fewer than fourteen of his students received 1851 Exhibition awards, as he had, and went to work in the Cavendish Laboratory.) Some of these students eventually returned and enriched Australian and New Zealand scientific life with their new-found research skills and orientation, but many of the best never came back. They constituted a permanent net drain on a scale that the still tiny Australasian physics community could ill afford.

ALTERNATIVES

Until the 1920s, the AAAS was the only organization in the Australasian region that provided national linkages of any kind in the physico-mathematical sciences. Without it, the tiny groups of practitioners in the different Australian states and in New Zealand would have been entirely on their own. With the formation of the ANRC, a second national scientific institution in which the physicists and mathematicians were represented came into being. The ANRC, however, did not sponsor conferences, and its specialist committees did their work mostly by mail.

More significant in the long run, so far as the Australian physics community was concerned, was the letter-writing campaign begun in 1923 by A. D. Ross, foundation professor of mathematics and physics at the University of Western Australia.[28] Ross was among the first Australians to join the London-based Institute of Physics, founded in 1919 with a view to enhancing the status of physics as a profession. Soon afterwards, he persuaded the Institute authorities to allow him to act as their 'local secretary for Australia', and he embarked on a recruiting campaign on the Institute's behalf. Within a few years he had convinced almost everybody resident in Australia who was eligible for membership to join. Moreover, by exploiting the Institute's need to have applications scrutinized by people familiar with the applicants and their qualifications, he had created a *de facto* local committee of the Institute that began to play a role in other matters of concern to Australian physicists. He also initiated the practice of calling a formal meeting of Australian members of the Institute during each AAAS congress, thus giving the informally constituted branch a visible identity of its own. It was a shrewd move since it enabled Ross to buttress arguments on matters of professional concern with the authority of the much larger British organization, compensating in this way for the want of local numbers to support him.

Quite soon, indeed, the local group struck out on a path not envisaged by the parent British organization (which tended to confine itself to matters of professional concern to physicists), and began to organize conferences of its own. For reasons that were never fully spelled out in public, by the late 1920s the AAAS meetings, previously so central a feature of the scientific life of the nation's physicists, had come to be seen as no longer satisfying the specialist needs of their discipline. At the meeting of Australian members of the Institute of Physics held during the AAAS Congress in Hobart in January 1928, it was agreed to hold an independent, specialist conference of Australian physicists and astronomers later in the same year. This took place in Canberra in August under the aegis of the Commonwealth Solar Observatory, and was a great success.[29] Thirty-four people attended and heard a variety of brief research reports and three highly topical symposium discussions. One, on radio research, was led by Laby. A second, on methods of geophysical prospecting, was prompted by the activities of the Imperial Geophysical Experimental Survey that had commenced operations in Australia a few months earlier.[30] The third, on the new quantum mechanics of Schrödinger and Heisenberg, was the first major discussion of this topic in Australia. Papers were presented by an outstanding research student from the University of Melbourne, H. S. W. Massey, and Edna Briggs, recently returned from two years in Cambridge, which she had evidently spent in absorbing the new ideas, while her physicist husband, G. H. Briggs, a lecturer at the University of Sydney, completed a PhD at the Cavendish Laboratory.

This conference was so successful that those attending agreed that others should be held regularly in the future, in the years between the AAAS (and later ANZAAS) congresses. Accordingly, further conferences were held in 1929, 1931, 1933, 1936 and 1939, the last on the eve of the Second World War.[31] The meetings provided welcome opportunities, uncluttered by the distractions of the much larger AAAS and ANZAAS congresses, for discussions of current work being done. In addition, they served as a focus for a growing collegial spirit among Australia's physicists. A remarkably large fraction of Australia's population of physicists and astronomers attended, along with many of their research students. Yet the total numbers involved were still quite small. Moreover, they were heavily concentrated in the two major centres of Melbourne and Sydney. Here, indeed, they were numerous enough to warrant the local Royal Societies forming specialist sections to cater for their interests (though in Melbourne this initiative quickly collapsed as a result of personal differences between some of those involved). For those from the smaller centres, the new series of meetings in effect doubled the number of opportunities for drawing intellectual sustenance from contact with other physicists. Where a choice had to be made, the decision often went to the new-style discipline-based conferences in preference to the traditional multi-disciplinary AAAS and ANZAAS congresses.

One result of this was an increasing separation between the physicists and the mathematicians. The attendance of the latter, even fewer in number as they were than the physicists and astronomers, at AAAS congresses had always been spasmodic, and some of the more senior people rarely went. While those whose interests lay at the 'applied' end of the mathematics spectrum sometimes went to the new, more specialized conferences, those who were interested in the 'purer' branches of the discipline invariably did not. As a result, Australian mathematicians remained more isolated from each other than did their physicist colleagues, a situation that was not redressed until the Australian Mathematical Society was founded in 1956.

The conference of physicists held in Melbourne in August 1939 saw the informal grouping that had been held together by Ross's letter-writing activities given formal standing as an Australian Branch of the Institute of Physics, with its own constitution and formally elected office-bearers. The New Zealanders, however, even fewer in number as they were, were not party to this development, since despite Ross's occasional attempts to include them in his network, the officials of the Institute in London had been unwilling to extend his brief as far as that. Not until may years afterwards, in the early 1960s, were there moves to follow the Australian example and form a New Zealand Branch of the Institute. Curiously, this occurred just as the Australians were cutting their links with the British body altogether and establishing their own, independent Australian Institute of Physics.

THE IMPACT OF WAR

The 1939 conference also saw one session devoted to a discussion of ways in which Australia's physicists might best contribute to the nation's effort in the war that was obviously coming. Already the government had been persuaded to set up two new physics-based divisions within CSIR to support a hoped-for expansion in Australian industrial capability, namely, the National Standards Laboratory and the Division of Aeronautics. As the course of the war drove Australia to depend more and more on its own resources, many Australian physicists would find employment in these laboratories.

Unknown to most of those attending the 1939 physics conference, a third physics-based CSIR division was about to be established in secret under the direction of D. F. Martyn to work on a new weapon that came to be known as radar. Located within the new National Standards Laboratory building on the campus of the University of Sydney, the enigmatically named Radiophysics Laboratory quickly became Australia's largest employer of physicists. Building on the extensive experience in radio research that had developed in Australia during the preceding fifteen years, it became, in time, a significant contributor to the nation's war effort. Several of the country's leading mathematicians also contributed to its work.³²

Meanwhile, in Melbourne, the outbreak of war saw Philip Bowden, an 1851 Exhibition Scholar from the University of Tasmania who had since become a lecturer at Cambridge, offering his services to the Australian government as an expert on the physics of surfaces. With an eye to the needs of the nation's infant aircraft industry, the government seized upon Bowden's offer and created another new unit within CSIR, called the Lubricants and Bearings Section, under his leadership. In this case the unit was located on the campus of the University of Melbourne. It, too, quickly created a demand for trained physicists.

The war generated many other demands for physicists and, in response, the numbers of trained physicists in the country increased rapidly. Some found employment in the meteorological service (transferred to Air Force control for the duration of the war), while others developed the Ionospheric Prediction Service, essential to the maintenance of Australia's long-range radio links. Electronics manufacturers such as AWA expanded their laboratory staffs dramatically to cope with wartime demands. Many physics graduates found employment at the vastly expanded Maribyrnong Munitions Laboratories. One, E. H. S. Burhop from the University of Melbourne, was seconded to the United States to join the Manhattan Project working to develop the atomic bomb. Many university physicists, and also the Mount Stromlo astronomers, became involved in the work of the Optical Munitions Panel that oversaw the creation, from nothing, of an optical components industry in Australia. Here, too, the mathematicians were also called into service.

These were exciting times for Australia's physical scientists. As in the First World War, however, ANZAAS had gone into hibernation for the duration, and so other forums had to be found to facilitate exchanges of scientific opinion. Some physicists devoted their energies to the affairs of the Australian Association of Scientific Workers, founded in 1939 to promote the social role of the scientist.³³ Within physics itself, however, the running was made by the Institute of Physics. The newly adopted constitution of the Australian branch envisaged the formation of local divisions in the various states, and several of these came into being during the war years and began organizing regular schedules of meetings at which one or more papers on physical subjects were read. The New South Wales division was first in the field and was active from 1940 on; Victoria, Western Australia, and South Australia followed later in the war. After the war these divisional programmes continued unabated (with a Queensland division soon being added to the roster as well), providing the majority of Australia's physicists with more regular opportunities for discussions with their fellows than ANZAAS or any other national organization could hope to provide.³⁴

In earlier times, only by organizing things on a national scale had it been possible to obtain a large enough body of physicists to sustain group meetings. The growth in numbers that occurred during and after the Second World War changed all that. By 1947 the membership of the

Victorian and New South Wales divisions had swollen to ninety-eight and sixty-six respectively, and though the numbers in the other states were much smaller, the deficiency could be made up by enrolling student members or at least encouraging them to attend the meetings. Ross in Western Australia was characteristically energetic in this regard, with the result that in 1947 the membership of his local division comprised two Fellows, six Associates, six subscribers, and no few than forty-three students. In Adelaide, even though there were only twelve 'corporate' members (i.e. Fellows or Associates) of the local division, and students did not enlist in such numbers, attendances at the five 'ordinary' meetings held that year averaged thirty-eight, while a special meeting with Richard Woolley as visiting speaker drew a crowd of over two hundred and fifty.

Physics went through a boom period in Australia, as elsewhere, during these years. Though the success of the atomic bomb project in the United States had unleashed the terrors of nuclear war on a previously unsuspecting world, it also seemed to promise untold riches in the form of cheap supplies of energy. At the same time, the achievements of the physicists in the wartime development of radar were at last made public. Physics suddenly became a glamour science in which the demand for trained researchers far outstripped supply. Student numbers swelled and so, too, in response, did university physics departments. To staff them and the new government laboratories, trained physicists were recruited in considerable numbers from elsewhere, chiefly from Britain, so reversing the direction of flow for the first time in two generations.

The first tentative Australian initiative in relation to atomic energy came in 1947, with the formation of a small atomic physics unit within CSIR, several members of which were seconded soon afterwards to the new British atomic research station at Harwell. The group, which included chemists and engineers as well as physicists, remained in Britain until 1955, when it returned to Australia to become the nucleus of the newly created Australian Atomic Energy Commission with its research establishment at Lucas Heights, near Sydney. Meanwhile, the Australian government had also encouraged the ambitions of L. H. Martin, Laby's successor in the chair of physics at Melbourne, to concentrate the research activities of his department on nuclear physics. Martin's appointment in 1948 as the Australian government's Defence Scientific Adviser testifies to the important place that physics, and atomic physics in particular, now held in government thinking.[35] Nuclear physics also figured large in the planning of the new, research-oriented Australian National University established in Canberra in the postwar years. One of the four founding schools that constituted the University was a Research School of Physical Sciences, and the man chosen to direct it was a leading authority on nuclear physics, the Australian-born and Rutherford-trained M. L. E. Oliphant.[36] At the University of Sydney, Harry Messel, appointed in 1952 to the long-vacant chair of physics, took advantage of the public interest in the subject to launch a Nuclear Research Foundation to raise additional funds for his department. The dramatic success he achieved quickly transformed the department's research capabilities and gave academic physics a new image throughout the country.[37]

INDEPENDENCE

Much the most dramatic developments in postwar Australian physics came, however, from CSIR's Radiophysics Division. At war's end, the large team of research physicists and engineers that had been assembled to work on radar was not dispersed in the way that the much larger United States group was. Some members of the team took up developmental work on civilian applications of radar, others for a few years pursued research on large-scale electronic computers before they were diverted to other tasks. A large group directed by the new chief of the Division, E. G. Bowen, applied their radar techniques to the study of clouds and the seeding of them in attempts, ultimately unsuccessful, to make rain. Another group, under J. L. Pawsey, turned their radar equipment towards the heavens and immediately made a series of fundamental and exciting discoveries that carried Australia to the forefront of the then new field of radio astronomy. Here, Australian physicists for the first time experienced the heady feeling of leading the world with their research.[38]

The other physics-based divisions within CSIR (and later the CSIRO) also continued to grow apace. Pre-eminent, perhaps, was the National Standards Laboratory, which divided in 1946 into three separate divisions of Physics, Metrology, and Electrotechnology. The wartime Lubricants and Bearings Section became a full-scale Division of Tribophysics, while a new Division of Chemical Physics grew out of the swollen Division of Industrial Chemistry. The Division of Aeronautics, however, was transferred in 1949 from CSIR to the Department of Supply where, together with the Maribyrnong laboratories and the newly established Long-Range Weapons Establishment at Salisbury and Woomera in South Australia, it provided a strong physics presence within the expanding Defence scientific complex. Physics was also well represented in another new government scientific organization, the Commonwealth Bureau of Mineral Resources, where a substantial geophysics group was brought together.

On Mount Stromlo, the Commonwealth Observatory entered a new and expansionary phase, even as the old Melbourne and Adelaide observatories finally expired. Its work was enhanced by a greatly expanded staff, substantial new equipment, strengthening links with the radio astronomers and, eventually, a new site on Siding Spring Mountain in northern New South Wales. With the establishment of the National University, the Observatory became a *de facto* department of the Research School of Physical Sciences, a move that greatly benefited both groups. By the 1960s Mount Stromlo had joined the Radiophysics Division as an institution of world standing, and Australian optical astronomers stood ready to take on a major new instrument that would consolidate their position in their field for years to come, the Anglo-Australian Telescope.[39]

Despite the great increase in numbers and research activity that these various developments encompassed, the prewar series of national conferences of physicists and astronomers was not resumed in the postwar world. For some years the nation's physicists rested content with the regular programmes of meetings now being sponsored by their local divisions of the Institute of Physics, together with the revived ANZAAS congresses (after 1946) as the only occasions for wider, national gatherings. At the latter, the Section A programme inevitably became more and more crowded. At the 1951 Brisbane Congress, for example, no fewer than forty-four papers were presented to the Section in the space of seven days, in addition to K. E. Bullen's presidential address and a further thirteen papers presented to a special optometry subsection that was linked to Section A at this period. The strain was obviously too great, and for the next meeting, at Sydney in 1952, the organizers bowed to the inevitable and divided the programme into subsections on related themes, with a sizeable cluster of 'miscellaneous papers' at the end.

Inevitable such a division of the programme may have been, but it also had the effect of institutionalizing and consolidating the different specializations represented within the Section. The subsequent history of the Section is of one attempt after another to maintain some overall coherence by means of broad-interest symposia and the like, while at the same time catering adequately for the needs of ever-growing numbers of researchers to discuss their work not merely in general terms but in detail with other specialists.

To an increasing extent during the 1950s and afterwards, these latter needs came to be catered for outside the ANZAAS umbrella. The experience of the mathematicians was typical in this regard. The 1955 ANZAAS Congress at Melbourne saw a discussion of 'the question of arranging a separate section for mathematics at future meetings of ANZAAS', the outcome of which was, however, the formation instead of a separate Australian Mathematical Society, which quickly began to organize its own, independent conferences.[40]

Among the physicists, the Institute of Physics likewise began to sponsor specialist conferences that attracted participants from around the country. Probably the earliest of these were a Heat Transfer Conference held in Sydney in 1948 and one entitled X-Rays in Industry held in Melbourne in the following year. More spectacular was an international conference on radio science held in Sydney in 1952 under the aegis of the international radio science union, URSI. This was the first occasion on which one of the international scientific unions had met outside Europe or North America, a striking tribute to the work of CSIRO's radiophysicists. From the mid-1950s, the trickle of specialist meetings began to turn into a flood, with conferences on Scientific Manpower in Melbourne and Contemporary Optics in Sydney in 1956, an Atomic Energy Symposium in Sydney in 1958, Solid State Physics and a second Contemporary Optics conference in Melbourne in 1959,

Theoretical Physics in Sydney in 1960, and Ionospheric Physics in Brisbane in 1961. In the late 1950s an agreement was forged to link these specialist conferences wherever possible with ANZAAS congresses, but in the event this was honoured more in the breach than in the observance.[41]

The trend continued under the newly constituted Australian Institute of Physics, which declared its independence from the parent British organization in 1962. Increasingly, Australian physicists came to see such local specialist conferences, the larger specialist conferences overseas made ever more accessible by improvements in air transport, or (from 1974) the biennial national physics congresses sponsored by the Australian Institute of Physics, as the meetings they preferred to attend. ANZAAS congresses became correspondingly less and less relevant to their research activities and more and more, from their point of view, exercises in scientific public relations. Even from the public relations point of view, however, for physics and mathematics other forums were at least as effective as ANZAAS. After the best part of a century, for the sciences that constituted the old Section A, the Association had lost its reason for being. If ANZAAS is to remain an organization in which these sciences have an effective presence, it must find a new role for itself, quite different from the one it has fulfilled during most of its history.

NOTES

Research for this paper was supported by grants under the Australian Research Grants Scheme.

1. W. A. Osborne, *William Sutherland: a Biography* (Melbourne: Lothian Press, 1920).
2. Russell, 'President's Address', *Report of the AAAS*, 1 (Sydney, 1888), 1–14, at 7.
3. Todd, 'Meteorological Work in Australia: a Review', *Report of the AAAS*, 5 (Adelaide, 1893), 246–70; M. E. Hoare, 'Science and Scientific Associations in Eastern Australia, 1820–1890', (Unpublished PhD thesis, Australian National University, 1974); W. J. Gibbs, *The Origins of Australian Meteorology* (Canberra: AGPS, 1975); J. Gentilli, 'A History of Meteorological and Climatological Studies in Australia', *University Studies in History*, 5 (1967), 54–79.
4. *University of Adelaide Calendar*, 1877, quoted by S. G. Tomlin, 'Horace Lamb, 1849–1934', *The Australian Physicist*, 12 (1975), 165–8, at 166.
5. H. Lamb, *A Treatise on the Mathematical Theory of the Motion of Fluids* (Cambridge: Cambridge University Press, 1879).
6. In Adelaide, too, a separate faculty of science was established at this time, but it was a further twenty years before additional professorial appointments were made.
7. R. W. Home, 'First Physicist of Australia: Richard Threlfall at the University of Sydney, 1886–1898', *Historical Records of Australian Science*, 6 (1986), 331–56.

8. J. G. Jenkin, 'The Appointment of W. H. Bragg, F. R. S., to the University of Adelaide', *Notes and Records of the Royal Society of London*, 40 (1985), 75–99.
9. *Bragg Papers, 37A* (Royal Institution, London), W. H. Bragg to his fiancée, Gwendoline Todd, 28–30 August 1888, quoted in R. W. Home, 'The Problem of Intellectual Isolation in Scientific Life: W. H. Bragg and the Australian Scientific Community, 1886–1909', *Historical Records of Australian Science*, 6 (1984), 19–30.
10. R. Threlfall, 'The Present State of Electrical Knowledge', *Report of the AAAS*, 2 (Melbourne, 1890), 27–54.
11. W. H. Bragg, 'On Some Recent Advances in the Theory of the Ionization of Gases', *Report of the AAAS*, 10 (Dunedin, 1904), 47–77.
12. Rosaleen Love, 'Science and Government in Australia, 1905–14: Geoffrey Duffield and the Foundation of the Commonwealth Solar Observatory', *Historical Records of Australian Science*, 6 (1985), 171–88.
13. W. H. Steele, 'On the Conductivity of a Solution of Copper Sulphate', *Report of the AAAS*, 4 (Hobart, 1892), 256–7.
14. Home, op. cit., note 7; R. W. Home, 'Lyle, Sir Thomas Ranken (1860–1944)', *Australian Dictionary of Biography*, 10 (1986), 172–4.
15. R. Threlfall and J. A. Pollock, 'On Some Experiments with Röntgen's Radiation', *Philosophical Magazine*, Series 5, 42 (1896), 453–63 (also published in *Proceedings of the Physical Society of London*, 15 (1896–7), 1–12); Hugh Hamersley, 'Radiation Science and Australian Medicine, 1896–1914', *Historical Records of Australian Science*, 5 (1982), 41–63.
16. E. N. da C. Andrade, 'William Henry Bragg, 1862–1942', *Obituary Notices of Fellows of the Royal Society of London*, 4 (1942–4), 277–92; *Dictionary of Scientific Biography*, 2, 397–400; *Australian Dictionary of Biography*, 7 (1979), 387–9; G. M. Caroe, *William Henry Bragg, 1862–1942: Man and Scientist* (Cambridge: Cambridge University Press, 1978); R. W. Home, 'W. H. Bragg and J. P. V. Madsen: Collaboration and Correspondence, 1905–1911', *Historical Records of Australian Science*, 5 (1981), 1–29.
17. W. H. Bragg, 'The Lessons of Radioactivity', *Report of the AAAS*, 12 (Brisbane, 1909), 1–30.
18. Home, op. cit., note 9, 25–6.
19. T. H. Laby, 'Some Recent Advances in Physics', *Report of the AAAS*, 13 (Sydney, 1911), 19–29.
20. 'Astronomy', *The Australian Encyclopaedia* (Sydney: Grolier Society, 1958), 1, 278–86.
21. Susan Davies, 'R. v. d. R. Woolley in Australia', *Historical Records of Australian Science*, 6 (1984), 59–69.
22. R[ichard] T[hrelfall], 'James Arthur Pollock, 1865–1922', *Proceedings of the Royal Society of London*, A, 104 (1923), xv–xxii.
23. *Laby Papers* (University of Melbourne Archives), Laby to E. Rutherford, 8 January 1918.
24. Jeffrey Crelinsten, 'William Wallace Campbell and the "Einstein Problem": an Observational Astronomer Confronts the Theory of Relativity', *Historical Studies in the Physical Sciences*, 14 (1983), 1–91, especially 61ff.
25. Isabel M. Kidson, (ed.), *Edward Kidson* (Christchurch: Whitcombe & Tombs Ltd, nd [1941?]).
26. W. F. Evans, *History of the Radio Research Board, 1926–1945* (Melbourne: CSIRO, 1973).
27. J. F. Richardson, *The Australian Radiation Laboratory: A Concise History, 1929–1979* (Canberra: AGPS, 1981).

28. R. W. Home, 'Origins of the Australian Physics Community', *Historical Studies*, 29 (1982–3), 383–400.
29. Conference of Australian Physicists, Canberra, 15th to 18th August, 1928: Proceedings and Abstracts of Papers.
30. A. B. Broughton Edge and T. H. Laby, (eds), *The Principles and Practice of Geophysical Prospecting: Being the Report of the Imperial Geophysical Experimental Survey* (Cambridge: Cambridge University Press, 1931); Barry W. Butcher, 'Science and the Imperial Vision: the Imperial Geophysical Experimental Survey, 1928–1930', *Historical Records of Australian Science*, 6 (1984), 31–43.
31. *Australian Institute of Physics Papers* (Basser Library, Australian Academy of Science, Canberra), Boxes 86/13 and 86/14.
32. W. F. Evans, *History of the Radiophysics Advisory Board, 1939–1945* (Melbourne: CSIRO, 1970).
33. Jean Moran, Scientists in the Political and Public Arena: a Social-Intellectual History of the Australian Association of Scientific Workers, 1939–49 (unpublished MPhil thesis, Griffith University, 1983).
34. AIP, op. cit., note 31, Box 86/2a.
35. Tim Sherratt, 'A Political Inconvenience: Australian Scientists at the British Atomic Weapons Tests, 1952–53', *Historical Records of Australian Science*, 6 (1985), 137–52.
36. S. Cockburn and D. Ellyard, *Oliphant: the Life and Times of Sir Mark Oliphant* (Adelaide: Axiom Books, 1981).
37. David Branagan and Graham Holland (eds), *Ever Reaping Something New: a Science Centenary* (Sydney: University of Sydney, Science Centenary Committee, 1985), ch. 4.
38. W. T. Sullivan, III, (ed.), *The Early Years of Radio Astronomy* (Cambridge: Cambridge University Press, 1984); A. C. B. Lovell, 'Joseph Lade Pawsey, 1908–1962', *Biographical Memoirs of Fellows of the Royal Society of London*, 10 (1964), 229–43.
39. Bernard Lovell, 'The Early History of the Anglo-Australian 150-inch Telescope (AAT)', *Quarterly Journal of the Royal Astronomical Society*, 26 (1985), 393–455.
40. A. L. Blakers, 'The Australian Mathematical Society: Foundation and Early Years', *Australian Mathematical Society Gazette*, 3 (1976), 67.
41. AIP, op. cit., note 31, Box 86/5.

7

Chemists at ANZAAS: Cabbages or Kings?

Ian D. Rae

'Only about 70 years ago was chemistry like a grain of seed from a ripe fruit, separated from other physical sciences', wrote Liebig in 1859.[1] An important figure in this 'separation' was Antoine Lavoisier[2] whose exciting work in the 1780s established European chemistry as a discipline that is thus approximately as old as European settlement of Australia. It would be almost another hundred years, however, before the discipline was firmly established in Australia, and if we wish to cite a particular date we might choose the centennial year, 1888, which saw the establishment of the Australasian Association for the Advancement of Science (AAAS).

Over the last century the AAAS and then ANZAAS have served the small community of Australasian chemists well, supplying a national forum for new research and collective action. Because of the great distances involved, however, it was only very senior people who could travel to meetings which were more than one state from their home. From the outset, Archibald Liversidge's influence was vital, but Orme Masson and his disciples played an increasingly important role and probably helped to maintain the AAAS very much at a 'professorial' level.

When local communities of chemists grew sufficiently numerous that state activities became possible, the AAAS declined to enter the field and the way was left open to other organizations, notably the Australian Chemical Institute. With increasing concentration on pure research in the 1920s and 1930s, Section B (Chemistry) continued in a very 'academic' style in the 1950s, and was well attended, but by the 1960s this was not the way in which ANZAAS as a whole was inclining. As the Association moved towards more social concerns, chemists turned to the more specialized offerings of the Chemical Institute, and in the 1980s they have made much smaller contributions to ANZAAS than one might expect.

A significant date is 1885, when E. H. Rennie (1852–1927) took up the chair in Adelaide and became the first Australian-born chemistry professor. In New Zealand the corresponding date was 1903, when W. P. Evans (1864–1959) succeeded Bickerton in the chair at Canterbury College.[3] However, academic chemistry remained weak for many years—there were only ten chemists in Australia's four university science faculties at the turn of the century and four New Zealand professors of chemistry—and much of the early vigour of Section B came from analytical chemists employed by governments. When the Australian Chemical Institute was formed in 1917, its main concerns were salaries and qualifications, but before long there were regular scientific meetings in state capitals. The AAAS (which became ANZAAS in 1930) was unchallenged as a national forum for the presentation of chemistry until the 1960s. The growth of specialist divisions within the Royal Australian Chemical Institute (RACI),[4] however, quickly eroded the role of ANZAAS, and in recent years RACI too has found it difficult to mount national meetings where a truly broad range of chemistry is presented to all participants. This change could not have taken place until critical numbers were achieved in specialist areas, but the concentration of ANZAAS on national gatherings meant that long-distance travel was involved, which mitigated against the participation of many chemists and left the way open for vigorous local groups such as Australian Chemical Institute branches and other professional societies, catering for a wider range of those people whose allegiance was primarily to chemistry as a profession.

In short, then, the AAAS and ANZAAS have concentrated on kings to the exclusion of cabbages. Nonetheless ANZAAS continues to play a valuable role as a shop window, especially where science is intelligently promoted and presented to the young people on whom the future of our nation will depend.[5] It is also a place where chemists can begin to dispel the negative image of their profession, in particular as it relates to chemical industry.

THE FIRST CENTURY

Colonial chemistry began with the needs of medicine (including pharmacy)[6] and such traditional industries as salt making, mining, tanning, and fermentation. Colonial medical men turned quite quickly to local flora, although they failed to profit from the experience of Aborigines, who had detailed knowledge of the medicinal properties of their native plants.[7] In November 1788, Dennis Considen, surgeon of the First Fleet, styled himself 'the colony's pharmaceutic pioneer' because of his medicinal use of eucalyptus oil and extracts of other plants.[8] Thus began a tradition of natural products chemistry which was strongly represented at AAAS meetings a century later and prospers still today.

Applications of chemistry multiplied rapidly in the second half of the nineteenth century, following the discovery of gold and other mineral wealth, and chemical manufacture was established by the Elliott brothers in Sydney,[9] Felton and Grimwade in Melbourne,[10] and Faulding in Adelaide.[11] All of these had begun as druggists, but by 1865 the Elliotts had also installed a factory at Balmain to produce sulfuric and nitric acids, superphosphate from bones, and sundry other chemicals. James Elliott, younger son of one of the three brothers who founded the company, studied under von Hoffmann at Berlin, where he took a PhD and taught for a time before returning to the family business in 1884 to work as a chemical engineer and pharmaceutical chemist.[12] Starting as a retail and manufacturing pharmacist in 1845, Francis Faulding later diversified into veterinary medicines, where he employed Joseph Bosisto, who shortly left for Melbourne to become famous for the distillation of eucalyptus oil and its medical applications.[13]

In the 1870s James Cuming developed the fertilizer industry in Melbourne with exquisite timing, as successive Land Acts permitted agriculture to develop at the expense of grazing. Cuming, son of a Scottish veterinarian, gained his knowledge of chemistry by working on the construction of an acid plant and by private study in the Melbourne Public Library,[14] before taking over an acid plant formerly operated by Robert Smith at Yarraville. Cuming Smith & Co. co-operated with Felton Grimwade's and the two companies later amalgamated some of their interests and joined in interstate developments such as the Adelaide Chemical Works.[15] The cost of shipping and judicious co-operation between local manufacturers gave them protection from import competition. The hazards of transporting acids and explosives half way round the world also favoured local production. Australian chemical technology, while refreshed by occasional visits by the owners to plants in the Northern Hemisphere, was marked by versatility and self-reliance and rewarded with good profits.

The early academic appointments gave further evidence of chemistry's roots in medicine and industry. At the University of Sydney, in 1851, one of the three foundation chairs was in Chemistry and Experimental Philosophy (Physics). The selection committee in England were instructed that the chemistry courses at Edinburgh, University College, and King's College, London, were to be taken as models.[16] John Smith of Aberdeen was appointed to the Sydney chair in 1852.[17] Sydney's choice of the medically qualified Smith was strongly influenced by his reputation as a lecturer and by his record of research in chemistry. During Smith's thirty years at Sydney University (1852–85) he did little research and left much of his demonstrating to his assistant. He did, however, assist the colonial government by analysing adulterated food and impure water, and his work for school education won him a CMG in 1878.

Smith's work at Sydney was greatly helped by A. M. Thomson[18] and later by Archibald Liversidge, who came to Sydney in 1872, having studied at the Royal School of Mines and the Royal College of Chemistry in London, and at the Chemical Laboratory in Cambridge.[19] He brought about major developments in the teaching of chemistry and was responsible for extensive new laboratories. Like Thomson, he was appointed as reader in geology and demonstrator in chemistry. In 1874 he became professor of geology and mineralogy, and in 1885 professor of chemistry and mineralogy, a title he also bestowed on Section B of the AAAS. The teaching of science in schools received strong support from both Smith and Liversidge, and both put much effort into the Royal Society of New South Wales.

When the University of Melbourne, founded in 1853, sought to elevate the teaching of chemistry above service courses designed for medical students, its selection committee in Britain included the incumbents of Cambridge, King's College (London), and Manchester.[20] They appointed David Orme Masson,[21] who came in 1886 from Edinburgh to continue the teaching of chemistry begun by Frederick McCoy—professor of natural science who had included chemistry in his repertoire[22]—John Macadam, and the first Melbourne chemistry professor, John Drummond Kirkland.[23] Some university appointees came from secondary school teaching. Thus Macadam,[24] a Scot with qualifications in medicine and chemistry, who was recruited by Scotch College (Melbourne) in 1855 as lecturer in chemistry and natural sciences, moved in 1865 to become lecturer in medical chemistry at the University as well as government analyst. Henry Andrew,[25] who taught chemistry at Wesley College, later became professor of physics at the University.

Vocational teaching of chemistry was also well established, having begun in response to the needs of the mining industry for analysts and assayers. The Ballarat School of Mines and Industries—the first technical college in Australia—was established in 1870, followed three years later by Bendigo and then by others in provincial cities.[26] In many of these institutions there were very strong chemistry departments. John Smith visited Sandhurst (Bendigo) in 1876 and noted that they had thirty-four chemistry students whereas his university had only six![27] Vocational teaching also took place outside universities and schools of mines. McCoy was director of the Museum of Natural and Applied Sciences in Melbourne (an honorary post)[28] and the museum–university connection was strengthened by the work of museum staffer and Harvard graduate James Cosmo Newbery.[29] His lectures at the Industrial and Technical Museum extended over seventeen years until they were taken over by the Working Men's College (later the Royal Melbourne Institute of Technology) in 1887. Newbery with Vautin developed a new chlorination process, adopted overseas for the extraction of gold, and he also did early work on Victorian brown coal, which included a technical mission to Germany in 1891. There was an American influence in Sydney, too, when R. K. Murphy came from Columbia University to the Sydney Technical College in 1915 to begin Australia's first course in chemical engineering.[30]

The third Australian chair of chemistry was established in Adelaide, to which E. H. Rennie was appointed in 1885.[31] Rennie graduated from Sydney University (BA 1870, MA 1876) before teaching at his old school, Sydney Grammar, for five years and for a shorter time at Brisbane Grammar School. He completed his DSc (the first Australian to do so), in London in 1882 and then worked in the Government Analyst's office in Sydney before his call to Adelaide.

The establishment of universities in New Zealand followed a more complicated path.[32] The four institutions founded in the nineteenth century—Otago University (1869), Canterbury College (1873), Auckland University College (1883), and Victoria University College, Wellington (1898)—together made up the University of New Zealand, and in each case chemistry had a foundation chair, although it was often allied with physics or geology.[33] J. G. Black, who came to Otago from Edinburgh in 1870 to a chair of natural sciences (later chemistry), was a very popular lecturer, who published little but worked extensively on mining problems, and encouraged the formation of schools of mines.[34] A. W. Bickerton was the initial appointee at Canterbury, coming from Southampton where he had been borough analyst as well as holding a position at the University College. His impact on New Zealand chemistry was limited, however, since most of his energy went into university politics and the study of cosmology.[35] Much more active in chemistry was the first professor at Auckland, F. D. Brown,[36] and his counterpart in Wellington, T. H. Easterfield.[37] Easterfield spoke strongly in favour of academic research in chemistry when delivering his inaugural address, but he spent much of his twenty-two year tenure on applied research and administration of his growing department. In 1920 he resigned to become the first director of the Cawthron Institute in Nelson, New Zealand, and took no further part in chemical activities.

The similarities between Australian and New Zealand chemistry in the latter part of the nineteenth century were very strong. Government analysts, schools of mines, and applied chemistry were prominent, but connections with chemical industry were weak. Chairs of chemistry at universities often embraced a range of disciplines when they were first established as they did at schools of mines, which were then strong. 'Mining chemistry' usually meant assaying by established methods, and there was little really original work. In government laboratories there was agricultural chemistry but, again, much of it consisted of quantitative and qualitative analyses.

Despite its much smaller population, there were as many university chemistry professors in New Zealand at the time the AAAS was founded (J. G. Black, F. D. Brown, A. W. Bickerton) as there were in Australia (A. Liversidge, D. O. Masson, E. H. Rennie). By the turn of the century the New Zealanders had gone one ahead, with the appointment of Easterfield, and that remained the position until B. D. Steele became the foundation professor of chemistry at the University of Queensland. N. T. M. Wilsmore's appointment to Western Australia in 1912 put the Australians ahead for good.

AUSTRALASIAN CHEMISTRY 1888-1914

The thirteen papers presented to Section B (Chemistry and Mineralogy) at the first AAAS meeting in Sydney were concerned mainly with minerals, ore bodies, natural products, and water analysis. The last topic was of particular concern to governments that had to deal with rapidly growing cities like Sydney and Melbourne, where population growth and overcrowding had outrun public health measures. It was a constant concern of the Victorian Board of Public Health, whose inspector, D. A. Creswell, a noted rower at Oxford, was said to have foresworn rowing when he first smelled the Yarra.[38] The pattern established in Sydney continued unbroken through the twelve more congresses that took place before the First World War. Thus the average was fourteen papers presented, in addition to the presidential address, and most of the contributions were of an applied nature, more akin to work published in chemical industry journals overseas than to that appearing in the journals of the chemical societies.

Although there were in all about eighty contributors, most names appeared only once with just a few making regular contributions. First among these was Liversidge himself with twenty-one papers, followed by Professor T. H. Easterfield from Wellington, New Zealand with eleven. Other prominent academics were David Masson and John Booth Kirkland (six each), Edward Rennie (five), and Norman Wilsmore[39] (four). Technical education was well represented by Donald Clark (four papers), the director of the Gippsland School of Mines in Bairnsdale, who took a special interest in gold deposits in that region,[40] and Professor Mica Smith, of the Ballarat School of Mines,[41] who was president of Section B in 1902. Government analysts were also well to the fore, with the New South Wales government analyst, W. M. Hamlet, giving five papers; his deputy, W. Doherty, nine; and Mines Department analyst, J. C. H. Mingaye, six. The New South Wales Agriculture Department's F. B. Guthrie, president of Section B for the Melbourne Congress in 1900, contributed two papers during this period.[42]

Mining and manufacturing were less well represented, but they did make some important contributions. Thus E. W. Knox,[43] the manager of Colonial Sugar, spoke in Melbourne in 1890 about the benefits of chemical control and materials balance in a manufacturing process, and Robert Sticht of the Mount Lyell Mining & Railway Company was president of Section B for the Adelaide meeting in 1907.[44] Explosives for munitions, for hard-rock mining, and for roadmaking were the special province of English expert C. N. Hake, who was Section president in Adelaide (1893); Masson had also had experience in this area.[45]

This emphasis on the 'practical' led W. M. Hamlet further into Shakespearean paths when delivering his presidential address to Section B at the Hobart meeting in January 1892:

the record of the year with regard to original research work, were it given in its entirety, would read very much like Falstaff's hotel bill, showing but a halfpennyworth of research to an intolerable deal of drudgery. This must be the case with most of us in a new country, who if not engaged in teaching or organisation are compelled to spend our time in assaying minerals or else in the pursuit of the agricultural, sanitary or criminal investigations.[46]

This emphasis was also reflected in the title of Section B, which was Chemistry and Mineralogy for the first four meetings, Chemistry for the next five, and Chemistry, Metallurgy and Mineralogy for two, before finally settling down to just Chemistry. The Section's coverage was extremely broad. From 1911 there was a subsection for Pharmacy, which contained a good deal of organic chemistry (notably absent in the parent Section B), and pharmacy eventually graduated to a Section of its own, at the nineteenth congress held in Hobart in 1928.

Organic chemistry was mainly in the hands of those who took an interest in Australian plants. At the first meeting, J. H. Maiden, the New South Wales government botanist, presented a bibliography of plant products,[47] and at the second meeting, E. H. Rennie spoke on the colouring matter of *Drosera whittakeri* and the occurrence of aesculin in *Bursaria spinosa*.[48] The pigment droserone was to occupy Australian organic chemists for years to come, and the aesculin of *Bursaria* has been commercially exploited in patent medicines for treatment of haemorrhoids. Maiden was followed by R. T. Baker and H. G. Smith, at the Technological Museum in Sydney, who contributed major papers on eucalypts to the fourteenth meeting in Melbourne in 1913.[49] Papers in this field also came from New Zealanders, particularly T. H. Easterfield at Victoria College, Wellington, who was president of the Section at Brisbane in 1909. Among his students who also presented papers in sessions that Easterfield's group dominated were the first women to do so before Section B, Miss A. I. Slowey and Clara M. Taylor.[50] Some Australian chemists had women assistants but women seemed to fare better in New Zealand.

Although New Zealanders experienced particular difficulties in travelling to many of the congresses, transcontinental travel for Australians was neither convenient nor cheap, and there were very few consistent attenders judging by the names of those presenting papers. Variations in attendance at the AAAS meetings do not permit a simple interpretation, except that attendances in Melbourne and Sydney generally outranked those at other venues.[51] Similarly the papers presented in Section B show the kinds of fluctuations expected with venue, although the totals are small enough that a rousing local effort often more than countered the decrease brought about by the difficulties of travel. The numbers in Section A (Physics) and Section C (various titles but nominally Geology) showed similar fluctuations, but there were always more papers in these sections, often as many as twice the number presented in Section B.

Overseas (i.e. non-Australasian) contributors were quite rare. C. N. Hake, mentioned above, held appointments in Australia, but some mystery surrounds the contribution by two Indian authors to the 1926 meeting in Perth.[52] A more significant visitor was Henry Dale, director of the Wellcome Physiological Research Laboratories in London and afterwards president of the Royal Society, who spoke at the Melbourne meeting in 1913 about the active principles of the fungus ergot.[53]

The growing strength of chemistry departments, which followed the institution of science degrees (Sydney, 1882: Melbourne, 1886), was reflected in the papers presented to Section B after 1890. The chemistry of minerals still occupied about 40 per cent of the programme, and many of the university contributions were descriptive or reported analytical results, but the quantity of original research began to increase. Masson, who was president of Section B at the Christchurch Congress in 1891, described his work on the physical chemistry of solutions, which attracted favourable attention overseas.[54] At the same meeting, Norman Wilsmore, who was then only the fourth Melbourne BSc graduate (1890: MSc, 1892), reported unsuccessful attempts to prepare magnesium ethyl,[55] another uncharacteristically 'academic' piece of chemistry for an Australian laboratory.

There were a number of contributions on the teaching of chemistry, and at Dunedin in 1904, Edgeworth David, who was general president, stressed the need for better science education. By way of example, he drew attention to the impressive industrial performance of Germany vis-à-vis Britain.[56] This meeting, at which he was president of Section A, launched William Bragg on his research career, after seventeen years in Adelaide during which 'it never entered my head that I should do any research work'.[57] At the Brisbane meeting of 1909, when Bragg was president of the AAAS, Thomas Easterfield described the paucity of chemical research in Australasia and put part of the blame at the door of the universities: 'for, with the exception of Orme Masson, none of us has succeeded in forming a recognised research school'.[58] It was to be a persistent criticism of chemistry in Australia for the next fifty years.

In contrast to the practically oriented chemists, William Sutherland, regarded as Australia's first physical chemist, pursued his work in almost complete isolation, although he did work as a lecturer in physics at Melbourne University between 1888 and 1897.[59] His papers dealing with intermolecular forces given in Section A of the first two AAAS meetings set him apart as a theoretician, well ahead of contemporary opinion. Sutherland also presented a paper on the molecular constitution of water at the Melbourne Congress in 1900, and several others during the prewar years.[60]

Even the pedestrian analytical work that Hamlet complained about eventually led to a worthwhile development—a sensitive microbalance in which weights were replaced by the effect of air pressure on a silica bulb in an airtight balance case. This was the invention of B. D. Steele, a gifted graduate of Masson's department,[61] and his collaborator, physicist Kerr Grant,[62] who were investigating the ionization of metals and required precision in mass of 10^{-9} g! Grant later became professor of physics in Adelaide, and Steele took a chair in Queensland, where he concentrated on solution chemistry, which was the topic of his presidential address to Section B at the Sydney meeting in 1911.[63]

The meeting of the British Association for the Advancement of

Science in Australia in 1914 fulfilled an ambition that Archibald Liversidge had cherished since the 1870s. The British Association had already held meetings in Canada (1884, 1897, and 1909) and South Africa (1905), and the leading role played by Masson in arranging the British visit was aided by the existence of the AAAS as a co-ordinating body. The arrival of the overseas delegates coincided with the outbreak of war, but the decision was taken to proceed with the conference and to permit the visiting German and Austrian scientists to take part in the meetings.

Australians—and not only scientists, for he gave several public lectures—were able to hear the New Zealander, Rutherford, at the height of his powers. Other notables were Moseley, whose researches on X-ray spectra came to a premature end with his death at Gallipoli in 1915, and the maverick H. E. Armstrong, who questioned the existence of isotopes. In Section B, the local contributions look quaintly colonial[64] beside the theoretical power of the Section president, William J. Pope (Cambridge), the physical organic chemistry of G. T. Morgan (Dublin) and H. McCombie (Birmingham), and the alkaloid syntheses of Robert Robinson, who had recently taken up the new chair of Organic Chemistry in Sydney, and his wife Gertrude.[65] The local British Association secretary was David Rivett, a former pupil of Masson, who had won a Rhodes Scholarship in 1907 and later worked with Arrhenius in Stockholm. He succeeded Steele in 1911 as lecturer and subsequently became Masson's successor in the chair.[66]

THE FIRST WORLD WAR AND THE ORGANIZATIONS OF AUSTRALIAN CHEMISTRY

The turn of the century marked a period, in scientific circles at least, when many new professional societies were founded. The AAAS survived this competition virtually unchanged, but its leadership in chemistry was to be deeply affected.

In Melbourne, Masson, with the support of a group of industrialists, founded the Society of Chemical Industry of Victoria (SCIV) in 1900. This was always a small organization (reaching a membership of 180 in 1917),[67] and perhaps for this reason its equivalent in New South Wales, under Liversidge's guidance, was established in 1902 as a branch of the Society of Chemical Industry of London instead.[68] Masson also founded the Melbourne University Chemical Society at this time, and it held occasional joint meetings with SCIV.[69] In 1915, Rennie founded the Society of Chemical Industry of South Australia,[70] and in the same year the Chemical Society of Western Australia was formed. This organization survived until 1933, for most of its life in parallel with the Australian Chemical Institute, to which body it eventually committed the balance of its funds.[71] A Chemical Society was also formed at Sydney Technical College in 1912, and one at the University of Sydney in 1929.

In Australia during the First World War three developments on the national scene affected chemists. First, the Commonwealth government appointed in 1916 an Advisory Council for Science and Industry, with Masson as one of its executive members. This was to lead eventually to the establishment of the Council for Scientific and Industrial Research (CSIR), a process hampered by poor funding and political vacillation[72] over the formation of the Institute of Science and Industry that preceded CSIR.

The second was the recruiting of Australian chemists to work in the munitions industry in Britain. In 1909, the English chemist A. E. Leighton was recruited by Hake to establish a cordite factory in Australia.[73] He quickly established himself and became the dominant figure in Australian munitions production for almost half a century. He was 'requisitioned' by the British government in 1915 for wartime work there. 'It very soon became clear to me', he wrote, 'that this country had already absorbed the British trained chemists, and that to staff the new explosives factories then under construction it would be essential to bring men from the Dominions.'[74] Thus seventy-five chemists selected by committees in Australia arrived in Britain in October 1916, to be followed by a further thirty in April 1917; compare this 105 with the foundation membership—372 in 1918—of the Australian Chemical Institute. Other British dominions also participated in this scheme, which was headed by South Africa chemical engineer K. B. Quinan.[75]

For this 'professional' version of military service, chemists were drawn from all states, with government and industry providing about 40 per cent each, and the remainder coming from educational institutions. Previous experience with explosives was not required. J. C. Earl, later to be professor at Sydney University,[76] and David Rivett[77] were the most highly qualified, but only about half of the munitions chemists had university degrees. Some had studied at technical colleges and others had no formal qualifications at all. They were assigned to duties in eleven factories, the largest of which was newly installed at Gretna, set up to produce 40 000 tons of cordite a year. Such was the need for their services that the Australian vanguard found themselves working twelve-hour shifts, seven days a week.[78]

David Rivett was part of the second group, and went to Brunner–Mond, where he developed improved methods for the production of ammonium nitrate. Although Brunner–Mond had vast experience in this process—a four-component system, with up to nine solid phases—Rivett tackled it late in the war and devised an improvement that was subsequently patented.[79] His expertise in the area was underlined by the postwar publication of his textbook on the phase rule.[80] Among the many Australian university men involved in the munitions effort, two—A. C. Cumming (TNT) and Bertram Steele (phenol)—were operating processes of their own design during the war. Both were Masson-trained, as was his own son, Irvine, whose researches at Woolwich resulted in major economies in cordite production.

Meeting in Melbourne in 1921 after an eight year break, the AAAS was much occupied with the recent war. Thus Wilsmore's presidential address to Section B was largely devoted to the wartime activities of chemists, and he concluded that 'on the whole, the effect of the war on the position of chemistry and chemists has been favorable'.[81] Chemists also met in joint session with Section A (Physics) to discuss 'The Application of Physical and Chemical Science in the Great War', and Section B recommended 'That this Association urge upon the Federal Government the necessity of taking steps to see that so far as possible all new chemical works and plant should be erected with a view to ready adaptability to war work in case of need.'[82] There is no evidence that government—or industry, for that matter—took any notice of their scientists' zealous advice, being forewarned perhaps 'that war leaves behind a minefield of dangerous rules and senseless orders'.[83]

The third significant wartime development in Australian chemistry was the formation in 1917 of the Australian Chemical Institute. This was modelled on the Institute of Chemistry of Great Britain and Ireland, founded in 1877 to secure official recognition of the status of chemists. By the turn of the century, of its total membership of about a thousand, there were about twenty Australian members, mainly academics and senior industrialists of British origin or experience. There was no formal organization in Australia, but the Institute was represented in four of the states by Fellows, appointed in 1908, who acted as honorary local secretaries (Liversidge was one such), advised prospective candidates, and reported to London on chemical matters of local interest.

This 'industrial' role was one that the AAAS had never sought to play. On the other hand, neither was there much local enthusiasm for the suggestion of F. M. Allan, in his paper to the Dunedin congress of 1904, that an Australasian branch of the Chemical Society of London be formed.[84] Masson spoke out in 1914 on behalf of those who were in favour of a move to raise the status of Australian chemists, but others with whom he discussed the project thought the move premature. By 1916, however, a group of chemists in Lithgow (a munitions centre) had formed themselves into an Australian Chemical Association with the purpose of improving their pay and prospects.[85] Other chemists were invited by circular letters to join, but Masson and his colleagues believed that the concept fell far short of the full range of professional qualities. Using the SCIV as his platform and enlisting the aid of Sydney chemists with whom he had close contacts because of their work with AAAS, Masson circulated all states with a draft of what ultimately became the constitution of the Australian Chemical Institute.[86] It had a foundation membership of 372 in 1918, expanding to 603 by 1920, and 802 by 1931.[87]

Despite 'early opinion that the Intitute would find no scope for holding meetings of its members for the reading and discussion of scientific papers',[88] the Institute sprang to life with regular meetings of its state branches, and before long two subgroups of the Victorian branch—Analytical and Chemical Engineering—were formed for more specialized regular dicussion. It took some time, and a good deal of often bitter discussion, before membership criteria were settled, but eventually the Institute agreed to admit graduates from recognized tertiary courses that it would accredit for that purpose.

Qualification for entry to the Institute had been a difficulty right from the beginning, and in pursuing it the Australian Chemical Institute found that Australia had no power to grant the Institute a charter. It was necessary to approach the Privy Council in London, which took advice from the Institute in Britain. The British body held that the standards adopted by the Australian Institute, that is, admission without separate examination, were inconsistent with aims to raise the status and level of education in Great Britain and the Dominions, where the Institute had been hoping to see uniform standards apply. It is worth noting that no technical qualification was needed to join either the British or Australian Associations for the Advancement of Science—they were clearly wise to keep out of such 'qualifying' activities! Indeed, the Australian Chemical Institute reminded their British colleagues that, in 1877 in the first six months of its own existence, the Institute of Chemistry had admitted whomever it deemed to be a competent chemist, evidence of specific training and experience being required only for applicants after that initial period. This battle introduced some strain between the Australian Institute and its British predecessor, which was finally allayed in the 1920s after discussions between Rivett, C. P. Callister, and their British counterparts.

BETWEEN THE WARS

Soon after the war, the Australian National Research Council (ANRC) was established. The AAAS was given responsibility for the election of the ANRC's permanent council at the 1921 congress, and it was to remain in close association until the ANRC was succeeded by the Australian Academy of Science in 1954. Soon after the formation of the ANRC, in which Masson had taken a leading part, his involvement with the Institute of Science and Industry came to an end.[89] Since its creation by the Hughes government in 1916, this body had worked as effectively as its meagre funds permitted, but legislation to secure its permanent existence was lacking. When the legislation finally came in 1920, it failed to incorporate important recommendations put forward by Masson and his colleagues on the advisory council that had guided the formation of the new Institute. In particular, the establishment of state advisory boards was not guaranteed; the scope of scientific direction was inadequate; the director's ability to appoint officers was subject to bureaucratic control; and finances for the ensuing four years were derisory. In fact, the budget was about one-third of the modest figure proposed by its director, George Knibbs.[90]

When Bruce succeeded Hughes as Prime Minister in 1923, the climate for Commonwealth scientific and industrial research improved. This began with Bruce's participation in the Imperial Economic Conference of that year, and continued through contact with Britain's Department of Scientific and Industrial Research. Finally, in 1925, Bruce decided upon reform and convened a conference to consider its shape. Masson was chairman of the fifteen man committee on reorganization, and through his advocacy nearly all of the features of structure and organization that he had been seeking since 1916 were agreed to. This led to the formation in the following year of the Council for Scientific and Industrial Research, with Rivett as chief executive,[91] G. A. Julius as chairman of the council, and W. J. Newbigin as the third member of the executive. Masson was chairman of the Victorian state committee.[92]

The AAAS congresses, while outwardly unchanged, provided evidence of a change in Australian chemistry that was not to become fully evident until after the Second World War. While Section B continued to attract about a dozen papers, half of which were delivered by professors or senior government scientists, there was a distinct shift towards basic chemistry at the expense of applications. Some of the best applied chemistry of this period, by I. H. Boas and by H. W. Gepp, never found its way onto a Section B agenda. The groundwork for the use of Australian hardwoods for pulp and paper manufacture were laid in Perth by I. H. Boas, working at his makeshift laboratory in the Technical School.[93] He later moved to Melbourne and, in 1928, became the first chief of CSIR's Forest Products Division. H. W. Gepp, who had attended Masson's chemistry lectures in the 1890s while working for the explosives company at Deer Park, became the first manager of the Electrolytic Zinc Company's refinery at Risdon, Tasmania, and mastered the problems of electrodeposition.[94]

Despite the importance of these works, the AAAS turned in other directions. The Section B committee was normally dominated by academics—mainly professors—with a few government chemists and very few (one or two in a committee of ten) industrial chemists. Interestingly, both Gepp (1931) and Boas (1937) served terms on the committee. The topics of the Adelaide congress of 1924 were valence and theories of atomic and molecular structure, carbohydrates, the influence of trace impurities on the properties of metals, and new ideas in laboratory practice. The first of these turned out to be an enduring topic, with contributions in Adelaide from E. J. Hartung, A. D. C. Rivett, and K. Grant,[95] and papers at later meetings by James Kenner,[96] H. G. Denham (a presidential address to Section B)[97] and N. V. Sidgwick. Sidgwick, at age 66, was at the Canberra meeting in 1939 as an invited guest of ANZAAS. It was not his first visit to Australia, since he had attended the 1914 British Association for the Advancement of Science meeting, travelling on the same ship as Rutherford and being inspired by his contact with the great physicist to pursue the chemical implications of atomic structure.[98] Colloid and surface chemistry was also strongly represented during these two decades, leading to discussions of the flotation processes that had been developed by the mineral industry. ANZAAS papers in this area were given by F. P. Worley, Erich Heymann, Thomas Iredale, S. W. Pennycuick and I. W. Wark.[99]

Organic chemistry was still not prominent, but the stalwarts J. C. Earl and A. K. Macbeth were joined by William Davies (later professor of organic chemistry at Melbourne) and by Victor Trikojus, at that time working in Sydney but later to become professor of biochemistry at the University of Melbourne. Perhaps the most exciting development in pure chemistry was the growth of co-ordination chemistry of metal ions,[100] mainly by chemists from the Sydney area: Francis Lions, George Burrows, David Mellor,[101] Ronald Nyholm,[102] and Francis Dwyer.[103] Commenting on this development, Branagan and Holland note that Burrows, Nyholm, and Dwyer all died in their early fifties;[104] since then we have added the name of another prominent Australian co-ordination chemist to that list—D. R. Stranks, who recently died, aged 57, some years after becoming vice-chancellor at Adelaide University.[105] Another star to rise in those two decades was that of Noel Bayliss, Rhodes Scholar for Victoria in 1927, and senior lecturer at Melbourne for four years before taking up the chair at Western Australia in 1938. The Sydney Congress of 1932 saw the advent of Adrien Albert, who presented papers in the Pharmacy Section (Section O).[106] After completing his PhD in London, Albert spoke about acridines in a joint symposium at the Canberra Congress of 1939 on the relationship between chemical constitution and biological activity. Albert's medicinal chemistry and his book *Selective Toxicity* were to earn for him the foundation chair of medicinal chemistry at the Australian National University.[107]

An indication of the new spirit of fundamental research at ANZAAS was the institution in 1930 of the Liversidge Lecture, a research presentation at each congress, given by a chemist to honour the wishes of its benefactor, who died in 1927. The list of speakers (appendix 6) makes interesting reading, because of their relative youth—very few were over fifty years of age—and because of their affiliations with fundamental research. In passing, we should note that the Royal Society of New South Wales instituted a biennial Liversidge Lecture in 1931, also with funds left by Liversidge. The purpose of the lectureship was the 'encouragement of research in chemistry',[108] and the list of lecturers includes some—G. M. Badger, L. E. Lyons, R. D. Brown, and Adrien Albert—who have given Liversidge Lectures (usually different ones!) in both series.[109]

Despite the expanded program of Section B in the two decades between the wars—even greater expansion was to come—ANZAAS itself had begun to change. While the Association still appointed research committees there were three significant changes, all in 1937. First, the ANRC founded the *Australian Journal of Science* which, because of strong links between the two bodies, became effectively the house journal for ANZAAS. Second, the Australian Chemical Institute held its first national conference in Adelaide that May; it was to be followed by others after the war.[110] And third, David Orme Masson, for many years the most powerful figure in Australian chemistry, died.[111]

At the same time, ANZAAS was evolving into a more 'socially' oriented organization which involved, among other things, the replacement of its original aims[112] by ones devoted to 'the advancement of knowledge' and the 'promotion of a spirit of cooperation' among those concerned or 'sympathetic'.[113] These social concerns embodied much more than the narrow professional concerns of earlier addresses by sectional presidents (for instance, Wilsmore's 'Present Position of Chemistry and Chemists' in 1921)[114] and were institutionalized with the formation of the Australian Association of Scientific Workers later in the year.[115]

CHEMICAL ORGANIZATION IN NEW ZEALAND

Both the formation of a chemical institute and the establishment of chemical research by the government proceeded more smoothly in New Zealand than was the case on the other side of the Tasman. The Chemistry Division of the Department of Scientific and Industrial Research (DSIR) had its foundation in the appointment in 1865 of William Skey to the position of analytical chemist at the Colonial Museum and Laboratory in Wellington. Its history has been extensively chronicled.[116] After successive reorganizations it became the Dominion Laboratory in 1909, when New Zealand achieved dominion status (i.e. was no longer technically a British colony), and then the Chemistry Division when DSIR was established in October 1926. The Laboratory has had only nine directors in this time, mainly because the first two, Skey and Maclaurin, occupied the post for thirty-five and thirty years respectively.

Shortly after this came the move to unite New Zealand chemists in a professional society. There were local societies based on the universities, often operating as branches of the Royal (previously Philosophical) Society of New Zealand, which even now plays a larger role than any of its counterparts in Australian states. In addition there was a New Zealand Section of the Institute of Chemistry of Great Britain and Ireland. The Auckland Chemical Society had sought to form a New Zealand Chemical Society in 1929, but the next year Professor H. G. Denham of Canterbury University College set out a scheme for New Zealand based largely on the constitution of the Australian Chemical Institute. Denham had been involved in the formation of the ACI while he was professor of chemistry at the University of Queensland; he moved to Christchurch in 1923 and led the department there for twenty years.[117] Thus the New Zealand Institute of Chemistry was formed in 1931 and registered as an incorporated society the next year. The first national conference was held in 1935 in conjunction with the local section of the British Institute of Chemistry, a practice that continued until the disbandment of the New Zealand Section in 1963. The membership of the New Zealand Institute of Chemistry grew from its foundation number of about 100, to 300 by 1946, 600 in 1955, and 1400 in 1980—a compound rate of 3.5 per cent a year.

New Zealand participation in Section B has diminished as ANZAAS has concentrated, through the postwar period, more on pure research and proportionally fewer ANZAAS congresses have been held in New Zealand. Chemistry at the DSIR is mostly applied chemistry[118] whereas

CSIRO chemistry, although it began as industrial chemistry in 1939, pursued more academic directions. In addition, ANZAAS had some competition from the science congresses organized by the Royal Society of New Zealand. These started after the First World War, but were held rather more often in postwar years than the roughly decadic ANZAAS meetings in New Zealand.[119]

CHEMISTS AT WAR AGAIN

With ANZAAS again 'in recess' during the Second World War, there was no opportunity for the chemistry of this period to be presented at a national forum, but a comprehensive account of the work done has been given by Mellor in his official war history.[120] Given the changing pattern of ANZAAS congresses, however, it is open to question that such applied chemistry would ever have surfaced there.

The natural development of the heavy chemical industry in Australia was accelerated by the approach of war, but it was 1940 before ICI's synthetic ammonia plant at Deer Park and soda ash factory on St Vincent's Gulf came on stream. Recognizing the need for pure sulfuric acid, stocks of imported sulfur had been built up by defence authorities in the late 1930s. The CSIR finally established a Division of Industrial Chemistry in 1940, with Ian Wark as chief. Wark had previously worked in the mineral processing industry, researching electrodeposition and flotation processes, leading to publication of his classic monograph in 1938.[121] Leighton[122] came out of retirement to act as consultant on explosives to Essington Lewis, director general of munitions, with special responsibility for the establishment by the private sector of factory annexes devoted to war work. Australian achievements were impressive, the more so because they were sometimes undertaken in the face of negative advice from overseas. Precision optical equipment for use, for example, in gun sights, was available, and local manufacture, when mooted in England, was coolly received, with the reply that four years would be needed to establish production, and the cost would be a million pounds.[123] The English advisors, however, had reckoned without the presence in Melbourne of a chemistry professor, E. J. Hartung, who as an amateur astronomer had a deep interest in optics.[124] Hartung's experiments established the formulations and the demanding process and production was undertaken in Sydney by Australian Window Glass Pty Ltd, a subsidiary of Australian Consolidated Industries. The first batch was produced in August 1941, just ten months after beginning on the project, and for less than one-tenth of the forecast cost.

Cut off from overseas supplies of drugs, Australia also had to produce its own vitamin C and sulpha drugs. This represented a sharp change of government policy, which until then had been to depend upon imported drugs. A high incidence of malaria among troops in Papua New Guinea caused the government to ask Professor A. K. Macbeth[125] of Adelaide to explore possible local manufacture of a new sulpha drug reputed to possess useful anti-malarial properties. Macbeth was already a consultant to a major chemical company (ICIANZ), which, towards the end of 1942, had sounded out its overseas principal about the likelihood of help if it were asked to undertake the synthesis.[126] In April 1943, however, the company accepted A1 priority from the government to build and operate an annexe for the production of twenty tons a year of sulphamerazine, and met the stipulation that the factory be in operation in seven months from that date. Work began on three fronts: rigorous investigation of Macbeth's eight-stage synthesis, building of a pilot plant, and construction of a factory. The company had already erected other annexes, but this one was completed with whirlwind speed between April and December, involving the preparation of 730 drawings and the expenditure of £262,000 ($5.7 million in today's currency).[127]

POSTWAR CHEMISTRY

The next four decades were to see ANZAAS replaced by the Royal Australian Chemical Institute (RACI) as a national meeting place for the chemical fraternity, just as the foundation of the Institute thirty years before had led to thriving state branches and mitigated against the development of such an infrastructure in the AAAS. Even ANZAAS's role as advisor to government, strong in Masson's day, appears to have weakened. Wark in 1951 noted that 'lacking any standards for admission "ANZAAS" could not speak for Science as a whole in Australia'.[128] When congresses resumed after the war, Section B continued much as before, with twelve papers at each of the 1946 and 1947 meetings. Then, however, came a flowering of 'academic' chemistry that was to last for almost twenty years: at Hobart in 1949, there were fifty-six papers, followed by fifty-one in Brisbane in 1951, and the average for the next ten meetings through to the mid-1960s was eighty.

Nonetheless, ANZAAS was moving away from this specialist emphasis. The aim had changed again, becoming 'to foster communication', and Professor M. L. E. Oliphant—then at the Research School of Sciences, Australian National University, and later to become Governor of South Australia—wrote in 1958, in an article published, ironically, in the *Proceedings of the RACI*, that ANZAAS

has become a cloak for the meeting of specialist societies, for the reading of specialist papers or for discussions of a specialized character. It is becoming difficult even for the scientist to be able to attend with any profit meetings of other than his own section. The ordinary man with interest in science is left out in the cold except for a certain number of public lectures of a more general character.[129]

Time was to show that most chemists preferred to attend specialist meetings of the divisions of the RACI rather than the more general offerings of an ANZAAS congress. Oliphant has had his wish that the 'ordinary man' (sic) should be catered for, but in the process most voluntary participation by scientists has been lost. Those who speak at ANZAAS do so strictly by invitation. Even the RACI has had to be careful to make its national convention—held at approximately four-yearly intervals since 1957, and with the eighth scheduled for Sydney in August 1987—a pastiche of divisional meetings, more ambitious ventures having failed to attract members.

The 1950s saw many other initiatives that changed the shape of Australian chemistry. At the beginning of the decade the New South Wales branch of the RACI had sponsored a lecture tour by Professor Alexander Todd (Nobel Laureate in 1957), who spoke in all state capitals and Canberra.[130] His visit was well received and facilitated the entry of many Australian chemists into the international community, not least through Cambridge, where many Australians completed their PhD degrees. Most subsequently returned to university teaching positions in Australia, and they form an identifiable group that assembles faithfully whenever their mentor visits Australia. The introduction of the PhD degree in Australia (1946) and New Zealand (1948, actually a 'reintroduction' since it also existed as a title there from 1922 to 1926)[131] meant that it was no longer necessary for graduates to proceed overseas to gain this research qualification, but many continued to do so until the early 1960s, aided by a system of scholarships and free passages on England–Australia shipping lines. In the 1980s hardly any Australian chemists take that route, nor for that matter do they take higher degrees in Europe or America.

Research and teaching absorbed most of the new doctoral graduates who now became available in increasing numbers, as ten universities were established between 1950 and 1970, doubling the number of such institutions. The CSIRO developed no fewer than seven chemical research laboratories from the base of the Industrial Chemistry Division, and by the mid-1960s their combined professional staff had increased from 44 to 359, rivalling the numbers of university chemistry faculty members. At Section B of the 1955 ANZAAS congress in Melbourne, twenty-four papers came from CSIRO chemists, twenty-three from Australian universities, three and one from their New Zealand counterparts, six from other government bodies in Australia, and two from the chemical industry. By 1965 CSIRO scientists gave seven and Australian university staff twenty-three of the papers at the Canberra Congress.[132]

Chemical industry was changed by the introduction of petrochemicals processing nurtured by state and federal governments, and developed largely by chemical giants from America and Germany. The established concerns, notably ICI, were mainly of British origin, and the new diversity sometimes brought with it the danger of producing too much for the small Australian market, from which local manufacturers had little experience of exporting.

An extensive search for medicinal materials among Australian plant products was conducted by the CSIRO in conjunction with the American pharmaceutical company Smith, Kline and French. This followed the exploitation of known medicinal plants during the Second World War. Under the leadership of J. R. Price (chairman of CSIRO 1970-7), a number of new alkaloids were isolated and their structures determined, a labour greatly aided by the introduction of nuclear magnetic resonance spectroscopy in the early 1960s.[133] A great many ANZAAS and RACI papers tell the story of this venture which, in the end, did not lead to the introduction of any new drugs but did much to develop organic chemistry in Australia.[134] The decade opened with a presidential address by L. H. Briggs[135] on Australasian plant products, and closed with the holding in Australia of a natural products symposium under the auspices of the International Union of Pure and Applied Chemistry (IUPAC), with Todd as President.[136] The symposium was a huge success and began a regular series of such meetings held successively in many different countries. The next IUPAC meeting in Australia (Sydney, 1969) was its twenty-second congress, held in conjunction with the meeting of the International Congress of Coordination Chemistry. This provided international recognition of another strong point in Australian chemistry, the practitioners of which had been numerous enough to organize a pre-ANZAAS symposium in Sydney in 1962, under the sponsorship of the New South Wales branch of the RACI, and with D. P. Mellor in the chair.[137]

By 1980, however, there was pressure on the CSIRO—and later on the universities—to resume their initial relationship with Australian industry and commerce. For many years there has been great strength in the area of solid-state and surface chemistry, which has progressed steadily from the early interests in mineralogy and ore treatment evident in early AAAS programmes in the 1890s. The development of flotation processes and the continued strength of the mining industry has been paralleled by the growth of two divisions of CSIRO, Mineral Chemistry and Chemical Physics, in the latter of which Alan Walsh developed atomic absorption spectroscopy.[138] In 1987 Chemical Physics was amalgamated with the Division of Materials Science (formerly Tribophysics) to form the Division of Materials Science and Technology.

A recent application of chemical skills has been the production of liquid fuels from Australian coals, a project well-funded for a decade from 1975 and involving the CSIRO, state governments, universities, and company researchers. As the energy crisis has receded, and governments face other pressing problems, activity in this area has been reduced but is likely to be resurrected when next the international market in petroleum products is interrupted. In the field of organic chemistry, the traditional interest in medicinal plant products has led, on the one hand, to syntheses and structure determinations of natural products, and, on the other, into drug design and production. The third strength, but one that has no obvious predecessor in colonial chemistry is in organometallic and coordination chemistry of metals. This has led to associated work in magnetochemistry and electrochemistry and, most recently, to the preparation of metal-containing drugs.

CHANGING PATTERNS IN AUSTRALIAN CHEMISTRY

Chemistry in Australia is strong in universities, as it is in Britain and the United States. The RACI has good coverage of the profession but its growth has slowed in the 1980s.[139] The weakness of industrial research in Australia (particularly in recent years),[140] compared with that in universities, is highlighted by the fact that Australia ranks ninth in the number of papers published but only fifteenth in ownership of industrial patents.[141]

In the late 1960s and early 1970s there was a severe job-crisis for graduate chemists, but this disappeared as chemistry enrolments in universities and colleges remained static for almost a decade.[142] Physics enrolments fell sharply in response to similar pressures and are now below the level of Australian self-sufficiency,[143] while the over-supply problems shifted to biology and biochemistry. Enrolments in geology have fluctuated over the last two decades as the mineral industry has gone through cycles of activity and quiescence.

In the first half of this century there was a net inflow of chemists, mainly from Britain, but since the Second World War a number of prominent Australian chemists have held chairs outside the country. In 1975 one of them, John Cornforth, became the first Australian Nobel Laureate in chemistry. He was among the University of Sydney's remarkable group of chemists—including Arthur Birch, A. David Buckingham, David Craig, and the late Ronald Nyholm—who have held or now hold distinguished posts in Britain. By way of interesting contrast, Adelaide graduates Roland Pettit (deceased), Lloyd Jackman, and Kevin Potts have held chairs in America, as has Sydney's Frank Anet and Melbourne's John Grutzner and Graeme Underwood, both of whom took their PhDs with Jackman while he held the Melbourne chair in the 1960s. It was not until the early 1950s that the number of chemistry chairs in Australia exceeded ten, so there were only limited opportunities for returning in mid-career to a suitable post. The postwar establishment of the Australian National University—Albert, Birch, and Craig returned there from England—and the burgeoning of the CSIRO have helped to stem this brain drain, and most Australian chemists now pursue their careers in Australia, even though many gain postdoctoral experience overseas.[144]

Foremost among the awards that took young Australian chemists abroad (mostly to England), were the 1851 Scholarships, funded by money remaining in London after the Great Exhibition of 1851. Birch, Cornforth, and Rita Harradance—soon to become Mrs Cornforth—all left Sydney that way in the late 1930s.[145] The primary objective of the original Commissioners was the improvement of industrial science in the manufacturing regions of Britain.[146] Up to the First World War, however, only 25 per cent of the 918 Scholars (117 of them Australians) entered industry: between the wars, 16 per cent, and in the period 1946–60, only 6 per cent. Since that time there have been a further 254

Scholars, of whom 76 have been Australian. Of these, only eleven (three Australians) have adopted careers in industry (about 4 per cent). While the Australian figures are small, they conform to the general pattern that outstanding scholars have increasingly preferred academic posts to those offered in industry.

CONCLUSION

Despite the fact that the most successful congress of recent years, the Festival of Science held in Melbourne in 1985, was organized by a chemist—Professor John Swan of Monash University—very few chemists were among the participants. Several chemists did make notable contributions to a session dealing with hazardous chemicals. There were also lectures by Professor W. Roy Jackson and Dr Patrick Perlmutter of Monash, and Professor Peter Andrews (Victorian College of Pharmacy), the Liversidge Address by Professor Tom Healy (Melbourne University), and a Youth ANZAAS[147] presentation by Associate Professor Ian Rae (Monash), but there were few 'chemical' faces in their audiences. The RACI had already experienced this effect when they included non-specialist sections in their national conventions. Few chemists, it seemed, would pay or were able to get institutional support to attend generalist meetings.

ANZAAS began one hundred years ago with three aims: to bring specialists together, to encourage communication between specialists in different disciplines, and to 'obtain more general attention for the objects of Science'—that is, to bring science to the public. Only the last objective survives. Barry Jones, Minister for Science in the Hawke (Labor) government, put it succinctly: 'scientists are now ultra-specialised, hermetically sealed off not only from the community, but often from other practitioners in related disciplines'.[148] But Jones is incorrect in supposing that scientists never communicate with the public. Prominent chemists, especially those with a flair for presentation of their work (and that of their colleagues) and those with strong views about the applications of chemistry, will continue to appear at ANZAAS, mainly by invitation to participate in Festivals of Science along the lines of those begun in Melbourne and continued in Palmerston North and Townsville in 1987. Specialist meetings will, henceforth, be the domain of the RACI and of smaller, more-specialist groupings, such as those convened each summer by the Australian Academy of Science. I would be surprised if, by the end of the century, there is any identifiable chemistry at ANZAAS at all.

NOTES

The title owes something to the presidential address of the Australian Chemical Institute for 1945 by R. D. Williams, 'Research Personnel in

Industry—Cabbages or Kings?', *Journal and Proceedings of the Australian Chemical Institute*, 13 (1946), 108–15.

The author wishes to express his appreciation for the help of Dr Joan Radford and the work of Dr L. W. Weickhardt; and the assistance of Monash University, the Royal Australian Chemical Institute, the New Zealand Institute of Chemistry, the CSIRO, and the New Zealand Department of Scientific and Industrial Research, Chemistry Division. The guidance and criticism of the editor of this volume, Professor Roy MacLeod, are acknowledged with gratitude.

1. Cited by Edward Mayhew in the first presidential address to Section O, Pharmaceutical Science, 'Pharmacy in Relation to Science', *Report of the AAAS*, 19 (Hobart, 1928), 665–73.
2. F. L. Holmes, *Lavoisier and the Chemistry of Life* (Madison: University of Wisconsin Press, 1985).
3. Anon., 'Obituary: William Percival Evans', *Journal of the New Zealand Institute of Chemistry*, 24 (1959), 193. Anon., 'Professor William Percival Evans, CBE, FRSNZ', *Transactions of the Royal Society of New Zealand*, 88 (1960), 59–61. Evans was born in Melbourne and brought to New Zealand as a small child. The first New Zealand born chemistry professor was P. W. Robertson, who held chairs at Rangoon (1909–11) and Wellington (1920–50). J. K. H. Inglis was the first New Zealander to hold a New Zealand chair of chemistry (Otago, 1912–35). See also H. N. Parton, *The First Eighty Years* (Christchurch: University of Canterbury, 1985).
4. The royal charter was granted in 1932, but the use of 'Royal' in the Institute's title was part of a supplemental charter in 1950.
5. The Melbourne Congress of 1985 featured the major involvement of 760 Year 11 students (mostly Victorians with some from New Zealand and Papua New Guinea) who for five days attended lectures by leading scientists and visited factories and laboratories. Youth ANZAAS has been particularly active over the years in Western Australia.
6. Mayhew, op. cit., note 1.
7. E. V. Lassak and T. McCarthy, *Australian Medicinal Plants* (Sydney: Methuen, 1983), 20.
8. L. A. Gilbert, 'Considen, Dennis (d. 1815)', *Australian Dictionary of Biography*, 1 (1966) 242–3.
9. 'Elliott's Chemical Works', *Sydney Morning Herald* (23 June, 1866), 9. 'Balmain Chemical Works', *Illustrated Sydney News* (23 November, 1872), 3. The site is now occupied by the Rozelle factory of Monsanto Australia Limited.
10. J. R. Poynter, *Russell Grimwade* (Melbourne: Melbourne University Press, 1967).
11. A. F. Scammell, 'Faulding, Francis Hardey (1816–1868)', *Australian Dictionary of Biography*, 4 (1972), 159–160.
12. Gregory Haines, 'Elliott, James Frederick (1858–1928)', *Australian Dictionary of Biography*, 8 (1981), 431.
13. James Griffin, 'Bosisto, Joseph (1824–1898)', *Australian Dictionary of Biography*, 3 (1969), 197–8.
14. John Lack, 'Cuming, James (1861–1920)', *Australian Dictionary of Biography*, 8 (1981), 172–3. This account of Cuming's eldest son also contains information about Cuming Snr and the family business Cuming Smith & Co.

15. Poynter, op. cit., note 10.
16. David Branagan and Graham Holland (eds), *Ever Reaping Something New* (Sydney: University of Sydney Press, 1985), 33.
17. Michael Hoare and Joan T. Radford, 'Smith, John (1821–1885)', *Australian Dictionary of Biography*, 6 (1976), 148–50.
18. Branagan and Holland, op. cit., note 16, 121–4.
19. D. P. Mellor, 'Liversidge, Archibald (1846–1927)', *Australian Dictionary of Biography*, 5 (1974), 93–4.
20. Joan T. Radford, *The Chemistry Department of the University of Melbourne. Its Contribution to Australian Science, 1854–1959* (Melbourne: Hawthorn Press, 1978), 38.
21. L. W. Weickhardt, 'Masson, Sir David Orme (1958–1937)', *Australian Dictionary of Biography*, 10 (1986), 433.
22. G. C. Fendly, 'McCoy, Sir Frederick (1817–1899)', *Australian Dictionary of Biography*, 5 (1974), 134–6.
23. Radford, op. cit., note 20, 28–34.
24. K. F. Russell, 'Macadam, John (1827–1865)', *Australian Dictionary of Biography*, 5 (1974), 118.
25. G. C. Fendley, 'Andrew, Henry Martyn (1845–1888)', *Australian Dictionary of Biography*, 3 (1969), 33–4.
26. Warren Perry, *The School of Mines and Industries Ballarat. A History of its First One Hundred and Twelve Years, 1870–1982* (Ballarat: BSM, 1984).
27. Frank Cusack, *Canvas to Campus. A History of the Bendigo Institute of Technology* (Melbourne: Hawthorn Press, 1973), 55.
28. Fendley, op. cit., note 22.
29. R. H. Fowler, 'Newbery, James Cosmo (1843–1895)', *Australian Dictionary of Biography*, 5 (1974), 331.
30. R. T. Fowler, 'A History of Chemical Engineering Education in Australia', *Chemical Engineering in Australia, ChE* 5 (1980–1981), 40–8.
31. P. Serle, 'Rennie, Edward Henry (1852–1927)', *Dictionary of Australian Biography*, 2 (1949), 267–8.
32. Hugh Parton, *The University of New Zealand* (Auckland: Auckland University Press for the University Grants Committee, 1979).
33. Joan T. Radford, 'Chemistry in 19th Century New Zealand', *Chemistry in New Zealand*, 49 (1984), 35–7, 60–2, 125–9.
34. J. K. Dixon, in F. R. Callaghan (ed.), *Science in New Zealand* (Wellington: A. H. & A. W. Reed, 1957), 48. This volume was prepared for the thirty-second ANZAAS Congress, held in Dunedin in 1957. P. P. Williams (ed.), *Chemistry in a Young Country* (Christchurch: New Zealand Institute of Chemistry, 1981).
35. Dixon, ibid., and Williams, ibid.
36. R. C. Cambie and B. R. Davis, *A Century of Chemistry at the University of Auckland* (Auckland: Percival Publishing Co., 1982), 2–6.
37. E. Marsden, 'Obituary Notice: Thomas Hill Easterfield, 1866–1949', *Journal of the Chemical Society* (1952), 1557. H. O. Askew, 'Obituary: Thomas Hill Easterfield (1866–1949)', *Transactions and Proceedings of the Royal Society of New Zealand*, 78 (1950), 381–4. D. R. MacFarlane, T. H. Easterfield: Science in Colonial New Zealand (MS submitted by the author as a third-year project in History, 1979). T. H. Easterfield, 'The Development of Science at Victoria College', *The Victoria University College Review*, Silver Jubilee Number (Easter 1924), 44–7.
38. Diana Dyason, 'Gresswell, Dan Astley (1853–1904)', *Australian Dictionary of Biography*, 9 (1983), 103–4.

39. Anon, 'Obituary: Norman Thomas Mortimer Wilsmore, 1868–1940', *Journal & Proceedings of the Australian Chemical Institute*, 7 (1940), 315–18.
40. S. Murray-Smith, 'Clark, Donald (1864–1932)', *Australian Dictionary of Biography*, 8 (1981), 5–6. Clark later became the first chief inspector of technical schools in Victoria; see also Perry, op. cit., note 26.
41. Perry, op. cit., note 26.
42. C. W. Wrigley, 'Guthrie, Frederick Bickell (1861–1927)', *Australian Dictionary of Biography*, 9 (1983), 143–44.
43. E. W. Knox, 'On an Application of Chemical Control to a Manufacturing Business', *Report of the AAAS*, 2 (Melbourne, 1890), 372–9.
44. R. C. Sticht, 'Progress in Rapid Oxidation Processes Applied to Copper Smelting', *Report of the AAAS*, 11 (Adelaide, 1907), 57–130. For further information about Sticht see Geoffrey Blainey, *The Peaks of Lyell* (Melbourne: Melbourne University Press, 1954).
45. D. O. Masson, Nitrous and Nitric Ethers of Glyceryl (unpublished DSc thesis, University of Edinburgh, 21 September 1883).
46. W. M. Hamlet, 'Presidential Address, Section B', *Report of the AAAS*, 4 (Hobart, 1892), 48–63.
47. J. H. Maiden, 'Bibliography of the Chemistry of Indigenous Australian Vegetable Products', *Report of the AAAS*, 1 (Sydney, 1888), 183–3.
48. E. H. Rennie, 'On the Colouring Matter of *Drosera whittakeri*', *Report of the AAAS*, 2 (Melbourne, 1890), 398–9, and 'On the Occurrence of Aesculin in *Bursaria spinosa*', ibid., 399.
49. Henry G. Smith, 'On the Kinos of Astringent Exudations of One Hundred Species of Eucalyptus', *Report of the AAAS*, 14 (Melbourne, 1913), 83–95, and 'Discussion on the Eucalypts and their Products', ibid., 116–25. R. T. Baker, 'A Census of Victorian Eucalypts and their Economics', ibid., 294–310, and 'Contribution to a Discussion on the Economics of the Eucalypts', ibid., 330–6.
50. A. I. Slowey, 'The Occurrence of Podocarpic Acid in New Zealand Red and White Pine', *Reports of the AAAS*, 12 (Brisbane, 1909), 153–4. Clara M. Taylor, 'The Phases of Sulphur', ibid., 158–9. Miss Taylor subsequently studied at Cambridge and had a long career as headmistress of girls' schools in England.
51. 'Table Showing Attendances and Receipts, and Sums Paid or Voted for Scientific Purposes (1888–1902)', *Report of the AAAS*, 9 (Hobart, 1902), xliv.
52. C. C. Palit and N. R. Dhar, 'Action of Nitric Acid on Metals in the Presence of Catalysts', *Report of the AAAS*, 18 (Perth, 1926), 162–70. The authors' address was Chemistry Department, University of Allahabad, India. Both Masson and Rivett had previously visited India, however: Radford, op. cit., note 20.
53. H. H. Dale, 'Ergot and its Active Principles', *Report of the AAAS*, 14 (Melbourne, 1913), 132–40. See also William F. Bynum, 'Dale, Henry Hallett (1875–1968)', *Dictionary of Scientific Biography*, Supplement (1978), 104–7.
54. D. O. Masson, 'The Gaseous Theory of Solution', *Report of the AAAS*, 3 (Christchurch, 1891), 84–103, and 'On Molecular Volumes and Boiling Points in Relation to Chemical Character', ibid., 103–6.
55. D. O. Masson, 'Does Magnesium form Alkyl Compounds?', ibid., 107–8. N. T. M. Wilsmore, 'Unsuccessful Attempts to Prepare Magnesium Ethyl', ibid., 108–15.
56. T. W. Edgeworth David, 'The Aims and Ideals of Australasian Science',

Report of the AAAS, 10 (Dunedin, 1904), 1-43.
57. G. M. Caroe, *William Henry Bragg, 1862-1942* (Cambridge: Cambridge University Press, 1978), 43. Bragg also wrote 'that I had had no laboratory training nor had I come in contact with research' (Caroe, ibid., 36). He had, however, published some theoretical work on electromagnetism: E. N. Da C. Andrade, 'William Henry Bragg, 1862-1942', *Obituary Notices of Fellows of the Royal Society (London)*, 4 (1943), 277-92 with bibliography (pp. 292-300) compiled by Dame Kathleen Lonsdale.
58. T. H. Easterfield, 'The Position of Chemical Research in Australasia', *Report of the AAAS*, 12 (Brisbane, 1909), 115-20.
59. T. H. Spurling, 'William Sutherland—Australia's First Physical Chemist', *Proceedings of the Royal Australian Chemical Institute*, 41 (1974), 313-14. W. Osborne, *William Sutherland, A Biography* (Melbourne: Lothian Book Publishing Co., 1920).
60. W. Sutherland, 'On the Law of Molecular Force', *Report of the AAAS*, 1 (Sydney, 1888), 39-41; 'On Molecular Refraction', ibid., 42-4. Further Investigations on the Laws of Molecular Force', *Report of the AAAS*, 2 (Melbourne, 1890), 368-70; 'The Molecular Constitution of Water', *Report of the AAAS*, 8 (Melbourne, 1900), 215.
61. Anon., 'Bertram Dillon Steele (1870-1934)', *Royal Society (London) Obituary Notices*, 3 (1934), 345. See also Radford, op. cit., note 20, 95-96.
62. S. G. Tomlin, 'Grant, Sir Kerr (1878-1967)', *Australian Dictionary of Biography*, 9 (1983), 77-9.
63. B. D. Steele, 'Presidential Address, Section B', *Report of the AAAS*, 13 (Sydney, 1911), 53-58.
64. Radford. op. cit., note 20, provides a pungent comment on rivalry at the British Association. A notable paper was presented by a young Sydney graduate who later became a prominent coordination chemist: G. J. Burrows, 'The Inversion of Cane-Sugar by Acids in Alcohol-Water Solutions', *Report of the BAAS*, 84 (Australia, 1914), 342-3; later published in *The Journal of the Chemical Society* (London), 105 (1914), 1260-70. Burrows found the rate of hydrolysis to be greater in 70 per cent alcohol than in water, with a minimum rate at 50 per cent ethanol, but not proportional to the hydrogen ion concentration nor to viscosity of the solvent. Subsequent work has shown the reaction is a third order one, depending on the concentrations of sucrose, water, and hydrogen ion, and that a better description is given when a more sophisticated acidity function is used (A. A. Frost and R. G. Pearson, *Kinetics and Mechanism* (New York: Wiley, 2nd edn, 1961), 11, 214, 333-4). A recent review of this reaction does not mention Burrows (Alison Rodger, 'The Kinetics of the Acid Catalysed Hydrolysis of Sucrose', *Chemistry in Australia*, 52 (1985), 70-1.
65. R. Robinson, 'The Synthesis of Isoquinoline Alkaloids', *Report of the BAAS*, 84 (Australia, 1914), 340-1. G. M. Robinson, 'The Condensation of Cotarnine and Hydrastinine with Aromatic Aldehydes', ibid., 341. Sir Robert Robinson, *Memoirs of a Minor Prophet* (Amsterdam: Elsevier, 1976).
66. Rohan Rivett, *David Rivett: Fighter for Australian Science* (Melbourne: Rohan Rivett, 1972). H. R. Marston, 'Albert Cherbury David Rivett, 1885-1961', *Biographical Memoirs of Fellows of the Royal Society (London)*, 12 (1966), 437-55.
67. 'Society of Chemical Industry of Victoria', *Chemical Engineering and Mining Review*, 10 (5 December 1917), 89. Radford, op. cit., note 20, provides much detail on SCIV.

68. Mellor, op. cit., note 19.
69. Radford, op. cit., note 20, 71–2.
70. L. D. Foley, History of the Australian Chemical Institute to 1945 (unpublished manuscript in the possession of RACI), 10.
71. 'Chemical Society of WA Disbands', *Chemical Engineering and Mining Review*, 25 (5 May 1933), 273.
72. G. Currie and J. Graham, *The Origins of CSIRO* (Melbourne: Melbourne University Press, 1966).
73. L. W. Weickhardt, 'Leighton, Arthur Edgar (1873–1961)', *Australian Dictionary of Biography*, 10 (1986), 69–70.
74. *Leighton Family papers* (in the possession of Miss Ann Leighton, Glen Iris, Victoria), letter from A. E. Leighton to the Official Secretary, Australia House, 12 July 1918. The existence of these papers was kindly made known to us by Dr L. W. Weickhardt.
75. N. T. M. Wilsmore, 'The Present Position of Chemistry and Chemists', *Report of the AAAS*, 15 (Melbourne, 1921), 19–42.
76. J. C. Earl, 'Being a Chemist', *Chemistry in Australia*, 45 (1978), 92–5. M. I. Bruce, 'John Campbell Earl 1890–1978', *Chemistry in Australia*, 46 (1979), 218–20.
77. Rivett, op. cit., and Marston, op. cit., note 66.
78. R. E. Summers, 'Recollections of a Munitions Chemist', *The Melbourne Technical School Magazine* (April 1933), 13–14.
79. A. C. D. Rivett, 'Ammonium Nitrate', *British Patent 131, 358* (16 May, 1918). *Chemical Abstracts*, 14 (1918), 99.
80. A. C. D. Rivett, *Phase Rule and Heterogeneous Equilibria: An Introductory Study* (Oxford: Clarendon Press, 1923).
81. Wilsmore, op. cit., note 75, 19.
82. *Report of the AAAS*, 15 (Melbourne, 1921), xxix.
83. Nicholas Hasluck, *The Bellarmine Jug* (Ringwood: Penguin Books, 1984), 21.
84. F. M. Allan, 'Proposal to Form in Australasia a Branch of the British Chemical Society', *Report of AAAS*, 10 (Dunedin, 1904), 162–3. Allan was manager of a cement works in Dunedin.
85. Foley, op. cit., note 70, 64. Radford, op. cit., note 20, 116.
86. A. E. Leighton, 'A History of the Australian Chemical Institute 1914–1932', *Proceedings of the Royal Australian Chemical Institute*, 21 (1954), 127–34, 145–54, 167–84. See also Foley, op. cit., note 70, and Daniel John O'Connor, The Royal Australian Chemical Institute 1917–1867 (unpublished MEcon thesis, University of Sydney, February 1968).
87. At this time the members were distributed 35 per cent in Victoria, 32 per cent in New South Wales, 12 per cent in South Australia, 10 per cent in Queensland, 7 per cent in WA and 4 per cent overseas, the Tasmanian branch not being established until 1938.
88. Foley, op. cit., note 70, 79.
89. Currie and Graham, op. cit., note 72.
90. Susan Bambrick, 'Knibbs, Sir George Handley (1858–1929)', *Australian Dictionary of Biography*, 9 (1983), 620–1.
91. Rivett, op. cit., note 66, 81–119.
92. Currie and Graham, op. cit., note 72, 152.
93. Newman Rosenthal, 'Boas, Isaac Herbert (1878–1955)', *Australian Dictionary of Biography*, 7 (1979), 332–333.
94. B. E. Kennedy, 'Gepp, Sir Herbert William (1877–1954)', *Australian Dictionary of Biography*, 8 (1981), 640–2.
95. *Report of the AAAS*, 17 (Adelaide, 1924).

96. Robert Robinson, 'James Kenner 1895–1974', *Chemistry in Britain*, 11 (1975), 146–7.
97. H. G. Denham, 'Modern Work on Molecular Structure', *Report of the AAAS*, 19 (Hobart, 1928), 125–46.
98. H. T. Tizard, 'Nevil Vincent Sidgwick 1873–1952', *Obituary Notices of Fellows of the Royal Society (London)*, 9 (1954), 237–55.
99. Alan Walsh, 'Sir Ian Wark, CMG, CBE', *The Australian Physicist*, 22 (1985), 169–70.
100. A. T. Baker and S. E. Livingstone, 'The Early History of Coordination Chemistry in Australia', *Polyhedron*, 4 (1985), 1337–51. D. P. Mellor, 'The Development of Coordination Chemistry in Australia', *Records of the Australian Academy of Science*, 3 (1976), part 2, 29–40.
101. S. E. Livingstone, 'David Paver Mellor 1903–1980', *Chemistry in Australia*, 47 (1980), 115.
102. Allan MacColl, 'Sir Ronald Nyholm, FRS, 1917–1971', *Chemistry in Britain*, 8 (1972), 341.
103. K. L. Sutherland, 'Francis Patrick John Dwyer', *Australian Academy of Science Yearbook 1963*, 30–41.
104. Branagan and Holland, op. cit., note 16, 51.
105. Dennis Mulcahy, 'Donald Richard Stranks, AO', *Chemistry in Australia*, 53 (1986), 437.
106. The Pharmacy Section became independent of Chemistry at the 1928 Hobart Congress, the necessary motion (of A. T. S. Sissons) having been accepted at the previous Congress: *Report of the AAAS*, 18 (Perth, 1926), xxv.
107. Adrien Albert, 'The Chemotherapy of Acridine Derivatives', *Report of ANZAAS*, 24 (Canberra, 1939), 48. The pharmaceutical papers were later published in the *Australian Journal of Pharmacy*, 20 (1939), 30–3, 173–4, 175–6, 369–71, 476–7. Adrien Albert, *Selective Toxicity* (London: Methuen, 3rd edn, 1965).
108. Anon., 'Awards of Liversidge Research Lectureship', *Journal and Proceedings of the Royal Society of New South Wales*, 91 (1957), xx.
109. A similar bequest was made to the University of Sydney. ANZAAS expressed concern that lectureships were not awakening an enthusiasm for, and providing a stimulus for, research work, as Liversidge had intended, and suggested to the University and to the Royal Society of New South Wales that they refrain for several years and then get a lecturer of really high standing from abroad! *Report of ANZAAS*, 22 (Melbourne, 1935), xxiii.
110. A. E. Scott, 'Institute Conference, Adelaide, May 1937', *Journal and Proceedings of the Australian Chemical Institute*, 4 (1937), 130–1. One of the major lectures given, by A. E. Leighton, was entitled 'Chemical Aspects of Defence'.
111. Anon., 'Obituary: Sir David Orme Masson, KBE, FRS, MA, DSc, LLD, FIC', ibid., 317–20. Masson was not only a professor himself, but also the son of a professor and of a professor's daughter, the father of a professor, the father-in-law of two professors, the brother-in-law of a professor, and the cousin of a professor. Five of his students became professors. N. T. M. Wilsmore, 'The Life and Work of Sir David Masson' (an address to the Western Australian branch of the RACI), *Journal and Proceedings of the Australian Chemical Institute*, 4 (1937), 410.
112. 'Objects and Rules of the Association', *Report of the AAAS*, 1 (Sydney, 1888), x.

113. Edward Wheller, 'The First Canberra Congress', *Search*, 15 (1984), 104–7.
114. Wilsmore, op. cit., note 75.
115. Jean Moran, 'Rhetoric and Representation in Australian Science in the 1940s and 1980s', *Prometheus*, 1 (1983), 271–89.
116. 'One Hundred Years of Chemical Research', *New Zealand Department of Scientific and Industrial Research Information No. 46* (Wellington: DSIR Chemistry Division, 1965); J. D. Atkinson, *'DSIR's First Fifty Years'* (Wellington: DSIR, 1976). W. G. M. Hughson and A. J. Ellis, *'A History of Chemistry Division'* (Wellington: NZDSIR, 1981). Williams, op. cit., note 34.
117. Information in this section is drawn from the Golden Jubilee issue of *Chemistry in New Zealand*, 45 (1981), no. 1, in which was repeated some material from the Silver Jubilee issue of 1955.
118. A. J. Ellis, 'Chemistry 1984: Design or Evolution', *Chemistry in New Zealand*, 43 (1979), 51–4. Ellis is the director of DSIR's Chemistry Division, and the article is adapted from his presidential address to Section 2 at the forty-ninth ANZAAS Congress in Auckland.
119. The first congress was held in 1919, and the twelfth in 1972 (*Proceedings of the Royal Society of New Zealand*, 100 (1971–1972), 51). The thirteenth congress, planned for Dunedin in 1976, was cancelled for lack of support (*Proceedings of the Royal Society of New Zealand*, 104 (1975–6), 45) and the series has not been revived.
120. D. P. Mellor, *The Role of Science and Industry, Australia in the War of 1939–45, Series 4 (Civil)*, vol. 5 (Canberra: Australian War Memorial, 1958).
121. I. W. Wark, *Principles of Flotation* (Melbourne: Australasian Institute for Mining and Metallurgy, 1938). Walsh, op. cit., note 99.
122. Weickhardt, op. cit., note 73.
123. Letter from the Australian High Commissioner, giving the opinion of Chance Brothers (UK), cited by H. C. Bolton, 'Some Themes in the Development of Optics in Australia', *Australian Physicist*, 24 (1984), 285–8. Bolton also supplied extra notes and references for the paper by Laurence Hartnett, 'Recollections of the Optical Munitions Panel in Australia', *Australian Physicist*, 22 (1985), 158–61.
124. J. T. Radford, 'Ernst Johannes Hartung 1893–1979', *Chemistry in Australia*, 46 (1979), 219–20.
125. G. H. Badger, 'Obituary: Alexander Killen Macbeth', *Proceedings of the Royal Australian Chemical Institute*, 24 (1937), 331–3.
126. Their reply was unequivocally negative, beginning: 'We are totally unable to appreciate what good purpose can be served by the proposal that you or even established drug manufacturers with some experience should undertake manufacture at the present time', cablegram no. 196 of 25 November, 1942, from ICI London to ICIANZ Melbourne, the existence of which was kindly revealed to us by Dr L. W. Weickhardt.
127. L. W. Weickhardt, 'The Production of Sulphamerazine in Australia 1943–1945', *Journal and Proceedings of the Australian Chemical Institute*, 14 (1947), 81–120. This is an account of a lecture given to the New South Wales branch of RACI on 21 August 1946, at the end of which Weickhardt commented that the talk 'to numbers of us . . . represent the sunset touch of part of our lives—days and months packed with incident'.
128. I. W. Wark, in *Science in Australia* (Canberra: ANU-Cheshire, 1952), 28–47. This was the opening address to the Physical Sciences section of a

seminar organized by the Australian National University on the occasion of Australia's Jubilee in 1951.
129. M. L. E. Oliphant, 'The Future of ANZAAS', *Proceedings of the Royal Australian Chemical Institute*, 25 (1958), 433–5.
130. Alexander Todd, *A Time to Remember* (Cambridge: Cambridge University Press, 1983), 118–23. Todd (born 1907) was the son-in-law of H. H. Dale (see note 53).
131. Parton, op. cit., note 32.
132. *Report of ANZAAS*, 31 (1955), 47–8; *Australian Journal of Science*, 28 (1965–1966), 148–9.
133. R. I. Willing and T. H. Spurling, 'Stanley R. Johns, 1935–1986', *Chemistry in Australia*, 53 (1986), 437.
134. An account of this work is being prepared for the Australian Academy of Science by J. R. Price and J. A. Lamberton. J. R. Price, 'Recent Developments in the Study of the Chemistry of Australian Plant Products' (Liversidge Lecture), *Report of ANZAAS*, 29 (Sydney, 1952), 67–82.
135. L. H. Briggs, 'Some Australasian Plant Products', *Report of the ANZAAS*, 28 (Brisbane, 1951), 15–26; B. R. Davis, 'Lindsay Heathcote Briggs 1905–1975', *Search*, 6 (1975), 200.
136. Todd, op. cit., note 130, 152–4.
137. 'Pre-ANZAAS Symposium on Coordination Chemistry', *Australian Journal of Science*, 24 (1962), 412; see also Baker and Livingstone, op. cit., note 100.
138. Alan Walsh, 'Atomic Absorption Spectroscopy—Stagnant or Pregnant', *Analytical Chemistry*, 46 (1947), 698A–704A. I. W. Wark, 'The CSIRO Division of Industrial Chemistry 1940–1952', *Records of the Australian Academy of Science*, 4 (1979), part 2, 7–41.
139. Figures kindly provided by Dr L. W. Weickhardt show that RACI membership grew between 1922 and 1975 at an exponential rate of 4 per cent a year, comparable to that of the NZIC (3.5 per cent) cited earlier. The British Institute grew at a similar rate, but in the United States ACS grew at 7 per cent until after the Second World War, then slowed (A.Thackray, J. L. Sturchia, P. T. Carroll and R. Bud, *Chemistry in America, 1876–1976* (Dordrecht: D. Reidel, 1985), 24). Similarly, university chemistry faculty increased at about 5.5 per cent a year until 1975, when growth abated somewhat (Weickhardt) while the Americans experienced 4.3 per cent over the period 1870 to 1978 (Thackray *et al.*, ibid., 140–1). An RACI survey in 1984, to which about half of its 5490 corporate members responded (J. T. van Gemert *et al.*, '1984 Remuneration Survey', *Chemistry in Australia*, 51(1984), 292–5, showed that 50.9 per cent were in industry, 21.9 per cent in education, and 24.5 per cent in government departments or institutions. The education component was divided among universities (11.4 per cent), CAEs (6.0 per cent), and schools (4.4 per cent), and of the government posts 12.8 per cent were federal and 11.7 per cent state.
140. John Cordner, 'Improving Scientific Innovation in Australia. Turning Good Research into Successful Commercial Applications', *Chemistry in Australia*, 52 (1985), 156–9. D. H. Solomon, 'Research for the Chemical Industry: Past, Present and Future', *Chemistry in Australia*, 53 (1986), 226–7 (Leighton Address of the RACI for 1985).
141. J. E. Kolm, 'Australian R and D Needs and Strategies for Industrial Growth', in *Manufacturing Resources of Australia* (Melbourne, Australian Academy of Technological Sciences, 1981), 271–301.

142. Total third year enrolments of 1058 in 1968 and 1053 in 1977 were reported by W. Stern, 'The Number of Chemistry Majors from Australian Tertiary Institutions', *Chemistry in Australia*, 45 (1978), 19–21.
143. J. R. Prescott, 'Jobs for Physicists—the 1985 Survey', *Australian Physicist*, 23 (1986), 230–2.
144. P. G. Lehman, 'Reemergence of the Brain Drain Bogey', *Chemistry in Australia*', 51 (1984), 251.
145. Branagan and Holland, op. cit., note 16, 60.
146. Roy M. MacLeod and E. K. Andrews, 'Scientific Careers of 1851 Exhibition Scholars', *Nature*, 218 (1968), 1001–16. I. W. Wark, '1851 Science Research Scholarship Awards to Australians', *Records of the Australian Academy of Science*, 3 (1977), part 3, 47–52.
147. Youth ANZAAS may well be the area where chemists have most impact on their 'public'. The RACI is increasingly active in promoting Youth lectures and visits by chemists to schools, and the Society for Chemical Industry has an annual careers guidance seminar, which in 1986 was attended by 340 Year 11 students (J. M. Stearne, *President's Report to Council, Society of Chemical Industry of Victoria*, 86 (1986), 2). In New Zealand the National Science Fair is an annual event organized by the Kiwanis with sponsorship by ICI(NZ)Ltd, *Proceedings of the Royal Society of New Zealand*, 104 (1984–5), 105.
148. Barry O. Jones, *'Opening Address to the National Science Summer School'* (Canberra: Canberra CAE, 1985). The focus of public comment has switched to the role of 'facilitators' who will bring industry and academia together: Bruce Cornell, 'Opportunity or Catastrophe: Australian Science in 1987', ABC Radio, Ockham's Razor, Sunday 8 February, 1987. Adrienne Clarke, professor of botany at Melbourne University, also commented on this in the ABC television programme Quantum (Wednesday 18 February, 1987).

8

Australasian Anthropology and ANZAAS
'Strictly Scientific and Critical'

D. J. Mulvaney

ANTHROPOLOGY AS SCIENCE

In retrospect, it was unduly optimistic to have launched anthropology as a foundation section at the 1888 inaugural meeting of the Australasian Association for the Advancement of Science (AAAS). After all, it was 1926 before Australia's first anthroplogy department was established at the University of Sydney, and thirty years more elapsed before a second teaching university followed suit, with a sub-department at Perth. (A research department was developed from 1948 in the Research School of Pacific Studies, the Australian National University.) The Australian School of Pacific Administration was established in Sydney in 1946, to train administration officials for territories. Across the Tasman, H. D. Skinner, the director of the University of Otago Museum, from 1919 doubled as lecturer (later reader) in ethnology. Ralph Piddington, a Sydney graduate, occupied the first New Zealand anthropology chair, at Auckland, in 1951.

The enthusiastic opening speaker at the 1888 sectional meeting, Dr J. J. Wild, was blissfully unaware of this uncertain future.[1] Wild proclaimed that, 'Anthropological science presents itself to us as the logical offspring of the new tendencies which in the course of the Nineteenth Century have invaded and revolutionized every branch of scientific inquiry.' Wild (1828–1900) circumnavigated the globe during the 1870s, doubling as secretary to the director and as official artist, on the HMS *Challenger* scientific expedition. Given this exposure to evolutionary biology, it is hardly surprising that Wild's optimistic new science of 'Anthropology may be more strictly defined as the critical examination of the intellectual and material progress of man from the earliest ages down to the present.' In keeping with this exalted purpose, anthropology 'professes to be strictly scientific and critical'. Many years passed before Section G (Anthropology) participants again attempted to define their discipline in terms of theory or of science.

Seven decades on, A.P. Elkin traced the history of Australian anthropology in numerous articles.[2] While these and his related, earlier essays constitute the most detailed version of history available, they are remarkable both for their unwavering thematic repetition and selective documentation, and for their Sydney bias.[3] Elkin paid little attention to research activities outside Sydney's sphere of influence, a fault which an error-prone, recent Sydney-based history does little to alter.[4] Neither does the biography of Elkin, because it was based on his personal collection of papers.[5]

The achievement of Adelaide fieldworkers in developing a different field methodology, while largely deprived of the financial support available to Sydney researchers, merits closer investigation than is presently available.[6] New Zealand's anthropological history has received minimal attention, mostly centred around the career of H.D. Skinner, at Otago.[7] Explanation is also necessary for the importance of Melbourne in the later nineteenth century, and for its subsequent decline. These are subjects requiring much further research. What follows is an attempt to re-examine the major landmarks highlighted by Elkin, but from a viewpoint not restricted by the Blue Mountains or the Tasman.

ANTHROPOLOGY BEFORE ANZAAS

It was a Dutch navigator who, on a Cape York beach in 1623, recorded the first facets of Aboriginal material culture. William Carstensz set a forlorn precedent by treating the indigenes like animals, shooting two and kidnapping another. His more illustrious countryman, Abel Tasman, had no more respect for savages when he obtained equally fleeting glimpses of Maori society a few years later, but on that occasion four members of his crew died in the encounter.[8]

In this sense, anthropology in Australia and New Zealand is as old as the first European landfalls. That sense has been termed ethnology by many modern anthropologists, meaning descriptive and disparate details concerning the customs, artefacts and other features of unfamiliar peoples, without attention to social relationships or other interrelated factors. In the ANZAAS region, the British term social anthropology was applied to studies integrating or comparing social structures, and analysing or interpreting institutions or symbolic systems. The raw material of ethnology was thereby utilized to test an explanatory model or to support a theory. As knowledge expanded, the prehistory of a society, the biology of its population, and its language all became subsumed within anthropology, and it was this complex fusion of disciplines that was embraced by the original AAAS Anthropology Section G in 1888.

For almost three centuries, chance observations, or more systematic attempts to describe poorly understood practices, feature as ethnology within the published accounts of explorers and colonists. One of the most comprehensive early descriptions was *An Account of the English Colony in New South Wales*, by David Collins, in 1798. Such records frequently were filtered through the mesh of prevailing philosophical notions or preconceptions of 'savage' life, such as noble savagery, unredeemed heathendom, or the perspective of unilinear physical and social evolution. Whichever eurocentric model was imposed, however, 'primitive' societies were judged inferior, while paternalistic Europeans provided dubious explanations of their origin and status. In an insightful critique, Sorrenson examined the implications for comprehending Maori history, while I traced the flaws in such interpretations in relation to Aboriginal Australians.[9]

Despite their methodological and conceptual errors, some Australian compilations dating from the later nineteenth century made systematic attempts to provide comparative data, although they offered minimal interpretation. Most of these ethnologists worked in Victoria and South Australia, where colonial government printing resources were made available. Some authors circulated printed questionnaires. E. M. Curr acknowledged the co-operation of every colonial secretary in distributing his material for *The Australian Race*.[10] Other compendiums published by government printers included R. Brough Smyth, *The Aborigines of Victoria* (1876), G. Taplin's two works, *The Narrinyeri* (1873) and *The Folklore, Manners, Customs and Languages of the South Australian Aborigines* (1879), and T. Worsnop, *The Prehistoric Arts, Manufactures, Works, Weapons, etc., of the Aborigines of Australia* (1897). Their titles betray their descriptive nature. Although the use of questionnaires was devised in Europe, and was utilized in America by L. H. Morgan, their intensive use and adaptation by several Australians over two decades formed an important benchmark in systematizing Aboriginal data, and emphasizing methods of direct observation.[11]

Some Australian colonists with experience of personal contact perceived that Aboriginal society possessed inner cohesion, and they hinted at relationships between social structure, economy, and ecological factors. Most notable amongst untutored settlers to publish was a Western District pioneer, James Dawson.[12] Outstanding amongst first generation direct observers were the two explorers E. J. Eyre and George Grey. Given their patent interest and participation in Aboriginal society, it is not surprising that both men fostered comparable activities in New Zealand, when holding administrative office there. Eyre was amongst the first Europeans to grasp the interconnection between environmental resources and social clustering, subjects of continuing research interest.[13] Grey proved even more percipient, for he was the first to discern firmly the bonds of kinship obligation, to recognize totemism as a spiritual entity (he termed it *Kobong*), and to comprehend the continent-wide occurrence of certain social practices, and the seasonal regulation of land and resource use.[14]

Australian and Oceanian anthropology entered the mainstream of contemporary evolutionary social theory through the efforts of A. W.

Howitt, a Victorian public servant, and a retired missionary from Fiji, Lorimer Fison. They corresponded with Lewis Henry Morgan in America during the 1870s. Through his encouragement they adapted his questionnaires and adopted a more stringent methodology, linked to models of social evolution. Through the collection of kinship terms, they analysed the classificatory system of kin relationships. The Victorian government printer produced several variants of questionnaire devised by Howitt, the questions proving unduly complicated for most recipients. Of 500 circulars posted during 1874, only 1 per cent 'yielded results' of value.[15] That the questions were erroneously predicated upon a model of unilinear social and institutional evolution is irrelevant to the reception of such work in the intellectual climate of Europe, during the ferment following those intellectual revolutions associated with Charles Darwin and Karl Marx. Howitt and Fison assembled geographically separated data, synthesized it and 'demonstrated' that it followed rules of social organization. Their first major collaboration was published soon after Morgan's *Ancient Society*, with a preface by, and a dedication to, Morgan. The Fison and Howitt version, on Morgan's coat-tails, became subsumed into classical Marxism via Frederick Engels's, *The Origin of the Family*.[16]

Upon Morgan's death, the two colonial social Darwinists switched their allegiance to E. B. Tylor, the Oxford foundation anthropologist. Under his patronage they contributed to British debates on theories of the origin of ritual, systems of marriage, and other group relations. This intellectual network ramified after Tylor's research assistant, Baldwin Spencer, took up the biology chair at the University of Melbourne in 1887. Howitt and Fison became Spencer's mentors before Spencer's association with the Alice Springs postmaster, F. J. Gillen. By the turn of the century, Howitt, Fison, and Spencer were linked in a question and answer postal dialogue with Tylor, Sir James Frazer, A. C. Haddon, and other British social theorists. Such relationships posed complex possibilities for circularity of armchair query and field response.[17]

Coincidentally, Spencer's Oxford classmate W. E. Roth also arrived in Australia and pursued ethnological studies in outback Queensland, independently of the Melbourne group. At AAAS congresses between 1891 and 1904, Roth, Spencer, Gillen, Fison, and Howitt (twice), all presided at anthropology section meetings. Unfortunately, few in their audiences shared their expertise or sympathy for indigenous mores. In any case, despite the importance of Fison and Howitt in 1888, the work of other inspired amateurs, including Roth, Spencer, and Gillen, lay in the future.

AAAS: THE FIRST HALF CENTURY

Ominously, the 1888 Section president was not the progressive Wild, but Dr Alan Carroll, a Sydney paediatrician. To judge from its title, his unpublished address skirted the lunatic fringe, 'on movements of races from Asia to America and Australia'. Three of the eight paper readers were missionaries from Oceania, and the immediate future of the section rested with their ethnological gleanings, rather than with social science. During the eight congresses down to 1900, clergymen, mostly missionaries, delivered fifty-one of the ninety-four contributed papers (table 8.1). The Reverend S. Ella set the negative theme in 1888, in his account of Polynesian 'customs and social habits'. He boasted that 'throughout Polynesia ... the Scriptures have been circulated in the various languages, and ... missionaries have laboured with increasing success, so that heathenism is rapidly dying out before the growth of Christian light and life'.[18] While a common thought at church fund-raising meetings, these were strange sentiments at a congress intended to study remote societies in the name of science.

Ella subsequently served as Section president at the 1893 Adelaide congress. His assessment of the purpose of anthropological research contrasted strikingly with Wild's prognostication.[19] Ella sought, 'Above the solution of scientific questions, an ethical and benevolent inquiry—How may these people be best reached by evangelical and civilizing

TABLE 8.1
Analysis of the themes of papers delivered in the Anthropology Section at AAAS and later ANZAAS congresses, 1888–1939. ('Academic' indicates those with relevant university degrees, including medical, where the subject is biological.)

	Focus of paper					Profession		
Year	Australia	Melanesia	Polynesia	New Zealand	Other	Academic	Missionary	Total papers
1888	3	2	1	1	1		3	8
1890	4	6	3		2		7	15
1891	1		2	2	1		2	6
1892	2	6	3		1		9	12
1893	6				2		1	8
1895	4		5		1		5	10
1898	7	5	5	3	1		9	21
1900	7	7				1	6	14
1902	5				1			6
1904	2			3	1	1	5	
1907	3	2				2	1	5
1909	3	2		1	5		3	11
1911	4	3		1	1		4	9
1913	5		1			1	1	6
1921	9	2	2		1	2	1	14
1923			1	1	1	1		3
1924	11		1	1	2	2	1	15
1926	6			1		3		7
1928	9	2				2		11
1930	5	2	1		3	8		11
1932	6	4	3		1	11		14
1935	18	6		1		11	1	25
1937	5	7	2	7	2	6	2	23
1939	12	4			2	8		18
Total	137	60	30	22	28	59	56	277

influences? Thus, our investigations will possess a tone and object of vast moment to the future well-being of the Polynesians.'

Already by 1889 Baldwin Spencer was less sanguine of anthropology's future than Wild. Spencer was a foundation member of the AAAS and a firm advocate of intercolonial scientific co-operation, serving on four Association Standing Committees of Investigation. When Tylor wrote from Oxford suggesting that Lorimer Fison merited a teaching post at the University of Melbourne, Spencer responded sympathetically. He regretted that the claims of certain other chairs needed priority. In any case, he observed, 'Anthropology is scarcely "practical" enough for the Australian mind in fact a preliminary lecture would probably be necessary to enlighten the general Melbourne public with regard to the meaning of the word and the aim of any individual calling himself an anthropologist.' Tylor had persisted with this idea, even though Howitt had informed him a few months earlier that he had lobbied the university chancellor, 'without warming him up in the least'.[20]

Despite Spencer's pessimism, or possibly because of it, he supported Howitt during the next year in establishing an ethnological section within the Royal Society of Victoria. As Spencer was secretary and editor for the Society, this was not difficult. In his inaugural chairman's address to the new section, Howitt felt purposefully optimistic: 'For the first time in Australia the study of Anthropology has taken up a definite position ... It seems remarkable that the science of man should have been the last to have attracted special attention here, where there are unrivalled opportunities for its presentation in a country where man still exists in as nearly a primitive condition as it is possible to find.'[21]

Howitt delivered his exhortation in May 1890, just as Australia's bubble of prosperity burst. The financial crisis devastated Royal Society attendance, so the vaunted sectional system terminated within two years.[22] Yet, despite the fact that no professionally trained anthropologists held posts in Australia, these unpropitious times heralded Australia's golden age of anthropology, which lasted until about 1905. Spencer met Gillen in 1894 and they completed the fieldwork for *The Native Tribes of Central Australia*, published by Macmillan by 1899. Nobody had attempted such intensive research with informants. In 1901 they crossed the continent and pioneered the use of movie film and wax cylinder sound recording. Their *The Northern Tribes of Central Australia* (1904) appeared in the same year in which Howitt, with Spencer's encouragement, published *The Native Tribes of South-East Australia* (1904). Roth's idiosyncratic but information-packed ethnology of western Queensland appeared in 1897. He shifted his operations to Cape York, and produced seven invaluable *North Queensland Ethnography Bulletins* by 1904. A work that stimulated later debate was *Eaglehawk and Crow, a Study of the Australian Aborigines*, published in 1899 by the Reverend John Mathew. Although it is seriously flawed, it was an original attempt at synthesis, utilizing evidence drawn from linguistics, social institutions, and religion. Pastor Carl Strehlow, a missionary at Hermannsburg, commenced his studies of Aranda religion in 1901, although the first volume was published in 1907.[23]

Studies of Aboriginal society written by, and on, women have been rare. Early this century, three women pioneered such work. Mrs K. Langloh Parker corresponded with the English social theorist Andrew Lang and published *The Euahlayi Tribe* in 1905, the year that Mrs Aeneas Gunn wrote *The Little Black Princess*. Daisy Bates commenced work on *The Native Tribes of Western Australia* during 1904, although it was not published until sympathetically edited by Isobel White in 1985.[24]

A. C. Haddon led the classic interdisciplinary expedition to Torres Strait in 1898. Bolger correctly observed that Australianists have not paid Haddon sufficient credit.[25] His team possessed unprecedented expertise, with W. H. R. Rivers (psychology and genealogical analysis), C. S. Myers (ethnomusicology), C. G. Seligman (ethnomedicine, pathology), S. H. Ray (linguistics), and A. Wilken (photography). The earliest field ethnographic movie films resulted, as also the documentation of Papuan–Australian cultural influences. Baldwin Spencer sought Haddon's advice on filming for his 1901 transcontinental journey, but did not have the resources to build a team project. Even so, Haddon later praised the first Spencer and Gillen book, which 'is reckoned by those whose opinion is worth having as being the best book of its kind about any people'.[26]

This period of intensive regional fieldwork and publication impinged upon European intellectual history. From early this century Aboriginal society became increasingly incorporated within the grand synthesis by European theorists, such as James Frazer, Emile Durkheim, Andrew Lang, and Edward Westermarck. Aboriginal art exerted great influence on contemporary interpretations of European cave art.

Speculation about the nature of primitive society also spawned the Anthropological Society of Australia in 1896. This society was the Sydney creation of Dr Alan Carroll. After some name changes, its journal was entitled *The Science of Man*; it folded in 1913, following Carroll's death. Today it provides a quarry for useful facts and idiotic theories, but at that time, it claimed to represent science. Carroll achieved a triumph of public relations, when the governors of all the colonies agreed to be Society patrons, while all Premiers accepted vice-patronage. This coup was literally crowned in 1899, by the addition of 'Royal' to the Society title, and a £100 subsidy from the New South Wales government.

The journal aimed to 'place the people of Australasia in touch with systematic anthropological research', but thereafter failed to do so.[27] Its notion of anthropology combined racism with Lamarckian sentiments and autocratic reformism: 'While from Anthropology comes an exact explanation of the cause ... If pauperism, crime, disease, dementia are increasing, then the cause will be found, and the remedy proposed and provided.'[28]

Despite patronage, the society was not patronized by leaders of the discipline seeking publication. None of those named above ever contributed to the journal. It may be a significant pointer to editorial views on acknowledged experts that, during the month in which Spencer assumed duties as Chief Protector of Aborigines in Darwin, *The Science of Man* urged the appointment of Daisy Bates to that post. In an interesting exchange, a former secretary of the Anthropological Society, W. R. Harper, wrote to Spencer concerning another person.[29] He added: 'But there is a worse foe of Australian Ethnology here in Sydney—I refer to Dr Carroll'. He stated his reasons for his resignation as secretary as revulsion from the 'dishonesty' and 'rot' involved.

The first Australian attempt to produce a professional periodical did not succeed like its New Zealand counterpart. The Polynesian Society was established in Wellington in 1892 'to promote the study of Anthropology, Ethnology, Philology, History and Antiquities of the Polynesian race.' Its journal continues today. There is perhaps a parallel during its first six decades, however, in the veneration shown by the Polynesian Society for assumed 'revealed' Maori and Polynesian traditions. This cannot be termed objective science, as demonstrated by papers published in that journal from the 1950s.[30] Some of the 'purest' traditions came from the lips of well-meaning but uncritical Europeans, who improved upon oral traditions to produce a travesty of history forced into a detailed chronological mould.

The turn of the century witnessed promising museum developments, although their potential soon stultified, and the Queensland Museum then occupied a new 'temporary' home, from which it only moved in 1986. Under E. C. Stirling, a member of the 1894 Horn expedition to Central Australia, the South Australian Museum obtained important ethnographic items. Baldwin Spencer became the honorary director of the National Museum of Victoria in 1899 and immediately arranged the move to its present site. He opened a major building extension and built up major Australian and Melanesian ethnographic collections. Spencer also arranged a massive display of artefacts, ordered in typological sequence, on principles similar to those used in the Pitt Rivers Museum in Oxford. He wrote a guidebook to explain the exhibition, which remained, for decades, the only museum guide to ethnographic collections in Australia.[31]

Because there were no anthropological teaching posts in Australia until 1926 and H. D. Skinner's New Zealand monopoly lasted from 1919 to 1951, AAAS congresses depended upon inspiration and leadership from amateurs. By 1912, however, Howitt, Fison, and Gillen were dead; Roth was in South America; and Spencer's concerns were chiefly administrative. Spencer even missed the 1911 Congress when, for the first time, contemporary Aboriginal welfare issues were a matter of serious debate. Archdeacon C. E. C. Lefroy vainly urged that Aboriginal affairs should become a federal government responsibility.[32]

Sectional leadership was lack-lustre and attendance fell, for no personality filled the void. At the six AAAS meetings held between 1907 and 1923, the three Section presidents from New Zealand even failed to submit their presidential addresses for publication, so their message was muted. Hubert Murray, a visitor from Papua, published his lecture, as did W. Ramsay Smith, a South Australian anatomist and medical health expert, who twice served as president during this dull period.

Reflecting the Section's low profile, only three papers were presented at the 1923 Wellington Congress, and only four speakers provided the five contributions at Adelaide in 1907. Except for Ramsay Smith, speakers with claims to expertise were rare: Elsdon Best, twice; Daisy Bates and John Mathew in 1913; Spencer, twice in 1921. The most frequent contributor was the former surveyor R. H. Mathews, who read eight papers at five congresses between 1898 and 1911. Despite Elkin's lengthy tribute to the volume and quality of his work, his contemporaries either ignored Mathews or accused him of plagiarism.[33] This does not suggest that he was a drawcard. The issue of his academic morality merits closer investigation. Accusations relate to different times and diverse incidents, and his detractors included Howitt (who published on the subject), Spencer, Harper, J. H. Maiden, R. Etheridge, and others.

Two activities enlivened the anthropological scene before the turning point of the early 1920s. In 1910 Daisy Bates joined the expedition led by Cambridge anthropologist A. R. Brown. Their academic relationship has become the stuff of legend, and his Australian fame transformed Brown into Radcliffe-Brown. Part of the legend also concerns plagiarism, so it is relevant to recapitulate. Bates and Brown were temperamentally incompatible and their field partnership soon terminated. He offered to help edit her proposed book, and in 1912 he left for England bearing a copy of her text. They met next at the 1914 British Association for the Advancement of Science congress in Melbourne. Asked to comment upon Radcliffe-Brown's paper, Bates declined: 'I said that Mr Brown had given my notes so nicely there was no occasion to add to them.'[34] Appropriately, the First World War had just begun.

THE BRITISH ASSOCIATION AND AUSTRALIA

The 1914 meeting of the British Association in Australia, however, has greater status in anthropological history than this riposte. Chief architects of the visit were the Sydney geologist Edgeworth David, and Spencer and Masson. David Orme Masson (1858–1937) was professor of chemistry at Melbourne and a future chairman of the Australian National Research Council. Masson and Spencer staged the Victorian session, and Spencer prepared the chapter on Aborigines for the congress's *Federal Handbook on Australia*. It constitutes one of the first concise, continental perspectives of Aboriginal society, although one subject to the erroneous preconceptions of that time. In his museum capacity Spencer also arranged a massive exhibit of 10 000 Aboriginal stone tools, which helped fix typological concepts for two generations in Victoria.[35]

The presence in Australia of eminent British anthropologists highlighted the importance of regional research. Haddon and Rivers from Cambridge and R. R. Marett of Oxford set the seal of imperial interest. Australian-born anatomist Grafton Elliot Smith (brother to S. Ramsay Smith) was an outspoken visitor. Their Australian experience confirmed

British Association members in their opinion that anthropological teaching was essential in key imperial universities. A committee already advocated an expansion of posts within Britain.[36]

This was August 1914, not a propitious time to recommend academic studies whose products were neither guns nor butter, so nothing positive eventuated. Only a year previously, upon his return from the Chief Protectorship of Aborigines in the Northern Territory, Spencer recommended sweeping changes in administration. His report was tabled in parliament and his recommendations conveniently forgotten, because they were costly.[37] While anthropologists have consulted Spencer's book on *The Native Tribes of the Northern Territory*, published expeditiously in 1914, historians have ignored his major government report. From the modern vantage point, this document is a failure, because it advocates applied social Darwinism and stern paternalism. Yet it ranks as the first comprehensive blueprint for administering indigenes, written by a person acknowledged as an expert. The federal Labor government, which appointed Spencer in 1911, soon after it assumed control of the Territory from South Australia, should be credited for its enlightened approach in seeking expert advice. His policies were enlightened and too advanced for the public opinion. By the time that Spencer recommended those policies, however, the Labor administration had fallen. At the 1913 AAAS Congress, Spencer delivered a public lecture, and seconded welfare recommendations discussed at the Anthropology Section, which were even more sweeping than his own report. These initiatives on public welfare issues lapsed with the war, and they were not renewed for over twenty years.[38]

One positive side-effect resulted from the British Association visit. Circumstances of war converted Bronislaw Malinowski, who travelled as Marett's secretary, into an enemy alien, so he spent the war years in Papua and Australia. Significant for the future of anthropological politics was Malinowski's marriage to Masson's daughter, Elsie. It is interesting to reflect that the crucial fieldwork experience that shaped the theoretical stance of the two founding fathers of British social anthropology, Malinowski and Radcliffe-Brown, was undertaken in this region. Both men attended the 1914 congress.[39]

During the Sydney stay of the congress caravan, a group of scientists led by Edgeworth David announced the discovery of the Talgai skull. Grafton Elliot Smith claimed its international evolutionary significance in a public lecture that even likened it to Piltdown. This represented one of the earliest influential scientific claims for the great antiquity of human settlement in Australia.

Archaeology developed slowly in Australia. Although a few excavations recognized the elements of stratigraphy, there was little systematic evidence. Around the turn of the century, two geologists investigated claims for the authenticity of isolated finds. Both Etheridge in New South Wales, and J. W. Gregory in Victoria, concluded that no evidence established great antiquity for human occupation.[40]

This conclusion was bolstered by stone tool collectors, who claimed that, despite differences in types of tools, their technology only reflected geological differences between rocks. They believed that the same basic industry extended throughout all Australian sites and that Aboriginal society possessed no great antiquity. As late as 1924, a group of Victorian collectors informed the Adelaide AAAS Congress members that 'we are faced with the fact that the classifications, so confidently relied upon by the European archaeologist, are quite inapplicable and that the use of terms implying a geological age as well as a stage of culture, cannot be sustained'. In advocating the same doctrine in 1949, S. R. Mitchell continued to refer to the shallowness of Australian sites and he failed totally to appreciate the relevance of stratigraphic excavation. It was an atmosphere that stifled archaeological enterprise.[41]

TOWARDS AN ANTHROPOLOGY CHAIR

The first postwar AAAS congress, held in Melbourne in January 1921, proved crucial for the future of academic anthropology. With the inauguration there of the Australian National Research Council, after an interim existence since 1919, Edgeworth David considered that this constituted 'the most important work that the AAAS has ever done.' His enthusiasm was justified in relation to the future role of the Australian National Research Council (ANRC) in anthropology.[42]

Under the presidency of Hubert Murray, Section F (Ethnology and Anthropology) sent three resolutions to the AAAS council. Appropriately, the congress president, Baldwin Spencer, chaired the council meeting that approved them:

1. That there be urged on the Federal Government the need for the formation of a Federal Museum of Australia and its Territories, and the immediate necessity for securing specimens, historical and ethnological, while they are yet available.
2. That there be urged upon the Federal Government the need for endowment of a chair in Anthropology, especially in view of its value in the government of subject races.
3. That there be ... notice of ... the desirability of at once investigating and recording the Ethnology of the northern part of Western Australia.[43]

The problem remained to involve the government—and sixty years passed before it moved to establish a Museum of Australia. In 1922, the ANRC unavailingly supported the need for 'the endowment of systematic scientific research in the Pacific Islands under Australian control'.[44] Presumably this tactic was designed to highlight Australia's responsibilities under its League of Nations mandate.

Elkin's version of these events, leading to the establishment of an anthropology chair at Sydney, emphasizes the personal role of Grafton Elliot Smith.[45] This minimizes the systematic international policy pursued by the Rockefeller Foundation, the eventual backers, and seriously undervalues the persistence and contacts of the ANRC executive, the Australian proponents. The ANRC grasped the propaganda value of the approaching 1923 Sydney and Melbourne meetings of the Pan-Pacific Science Congress. Some months before that congress, it sought expert overseas opinion to bolster their case. Sir James Frazer, a single-minded researcher, favoured a Smithsonian Institution model, with a Bureau of Ethnology.[46]

A national bureau was a concept current in Australia, winning support from some state Royal Societies in that year. Its chief publicist was W. D. Campbell, who was concerned to record art sites in the Sydney area. His interest was long-standing, for in 1914 he had addressed the British Association Congress on the need 'for systematic ethnological research in Australia'. Campbell wrote to the Pan-Pacific Congress secretary, voicing the same terms of urgent salvage that four decades later were used to promote an Institute of Aboriginal Studies.[47]

Haddon attended the 1923 congress, as did Peter Buck and H. D. Skinner from New Zealand. They all agreed that a university department of anthropology was essential, both to train administrators and others for work in territories, and to engage in research. Significantly, the congress president who steered the debate was Malinowski's father-in-law, Masson. On 7 December 1923, Masson headed an ANRC deputation, including Spencer and David, to meet the acting Prime Minister, Earle Page. 'The definite suggestion was advanced that the Commonwealth Government should found and maintain a Chair of Anthropology at the University of Sydney at an approximate cost of £1700 per annum.'[48] Cabinet approved the general concept later that day.

Because the Minister reversed his opinion early in 1924, it was June 1925 before the University was in a position to accept the offer, but the mobile A. R. Radcliffe-Brown was there to occupy it early in 1926. The problem was an attack mounted by a British colonial administration 'expert', Colonel John Ainsworth, whom the vacillating government brought out as a consultant. He advised against university-trained officials, preferring 'men of good tone, character, personality and initiative, and a tolerant and patient disposition',[49] virtues presumably lacking in graduates. The ANRC effectively counter-attacked the Colonel's position, using as ammunition the opinions of several renowned colonial governors, all pro-university men. The Minister wavered.

AMERICAN PHILANTHROPY

It was at this stage, in August 1924, that Grafton Elliot Smith arrived, to 'discuss problems of anthropological research'. He came with a brief from the Rockefeller Foundation of New York, empowered to inform the government that, provided it established a teaching department, ample research funds would follow. It was the persuasive rustle of

American dollars, therefore, rather than the smoothness of Smith's tongue, that ensured government assent.[50]

The federal government contributed £1,000 a year, while most states agreed to varying contributions, the largest shares coming from New South Wales and Victoria. It also was agreed that, in return for funding, several free places would be allocated to train native affairs adminstrators and other officials. During November 1925, E. R. Embree, a Rockefeller Foundation director, and Clark Wissler, American Museum of Natural History anthropologist, arrived on a fact-finding university tour. By the following May, the ANRC received Rockefeller Foundation assurances of field research and fellowship funding over five years, in the general field of 'human biology.'

During the inter-war years, the Rockefeller Foundation pursued a global policy of improving living conditions by promoting the efficiency and objectivity of the social sciences, assuming that a methodology akin to the natural sciences offered the solution. Institutional centres were funded to focus development, and from 1923 a fellowship programme complemented university training. In Britain, over one million pounds passed to universities and affiliated institutions. The Foundation directors emphasized anthropology as one of those disciplines most amenable to scientific methodology. It is ironic that they employed Grafton Elliot Smith as an intermediary, because the Rockefeller philanthropists judged his traditional 'diffusionist' school to be unscientific and starved it of funding. Malinowski's 'functionalist' approach impressed them by its apparent objectivity, and his London School of Economics department flourished through the Depression era. Malinowski became the Rockefeller Foundation's chief European consultant.[51]

Australia was assessed as a useful unit in their scheme for anthropological improvement, and Radcliffe-Brown was a ideal person to implement the Foundation's philosophy of anthropology as science. During his Sydney tenure (1926–31), Radcliffe-Brown dismissed diffusionism. Anthropology, he asserted in 1930, 'will make little progress until we abandon these attempts at conjectural reconstructions of a past about which we can obtain no direct knowledge in favour of a systematic study of the culture as it exists in the present'.

In that same year, he took as the text for his ANZAAS sectional presidential address, 'Applied anthropology'. Although he defined anthropology as 'the scientific study of mankind and of all aspects of human life', he ignored archaeology and history. He was laying down the rules whereby knowledge was categorized into 'acceptable' and 'unacceptable'. There was a risk that 'acceptable' students conformed to set rules and abhorred anthropologists who adhered to other values. Such trends were approved by the Rockefeller Foundation, however, as a fixed methodology added apparent objectivity. As Fisher observed of British anthropology supported by the Foundation, it became 'in their terms, more empirical, more realistic, more practical', and more intolerant of different methodological approaches.[52] This may explain why prehistory and human biology were not taught in depth at Sydney for thirty-six years, and why interstate anthropologists from different traditions received so little encouragement from that pioneer department.

Initially, however, the Foundation instructed the ANRC that funding should flow towards the following areas of what it termed as 'human biology: anatomy, archaeology, ethnology, geography, pathology, physiology, psychology and sociology'. Anthropology was absent from the list. To advise it, the ANRC established a Committee on Anthropological Research, under Radcliffe-Brown's chairmanship. At its initial meeting in October 1926, a memorandum detailing these disciplines added linguistics to the list and was approved. 'Anthropology' was not listed; urgency of salvage research amongst vanishing communities was given as a criterion for awarding grants.[53]

Between 1926 and 1940, the Rockefeller Foundation provided £52,500 for research in Australia and Melanesia, while the Carnegie Corporation granted £3,000 in 1940. Thirty researchers or institutions were involved in the forty-two Rockefeller-funded projects. Their names constitute an honour role in anthropological history: Capell, Elkin, Firth, Fortune, Hart, Hogbin, Kaberry, Piddington, Porteus, Powdermaker, Radcliffe-Brown, Sharp, Stanner, Strehlow, Thomson, Warner, Wedgewood. The journal *Oceania* depended upon these funds throughout.[54]

DEPRIVATION IN ADELAIDE

It is a striking fact that, despite the Foundation's brief, most grantees were social anthropologists and archaeology was unrepresented. So were some other fields, except for the contribution offered by the South Australian Board for Anthropological Research. Between 1927 and 1937 that body evidently received less than £3,500 for interdisciplinary research across areas favoured by the Foundation, but, it seems, little favoured by Sydney anthropologists. The role of Adelaide in research activities merits closer investigation than is possible here, but the annual expeditions into the interior were of greater importance than history has credited.

Anthropological studies in Adelaide by 1926 were buoyant. The University Board for Anthropological Research was formed that year, and the Anthropological Society of South Australia was founded. This occurred two years before New South Wales established its own anthropological society in 1928 and its journal, *Mankind*. The South Australian Museum was the most active field research museum in Australia, and a major expedition from Adelaide spent 1921-2 on Groote Eylandt. The museum was soon to pursue significant archaeological and ethnological research along the Murray River and in Cape York. Its *Records* were an important series, and like the *Transactions of the Royal Society of South Australia* published frequently in the Aboriginal area.

The professor of anatomy at the University of Adelaide, F. Wood Jones, was the driving force, but a wide spectrum of scientific expertise already had undertaken planned expeditions into the interior of the state. Wood Jones delivered the 1926 AAAS Anthropology Section presidential address.[55] It was a trenchant criticism of current white Australian attitudes to 'the claims of the Australian Aborigine'. He was optimistic that the new Sydney department would produce humane graduates to serve as able native affairs administrators. He expected that the ANRC would use its new wealth which, 'pathetically enough, comes not from Australian sources', to fund fieldwork 'organized on the best modern lines of anthropological team-work'. This assumption of group research was a significant one.

Wood Jones had visited New York earlier that year to lobby the Rockefeller Foundation, but so had Radcliffe-Brown. Wood Jones suffered disappointment, therefore, when the Foundation presented the ANRC with a Sydney monopoly: 'Although definite applications have been made to us for direct support to Adelaide, we feel ... that decisions should be made by the group in Australia rather than by us'.[56] The ANRC was to be advised by an anthropological committee chaired by Radcliffe-Brown.

The Sydney department was distinguished by some outstanding graduates, but its first three heads—Radcliffe-Brown, Raymond Firth, and A. P. Elkin—were individualists, practising the British tradition of the solitary fieldworker in prolonged isolation with his informants. Despite the precedent set by A. C. Haddon in Torres Strait, the idea of promoting interdisciplinary projects, to bring together diverse specialisms, was alien to them. The relatively brief Adelaide expeditions, involving several specialists in contact with over a hundred 'informants', must have been judged superficial, and dismissed as 'not anthropology.' This became obvious, when an application was rejected by the ANRC Committee for Anthropological Research. J. B. Cleland wished to analyse the chemistry of the narcotic plant pituri, a basic commodity in widespread ceremonial exchange systems. 'Such investigation does not come within the scope of anthropology and human biology', was the stern response.[57] It would be likely to gain acceptance today. Even then, it probably fell within the Rockefeller guidelines.

Under Radcliffe-Brown's chairmanship until 1931, South Australian applications for moderate funds to assist annual expeditions were approved. But policy disagreements occurred from 1932, by which time expeditions were ranging further. The South Australian Board for Anthropological Research was requested by the ANRC executive committee to add a Sydney pathologist to its 1932 party. This request was adamantly refused over several weeks, on the grounds that the Board had sole right to determine membership, and that others with strong claims already had been rejected. Before this acrimonious correspondence, which followed the ANRC request, the executive committee already had approved the fieldwork grant. In view of the executive's later attempt to add a member to the party, its accompanying advice seems contradictory, but revealing: 'the Council would favour longer expeditions with a smaller personnel'.[58]

Once Elkin became chairman of the ANRC Anthropological Committee, in 1934, Sydney control tightened even further. Initially the South Australian Board was asked to furnish details of work undertaken in social anthropology. Rockefeller funding for 1935 was made conditional upon duplicates of films and photographs being supplied to the ANRC, while Elkin was to receive 'duplicate notes of any matters bearing on social anthropology'. By 1936 that condition was extended to copies of all field notes. As these requirements were framed in the name of the ANRC, it is striking that the executive minutes record that the duplicate films submitted 'are at the Department of Anthropology'. The screw was turning on a group that worked voluntarily and also contributed to expedition costs. Today, these demands seem totally unreasonable.[59]

The same meeting recorded further cultural imperialism, with the appointment to the Anthropology Research Committee of two members of Elkin's own staff and another faculty colleague. The explanation offered was that it would assist administration between formal committee meetings. Reading of the ANRC executive minutes indicates, however, that over subsequent years, the full committee simply confirmed decisions already made. After 1936 the South Australian Board received no further funds, although probably little was requested. It is interesting that one of the team, C. P. Mountford, was turned down. A decade later, Elkin strenuously opposed Mountford's leadership of the Australian–American expedition to Arnhem Land.[60]

The results of Rockefeller philanthropy were surveyed by Elkin in *Oceania*. Resulting publications totalled around 160 items, of which South Australian activities accounted for 30. Elkin's explanatory text describing this work totalled thirty-nine pages, but only two concerned the South Australian expeditions. Although 15,000 feet of invaluable film was one consequence, Elkin barely paused to mention it. H. K. Fry, a medical scientist who accompanied the expedition as social anthropologist/psychologist, was dismissed in two phrases, although he had published eight papers (one in *Oceania*, before Elkin's editorship). South Australian research indeed received faint praise. Their expeditions, observed Elkin, 'have always added something to our knowledge'.

Taking quantity alone as the measure, apart from the films and their documentation, South Australian research accounted for 18 per cent of Rockefeller research publication by 1940, achieved from about 6 per cent of the funds ('assisted, sometimes very considerably', Elkin primly observed). In fact South Australian Museum records establish that, during 1927–38, a total of sixty-seven scientific papers resulted from these combined operations, so Elkin's bibliography was not a full one.[61] The proportion of South Australian publications was over one-third of the total.

The traumas before Radcliffe-Brown's departure for Chicago in 1931, over uncertainties concerning government subvention for the chair, and whether *Oceania* could continue, have been discussed by Elkin and Wise. Documentation in the ANRC papers demonstrates that their explanations of various incidents were simplistic. Because Wise based her version on Elkin's private papers, she was unaware of the more complete documentation in the ANRC files. Consequently, she caricatured the attitude of Melbourne members of the ANRC executive towards the Sydney department, while Elkin once again attributed undue influence to Grafton Elliot Smith during the events of 1932–3.[62] Elliot Smith certainly interceded with the Prime Minister for renewed commitment to the department, yet ANRC executive member links with government and, through Masson, with Malinowski, were influential. During the 1933 crisis over departmental continuity, when its acting head, Raymond Firth, left Sydney and indications were that he would not be replaced, Malinowski was in America. Based in Rockefeller headquarters, he maintained contact with Masson. Malinowski seemed assured that, if the Australian government did not renew the Sydney Anthropology departmental budget, the Foundation would have compromised. In order to maintain its Australian programme, it would have authorized the use of research funds to provide teaching on a temporary basis.[63]

ANTHROPOLOGY EXPANDS

From the establishment of the Sydney chair to 1939, by which time the permanent future of a well-staffed department was assured, participation in AAAS congresses altered. Graduate researchers and Rockefeller fellows sought to participate and so eased out the amateurs. One-third of the anthropology contributions in 1926 and 1928 were from professionals. At the five meetings between 1930 and 1939, professionals delivered forty-four of the eighty-one papers. Only three clergymen contributed to these seven congresses. The trend continued after the war. In 1951 seven of the nine papers were by professionals, and twenty-seven of the twenty-nine contributors at Sydney in 1952 were academics. With professionalization came specialization. Between 1926 and 1939, the emphasis upon Aboriginal and Melanesian research became even more prominent. Despite the New Zealand connection, total ANZAAS contributions on the anthropology of its people were few (twenty-two), and this feature persisted after the war[64] (see table 8.1).

The expansive phase of Rockefeller support ended with the Second World War. Elkin successfully approached the Carnegie Corporation in 1940, and its £3,000 grant facilitated a trickle of wartime research, most notably that by Ronald and Catherine Berndt. As late as 1951 Elkin bleakly informed the ANRC that its Anthropological Research Committee was not functioning: 'It has no money to enable it to plan and carry out research.'[65] By this time, however, the Commonwealth Government had committed itself to establishing the Australian National University, whose Research School of Pacific Studies included research in social anthropology and linguistics. During the Elkin era, its chief research concerns were in Melanesia. The Sydney-based Australian School of Pacific Administration also instructed territorial officials.

The University of Auckland established New Zealand's first anthropology department in 1951. Under Ralph Piddington, it adopted a broad approach, soon teaching linguistics, prehistory, and biological anthropology. With prehistorian Jack Golson, from 1954, it was the first university in New Zealand or Australia to teach Pacific prehistory. Australia's second university to offer undergraduate anthropology was the University of Western Australia, from 1956, as a sub-department of Psychology. Significantly, the Carnegie Corporation also funded this initiative during its first three years. About this time, also, Carnegie support enabled the newly established (Australian) Social Sciences Research Council (SSRC) to sponsor research. Of £40,000 provided to the SSRC by 1958, two anthropologists benefited by £1,900.[66]

The fifties continued as lean years for anthropology in Australia, although universities in Melbourne, Adelaide, and Sydney all contributed small sums to fund research by staff members. The Carnegie Corporation sponsored an investigation into the conditions in Australian museums and art galleries during 1933. Its findings were startling, although penurious governments chose to ignore them. Staff conditions were deplorable, collection storage and exhibition were inadequate, and scientific conservation facilities virtually non-existent. Buildings were often drab and unsuitable, the Queensland Museum constituting 'a positive fire trap' then, and for the following half century.[67] Australian museums received so little support that it is remarkable how much research into material culture, human biology, and ecology was undertaken by dedicated and underpaid curators, such as N. B. Tindale in Adelaide, F. D. McCarthy in Sydney, and, in Melbourne, E. D. Gill.

Forty years on, the 1975 Committee of Inquiry on Museums and National Collections found the situation similar, but aggravated by years of further neglect. Since 1975 one of the most significant changes in Australian culture has been the comprehensive attempt to improve standards in museums and galleries. Ethnographic collections have received particular attention. The provision of conservation facilities at many institutions and successful courses in materials conservation, chiefly that offered by the Canberra College of Advanced Education, represent constructive efforts to ensure the preservation of the indigenous cultural heritage. Despite some positive developments, further action is essential, however, to conserve relics in the field, including rock art sites.[68]

Australian anthropology's absolute dependence on foreign philanthropy ended with the 1960s, when the nation embarked upon cultural self-help. The Murray committee in 1957 pointed the way, emphasizing the financial needs of tertiary education and the fundamental importance of research. The resulting Australian Universities Commission provided the means for a dramatic expansion in the number of universities. This was accompanied by the establishment of several new anthropology departments and the development of linguistics as a separate discipline. While the archaeology of classical and biblical lands had been taught at the University of Sydney since the last century, the prehistory of the Pacific region was taught nowhere until the author presented the first course on Australia and the Pacific as a history honours option at Melbourne, in 1957. By 1962 some prehistory was also being taught at Sydney, the University of New England, and the University of Queensland, while the Australian National University appointed J. Golson to a research post.

In 1979 a survey established that about 4000 students in thirty-four Australian tertiary institutions attended introductory courses in 'anthropology', taught by 171 staff.[69] At university level, in 1986, there were eight departments with anthropology chairs, offering comprehensive courses, and others at sub-department level. Six departments of prehistory/archaeology have chairs (including one with a Mediterranean-Near Eastern focus) and three other universities offer some options. There are two chairs of linguistics in teaching departments, but that discipline is offered also within other contexts, including anthropology. There is a chair of material culture at James Cook University. Biological anthropology lacks the status of a chair, although it is variously offered within prehistory contexts, or by arrangement with relevant science departments. The past quarter century has also witnessed an expansion of 'anthropology' within New Zealand. Four departments have comprehensive social anthropology offerings, and another teaches a Maori Studies programme; there are two prehistory chairs; and linguistics and human biology are taught within most of these departments.

Australian-based foundations made substantial contributions towards research during this same period. The Nuffield Foundation assisted archaeology around 1960. The Myer Foundation and the Myer Charity Trust provided unprecedented funding in 1964, with $78,000 towards the SSRC project directed by Charles Rowley, 'Aborigines in Australian Society', which produced a small shelf of influential books.

Aboriginal Australian research benefited most, however, from the establishment, during 1964, of the Australian Institute of Aboriginal Studies. Commencing with a budget of $135,000, it reached $400,000 in 1969, rising to $2,000,000 during the seventies; it now exceeds $3,000,000. Over 1300 research projects have been sponsored, an increasing number within recent years being conducted by Aboriginal researchers. The Institute houses a major research library, an invaluable tape archive, and it has made many ethnographic films. It is an important publisher of research, through books and its journal.[70]

ANZAAS AND ANTHROPOLOGY TODAY

The flood of graduates after the expansion of the sixties and anthropology's increasing complexity had repercussions for the structure of ANZAAS. As there were too many potential contributors from the various specialisms within anthropology, experiments with parallel sessions gave way to demands for separate sectional status for linguistics, and then for prehistory.

The titling and scope of anthropology was not a new problem. Section F, Anthropology, changed to Ethnology and Anthropology at the eighth congress, in 1900, and retained that title until 1926, except for the 1907 and 1911 meetings, when it became Anthropology and Philology. As amateur contributions declined with the advent of eager Sydney graduates, the original title of Anthropology was restored in 1928. Participant philologists were so numerous in 1946 and 1947, however, that Speech Science was added. By 1975, both Linguistics and Archaeology had become separate sections (25A/25B).

By that period, however, the social anthropologists had formed their own special Australian Anthropological Society (1973), and the Australian Linguistic Society already existed. The Australian Archaeological Association followed in 1975. Ironically, the last was constituted formally at the first ANZAAS meeting after archaeology achieved sectional independence. Each of these societies holds annual meetings, and societies of consultant archaeologists and anthropologists have been spawned since. Historical archaeologists also have a national organization and hold conferences. A considerable overlap exists between the membership of these societies and regular meetings of the international conservation body, Australia ICOMOS. Most practitioners opt to spend their available time and money attending these 'trade' meetings rather than ANZAAS. The creation of the Australian Institute of Aboriginal Studies, which sponsored many conferences, also siphoned off many of those whose concerns lay with Aboriginal Australians. The trend towards specialist societies was similar in New Zealand, where the active New Zealand Archaeological Society was established in the 1950s. In both countries, there has been a growth of journals and other serials. A 1982 survey found that, in Australia alone, at least twenty-six periodicals were current, not including a number of other serials. Of this list, only *Mankind* and *Oceania* existed before the 1960s.[71]

During the past decade, the public visibility of anthropologists and archaeologists has increased dramatically, even if not through their ANZAAS presence. Their disciplines have assumed political, economic, and social dimensions unthought of when Radcliffe-Brown chose Applied Anthropology as the subject for his Section F address at the 1930 ANZAAS meeting.[72] By this term, Radcliffe-Brown meant, 'the deliberate attempt made to control and alter the civilisation of other peoples in administration and education ... based on the application of the discovered laws of anthropological science'. He hoped for a time when, through his Sydney department, 'all the officers of the administration who are in actual control of natives will have a knowledge of anthropological theory'.

Radcliffe-Brown's natives were passive recipients of a paternalism based on scientific rules interpreted and applied by Europeans. Any personal sense of humanity or cultural richness was lacking. He classified Aborigines as 'a highly specialised variety of our species', in process of rapid extinction.[73] His successor, A. P. Elkin, inherited similar social Darwinian assumptions, but he is rightly praised for his championing of unpopular Aboriginal causes during the unsympathetic

1930s and 1940s. Yet his humanitarianism was founded on stern, authoritarian principles. He expected neither anthropologists nor administrators to ask Aboriginal people what they wanted. Anthropology held the answer, that a conscious policy of assimilation would produce a just world, in which 'they' became 'us'. In the meantime, rather than dying out, the Aboriginal population increased.

For about twenty years, however, anthropologists have worked closely with, and for, Aboriginal communities. On many public issues since the Gove bauxite mining case and subsequent land rights controversies, many anthropologists have been outspoken in support of Aboriginal self-determination, even though such views are unpopular in the general community. Many have acted as advisers to Aboriginal Land Councils or local communities and assisted in documenting land claims. Even access to their private field notebooks has become a legal issue in the Northern Territory. This is an applied form of anthropology inconceivable half a century ago.

In archaeology, also, what most people dismissed as obscure findings about the remote Australian past have assumed political proportions. Aboriginal leaders quantify their land claims by appeals to their ancestral forty millennia of occupation; environmental impact assessments must add the new dimension of past cultures to complicate their statements; to the evaluation of conflicting land-use options, whether minerals, tourism, forestry or urban spread, is added the presence of places and relics of significance to present and future black or white Australians. It was the public recognition of the importance of archaeological resources, as much as respect for environmental integrity, that made the proposed Tasmanian dam on the Franklin River such a landmark in the emergence of a new cultural nationalism.[74]

The anthropological sciences have demonstrated that evidence may be assembled and interpreted, using methods appropriate to basic social or natural science practice. That does not mean that the implicit objectivity of scientific research requires that its practitioners remain silent when their work poses political or social issues impacting upon the public arena. It always was a fiction that science stands above and outside politics. When the results of scientific research are ignored or distorted by governments or by sectional interests, the lesson of much contemporary anthropology and anthropology is clear—that silence is neither golden nor objective.

A CENTURY OF ANZAAS

During its first quarter century, the AAAS meetings bridged immense distances and provided opportunities for leading, but isolated, practitioners of anthropology to converse. This shared experience may have proved more significant than were the papers they presented. At the Melbourne Congress in 1900, Gillen met Howitt and Fison; these latter visited Hobart in 1902, where Roth presided over the meeting; Spencer regularly attended New Zealand congresses. Between the wars, attendance at meetings by Hubert Murray, E. W. P. Chinnery and F. L. Williams, in particular, served to emphasize the importance of research and administration in Papua New Guinea.

The era of academic anthropology dates from the 1930 Congress, symbolized by the presence of Section president Radcliffe-Brown. It became not only a matter of pooling information, or of meeting colleagues, but also a means whereby young researchers sought to gain academic recognition or preferment. For almost half a century, ANZAAS provided the basic public venue for academic anthropology. As the human sciences diversified and specialized, such general meetings held less appeal. This was an inevitable outcome.

ANZAAS provided another vital co-ordinating role in acting as the agent for interface with government and international agencies. The history of anthropology in Australia might have been very different if the British Association had not ventured here in 1914. Without the creation of a strong ANRC in 1921, with positive sympathy for anthropological teaching and research, Australia would not have achieved one of the first chairs of anthropology in the British Empire. Its role in attracting American Foundation funding proved of inestimable importance.

A century on, anthropology (used in its broadest sense) is a major discipline in Australian tertiary education. Whether anthropology or anthropologists across the last century have been as 'strictly scientific and critical', as Wild predicted in 1888, is a matter of doubt. Credit is due to ANZAAS, however, for providing the opportunities that Wild anticipated: 'to press the claims of this youngest of sciences upon the attention of all those who are disposed to encourage research within the limits of the Australian colonies'.[75]

NOTES

1. J. J. Wild, 'Outlines of Anthropology', *Report of the AAAS*, 1 (Sydney, 1888), 442–6. Wild was at that time the assistant secretary of the Royal Society of Victoria.
2. Amongst other historical surveys see, A. P. Elkin, 'Anthropology in Australia: One Chapter', *Mankind*, 5 (1958), 225–42; 'A Darwin Centenary and Highlights of Fieldwork in Australia', *Mankind*, 5 (1959), 321–33; 'The Development of Scientific Knowledge of the Aborigines' in H. Sheils (ed.), *Australian Aboriginal Studies* (Melbourne: Oxford University Press, 1963), 3–28; 'The Journal Oceania: 1930–1970', *Oceania*, 40 (1970), 245–79; 'R. H. Mathews: His Contribution to Aboriginal Studies', *Oceania*, 46 (1975–6) 1–24, 126–52, 206–34.
3. A. P. Elkin, 'Anthropological Research in Australia and the Western Pacific, 1927–1937', *Oceania*, 8 (1938), 306–27; 'Anthropology in Australia, 1939', *Oceania*, 10 (1939), 1–29, 'Anthropology and the Peoples of the South-West Pacific: the Past, Present and Future', *Oceania*, 14 (1943), 1–19; 'A. R. Radcliffe-Brown, 1880–1955', *Oceania*, 26 (1956), 239–51.
4. G. McCall (ed.), *Anthropology in Australia: Essays to Honour 50 Years of*

'*Mankind*' (Sydney: Anthropological Society of New South Wales, 1982), chapters by McCall and A. Hamilton.
5. T. Wise, *The Self-Made Anthropologist* (Sydney: George Allen & Unwin, 1985).
6. Proceedings of a seminar to honour C. P. Mountford, *Anthropology in Australia* (Adelaide: Anthropological Society of South Australia, 1976); H. M. Hale, 'The First Hundred Years of the Museum—1856-1956', *Records of the South Australian Museum*, 12 (1956); N. B. Tindale, 'A South Australian Looks at Some Beginnings of Archaeological Research in Australia', *Aboriginal History*, 6 (1982), 93–110; P. Sutton 'Anthropological History and the South Australian Museum', *Australian Aboriginal Studies* (1986/1), 45–51.
7. J. D. Freeman and W. R. Geddes (eds), *Anthropology in the South Seas* (New Plymouth: Avery, 1959). But see the interesting study by M. P. K. Sorenson, 'Polynesian Corpuscles and Pacific Anthropology: the Home-Made Anthropology of Sir Apirana Ngata and Sir Peter Buck, *Journal of the Polynesian Society*, 91 (1982), 7–28. On archaeology, see J. Sanders, Historiography or Prehistory (unpublished MA thesis, University of Auckland, 1968).
8. J. E. Heeres, *The Part Borne by the Dutch in the Discovery of Australia* (London: Luzac, 1899), 36–41; A. Sharp (ed.), *The Voyages of Abel Janszoon Tasman* (Oxford: Clarendon Press 1969), 120–4.
9. M. P. K. Sorenson, *Maori Origins and Migrations* (Auckland: Auckland University Press, 1979); D. J. Mulvaney, 'The Australian Aborigines, 1606-1929: Opinion and Fieldwork', *Historical Studies*, 8 (1958), 131–51, 297–315; 'The Ascent of Aboriginal Man: Howitt as Anthropologist', in M. H. Walker, *Come Wind, Come Weather* (Melbourne: Melbourne University Press, 1971), 285–324; 'Gum Leaves on the Golden Bough: Australia's Palaeolithic Survivals Discovered', in J. D. Evans *et al.*, *Antiquity and Man* (London: Thames and Hudson, 1981), 52–64.
10. (Melbourne, 1886), XI. For elaboration of this subject, see D. J. Mulvaney, 'Patron and Client: the Web of Intellectual Kinship in Australian Anthropology' in N. Reingold and M. Rothenberg (eds), *Scientific Colonialism: A Cross-Cultural Comparison*, (Washington: Smithsonian Institution, 1987).
11. J. Urry, 'A History of Field Methods' in R. Ellen (ed.), *Ethnographic Research* (London: Academic Press, 1984), 35–61.
12. J. Dawson, *The Australian Aborigines* (Melbourne: Robertson, 1881).
13. E. J. Eyre, *Journals of Expeditions of Discovery into Central Australia* (London: Boone, 1845), e.g. vol. 11, 151–4, 218.
14. G. Grey, *Journals of Two Expeditions of Discovery in North-West and Western Australia* (London: Boone, 1841), e.g. vol. 11, 225–9, 219, 259–63.
15. Mulvaney, in Walker, op. cit., note 9, 301.
16. L. Fison and A. W. Howitt, *Kamilaroi and Kurnai* (Melbourne: Robertson, 1880); Mulvaney, 'The Australian Aborigines', op. cit., note 9, 309.
17. D. J. Mulvaney and J. H. Calaby, '*So Much That is New*'. *Baldwin Spencer, 1860–1929: A Biography* (Melbourne: Melbourne University Press, 1985).
18. S. Ella, 'A Comparative View of Some of the Customs and Social Habits of the Malayan and Papuan Races of Polynesia', *Report of the AAAS*, 1 (Sydney, 1888), 493.
19. S. Ella, 'The Origin of the Polynesian Races', *Report of the AAAS*, 5 (Adelaide, 1893), 143. For Spencer's AAAS committee membership see Mulvaney and Calaby, op. cit., note 17, 108.

20. *Tylor Papers* (Pitt Rivers Museum, Oxford), W. B. Spencer to E. B. Tylor, 23 May 1889 and A. W. Howitt to Tylor, 28 January 1889.
21. A. W. Howitt, 'Anthropology in Australia', *Proc. Roy. Soc. Vict.*, 3 (1891), 15–22.
22. Mulvaney and Calaby, op. cit., note 17, 100.
23. W. E. Roth, *Ethnological Studies Among the North-West-Central Queensland Aborigines* (Brisbane: Government Printer, 1897). C. Strehlow, *Die Aranda- und Loritja-Stamme in Zentral-Australien*, (Frankfurt: Frankfurter Museum für Völkerkunde, 1907–20).
24. I. White (ed.), *Daisy Bates. The Native Tribes of Western Australia* (Canberra: National Library, 1985).
25. P. Bolger, 'Anthropology and History in Australia: the Place of A. C. Haddon', *Journal of Australian Studies*, 2 (1977), 93–106.
26. *Spencer Collection*, (Pitt Rivers Museum, Oxford). Box 1, A. C. Haddon to W. B. Spencer, 5 May 1902.
27. *Australasian Anthropological Journal*, 10 (August 1896), 3.
28. *The Science of Man* (September 1906), 10; (April 1897), 98, 110; (September 1896), 4.
29. *The Science of Man* (1 January 1912), 84; *Spencer Papers* Mitchell Library MSS. 29/7, W. R. Harper to W B. Spencer, nd (probably 1898).
30. R. Piddington, 'A Note on the Validity and Significance of Polynesian Traditions', *Journal of the Polynesian Society*, 65 (1956), 200–3; J. Golson, 'Archaeology, Tradition, and Myth in New Zealand Prehistory', *Journal of the Polynesian Society*, 69 (1960), 380–402.
31. W. B. Spencer, *Guide to the Australian Ethnographical Collection* (Melbourne: Government Printer, 1901); Mulvaney and Calaby, op. cit., note 17, 243–53.
32. Lefroy, 'The Future of the Australian Aborigines', *Report of the AAAS*, 13 (Sydney, 1911), 453–4. Report of Research Committee, 'Export of Anthropological and Ethnological Specimens', *Report of the AAAS*, 14 (Melbourne, 1913), 453; Mulvaney and Calaby, op. cit., note 17, 274–5.
33. A. P. Elkin's 'R. H. Mathews: His Contribution to Aboriginal Studies', *Oceania*, 46 (1975–6), 1–24, 126–52, 206–34; Walker op. cit., note 9, 248; Mulvaney and Calaby, op. cit., note 17, 215, 369; Mitchell Library MSS 29/7. W. R. Harper to W. B. Spencer, 1898. See also D. E. Barwick, 'Mapping the Past', *Aboriginal Studies*, 8 (1984), 102.
34. E. Salter, *Daisy Bates* (Sydney: Angus & Robertson, 1971), 176.
35. *Federal Handbook of Australia* (Melbourne: Government Printer, 1914), 33–85; Mulvaney and Calaby, op. cit., note 17, 318.
36. Report of the Committee, 'The Teaching of Anthropology', *Report of the BAAS*, 84 (London, 1914), 235–6.
37. 'Preliminary Report on the Aboriginals of the Northern Territory', *Bulletin of the Northern Territory*, 7 (Melbourne: Department of External Affairs, 1913).
38. Report of Research Committee, 'Welfare of Aborigines Committee', *Report of the AAAS*, 14 (Melbourne, 1913), 450–3; Mulvaney and Calaby, op. cit., note 17, 305–14.
39. For further information on the 1914 Congress, and on Malinowski, see Mulvaney and Calaby, op. cit., note 17, 228–9, 252, 257, 317–8, 322.
40. R. Etheridge, 'Has Man a Geological History in Australia?', *Proc. Linnean Soc. NSW*, 5 (1891), 259–66; J. W. Gregory, 'The Antiquity of Man in Victoria', *Proc. Roy. Soc. Vict.*, 17 (1904), 120–44.
41. A. S. Kenyon, D. J. Mahony and S. F. Mann, 'Evidence of Outside Culture Innoculations', *Report of the AAAS*, 17 (Sydney, 1924), 464–6;

S. R. Mitchell, *Stone Age Craftsmen* (Melbourne: Tait, 1949), 5, 107–8.
42. David is quoted in Mulvaney and Calaby, op. cit., note 17, 371. For a historical outline, see A. P. Elkin, 'The Australian National Research Council', *Australian Journal of Science*, 16 (1954), 203–11.
43. 'Summary of Resolutions Affecting Committees of the Various Sections', Section F *Report of the AAAS*, 15 (Melbourne, 1921), xxxiii.
44. *ANRC Papers* (National Library of Australia) NL MS 482, ANRC Minute Book 49, 17 August 1922.
45. e.g. *Oceania*, 40 (1970), 252; *Mankind*, 5 (1958), 231.
46. *Spencer Papers* (Museum of Victoria), Box 25, file 3/10. (A copy of Frazer's submission is held and is dated 27 June 1923.) *Haddon Collection* (Cambridge University Library). Haddon wrote to Spencer on the subject on 13 May 1923. NL MS 482, ANRC Executive Minutes, 19 March 1923, 68.
47. Report of the BAAS, op. cit., note 36, 534. Campbell's 1923 activities are inferred from *ANRC Papers* NL MSS 482, Box 58/828. During 1923 the Royal Societies of Queensland, South Australia and Western Australia, and the Historical Society of Victoria, all supported a Federal Bureau of Ethnology.
48. NL MS 482 ANRC. Executive Minutes, 7 December 1923.
49. The ANRC prepared a comprehensive document on the proposed chair in April 1925, which was submitted under Masson's signature to the University of Sydney. It outlined Ainsworth's views and counter-arguments. It cited the support of colonial administrators. These included Lugard, Swettenham, Wingate, Temple, Im Thurn and Hubert Murray. NL MS 482, ANRC Executive Minutes, 17 April, 5 June 1925. The printed report is included pp. 148–51. A more informal account of the Ainsworth affair is quoted in the text. It was given by Masson, in a letter to Haddon (*Haddon Papers*, op. cit., note 46, Box 4/M, nd, but 1925). In an earlier letter, certainly 1923, Masson thanked Haddon for his support for the chair—'we owe it to you'.
50. NL MS 482, Box 61/853B, 2 September, 13 November 1924; Executive Minutes, 22 August 1924.
51. For the policy of the Rockefeller Foundation and its elevation of anthropology, see D. Fisher, 'American Philanthropy and the Social Sciences in Britain, 1919–1939;' The Reproduction of a Conservative Ideology', *Sociological Review*, 28 (1980), 277–315.
52. A. R. Radcliffe-Brown, 'The Diffusion of Culture in Australia', *Oceania*, 1 (1930), 366–70; 'Applied Anthropology', *Report of ANZAAS*, 20 (Brisbane, 1930), 267–80. See Fisher's discussion of Rockefeller policy, op. cit., note 51, 300–6.
53. NL MS 482/853c, E. R. Embree to D. O. Masson, 27 May 1926 (two letters). NL MS 482/862, Box 62—the first meeting of the advisory committee on anthropological research.
54. Details are provided by Elkin in *Oceania*, 8 (1938), 306–27; 10 (1939), 2–29; 40 (1970), 245–79.
55. F. Wood Jones, 'The Claims of the Australian Aborigine', *Report of the AAAS*, 18 (Perth, 1926), 497–519; quotation from 519.
56. NL MS 482/853c. Embree to Masson, 27 May 1926.
57. NL MS 482/848A, Box 60, 17 October 1932.
58. ibid., correspondence June–July 1932; advice, 7 March 1932.
59. NL MS 482, Minutes Book 1926–35, November 1934, 27 March 1935; Minute Book 1935–41, 7 April, 25 September 1936.
60. NL MS 482, Minutes Book 1935–41, 25 September 1936, 23 April, 3

December 1937; Minutes Book 1941–48, 19 February 1943. Wise, op. cit., note 5, 204–5.
61. *Oceania*, 8 (1938), 10 (1939). Fry is mentioned on 312 and 22 respectively. The quotation, 10 (1939), 19; considerable assistance, 8 (1938), 312. Some details of South Australian activities between 1927 and 1940 are given by H. M. Hale, *Records of the South Australian Museum*, 12 (1956), 149–55. The Museum bibliography was supplied by Philip Jones, South Australian Museum.
62. Wise, op. cit., note 5. The ANRC collection contained many documents not available to Elkin. Wise's account of Donald Thomson's resignation from a Rockefeller fellowship in 1930 omits some crucial documents (96–100); her reference to R. Piddington is facetious (115), for the episode was a major incident concerned with academic freedom and civil rights. The ANRC is accused of censuring Radcliffe-Brown, 'like a schoolboy' (102), but the latter had embarked on major policy initiatives without any reference to them (NL MS 482/855, 859B). For Elkin on the role of Elliot Smith during 1932–3, see *Oceania*, 40 (1970), 263.
63. NL MS 482/853c, Masson to ANRC, 30 March 1933; Malinowski to Masson, 3 May 1933.
64. I make no claim to accuracy in these totals, as subjective judgements are involved in assigning categories, but the proportions are significant. Some joint symposia in the 1930s are omitted.
65. NL MS 482/539, 19 March 1951.
66. Social Science Research Council of Australia, *Report* 1956 (Canberra, 1957).
67. S. F. Markham and H. C. Richards, *A Report of the Museums and Art Galleries of Australia* (London: Museums Association, 1933). D. J. Mulvaney, 'Museums', *Australian Cultural History*, 2 (1982–3), 38–45.
68. A. Rosenfeld, *Rock Art. Conservation in Australia* (Canberra: Australian Heritage Commission, 1985).
69. G. M. McCall, 'Teaching Anthropology in Australia', *Australian Anthropolgcial Society Newsletter*, 8 (1980), 16–20.
70. For a more detailed review of AIAS history, see Mulvaney, 'A Sense of Making History': Australian Aboriginal Studies, 1961–1986', *Australian Aboriginal Studies*, 2 (1986), 48–56.
71. P. Hinton and G. McCall, 'The Great Australian Anthropological Periodicals Explosion', in McCall, op. cit., note 4, 109–34.
72. A. R. Radcliffe-Brown, 'Applied Anthropology', *Report of ANZAAS*, 20 (Brisbane, 1930), 267–80.
73. ibid., 268.
74. D. J. Mulvaney. 'Towards a New National Consciousness', *Australian Natural History*, 21 (1983), 88–9.
75. *Report of the AAAS*, 1 (Sydney, 1888), 446.

9

Education, Social Science and the 'Common Weal'

Alison M. Turtle

Around the nexus of education and psychology have clustered a variety of concepts relating to the physical and psychological welfare and betterment of Australians, both individually and collectively. Such concerns emerged at the level of formally organized study and discussion in the late 1880s amongst theorists and practioners of education and what was then termed 'mental science'. In Australia the emergence of the first universities barely preceded the beginnings of the formal study of education and the new scientific psychology, so that the forum provided by the Australasian Association for the Advancement of Science (AAAS) for the development and dissemination of new ideas of local relevance within these fields was, in the late nineteenth and early twentieth centuries, of particular significance. The first meeting of the AAAS was held, in fact, less than a decade after the establishment in Germany of the first laboratory of experimental psychology.

The present chapter will discuss aspects of such developmental and dissemination within Australia, with passing reference to Australia's near neighbour New Zealand, in relation to the activities of Section J of the AAAS, Mental Science and Education. Of particular interest is the psychology/education interface in the early stages, with practitioners of both disciplines perceiving the possibilities for fruitful interaction in filling the needs of the adolescent society through ambitious programmes of measurement and testing. Many of their recommendations became firmly entrenched in the practices of government departments, though in later stages massive expansion of educationists and psychologists in numbers, and, particularly of the psychologists, in interest, led to a gradual loosening of the early tie between them. In the background was the constant presence of the British model, influential through personal interaction as well as emulation, the slow consolidation of these disciplines within the universities, and the formation of specialist organizations to meet the requirements of psychologists and educationists thinly spread over vast distances for professional interaction and public platforms, and, later, public funding.

During its three-quarters of a century of existence, Section J sponsored seven committees, all during the first half of this period. Four of these were narrow in focus, concerned with matters of the school curriculum and not involving the psychologists. The other three, while similarly pertaining to the affairs of children, were much broader in perspective and reflected the joint interests of educationists and psychologists. The central part of this chapter will be organized around discussion of the context in which these three committees were conceived and operated.

INSTITUTIONALIZATION OF EDUCATION AND PSYCHOLOGY—BACKGROUND AND BEGINNINGS

The history of the scholarly disciplines of education and psychology, the study of the human mind and behaviour, necessarily bears an intimate relationship to that of the development of formal systems of education. In Australia, where the control of education became and remained far more centralized than in Britain, this relationship is particularly close. As one historian remarked (on the occasion of a previous ANZAAS jubilee), 'The central feature of Australian education in the nineteenth century was the struggle for State control of education, resulting in the emergence of free and compulsory education for every child.'[1]

Victoria set the pace in 1872, making primary education free, compulsory, and secular; New South Wales followed suit in 1880 for children aged from 7 to 14 years and set up a Department of Public Instruction; by 1895 all the other states had passed similar legislation. Apart from New South Wales, where a small-scale attempt was made in the 1880s, no steps towards the extension of state education to the secondary level were taken until the turn of the century. Having conducted commissions of inquiry just after this time, New South Wales and Victoria each appointed directors of education, respectively Peter Board and Frank Tate, who inaugurated wide-scale reforms, including liberalization of the primary school curricula, introduction of state systems of secondary education, and creation of teacher-training systems. The Melbourne Teachers College opened in 1903, that at Sydney in 1906. Their programmes served as models for Tasmania and South and Western Australia. More students thus became available for the universities, the first of which had been founded at Sydney in 1850; over the next six decades this was followed by the establishment of one in each of the other colonial and state capitals.

The first public Australian platform of education and psychology, however, was within the AAAS, where both appeared remarkably early.

At the second meeting of the AAAS in Melbourne in 1890, Mr W. Sutherland (of Victoria) moved that a new section be added for the science of education, saying that 'he thought educational enthusiasm here was more general than in the mother country'. The matter was deferred, and Henry Laurie, professor of philosophy at the University of Melbourne, gave notice of motion 'that a new section be added under the head of Mental and Moral Science'.[2] At the fifth meeting in 1893, Section J, Mental Science and Education, was inaugurated with Laurie as chairman. Sociology, on the other hand, made a belated appearance. The second chairman was Francis Anderson, professor of philosophy at the University of Sydney and for years an outspoken advocate of the introduction of education, psychology, and sociology to the university curriculum. Although Anderson in 1911 gave a paper at the AAAS on 'Sociology in Australia: a Plea for its Teaching', and a resolution was passed supporting him,[3] sociology appeared as a Section only in 1972, at the forty-fourth meeting. The affiliation of education and psychology within the AAAS was an uneasy one throughout. Upon their very first appearance in 1893, notice of motion was given to separate mental science and education into separate Sections, but this came to nothing.[4] In 1926 Philosophy was added to Section J, which became Education, Psychology, and Philosophy. Immediately after the Second World War, in 1946, a proposal was again put forward to create separate sections for education and psychology, but it was withdrawn.[5] In 1969, after a series of preliminary manoeuvres including the dropping of philosophy from the group in 1968, education and psychology finally separated, to form Sections 22 and 23.

In Britain the appearance of education and psychology in the British Association for the Advancement of Science came rather later, education in 1901 and psychology not until 1921, after having been a minor segment of the physiology section since 1896. Presumably its Australasian counterpart may be seen as a relatively more significant protagonist in the early regional development of these disciplines. Simple paucity of numbers probably caused the educationists and mental scientists to take advantage of the opportunity to come together provided by such an umbrella organization as the AAAS. The comparative situations of the disciplines within universities in the colonies and the mother country likewise supports this theory of the relatively greater importance of the AAAS in their early stages. The University of Sydney was the first in the area to be incorporated in 1850, and it was not until 1881 that Australia saw the first appointment of a philosopher, that of Henry Laurie to a lectureship in logic at the University of Melbourne, which was converted to a chair of mental and moral philosophy in 1886. Otago University in New Zealand was somewhat in advance here, having among its first three chairs in 1871 one in mental and moral science. The first full-time post in psychology in Australia was within the philosophy department at Sydney, where Henry Tasman Lovell became associate professor in psychology in 1920; in 1929 his position was elevated to a full chair within a separate department. This remained the only one in either country until 1946, when the University of

PLATE 11
The AAAS meeting, Brisbane, 1909, cover of train brochure (*Liversidge Papers*, University of Sydney Archives).

PLATE 12
'Members of the General Committee of the Congress, Adelaide, 1911' (*Daily Telegraph*, 10 January 1911; *Liversidge Papers*, University of Sydney Archives).

PLATE 13
1923 Congress, Victoria University College, Wellington (Courtesy of Peter Lever-Naylor).

PLATE 14
Left to right: Mr David Carment, Sir Hubert Murray and Sir Edgeworth David (*Sun*, 17 August 1932).

PLATE 15
'Will science lend a hand?' (Stan Cross, in *Smith's Weekly*, 15 September 1923).

PLATE 16
Mueller Memorial Medal, first awarded in 1904.

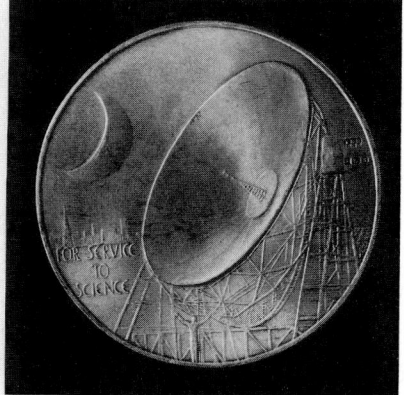

PLATE 17
ANZAAS Medal. Designed by Andor Meszaros, it depicts the progress of knowledge from the pyramids of Egypt to the radiotelescope: 'only the moon and horizon have remained constant' (*Australian Journal of Science* 28, September 1965, 98. Courtesy of Peter Lever-Naylor).

Melbourne appointed as professor Oscar Oeser, formerly of the University of St Andrews in Scotland. New Zealand set up joint chairs in psychology and philosophy at both Victoria University and Canterbury Colleges in 1949, the chairs being separated in 1951 and 1953 respectively. Chairs of education were established at Sydney in 1910 and at Melbourne in 1918, and at all four New Zealand university colleges between 1920 and 1924. In Britain, on the other hand, philosophy had been entrenched as a teaching subject for centuries, and mental philosophy specifically since the first half of the nineteenth century. The first chairs of education, at Edinburgh and St Andrews, were set up in 1876, and by 1899 there were seven such throughout the British Isles. The first lectureship in psychology was occupied by G. F. Stout at Cambridge from 1887, and after the creation of the first chair at Manchester in 1919 others proliferated.

In view of the alliance between physiology and psychology in the British Association, and of psychology with philosophy in most of the universities in Australia and New Zealand throughout the early decades of the twentieth century, the alliance of psychology and education within the AAAS requires explanation. The Scottish influence on Australian intellectual life in the second half of the nineteenth century, in terms of formal academic structures and of personnel must be remembered here. In Scotland the tie between education and psychology within the universities and teachers colleges in the years leading up to the twentieth century was much stronger than it was in England, where philosophy and, in the case of Cambridge where the first psychological laboratory was established, physiology were the foundation blocks on which the new psychology was built. It is noteworthy that the first two chairmen of Section J, Laurie and Anderson, were graduates of Scottish universities, Laurie of Edinburgh and Anderson of Glasgow. A good example of a product of the prevailing interdisciplinary mentality is the career of Henry Tasman Lovell, first professor of psychology in Australia. Lovell studied philosophy under Anderson at Sydney, became a lecturer at the new Sydney Teachers College, went to Jena in 1907 to take a doctorate with Wilhelm Rein, the renowned educator of the day, and returned to the Sydney philosophy department in 1910 where he established an experimental psychology laboratory. In 1920 he was appointed associate professor in a newly independent psychology department, and in 1929 he became a full professor.

Notwithstanding these different patterns of organizational beginnings, in later stages of planned growth the Australasians constantly aspired to emulate their British colleagues. At the first meeting of Section J in 1893, a committee was appointed to consider the best means of encouraging psychophysical and psychometrical investigation in the Australasian colonies, with Laurie as its chairman. This committee decided to write to two illustrious psychologists, Francis Galton in London and the British-born Titchener at Cornell, for advice on how to proceed.[6] This deferential posture is reflected in a train of subsequent developments. For example, the National Institute of Industrial Psychology was formed in Britain in 1918 and its counterpart, the Australian

Institute of Psychology, only in 1927, even though the founder and director of the former, C. S. Myers, paid explicit tribute to the influence of the Australian philosopher/psychologist, Professor Bernard Muscio of the University of Sydney, on British thinking in these directions.[7] Only in 1965 was the Australian Psychological Society established as a body independent of the British Psychological Society, of which it had had branch status for twenty years. Likewise the New Zealand Psychological Society, founded in 1968, had been the New Zealand branch of the British Psychological Society since 1947.

THE EARLY DOMINANCE OF CHILD RESEARCH IN THE ACTIVITIES OF SECTION J

The first three decades of Section J produced a plethora of papers on formal structures of education, such as curricula and teaching training and on the history of education, a handful on traditional problems of mental philosophy and pathology, and a small but increasing number concerned with the definition and role of the new psychology. Most conspicuously, however, these years saw two major thrusts in the direction of betterment both of the race and the individual, through anthropometric measurement and mental testing. Each discourse spawned its own progeny of problems, techniques and theory, the overlap of interest as well as of origins being considerable, and each began with the appointment of a committee whose brief included these matters. Whilst anthropometry in its later stages came to include mental testing, chronologically it preceded the development of modern tools of mental measurement.

Both movements placed major emphasis on the study of children, and both were considerably influenced by the child study movement. In fact some historians, such as Roselyn Gillespie, incorporate the whole early development of the scientific movement in education into this movement. According to Gillespie the organized form of child study was probably initiated in the United States in 1894 with the National Educational Association Child Study Department under G. Stanley Hall, followed by the British Child-Study Association founded by James Sully in 1896.[8]

In 1894 the Anthropological Society of New South Wales was established with Dr Alan Carroll as secretary, and for several years it received a state subsidy of £150 a year. One of its main objects was to arouse interest in child study, and to set up a laboratory where teachers and others involved in child care could learn the principles of measurement.[9] The largest child-study organization in Australia was, however, the Child-Study Association of Australia, founded in 1902 with Dr Carroll as president. Its objects emphasized both the measurement and grading of children and assistance for defective children; the study of children was not seen as being the exclusive province of specialists.[10] As time went on, matters of child nutrition came to the fore in Dr Carroll's interest, and in 1904 he severed his connection with the first Association and founded a second with the same name,[11] while the first became transformed into a Parents and Teachers Union.

As the teachers colleges, Departments of Education and universities gradually took upon themselves many of the tasks originally self-allotted by the various child-study bodies, the latter increasingly distanced themselves from scientific activities. Although the occasional paper on child study continued to be given at meetings of the AAAS until the end of the first decade of the new century, by 1909, when a Child Study Association was formed in Brisbane, the questions addressed related far more to custodial care than to scientific measurement—food, clothing, sleep, moral development, habit formation, and the like.[12] One may speculate, however, that the original movement, bringing to public awareness the possibility of a scientific approach to the study of children, who had hitherto been neglected by mental philosophers and the new experimental psychologists alike, was one of the factors facilitating the early liaison in the AAAS between education and mental science.

STAGE ONE: ANTHROPOMETRIC MEASUREMENT

Anthropometric surveys in Australia have been conducted almost entirely upon children, usually within schools, where such surveys have served the additional purpose of medical inspection. The first set of anthropometric data collected dates back to the 1880s, when a Dr F. N. Manning of New South Wales reportedly measured the heights of a number of schoolchildren.[13] Surveys by government officials in New South Wales and Tasmania were made in 1902, and in 1908 another followed in New South Wales, and in 1909 one in South Australia. Within the Departments of Education of these three states and Victoria, School Medical Services were established between 1907 and 1914. The coming-of-age of anthropometric activities occurred in 1908, when a conference of state premiers in Melbourne agreed to the adoption of a uniform system of medical inspection, including anthropometric measurement, for schoolchildren, and to accept the offer of the Commonwealth government to co-operate with them by providing the services of the Commonwealth Statistician.[14] At its congress in Sydney in January 1911, as the result of a paper given in Section I (Sanitary Science and Hygiene),[15] the AAAS appointed an Anthropometric Committee comprised of representatives of the Commonwealth, each of the Australian states, and New Zealand, to promote the study of anthropometry. This undertaking won the support of the Medical Congress held in Sydney the following October[16] and had the effect of persuading the Prime Minister, Andrew Fisher, to prod the state premiers into renewed activity.[17]

The Association itself retained a long-term interest in the area. Such interest, it will be recalled, began back in 1893 when a committee had been appointed to report 'on the best means of encouraging psychophysical and psychometrical investigation in Australasia'. In its report to the sixth meeting, in 1895, this committee concluded regretfully that the

Australasian colonies could not as yet afford chairs and laboratories for experimental psychology, but it aimed 'to devise a scheme of statistical inquiries of theoretical and practical value, in connection with the State school systems of the Australasian colonies'.[18] In his letter soliciting advice from Sir Francis Galton, renowned British anthropometrist and pioneer of the study of individual psychological differences, E. F. J. Love, the secretary of this committee, wrote that the group was of the opinion that 'a statistical enquiry into the development of psychical faculty among the children of these colonies is at present within the compass of possibility', owing to the universal and compulsory system of state education. He also mentioned that the support of officers of the Education Departments in each state had already been secured.[19] No later reports of the committee are available. In 1932 the Section I Committee, which had been revived after the war, finally reported to ANZAAS on the 'Range of Physiological Variables in the Inhabitants of Australia and New Zealand', summarizing a dozen anthropometric surveys dating back to 1902, and comparing results from the two countries.[20]

From its history within the AAAS, it may be discerned that anthropometry came under the umbrella of a number of disciplines, the most conspicuous being medicine, statistics, psychology and education. Anthropometry provided a useful vehicle for interaction of psychology and education with other social and medical sciences. For example, although the 1932 report was made to the combined ANZAAS Sections of Medical Science and National Health, and Physiology and Experimental Biology, it included measures of intelligence based on formal psychological tests. Scheduled interactions between Section J and others were to become commonplace in later years.

Probably nowhere was the emulation of the British model more prominent among the Australian scientists than in this field of anthropometric measurement, where the major goal of the Australians was the comparison of their data with those of the British. At the meeting of the British Association for the Advancement of Science in Bristol in 1875 an Anthropometric Committee had been appointed, with Galton as its secretary; it conducted an extensive survey continuing until 1883 of some 53 000 persons of all ages throughout the British Isles. Measurements were taken not only of bodily structures but also of a range of physical capacities, such as breathing and vision, and related to demographic variables such as occupation and social class.[21] In 1884 Galton established his own Anthropometric Laboratory at South Kensington to record measurements of such of the British race as chose to pay threepence for the privilege. Between 1887 and 1893 anthropometric laboratories were set up at meetings of the British Association to measure the physical characteristics of the class of person attending such meetings. In 1892 the British Association appointed a committee to investigate mental and physical deviations from the norm among schoolchildren, which operated till about 1900.[22] Another committee was appointed in 1902, and it reported in 1908; this was the first explicitly to suggest the inclusion of psychological measures, although it recognized that the scheme of mental measurement that it proposed, relying on ratings of attributes of character and capacity by trained observers, was only tentative and experimental.[23] The citing of British precedents in methodology was commonplace in both proposals for and reports of actual anthropometric surveys in Australia,[24] and the AAAS in 1911 went so far as to recommend the wholesale adoption by governments in Australia and New Zealand of the British standard of 1908.[25] Interest in anthropometric matters however was sustained much longer by the AAAS and ANZAAS than by the British Association, where it seems to have fizzled out for lack of funds after 1913.[26]

Apparently the colonial mentality lingered on, with a need for continuing reassurance as to the adequacy of the racial materials on which the new nation was being erected. The enunciation of intent behind proposals for anthropometric studies in Australia and New Zealand remained remarkably stable for at least four decades from the 1890s. In 1892 F. B. Suttor, New South Wales Minister for Public Instruction, approved the suggestion of Charles MacKellar, MP, that Drs Graham and Jamison be employed (without payment) to undertake an anthropometrical survey of the state's schoolchildren. This produced a long statement regarding the nature and aims of such a survey from Dr Graham; in an accompanying letter to the Minister he wrote that 'The subject is largely a scientific one and would do much to answer the question whether the Australian born race is physically superior or inferior to the English speaking people in other parts of the world, and what causes are at work in producing such changes, if such are found to exist.'[27]

This theme, of the need for preservation of the physical standards of the parent race, was sustained and elaborated upon throughout subsequent utterances. By the next decade it included reference to state educational policies, and, as mental testing got underway, standards of mental as well as physical ability received mention. Thus at the Hobart meeting of the AAAS in 1902, the New South Wales government statistician, T. A. Coghlan, reported to the Economics and Social Science Section the results of a study that he described as the first of anthropometric importance in Australia, which he himself had conducted on 2000 Sydney schoolchildren. He described such studies as serving the purposes of education in the following terms:

The object of all worthy educational systems is the freest and most symmetrical development of individual minds and bodies, and the training of our future citizens, so that they may approximate, as nearly as possible, the ideal human type. As a first step towards this, it is necessary to determine a standard by which physical development may be tested. This can be done by taking measurements of a large number of children, and summarizing the results according to approved methods.[28]

Coghlan went on to make detailed comparisons of the data for the Australian children with that available for those of North America and England.

An alternative to Coghlan's concept of the ideal human type was that of efficiency. In 1926 E. Morris Miller, director of the State Psychological Clinic in Tasmania, published a monograph on 'Brain Capacity and Intelligence' under the auspices of the Australasian Association of Psychology and Philosophy, in which he worked out norms of brain growth of Tasmanian children. In a foreword to this study the state chief secretary, J. Allan Guy, declared:

The preservation of the high standard of efficiency of the British peoples in all parts of the Empire is a pressing demand upon all Governments, and no portion of the Britannic Commonwealth of Nations should be permitted to deteriorate in physical and educational development. It is necessary that scientific workers should prepare various norms of growth, both of physique and intelligence, so that Governments may have information to guide them in matters of policy.[29]

As recently as 1932 the report of the Anthropometric Committee of the Association asserted 'Is our race improving or deteriorating in these new lands is a question of first importance, and frequently asked about tropical Australia.'[30]

Anthropometry offered a means of scientific inquiry into the idea of the 'national type', a concept identified by Richard White as being conspicuous on the late nineteenth century intellectual landscape of Australia and elsewhere. In this concept notions of natural selection, social Darwinism, and the supposed superiority of the British racial type blended, to produce questions in the transplanted colonial communities as to whether the Anglo-Saxon racial type would continue to progress or commence to degenerate in foreign climes. Underlying this concept is the quest to improve 'national efficiency' described by G. R. Searle as characterizing much of British political thought in the early years of the twentieth century.[31] But the goals of the anthropometrists were far from realized. Today their statistical results are considered either obvious or dubious,[32] and even had it been otherwise, to use such data in the cause of racial betterment posed no mean challenge. What, for instance, was to be identified as improvement and what as degeneration, in terms of departures from the physical and mental norms of the parent country? Interestingly, reports of anthropometric studies in Australia and New Zealand most usually simply record the data and rarely evaluate them.

Although nourished by ambitions for racial betterment, anthropometry in Australia was not to any noticeable extent tuned to the ends of the eugenecists and used as ammunition for a programme of selective breeding of the population, as had been recommended by Galton in Britain.[33] Nor is there any evidence that anthropometrists were involved in the debates over immigration policy at the time of the formation of the White Australia policy, as was the case with psychologists in the United States after the First World War.[34] In Australia the proponents of anthropometry widely and consistently attributed the differences they observed among those measured to differences in environment. Thus when the first survey carried out in 1908 by the New South Wales Education Department found marked differences in physique between children from poorer and industrial suburbs and those from less densely populated and seaside areas, warnings were issued as to the need for precautions in urban planning to protect the future.[35] In 1901 Coghlan wrote to the Under Secretary for Public Instruction in New South Wales, 'it is now generally admitted that...the causes of physical variation must be sought in the life-conditions of individuals...By means of Anthropometry, all deviations from the laws of normal growth may very quickly be recognized at a stage when timely treatment may overcome them...Anthropometry enables us to detect the evil effects of study upon eyesight, chest capacity and muscular powers no less than upon the brain.'[36] Similarly, in 1913 the AAAS Anthropometric Committee reported that: 'The objects of Anthropometry are primarily to measure the structural features of the activities of the human body, and secondly from the changes in structure and in activities disclosed in successive measurements to estimate the influence of various differences of environment.'[37]

The administrative tie that came most easily to anthropometry was with medical hygiene in schools. To the Departments of Education, which have been the main agencies conducting these surveys, the anthropometric aspects have from the beginning been only a part of what they hoped to achieve. In submitting a proposal to the New South Wales Department for Public Instruction in 1906 for regular medical inspection of schools, the director, Peter Board, listed six other items of interest besides anthropometric survey, these being defects of eyesight, hearing, other physical defects, infectious disease, school hygiene, and mentally defective children. Writing to the premiers in 1909 about the need for uniformity in lines of investigation, Andrew Fisher commented that the state governments 'are not agreed as to how far any scheme should lean to the purely hygienic or the purely anthropometric side of the work'.[38] It was this tie with medical hygiene that sustained anthropometric measurement for so long. Whilst the Association maintained its enthusiasm for these surveys into the 1930s, and undoubtedly acted as a catalyst in the early stages, it lacked the finances for such expensive undertakings. Had the recommendations of the AAAS not been endorsed by the School Medical Services, anthropometry could not have flourished as it did. As it is, government departments still from time to time assemble anthropometric data on Australian children.[39]

STAGE TWO: THE START OF MENTAL TESTING

Mental testing reached Australia and New Zealand early. Its development was nurtured by two predecessors besides anthropometry. The advance of the scientific movement generally in the nineteenth century stimulated the application of the new psychology to educational matters, and passage of the various Education Acts caused the problem of identification and management of the backward child to surface shortly thereafter.

The enthusiasm for anthropometric measurement had helped to pave the way for the acceptance of mental testing. While the bulk of the anthropometrical measures taken were physical, ranging from the basic recording of height and weight to that of literally dozens of measures of physical structures and capacities, an awareness of the potential for inclusion of psychological material was there from at least 1894, as may be seen in the Association's letter to Galton. The first published report from the New South Wales Department of Public Instruction in 1908 contained a table comparing the physical development of pupils with their mental progress as indicated by class in school,[40] while that from South Australia the following year had a section, strongly emphasized in its conclusions, on ' "Dullards" and their Physical Defects'.[41] In her 1911 address to the AAAS Mary Booth, subsequently the secretary of the Association's Anthropometric Committee, described the enterprise as having 'for its object the exact measurement of the anatomical, physiological and psychological characters of man'.[42] By the time Morris Miller conducted his 1926 survey in Hobart, psychological testing was well underway in Australia, and he used formal tests of mental development in order to obtain figures on mental levels to compare with those on brain capacity.[43]

During the 1880s the new scientific psychology was vaunted in Australia by philosophy professors Anderson and Laurie and, as has been seen, its introduction to Australasia formed one of the topics of the very first committee of Section J in 1893. The first laboratory along Wundtian lines was set up by John Smyth when he became principal of Melbourne Teachers College in 1903;[44] the venture failed to flourish, and the next one in Australia was established a decade later by Lovell within the philosophy department at Sydney. Thomas Hunter meanwhile began one at Victoria College, New Zealand, in 1908, on his return from a visit to Titchener's Wundtian laboratory at Cornell. In 1923 K. S. Cunningham revived Smyth's undertaking with a research programme oriented towards problems in education,[45] and in 1928 Hunter converted his laboratory at Victoria in New Zealand to similar ends. About the same time Clarence Beeby of Canterbury College developed a laboratory primarily servicing the psychological components of courses in education there.[46]

Mental deficiency was initially claimed as their province by the medical profession. As early as 1892 a paper was delivered at the Intercolonial Medical Congress of Australasia on 'Idiocy and Juvenile Insanity in Victoria'.[47] The medics sustained an interest in the area, and in 1911 the same congressional group appointed a Committee on the Care of the Feeble-Minded in Australasia.[48] Meanwhile, on the recommendation of Section J, the AAAS in 1900 appointed a committee 'to consider and report upon the need of separate State education of defective children in Australasia';[49] its report was presented in 1902 but unfortunately has not survived. Once again, however, it seems that an early Section J committee may have acted as a catalyst to official action. In 1904 a New South Wales Commission on Primary Education recommended the establishment of special schools for the feeble-minded, on the principle that 'uniformity of result is not an aim of education... The aim of education is to make the most of each child as he stands.'[50] This initial resolve, however, was not implemented for some time. The first move was in Victoria, where between 1911 and 1912 the Bell Street School at Fitzroy was set up for feeble-minded and maladjusted children.[51] Interest was strengthened by the publication in Britain in 1908 of the report of the Royal Commission on the Care and Control of the Feebleminded; in 1911 New Zealand passed a Mental Defectives Act based largely on this report.

Gradually the Victorian example was followed by other states, and during the 1920s the school psychological services began. Psychologist H. T. Parker was appointed in Tasmania in 1922; Lorna Hodgkinson in 1922 as Superintendent of the Education of Mental Defectives in New South Wales; Constance Davey in South Australia in 1924; and Ethel Stoneman in West Australia in 1926.[52] The major concern of all these new appointees was the classification and educational placement of children identified as different; while some attention was given to supernormal and delinquent children, most went to the subnormals. By the 1920s the new tools of the mental testing movement were available for use, and were widely used by the state psychologists.

Mental testing stemmed from Binet's work in France. Together with Simon, Binet devised the first scale of intelligence which was published in 1905; revised versions appeared in 1908 and 1911. The scale was used almost immediately by staff of the Sydney Teachers College; R. G. Cameron tried out the 1908 version on 71 children and switched to the 1911 revision for another 177.[53] In 1913 Elizabeth Skillen produced her own modification of the 1911 scale for children attending the Sydney Blackfriars School.[54] In Melbourne, Stanley Porteus commenced in 1913 as head of the newly established Special School for subnormal and maladjusted children at Bell Street and, though lacking any formal psychological training, promptly proceeded to place Australia's name on the international map of testing technology. Finding the Binet tests of intelligence inadequate for use with these children, as well as in their coverage of prudence, foresight, initiative, and purpose, he devised his own series of tests, the Porteus Maze Tests. At John Smyth's instigation he presented his results to the 1914 overseas meeting of the BAAS in Melbourne, where they met with an excellent reception. In 1916 he was appointed part-time lecturer in experimental education at the University of Melbourne and worked with various students, including K. S. Cunningham, on standardizing the maze tests against the Binet. In 1919 he left for an appointment in the United States, where he remained.[55] Other revisions of the Binet tests for Australian use were undertaken in the 1920s, one by H. T. Parker (who had trained at Sydney Teachers College) in Tasmania in 1921, and a major one by G. E. Phillips at Sydney Teachers College in 1924. [56] Over the next few years the staff of Melbourne Teachers College produced their own group tests of intelligence, complementing these others intended for individual administration.[57] Thus by the end of the decade Australia was well equipped with tests normed to local standards.

Both Parker and Philips reported on their work to the first AAAS meeting after the war, in 1921, as also did E. C. Blackwell on hers on the application of Burt's tests to Australian children.[58] At the next meeting, in 1923, Section J passed a resolution that paid tribute to the value of such work.[59] Clearly by this stage the AAAS had abandoned the active role of instigator and was serving more as a transmitter. Had its meetings not been suspended during and immediately after the war years (1914–20), a crucial period in the development of basic techniques of mental testing, the AAAS might have played a more active role in this area. In its absence, other organizations filled this function. Although there existed no scholarly research journal in education in the first decade,[60] the Education Society set up at Sydney Teachers College in 1908 was publishing its own monograph series, the *Records*, which over the two decades of its appearance ran to forty-five issues. In 1917 the College commenced publication of *Schooling*, and there was also the *Education Gazette and Teachers Aid* put out by the Victorian Education Department. Furthermore, in 1922 the Australasian Association of Psychology and Philosophy was formed in Sydney at the instigation of the recently appointed Professor of Philosophy, Bernard Muscio, and in 1923 it produced the first issue of the *Australasian Journal of Psychology and Philosophy*. Both this Association and its journal survived until the philosophers and psychologists went their separate ways after the Second World War, and its meetings and journal provided a vehicle for the communication of the new work in mental testing. Although these tiny societies did not bring educationists and psychologists together as did the AAAS, their existence meant that the AAAS *Reports* were no longer the only outlet for research publications in their areas.

BETWEEN THE WARS: OFFICIAL APPLICATION OF TESTING TECHNIQUES AND THE ARRIVAL OF ACER

Shortly after the First World War moves were afoot within the Association and government circles to bring about widespread application of the newly devised tools of testing and guidance. Military exigencies had already consolidated their position overseas, and while fundamental research into the nature of the characteristics being measured continued at an active level in the United States and Britain alongside the development of public testing services, the Australians appeared convinced of the utility of the new approach. In 1924, at the request of Section J, the AAAS resolved to appoint a 'small committee of psychologists and men engaged in industry or commerce', to draw up a series of measures of vocational aptitudes, 'which may be of use to teachers, and others'.[61] This proved to be the last committee proposed by this Section. In 1928 Section J recommended the formation of State Vocational Guidance Bureaux, and to this end requested the Commonwealth Statistical Bureau to compile the number of entries yearly in different occupations in each state. A small sub-committee of the Section was formed in each state to monitor progress.[62]

At the 1930 meeting of the renamed Association, ANZAAS, K. S. Cunningham, who had been appointed director of the newly constituted Australian Council of Educational Research (ACER), reported on the steps taken since the 1928 meeting, and on developments to date.[63] The first part of Cunningham's account was unfortunately not recorded, so that the precise role of the AAAS in achieving results must remain conjectural. But the results themselves are clear enough. By 1930 New South Wales, Victoria, and South Australia each had provision for vocational guidance work under the auspices of their Education Departments, though in New South Wales the establishment of a special bureau for guidance purposes predated the AAAS recommendation to that end in 1928. Additional reports were given to the same meeting about progress in South Australia and Tasmania, although their efforts had to date achieved no structural changes within the Education Department. A detailed account is given of the steps taken by the Vocational Guidance Committee in Tasmania, probably representative of those taken elsewhere. These involved considerable time and hard work, presumably unpaid, and included examination of schoolchildren, including the administration of intelligence tests, an economic survey of Launceston in order to estimate the scope of the various occupations, as well as study of the relationship between school record, sport and hobbies, and suitable careers.[64]

By 1924, then, Section J was foreseeing the possibility of extending the use of mental tests to the sphere of vocational guidance, and of converting the latter to a professional arena. The formation in that year of the Vocational Guidance Committee was undoubtedly prompted by the presidential address to Section J delivered by J. Nangle, superintendent of Technical Education in New South Wales. Nangle described how over the previous two decades he had interviewed hundreds of boys seeking vocational advice, and he referred to the new scientifically based testing for vocational aptitudes, and particularly to the norming enterprises for tests of intelligence being undertaken in Australian universities.[65] In fact in 1914 a Boys Employment Bureau had been established for junior technical school graduates by the New South Wales Department of Public Instruction at the behest of Peter Board,[66] and this evolved into the Vocational Guidance Bureau, available to all children passing through New South Wales schools, officially constituted in 1926. In this area it seems that the AAAS, while not setting the ball rolling, accelerated its passage, and was deliberately utilized as an agent for this purpose by a senior government officer.

From the late 1920s, sufficient local graduates were being produced in psychology to advance their own claims. The University of Sydney had separated psychology from philosophy in 1920 and created a full chair in 1929; the University of Western Australia established the country's second independent department in 1928, with the appointment of H. L. Fowler as lecturer-in-charge. Increasingly the other universities were including courses in psychology within their philosophy curricula.

Graduates in education had been available rather earlier, the products of the Sydney and Melbourne teacher training colleges. A. H. Martin, lecturer in psychology at the University of Sydney, was instrumental in the formation in 1927 of a new organization, the Australian Institute of Industrial Psychology, directed towards conducting research for industry and providing guidance for young people. This body was financed in the first place by business, and individual projects were funded later by the Australian Council for Educational Research (ACER); its first major project was the construction of a battery of tests for the New South Wales Vocational Guidance Bureau.[67]

Consolidation of this application of the new psychological techniques within the educational sphere was rapidly followed by diversification of their role. During the 1920s and 1930s New South Wales saw the widest range of techniques within government programmes. While by 1932 all states had vocational guidance services, only the New South Wales Bureau was using psychological tests.[68] In that year the Bureau was transferred to the Department of Labour and Industry as a result of the changing employment scene consequent on the Depression, but interest in matters psychological was maintained within the Education Department by its first research officer, H. S. Wyndham, appointed in 1935. Wyndham established the state's school counselling service; H. T. Parker was working along similar lines in Tasmania. In 1936 a Child Guidance Clinic was set up within the New South Wales School Medical Service, and a vocational guidance and welfare officer was appointed to the Sydney Technical College. Meanwhile child clinics were established in Melbourne (by Bachelard), in Perth (by Stoneman), and as early as 1922 in Hobart (by Miller). Supporting the work of these clinics was a number of mental health bodies. In 1922 a Mental Health Section of the Public Health Association of Australasia was formed, with Miller as chairman, and in 1930 the Victorian Council for Mental Hygiene was established, followed in 1932 by a similar one in New South Wales. The emphasis of these groups was on mental disorder, but the Victorian organization in addition established a Vocational and Child Guidance Centre in 1931.[69]

Whilst the application of the techniques of educational psychology was being thus extended between the wars, a major consolidation of the research activities underpinning them took place with the advent of ACER in 1930. This body was founded on the initiative of the Carnegie Corporation of New York; the history of these moves is well documented elsewhere.[70] A special study of the Australian educational scene was conducted in 1928 by a trustee of the Corporation, James E. Russell, whose report ran thus:

The one greatest need in Australia—a need voiced by every leader that I have met—is for some means of checking impartially the work of the school and of supplying reliable information for continuous development...An Institute of Child Welfare Research...together with a Bureau of Reference and Research, such as many of our cities maintain, would be a godsend to Australia and New Zealand.[71]

In Australia Russell had lengthy discussions with Lovell and Mackie, professors respectively of psychology and education at Sydney, and Frank Tate, Victorian director of education. Detailed planning was left to the locals, with Tate taking particular responsibility for the proposed research bureau, though the resulting ACER was wholly sustained by American funds for its first thirteen years. The Institute of Child Welfare Research failed to materialize.

K. S. Cunningham, at that stage in charge of the psychological laboratory at Melbourne Teachers College, was appointed ACER's first executive officer. Cunningham saw the new body as being, first, a centre devoted to the scientific study of education; second, an information resource on the latest educational thought and practice; third, a supporter of progressive education; and fourth, a body with an Australia-wide outlook.[72] Research was both fostered through grants and undertaken directly; one of the first two grants went to H. T. Parker for investigations into variations of intelligence among subnormal children, and the Council decided that its own first research would include the standardization of scholastic and mental testing throughout Australia.[73]

From its inception, ACER performed many of the same functions as ANZAAS. During the 1930s, as well as funding research, it financed publication of about sixty accepted manuscripts in education and allied fields, and assisted the holding of conferences. Cunningham later justified this policy: 'In the 1930s facilities of these types were much less developed in Australia than they are today. The ACER felt it appropriate at that time to assist activities somewhat peripheral to its central task.'[74] In 1934, with the direct encouragement of ACER, a New Zealand Council for Educational Research was set up, similarly constituted and funded by the Carnegie Corporation. From the start there were some research and conference contacts between the two bodies, the rate of which was stepped up after a direct grant for the purpose was made by the Carnegie Corporation in 1950. The only remaining function of Section J of ANZAAS not already covered by ACER, that of bringing together Australians and New Zealanders, was thereby handled. Whilst the Section survived intact until 1968–9, when first philosophy, added in 1926, was dropped, and then psychology and education finally split up, its prime was past. A good indicator of this is the fact that Section J appointed no new committees after the state subcommittees for the vocational guidance committee were set up in 1928; the decline in committees in other Sections set in only after the Second World War.

THE POSTWAR YEARS

The onset of the Second World War and the ensuing suspension of the meetings of ANZAAS from 1940 to 1945 inclusive made the decline in significance of Section J less apparent than it might otherwise have been.

At the same time the war led to an enormous expansion in the demand for trained psychologists. Both ACER and the Australian Institute of Industrial Psychology contacted the federal government in the early stages of the war, suggesting use of their expertise in the selection of personnel, and in the event various government departments and the armed forces employed psychologists for selection of industrial and military personnel during the war years, and for rehabilitation purposes both during and after the war.

One result of this increased recognition was the birth of the formally independent local organizations for trained psychologists of whatever persuasion, the Australian and New Zealand Branches of the British Psychological Society in 1945 and 1947. Another result was the rapid proliferation of psychology departments throughout the universities of both countries. Within these departments there was now much more room for the development of areas such as the experimental psychology of learning and perception than had existed hitherto, and this trend accelerated as more government funds began to flow into pure research. The move within ANZAAS in 1946 to separate psychology and education reflected the increasing separation of interest between the academic practitioners of the two disciplines.

One obvious sign of this separation was a gradual withdrawal of the psychologists from the activities of Section J, to the extent that Elkin, writing a brief history of the Association in 1962, could state that 'only Education is active in the Congresses'.[75] This was an overstatement, but the fact is that the papers in Section J in 1962 and 1965 were given almost entirely by educationists. In 1962 the Section created its first prize, the Mackie Medal. Although this was to be awarded to an outstanding scholar in education, philosophy or psychology, it is noteworthy that the medal was named after an early professor of education. In 1967 and 1968 Section J operated through two subsections, Education and Psychology (Philosophy being dropped altogether), before dividing in 1969 into two separate Sections, 22 and 23. These continued to meet throughout the 1970s and early 1980s, with Section 23 (Psychology) absenting itself altogether in 1977 and inclining increasingly towards joint symposia with various other sections. As more and more specialist societies grew up and organized conferences of their own, the presentation of a paper at ANZAAS became less and less a useful exercise.

The postwar dominance of Section J by education can be explained by a number of factors. For a start, it was nothing new. From 1893 far more of the presidents of the Section inevitably held formal posts in education than in psychology. But even after psychological posts were created, the custom continued, so that between 1946 and 1965 there were nine presidents from education and only four from psychology (as well as two philosophers). Furthermore, while the psychologists now had their own professional body, nothing comparable to the Australian branch of the British Psychological Society existed for the educationists. A range of groups oriented towards research in education appeared from the early 1960s, but no one professional group has as yet exerted a cohesive pull.

Perhaps the existence of ACER, a powerful monitor of professional standards, is of relevance here.

Much of the activity of the psychologists within Section J during the 1950s and 1960s consisted of symposia and the presentation of papers on topics of broad general interest with a decided sociological slant. Before this Section J had participated in only two symposia, in 1932 and 1935. The pattern was set when Professor O.A. Oeser of Melbourne's psychology department gave a paper at the 1954 Congress entitled 'Culture Patterns and Social Tensions'. The same meeting saw a symposium on Scientific Study of Industrial Relations (together with Section F, Anthropology), and 1957 and 1960 two on Assimilation of Migrants (the first with Section F). In 1961 it participated in two more with Section F, Sociology of Disease and Aborigine Research, and the trend continued throughout the decade. In 1969 the new Section 23 (Psychology) included groups of papers on Drug Addiction, Psychology and the Problem of Crime, Psychology and Technological Change, and The Effects of Mass-Media of Communication. In 1972 a new Section was created for sociology, and for a time the number of sessions in Section 23 on experimental topics was stepped up.

The role of the Association *vis-à-vis* sociology was the reverse of the one it had played for education and psychology, where the establishment of Section J preceded by many years the creation of university departments in these subjects. Australia's first undergraduate department of sociology was established within the University of New South Wales (then the New South Wales Institute of Technology) in 1959, barely anticipated by the School of Social Science at Victoria University, New Zealand, in 1957. By 1972 there were thirteen such departments in the two countries. Relative to overseas developments this itself was a belated arrival, and one that has provoked considerable discussion among historians of the discipline.[76] Before the 1950s there had been sporadic bursts of sociological activity by academics: sociology appeared as part of the philosophy syllabus at the University of Sydney in 1909 and of that of Melbourne ten years later, while in New Zealand a course in sociology became a compulsory requirement for the Diploma of Social Science established by the federal university in 1921. These attempts were short-lived, and the resolution passed by Section J in 1911 in support of Francis Anderson's plea for the teaching of sociology in Australian universities[77] brought no results. A later sign of interest in the discipline was the formation of the Australian Institute of Sociology in 1942, under the chairmanship of anthropologist Arthur Elkin, a former student of Francis Anderson. This body had a clear postwar orientation towards 'the changes which lie ahead and to all that is denoted by the word "reconstruction"', and it achieved a membership of 132 before it lapsed in 1956.[78] In 1963 the Sociological Association of Australia and New Zealand was formed. This story may serve as a comment on the declining status of ANZAAS in the eyes of the academic community, in that the practitioners of the newly established discipline did not see it as imperative to gain immediate independent recognition within the Association. Less directly, it may have been felt by the organizers of ANZAAS that psychology could not or would not sustain a programme of sufficient general interest without sociological topics.

A broader external influence than the rise of ACER and of established psychology was conspiring with these two to weaken the role of ANZAAS, as far as Section J was concerned, in the postwar years. From 1941 to 1945 the Commonwealth government sponsored individual projects in the social sciences, firstly through the Reconstruction Division of the Department of Labour and National Service (1941–2), and then through the Department of Post War Reconstruction. As early as November 1942, certain members of Prime Minister John Curtin's Committee on National Morale met with members of the Australian National Research Council (ANRC) and leading philosophers Alan Stout and John Passmore to discuss a proposal for a Social Science Research Council. This proposal resulted in the formation in April 1943 of the Provisional Social Sciences Research Committee of the ANRC (the SSRC), presided over by the director of ACER, K. S. Cunningham.[79]

The term 'Provisional' was dropped early in 1945, but nonetheless a certain uncertainty accompanied the early stages of the SSRC. Despite an initial intention to establish a general Australian Association for the Social Sciences open to all qualified workers, no steps were taken in this direction, and membership began and continued to be by invitation. Funding was a problem: the ANRC was able to provide only a small grant for clerical purposes; ACER made an inaugural grant that enabled the first four meetings to be held; and from 1947 the Commonwealth government came to its assistance with UNESCO funds. A movement to reorganize the ANRC in such a way that the physical, biological, and social sciences would each be represented by committees of equal status never came to fruition. The SSRC did, however, from the beginning control its own membership and was responsible for its own programme, much of which was devoted to publicizing the activities of social scientists in Australia and to canvassing support for their funding.[80] In 1952 it decided to become fully independent of the ANRC, and August of that year saw the establishment of the new Social Science Research Council of Australia; the title was changed in 1971 to the Academy of the Social Sciences in Australia.

Over the years the Council and then the Academy have raised funds from public and private bodies for a range of major research projects, such as on the status of women, Aborigines, and immigrants in Australia, as well as providing support for publication and conference activities. Its constant focus has been towards applied rather than laboratory-based social science, and over the last two decades it has developed a strong orientation towards Asia.[81]

Although the Academy draws upon many other disciplines besides psychology and education for its membership, and the concept of social science is unmanageably broad, it is almost certain to have been perceived by its members as taking over at least some of the functions of Section J of ANZAAS (and doubtless of other sections as well). Far

PLATE 18
'Looking for a Leak at the C.S.I.R.' (*Bulletin*, 13 October 1948).

PLATE 19
'The Double Helix' (*Search* 2, 1971, 214).

PLATE 20
Harnessing the five-in-hand waggonette during the Glacial Committee expedition by Edgeworth David and Walter Howchin to Crown Point, Central Australia, July 1921 (*Edgeworth David Papers*, University of Sydney Archives).

PLATE 21
'A Melbourne University Professor'
Professor Henry Laurie (*Melbourne Punch*, 24 July 1890).

PLATE 22
'Professor Francis Anderson, M.A.'
(*Arts Journal* 4(3), University of Sydney 1921).

PLATE 23
Department of Public Instruction (NSW). *Report upon the Physical Condition of Children Attending Public Schools in NSW* (Sydney: William Applegate Gullick, (G.P.), 1908).

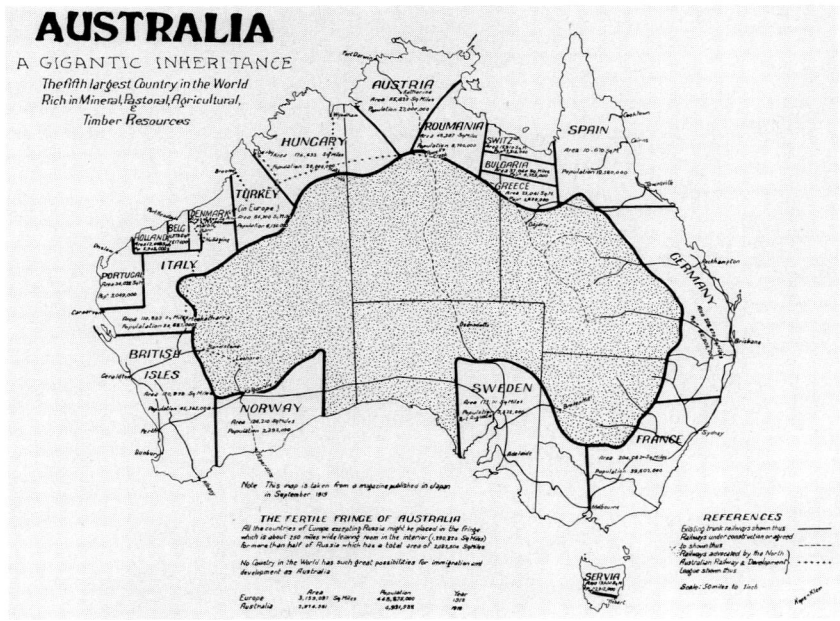

PLATE 24
The 'Gigantic Inheritance' of Australia (redrawn from E. J. Stuart, *A Land of Opportunities*, London: 1923).

PLATE 25
Griffith Taylor as interfering marriage broker (*Daily Telegraph*, 25 June 1923).

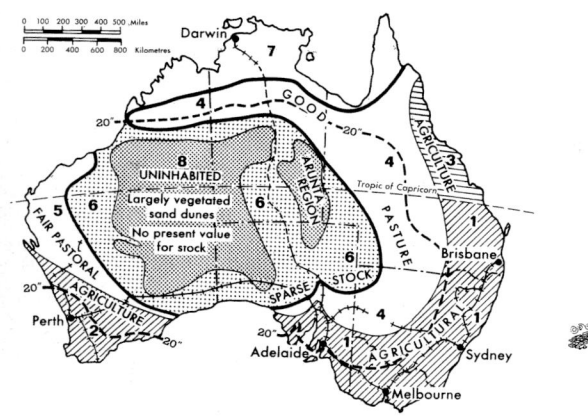

PLATE 26
Taylor's preferred sequence of settlement (adapted from his map in the *Sydney Morning Herald*, 28 February 1925).

PLATE 27
'The Scientist Seen as Villain' (*Search*, 1 October 1970).

more directly than ANZAAS, it has been concerned with the raising of funds for research. Like ANZAAS, its pronouncements carry the signature of a variety of disciplines, but they bear as well the underlining of elitism, with its membership-by-invitation-only. Whether the Academy has performed any better than its non-elitist peer in the public impact of its statements is questionable, but it is probable that its existence has served to divide forces that were in any case no longer strongly united.

IN SUMMARY

Analysis of the role of the AAAS and later ANZAAS in the development of psychology and education in Australia and New Zealand is complicated by two sets of factors. First, there was the intervention of two world wars and the ensuing suspension of meetings at times critical in the history of mental testing. The first interruption came just when major innovatory techniques were being developed overseas, and the second when, owing almost entirely to external influences, a powerful new institution, ACER, dedicated to most of the goals ANZAAS had been pursuing through Section J, had recently arrived on the local scene. Second, there is the lack of a good archival collection of the Section's records. As has been indicated at various points throughout this chapter, not only are the documents of the early committees for the most part unavailable, but even their reports were sometimes not printed and no indication was given of whether or how these were followed up.

Nonetheless, to attribute considerable importance to the existence and activities of Section J in its early stages is more than mere conjecture. As a meeting place and publication outlet for the practitioners of the new psychology and education it had no rival. Two of its three committees involving both educationists and psychologists can be seen as having had definite catalytic properties. The 1893 committee was a very early pioneer of the anthropometric survey in Australia and New Zealand; particularly noteworthy was its eagerness to include psychological measurements in such surveys over a decade before Binet devised his scale of intelligence. Like it, the 1900 committee on the need for separate state education for defective children was followed by relevant action by the state Departments of Education. The same is true of the 1924 committee with its 1928 sub-committees on vocational guidance, but events moved so rapidly in this area that it is impossible without better records to know to what extent the Association's influence was crucial. Clearly the situation varied from state to state. At the very least, it is apparent that the AAAS served if not as an instigator then certainly as an accelerator.

The only area of importance to both education and psychology where the AAAS did not play a conspicuous part in these early decades was the norming of mental tests for Australian conditions, and its mere token presence on this scene was probably due to the suspension of meetings during the First World War. At a different level of analysis, it may be seen that the Association in the early decades facilitated the interaction between education and psychology, for which the conditions were already appropriate in terms of the dependence on the Scottish model of university structure and employment of Scottish personnel in relevant disciplines.

Virtually a *deus ex machina* in the early 1930s, ACER took over a whole range of functions of Section J. The massive postwar expansion in the numbers of psychologists and the range of psychological interests led to an eventual separation between the original partners of Section J, an event not surprising in terms of psychology's numerous other liaisons and the dominance of education throughout the relationship. The creation of the Academy of the Social Sciences in Australia weakened the position of ANZAAS from without, while these sectional problems weakened it from within. The very success of Section J in its early years was not conducive to a stable role for AAAS in terms of the disciplines it represented. With their existence publicly justified, their practitioners became more firmly entrenched within the academies and the government services, formed their own specialist organizations, and forgot about using the Association to inform the public about their scientific activities.

From the 1950s onwards, one function that ANZAAS can fulfil better than any other forum has remained. It has allowed public, interdisciplinary contact (of particular importance for such eclectic disciplines), which in turn has perpetuated another property of the early papers and committee reports. The optimism and enthusiasm for the various causes of social welfare and individual betterment that these espoused cannot be missed by the reader of today, and it is just these qualities that fail to flourish within the confines of highly structured and complex modern disciplines. It is in the interdisciplinary symposia of the postwar period of ANZAAS's existence that they are most likely to survive.

NOTES

The assistance of Marilyn Orr in collecting and organizing background material for this chapter is gratefully acknowledged.

1. A. G. MacLain, 'Education in Australia' in A. P. Elkin (ed.), *A Goodly Heritage* (Sydney: ANZAAS, 1962), 130–46. Numerous histories of this period are available; see for instance C. Turney (ed.), *Pioneers of Australian Education*, vols 1 and 2 (Sydney: Sydney University Press, 1969, 1972).
2. *Report of the AAAS*, 2 (Melbourne, 1890), xvii–xix.
3. Proceedings of Section J of AAAS, 13 January 1911, reported in AAAS Minute Book.
4. *Report of the AAAS*, 5 (Adelaide, 1893), xxiii.
5. *Report of ANZAAS* 25 (Adelaide, 1946), xx.
6. Report of Committee on the Best Means of Encouraging Psychophysical

and Psychometrical Investigation in Australasia, *Report of the AAAS*, 6 (Brisbane, 1895), 842–3.
7. In a personal letter to Mrs Muscio after her husband's death in 1925, Myers wrote: 'I can never forget that it was through your husband's pioneer lectures that I got first to know anything about industrial psychology...Hence Bernard Muscio was responsible for the development of the subject throughout the British Empire.' Quoted in 'In memoriam—Bernard Muscio', *Australasian Journal of Psychology and Philosophy*, 4 (1926), 157–9.
8. R. R. Gillespie, 'The Early Development of the Scientific Movement in Australian Education—Child Study', *ANZHES Journal*, 11 (1982), 1–14.
9. T. A. Coghlan, 'Child Measurement', Presidential Address to Section G, Economic and Social Science, *Report of the AAAS*, 9 (Hobart, 1902), 541–53, esp. 541–2.
10. *New South Wales Educational Gazette* (1 August 1902), 66.
11. Anon., *Health and Longevity According to the Theories of the Late Dr Alan Carroll* (Sydney: Epworth Printing and Publishing House, 1915).
12. A. W. Rudd, 'The Scientific Study of the Child', *Report of the AAAS*, 12 (Brisbane, 1909), 763–74.
13. Reported by Coghlan, op. cit., note 9, 542.
14. *Department of Education Subject Files, 1875–1948* (NSW State Archives), Box 20/12864, Bundle: Medical Inspection, 1912; Minute from Dept. Attorney General and Justice, 1 June 1908.
15. M. Booth, 'School Anthropometrics: the Importance of Australasian Measurements Conforming to the Schedule of the British Anthropometric Committee, 1908', *Report of the AAAS*, 13 (Sydney, 1911), 689–96.
16. New South Wales Department of Education, op. cit., note 14, Mary Booth, Hon. Sec. Anthropometric Committee, to Premier of New South Wales, October 1911.
17. ibid., Andrew Fisher to state premiers, 13 October 1911.
18. See note 6.
19. *Galton Papers* (University College London Archives), E. F. J. Love to F. Galton, 14 April 1894.
20. 'First Report of Committee on Range of Physiological Variables in the Inhabitants of Australia and New Zealand', *Report of ANZAAS*, 21 (Sydney, 1932), 477–509.
21. 'Report of the Anthropometric Committee', *Report of the BAAS*, 51 (York, 1881), 225–302.
22. *Reports of the BAAS*, 58 (Bath, 1888), 854–55; 59 (Newcastle, 1889), 423–5; 60 (Leeds, 1890), 549–52; 63 (Nottingham, 1893), 614–20; 70 (Bradford, 1900), 461–6.
23. *Anthropometric Investigation in the British Isles—Report of the Committee* (London: The Royal Anthropological Institute, 1909), esp. 43–6.
24. See for example Department of Education, op. cit., note 14, James Graham on 'Nature and Scope of the Enquiry into the Physical Characters of Australian Born School Children', June, 1892, and letter from the Premier of Tasmania to the Premier of New South Wales, 27 November 1906.
25. *Report of the AAAS*, 13 (Sydney, 1911), vix.
26. 'The Organization of Anthropometric Investigation in the British Isles—Report of the Committee', *Report of the BAAS*, 83 (Birmingham, 1913), 230–1.
27. Department of Education, op. cit., note 14, Charles MacKellar to the Hon. F. B. Suttor, 11 May, 1892, with Minister's note attached. James Graham to the Hon. F. B. Suttor, 9 June 1892.

28. Coghlan, op. cit., note 9, esp. 542–3.
29. E. M. Miller, *Brain Capacity and Intelligence* (Sydney: Australasian Assocn for Psychology and Philosophy, Monograph Series no. 4, 1926), esp. 3–4.
30. *Report of ANZAAS*, op. cit., note 20, esp. 477.
31. See R. White, *Inventing Australia* (Sydney: George Allen & Unwin, 1981), esp. ch. 5, and G. R. Searle, *The Quest for National Efficiency. Study in British Politics and Political Thought, 1899–1914* (Oxford: Basil Blackwell, 1971).
32. Gillespie, op. cit., note 8, esp. 8.
33. See D. W. Forrest, *Francis Galton: the Life and Work of a Victorian Genius* (London: Paul Elek, 1974).
34. See L. J. Kamin, *The Science and Politics of I. Q.* (New York: John Wiley & Sons, 1974), esp. ch. 1.
35. 'Education and Physique—Development of Australian Youth—Interesting Medical Investigation—Remarkable Comparisons', *Daily Telegraph* (7 January 1909).
36. Department of Education, op. cit., note 14, T. A. Coghlan to Under Secretary for Public Instruction, New South Wales, 25 January 1901.
37. 'Report of Anthropometric Committee', *Report of the AAAS*, 14 (Melbourne, 1913), 605–21, esp. 607.
38. Department of Education, op. cit., note 14, Peter Board to the Minister for Education, 23 October 1906; Andrew Fisher to the Premier of New South Wales, 29 March 1909.
39. An example of a recent study is that by New South Wales Department of Health, *Height, Weight and Other Physical Characteristics of New South Wales Children* (Sydney: Government Printer, 1973).
40. New South Wales Department of Public Instruction, *Report upon the Physical Condition of Children Attending Public Schools in New South Wales* (Sydney: Government Printer, 1908), esp. 56.
41. *Report on the Medical Inspection of One Thousand Public Schools of South Australia During 1909* (Adelaide: Government Printer, 1910), esp. 62–4.
42. Booth, op. cit., note 14, esp. 684.
43. Miller, op. cit., note 29.
44. K. S. Cunningham, 'Ideas, Theories, and Assumptions in Australian Education' in J. Cleverley and J. Lawry (eds), *Australian Education in the Twentieth Century* (Victoria: Longmans, 1973), 101–23, esp. 108.
45. K. S. Cunningham, 'Experimental Education: 1. Individual Differences', *Education Gazette and Teachers Aid*, 23 (June, 1923), 101–2.
46. C. E. Beeby, 'Psychology in New Zealand Fifty Years Ago', in R. St George (ed.), *The Beginnings of Psychology in New Zealand* (Palmerston North, New Zealand: Dept Education, Massey University, Delta Research Monograph, 1979), 1–6.
47. J. V. McCreery, 'Idiocy and Juvenile Insanity in Victoria', *Intercolonial Medical Congress of Australasia: Transactions of the Third Session* (Sydney, 1892), 665–8.
48. *Australasian Medical Congress: Transactions of the Tenth Session* (Auckland, 1914), 701–31.
49. *Report of the AAAS*, 8 (Melbourne, 1900), xxix.
50. Legislative Assembly of New South Wales, *Interim Report of the Commissioners on Certain Parts of Primary Education* (Sydney: Government Printer, 1904), esp. 56, 65.
51. Report of the Victorian Minister of Public Instruction 1911–12, esp. 28.

52. In the early years these positions were subject to interruption; Hodgkinson's for instance was terminated in 1923, and Stoneman's in 1930.
53. R. G. Cameron, *The Measurement of Intelligence (The Binet Tests Applied to Australian Children)*, (The Teachers College, Sydney: Records of the Education Society, no. 14, 1913).
54. E. Skillen, *The Measurement of Intelligence. The Application of Some of Binet's Tests at Blackfriars* (The Teachers College, Sydney: Records of the Education Society, no. 16, 1913).
55. S. D. Porteus, *A Psychologist of Sorts: The Autobiography and Publication of the Inventor of the Porteus Maze Tests* (Palo Alto: Pacific Books), 1969.
56. W. F. Connell, *The Australian Council for Educational Research 1930–1980* (Hawthorn: Australian Council for Educational Research, 1980), esp. 32.
57. ibid., esp. 35.
58. *Report of the AAAS*, 15 (Melbourne, 1921), 364.
59. *Report of the AAAS*, 16 (Wellington, 1923), xvi.
60. The *Australian Journal of Education*, which appeared in New South Wales between 1903 and 1912, was aimed at a popular readership and was journalistic in style.
61. *Report of the AAAS*, 17 (Adelaide, 1924), xxviii.
62. *Report of the AAAS*, 19 (Hobart, 1928), xxiv–xxv.
63. K. S. Cunningham, 'Vocational Guidance: with Special Reference to General Principles and to Recent Developments in Australia', *Report of the AAAS*, 20 (Brisbane, 1930), 92–8.
64. W. Gibson, 'Vocational Guidance in Tasmania', ibid., 99–100.
65. J. Nangle, 'Vocational Guidance', *Report of the AAAS*, 17 (Adelaide, 1924), 617–27.
66. *Department of Education Subject Files, 1875–1948* (New South Wales State Archives), Box 20/13373; Bundle: Vocational Guidance, 1914–1915, 1918–1932; Memorandum by P. B. dated 27 October 1914.
67. Australian Institute of Industrial Psychology, *Seventh Annual Report, 1933–34*.
68. G. R. Giles, 'Vocational Guidance in Australia in 1932', *International Labour Review*, 26 (1932), 530–43.
69. Australian National Association for Mental Health, *A History of Mental Associations in Australia* (Sydney: ANAMH, 1969).
70. W. F. Connell, op. cit., note 56.
71. Quoted ibid., 1.
72. ibid., 49–52.
73. ibid., 22.
74. K. S. Cunningham, The Social Science Research Council of Australia, 1942–1952 (Canberra: Social Science Research Council of Australia, 1967), 7.
75. A. P. Elkin, 'ANZAAS: A History', *Australian Journal of Science*, 25 (1962), 2–4.
76. See, for example, C. V. Baldock and J. Lally, *Sociology in Australia and New Zealand* (Westport: Greenwood Press, 1974); J. Zubrzycki, 'The Teaching of Sociology in Australian Universities, Past and Present', in J. Zubrzycki (ed.), *The Teaching of Sociology in Australia and New Zealand* (Melbourne: Cheshire Publishing, 1970), 1–25; C. Veliz, 'Connolly's Decade', *Quadrant*, 24 (1980), 5–9.
77. See note 3.
78. K. S. Cunningham, op. cit., note 74, esp. 11.

79. K. S. Cunningham, *Social Science Research in Australia: History and Functions of the Committee on Research in the Social Sciences* (Sydney: Australian National Research Council, 1945).
80. K. S. Cunningham, 1967, op. cit., note 74.
81. Australian Social Sciences Research Council, *Annual Report 1972-3* (Canberra: ASSRC, 1973), esp. 7-10.

PART III

SERVING SOCIETY

10

Protracted Reconciliation: Society and the Environment

J. M. Powell

With proofs of the 'ecological agency' of industrial society accumulating daily about them, some nineteenth century American and European *savants* held out brighter prospects for our part of the world.[1] Until the last quarter of the century, however, predominantly exploitative attitudes towards the environment were not seriously challenged in Australia and New Zealand: there were, instead, massive landscape transformations, and costly reverberations that would torment succeeding generations. However, important changes seemed to be signalled in the 1880s, when there was an increasing interest in the 'management', as opposed to the 'development', of resources and a strengthened emphasis on the associated appraisal and protection of the environmental base. Scientists were expected to lend a hand, and as they did so they began to build up distinctive institutions and research preferences.

The early efforts of the champions of the Australasian Association for the Advancement of Science (AAAS) reflected this transition, not least in the provision of a moving forum on what we would call today resource evaluation, regional development, and environmental quality. Indeed, in some senses 'accountability' and 'service' were probably better appreciated during the first fifty years of the Association's history than they have been in our own narcissistic age, and the following notes concentrate on that early engagement. But the large direct influence of the science fraternity is easily over-estimated. Like other specialists at any time over the past century, scientists have mainly addressed one another, while much of the actual business of environmental appraisal and landscape modification has been conducted at a more public level, encompassing competing 'vernacular' evaluations and aspirations.

'GOSPELS OF EFFICIENCY' BEFORE THE FIRST WORLD WAR

While the colonial economies continued their customary booming and busting, a loud antiphonal was heard, especially in Australian society. It drew partly on a nascent, if still ill-defined, nationalism and on an urgent intellectual current that promoted science and technology in the polity.

The first was related to the continued rise in sensitivity towards uniquely southern landscapes, fauna, flora, and ways of living, and encouraged more expansive visions of the challenges and potentials of each country taken as a whole. The second influence was mainly exhibited in the labours of increasing numbers of specialists whose evangelistic zeal established the need for 'expert' interpretations and the determination of coherent management strategies for the built and natural environments. In the United States this impulse was later characterized as 'progressive', and the charismatic missionaries were said to have proclaimed a 'Gospel of Efficiency' with discernible roots in the mid-nineteenth century and earlier.[2] So, too, in Australia and New Zealand, though there has been less trumpeting here. For example, there were lively debates in the 1880s on the need for land tenure reform, and while the hallowed freehold ideal was generally retained, the leasehold principle was more widely adopted. Contemporaneously, there were early indications of a comprehensive refashioning of colonial water management systems, traditional approaches to land classification and land reservation were quite carefully reviewed, and the management of urban environments benefited from the application of more elaborate statistical monitoring and the diffusion of novel overseas interpretations.[3]

Neither the initiation nor the consolidation of major scientific reorientations should be simplistically ascribed to any kind of tightly formalized reasoning. In rural Australia vast regions were 'marginal', and they remained so. Extensive farming and grazing regions, which mirrored contemporary understanding of climate, soils, topography, crop and livestock tolerances, were also gigantic practical laboratories in which vernacular practices were established and transmitted. That is not to under-estimate the singular contributions of government agriculturalists, botanists, chemists, engineers, statisticians, and other specialist officers whose interpretative works are appropriately considered in their connections with rural expansion. But a reasonable claim may be made for the contributions of authentic 'agricultural revolutions' in which orthodox science occasionally played only a secondary role.

The early emergence of Australia's south-eastern wheat belt, for instance, was based to a large extent on *in situ* innovations in novel agricultural implements, complex adaptations of landholding patterns and credit arrangements, the determination of secure cropping rotations, and the provision of road and rail transport and other highly politicized infrastructural improvements by the colonial authorities— even if they must be assessed *en bloc*, rather than individually.[4] Similarly, in the grazing regions, the several regional variations in the sizes of 'living areas' and 'homestead maintenance areas'—celebrated bases for equitable working units, especially in the leasehold territories of Queensland and New South Wales—were in part the products of recurrent official–popular dialogue in good and bad times.[5] Detailed official advice on grazing and cropping practices was not always well regarded or well understood in the rural communities until more supportive 'extension' services were introduced in the twentieth century, and the identification of government and non-government origins of rural innovations continues to present innumerable puzzles.[6]

In his inaugural address, H. C. Russell (1836–1907), government astronomer in New South Wales, argued that the AAAS might take up the lead of its British parent by examining the special environmental problems and resource endowments of 'new' countries.[7] Following the mixed arts–sciences tradition of the earlier learned societies in the colonies, this grand opening statement included felicitous references to the comprehensive character of science—'man and nature are correlative, and, therefore, true science must embrace both studies'.[8] Yet science was still repeatedly threatened within and beyond the colonial bureaucracy. Thus his Victorian colleague, the renowned botanist Baron Ferdinand von Mueller (1825–96), had felt insecure during most of his career. Internationally, Mueller was probably best known for his prolific taxonomic writings and sustained contributions to basic exploration and the co-ordination of plant identifications, but he was also an adviser to various colonial authorities in the field of economic botany—crop diseases, the selection and acclimatization of exotic commercial species, and the use of native and introduced vegetation to combat drought and temper troublesome geomorphic processes. Government employment brought extraordinarily fortunate research openings—his directorship of the Melbourne Botanic Gardens alone provided unusual opportunities—but his treasured correspondence with Bentham and the Hookers showed an acute anxiety about the need to justify a plum position to supposedly philistine paymasters.

Mueller's loss of the Botanic Gardens to the gifted landscape-gardener-cum-nurseryman W. R. Guilfoyle (1840–1912) was intensely felt, but his austere confusions had become irritating to the growing class of leisure-seeking Melbournians who put beauty above instruction. The episode offers an interesting metaphor for the predicament of colonial science.[9] Like so many talented immigrants, Mueller was seldom in step with his contemporaries. But it was, perhaps, that same Gardens controversy and urgent reflections on his prodigious treks in unique wilderness areas that were subsequently engulfed by waves of enterprising settlers that lent an unusual incisiveness to some parts of his own AAAS presidential address at Melbourne in 1890. It included what is now seen as the first great public appeal for the reservation of selected areas for the benefit of future generations.

I have argued elsewhere that modern conservationism may be traced to the focusing of changing aesthetic, ecological and utilitarian motivations in the latter half of the nineteenth century. 'The Baron' was one of a small number of scientists who seemed prepared to recognize the indivisibility of this crucial trinity. Writers, poets, painters, and balladists brought an aesthetic appreciation of nature and landscape to young communities that were beginning to set enduring, non-material values on their national heritage. But Mueller and a handful of bureaucrats in the colonial lands, forests and water management agencies made good advances in their articulations of utilitarian or 'wise-use' approaches. Some of these early champions are best described as non-specialized administrators, but none was oblivious to prominent scientific arguments and all employed commonsense economics to protest the absence of authoritative frameworks to guide development and preservation programmes.[10]

Conservation was frequently discussed at the AAAS congresses before the First World War, and the New Zealand Institute and the Royal Societies occasionally accommodated the theme in their published proceedings. A few of the beleaguered foresters may be met therein, going a good deal further than Mueller in highlighting the problem of 'timber slaughtering' and the demand for comprehensive national policies. But forests remained the poor relations, especially in the hierarchies of Australian bureaucracy, and land administration and water management continued to hold the reins. Law and engineering, combined with basic surveying and accountancy, appeared to be the most sought-after skills, and for the most part the sciences proper were only grudgingly admitted. In the interim, a consuming passion for 'closer settlement' was indulged by every Australian and New Zealand government, and despite many shortcomings the stock of environmental information was markedly augmented through the inspection, design, subdivision, and preparation of each of the great purchased estates and selected Crown lands. The resultant 'intensified' landscapes chiefly reflected social and economic goals—especially welfare state notions regarding the redistribution of national resources and 'efficiency' emphases on increased productivity.[11]

No separate role need be identified for science and technology in any of this. The diffusion of refrigeration and various new production techniques accelerated the expansion of dairy farming—a godsend for the precarious 'yeoman' imagery—and the introduction of superphosphate and successful experiments with improved wheats and grasses helped to extend and stabilize some refashioned frontiers. But the contextual foundation for the making of rural Australia was undeniably 'progressive' and was, therefore, firmly set in social–political philosophy, the complex textures of legislation, refinements of bureaucratic management procedures, and a battery of improvements in transport and communications, tenure arrangements, credit provisions, and bank facilities, and the establishment of rural co-operatives. For this first period it would be misleading to sketch a high profile for scientific applications in any of these matters unless, following Russell's generous lead, we make room for the social sciences and humanities. And it is no less important to remember that, with some qualifications, it is fair to say that the predominating rural distributions—the great regional belts—were quite well established by 1918.

PROCLAIMING THE 'ECUMENE', 1918–1950

In the inter-war years a resurgence of the settlement imperative, partly reflecting nationalist–imperialist ambitions, resulted in a good deal of ill-founded boosterism in political circles. 'Population capacities' of between 100 and 500 million were regularly conjured by those who portrayed Australia as the future pivot of white settlement in a revivified, integrated Empire. Their naivety was compounded by the pursuit of intoxicating comparisons with Europe and the United States, foolishly based on size alone, and closer settlement schemes were continued in greater diversity. It was decided that much of the debt owed to returning veterans should be paid in land, and eventually 40 000 applicants in Australia and 17 000 in New Zealand received special assistance. Later, the British Empire Settlement Act and the £34 million Agreement, which were reminiscent of the 'systematic colonization' of early nineteenth century planning, appealed to Britain's liberal imperialists and their Dominion associates.[12] Before the end of the 1920s, however, all such ventures were being roundly condemned on economic and social grounds, and the most vigorous critiques had already been delivered by the environmentalist T. Griffith Taylor (1880–1963), founder of Australia's first full department of geography at the University of Sydney.[13]

Taylor insisted that the contemporary margins of settlement already approximated the limits 'set' by incontrovertible nature; the real challenge in the search for collective identity did not reside in romantic nineteenth century frontiering, but in the clarification of more 'rational' objectives. Attacking the popularized illusion of an 'Australia Unlimited', he employed elementary resource inventories and latitudinal analogies (with Northern Africa and Spain, for example) to predict a total population of 19 or 20 million at the end of the twentieth century, with an unacceptable 'saturation' capacity of about 60 million.[14]

British and Australian imperialists reviled him: a 'croaking pessimist', a miserable Jeremiah calling from the black depths of 'environmental determinism'. In 1921 his introductory geography of Australia was banned by Western Australia's education authorities because of its impudent reference to the wide influence of aridity, and in 1923 his address 'Geography and Australian National Problems', at the Wellington meeting of the AAAS, soon caused a furore.[15] Every sacred cow was ridiculed—the White Australia policy, the extension of white settlement in the tropics, and railway construction into the interior and the far north. Large areas of Western Australia were assessed as 'almost useless' and northern development was brusquely rejected. Taylor declared an end to exploration in the traditional Australian sense. Instead, geographical analysis was the best guide for political leaders and immigrants alike, and was infinitely superior to the mere opinion of the 'prescientific period'. Similarly, he dismissed the courage of the early settlers as the 'valour of ignorance'. His use of 'homoclimes', a convenient teaching device that compared popularly accepted 'difficult' regions overseas with Australia, confirmed the absurdity of optimistic predictions for our 'vast open spaces'. And there was more. Those who had framed Australia's immigration policies were equally misguided: 'almost all the despised "coloured people" of Southeast Asia are our own superiors in an ethnic sense', and the Chinese especially need not and should not be excluded.[16]

The message was clear: the fundamental geography of settlement was completed, fixed, and 'the wise statesman is he who moulds his policy in harmony with the varying environment for which it is his privilege to legislate'.[17] But Taylor roared too loudly for anyone's comfort, including his own, and his reminiscences complain bitterly of the apathy of senior colleagues. As the distinguished botanist J. H. Maiden (1959–1925) remarked to him, 'An undesirable side of human nature is that scientific men do not pull together, are afraid of showing their hands, so to speak': 'Many scientific men know perfectly well that you are right, but will not say so, and so you have to fight against the hordes of ignorance practically alone.'[18] The geographer's North American associates were kinder. Above all, he was drawn inexorably to Isaiah Bowman's (1878–1950) international vision of a 'science of settlement' for the investigation of an apparently diminishing *ecumene*, and in 1928 he decided to move from Sydney to Chicago.

Taylor had been mercilessly lampooned in Australia's anti-intellectual press, and the *Sydney Morning Herald* and other reputable dailies had fanned the flames. For example, the *Herald* published a stirring protest from Daisy Bates, 'the great white queen of the Never-Never', who accused him of slandering pioneering spirit and creativity. For her, empirical testing and national character were indivisible, and probably preferable to the speculations and pessimisms of jumped-up science. Furthermore, the maligned outback still presented an admirable challenge to a youthful nation; perhaps it was also a *necessary* frontier. And 'the early British pioneer knew nothing of "physical controls", and geology, rainfall, and temperature were, scientifically speaking, sealed books to him... [he] was not, and is not, the squealing kind, and he went on learning about the land all the time'.[19]

As I have attempted to explain elsewhere, powerful forces were marshalled to spike Taylor's guns, hastening his departure for a less parochial milieu. One strange variant of the old gunboat tactic may be cited here. Vilhjamur Stefansson (1879–1962), Canada's celebrated 'possibilist' with grand ambitions for the world's deserts, was brought out with much fanfare to challenge Australia's 'environmental determinist'. In the event, Taylor came out on top, because he drove home his main argument in the popular media, while communicating a 'preferred sequence of settlement' which now seems perfectly straightforward. But by temperament and inclination he was unsuited to the delicate business of administering unpalatable home-truths. In contrast, Stefansson was the darling of Australia's politicians, and his infectious optimism pleased the imperialists.[20] The tragedy of it all is that Taylor concentrated almost exclusively on macro-scale issues for the country as a whole: that was unfamiliar territory for most of his contemporaries, whereas the experience of local and regional analyses during the long history of closer settlement offered an area of common ground. 'Australia Unlimited' eventually evaporated with the rest of the rhetoric of the 1920s, and by the end of the next decade, Australians were more willing to attend to Taylor's sober predictions.

In fat and lean times, similar 'progressive' forces repeatedly rejected public and official perceptions of the pace and direction of future development in Australia. Conservation and planning maintained a significant if fluctuating presence in the activities of the AAAS and later the ANZAAS congresses and research committees, and Taylor's successor at Sydney, J. Macdonald Holmes (1896–1966) offered enterprising contributions to debates on land-use planning and the 'new states' movement.[21] In 1925, the prominent Victorian progressive James Barrett (1862–1945) edited a popular collection of essays to make a 'Plea for the Right Use of our Flora and Fauna', under the title *Save Australia*. Habitat destruction through forest clearances for settlement was a favoured target; so was the fur trade, which had taken approximately six million animals in 1919–21. Other essays covered such topics as the states' varied reservation and sanctuary policies, the problems caused by introduced weeds and animals, the co-ordination of popular local history projects, and the official recognition and protection of sites and places of historical significance. Less happy proposals suggested a virtually unrestricted assault on forest fires—a familiar *idée fixe* of that time, abandoned with some reluctance by later researchers. The AAAS made room for these concerns, and occasionally highlighted soil science, biological control procedures, and forestry. In the process, an invaluable interchange between academic and public service scientists was facilitated.[22]

A minor reformation in soil management was instigated in Australia by the publicity given to America's Dust Bowl and the spectacular arrivals of wheat-belt dust in the coastal capitals. A few enlightened measures were proposed in most states after the presentation of competent reports in the late 1930s and early 1940s, but the cash-grain fixation proved to be a stubborn handicap at both official and popular levels. Certainly the continued, traditional Australian dependence on empirical testing was intolerant towards untried or intangible scientific and technical advice, but by the 1930s the simplicity, cheapness, and 'natural' appeal of fallowing had won many supporters; so, when yields declined and the soils were blown or washed away, government advisers and farmers' organizations sermonized wheat farmers on the need for improved techniques and fallowing in the drier country. The chastened response brought ecological traumas—inevitably, as we now say all too smugly.

Of course, the trouble was that prolonged and repetitive fallowing after the New World fashion—that is, by means of an incorporation of several manipulations of the surface layers, including ploughing and harrowing—had eroded Australia's farming soils. But soil erosion was conceived as a *natural* hazard, producing a 'nuisance' effect of blocked roads, drains, and railway lines, and calling only for muscle and coarse maintenance engineering.[23] Following the lead of Prescott and others, however, Australia's scientists explained the situation: much of the reliable rainfall country was not commensurately endowed with fertile soils; moisture deficits often occurred in well-favoured soil districts; the preferred southern third of the continent was deficient in phosphate, and there might be an even more general deficiency of nitrogen.[24] And yet by the time the scientists presented their verdict, collective popular experience had probably yielded as much new information.

By the early 1940s the accumulated evidence seemed overwhelming, and specialized soil conservation and related agencies were established. But the preferred disciplinary structure of ANZAAS was not well adapted to cope with such many-sided themes; a few thorough theme-based congresses might have been more useful. More or less by default, the task was taken up by a handful of outspoken individuals. Data were usefully summarized for broader consumption by prominent geographers in both countries—especially by Sydney's J. M. Holmes, and New Zealand's K. B. Cumberland (1913—)—but in the public history of conservationism the classic reference is undoubtedly Francis Ratcliffe's *Flying Fox and Drifting Sand* (1938).[25] Ratcliffe (1904–70) was one of a small number of talented public servants who carried the flag of applied science and technology in the Council for Scientific and Industrial Research (CSIR) and related state agencies in the inter-war years, and his memoir has found an honoured niche in our literary heritage; his compact in-house publications merit more attention than they usually receive. In most parts of the country technical expertise was becoming agreeably proficient, but Ratcliffe's biologist's vision enabled him to understand that a widening gap between the development and diffusion of innovations measured the distance between the lay community and the subculture of Australian science. So, for example, the demonstration of copper–cobalt deficiencies in coastal soils was applauded in Western Australia and South Australia, but marginal wheat farmers in the west refused to accept the scientists' observations on the threat of increasing salinity. In the same state, the impressive soil surveys of L. J. Teakle, S. J. Stokes and others had little immediate impact, though they were to form the basis of important rationalization schemes in later years. Similarly, the great burst of 'light land' development after the Second World War was promoted by successful investigations into trace element deficiencies during the 1920s and 1930s.[26]

Between 1901 and 1950 Australian forestry was guided by a surprisingly small group, including a few very strong personalities with research and administrative experience in more than one state.[27] Too little is known of the role of the personality factor in this chapter of the story, and still less of the place of inter-personality relationships that were probably critical. Agreement on certain common goals was allowed top priority, and to this end seven Interstate Forestry Conferences were held in the capital cities between 1911 and 1924. They emphasized British-led ambitions for national and imperial self-sufficiency, as well as the need for federal initiative and co-ordination, advanced technical training, afforestation with natives and exotics, the reservation of forest cover for water conservation and erosion control, and fire monitoring and prevention measures. Their regular complaints also criticized continued state sovereignty over land resources, and the ossifications of bureaucratic structures that blocked progressive forestry.

At the Commonwealth level, C. E. Lane-Poole (1885–1970) was an unflinching advocate of forest conservation whose views did not always coincide with those of his state colleagues. His 1926 AAAS address caught their mood well enough, however, when he drew on his own earlier predicaments in Western Australia to complain of the typical Minister of Lands who 'is, as a rule, selected because of his ability to promote settlement'.[28] The venue was Perth itself (the first of the AAAS meetings there), and Lane-Poole was the newly appointed Commonwealth Forestry Advocate. Western Australia was said to have 'the worst record of all', and he suggested that its 'most purblind of land settlement enthusiasts' should have foreseen the miserable results of the sacrifice of the jarrah and karri regions. But the forester also struck hard at equally entrenched attitudes, official and popular, that continued to rule land management practices in all the states:

Except for a small number, who know the truth and are not afraid to expose the idiocy of the system, the general public encourage the Minister to pursue his primrose path. His electors are tickled by such phrases as 'we want men not trees'. . . To the Minister all land is potentially agricultural, all is too valuable to devote to forestry.[29]

In the sphere of resource appraisal and environmental management, the axe and the pen had been potent colonial symbols. The exigencies of pioneer settlement had monopolized the best energies of politicians and bureaucrats and, despite the efforts of scholars, scientists, and public servants, most public attention was focused on development-orientated land administration and water management. For our purposes, the vehemence of Lane-Poole's attack is less important than his perception of a definite mandate to interrogate the oldest and most influential foe in the bureaucracy. State-controlled or state-supervised timber production under a system of 'sustained yield management' had become the foresters' unifying goal, and it was aiding their steady climb on the ladder of public and official esteem. Whatever their private thoughts on 'softer' ecological and aesthetic values, the foresters' adoption of a utilitarian stance with a heavy emphasis on secure timber yields, and self-funding arrangements through sales and royalties and the like, established a comprehensible *raison d'être* in the polity and threatened the old monopolies over ministerial interest and budgetary consideration. And yet in Tasmania and elsewhere in Australia, the wheel was turning full circle. Colonial foresters had been blocked in their efforts to found large-scale government enterprises. Their successors were beginning to fashion an accord with the same industrial interests that had been bitterly condemned in the past; in particular their *rapprochement* with the politicians meant that compromised governments granted monopolistic concessions over enormous districts to the sawmillers and pulp and paper manufacturers.

If wind erosion delivered the more spectacular environmental warnings in Australia during the 1930s, in New Zealand extensive flooding in the lowlands and savage landslips and gulleying in the hill country probably made the clearest case for rigorous conservation measures.[30] The notion of 'protection forests' was applied on both islands, but exotic plantations took up most of the small forestry funds at the cost of urgent plans for the conservation of indigenous species. As in Australia, a responsible popularization of vital concepts and information condemned the narrow pretensions of the public works enthusiasts for piecemeal engineering solutions. The protest was strengthened by some devastating natural events, including avalanches, earthquakes, and floods, and by the keen evangelism of Lance McCaskill (1900–85) and his colleagues.

In the 1880s the New Zealand government seemed determined to implement enlightened forest policies. Sustained yield approaches were proposed in 1885 by the first chief conservator of forests (T. Kirk) as part of a general conservation scheme for indigenous forests. At popular and parliamentary levels, however, tree cover was still considered an obstacle to settlement, and vast expanses were simply destroyed, especially in the North Island. Kirk's plans were effectively shelved after the onset of depressed economic conditions in the later 1880s and 1890s, but a few large reserves were set aside, notably in the mountains. The destructive forces of pioneer settlement are seldom analysed in the parsimonious historical accounts of either country, yet in New Zealand the bulk of the clearance activity in modern times was probably concentrated in the relatively recent period of 1870–1914, when almost one-third of the total pre-European cover was removed. Most of these forests were axed and put to the torch, left to rot, or wastefully exploited: one broad indictment insists that the sawmills were then using an estimated 12 per cent of each tree they received.[31] Surely the darkest chapter in the chronicle of recent landscape change should be meticulously sifted for insights into obstinate environmental attitudes.

A change of heart in the early inter-war period saw the establishment of the New Zealand Forest Service, in 1919, and the reformulation of sustained yield frameworks assisted by selective logging strategies. Once again, planning was defeated by a return to depressed conditions, followed by the special demands of the wartime economy. On the other hand, although the conservation of indigenous forests was low in the pecking order, contemporary hopes for long-term investments in exotics led to landscape transformations on an astonishing scale. Open districts, such as the Canterbury Plains, and forested or recently cleared areas where the soils were considered unsuited to grazing were made over to plantations. Radiata pine soon proved to be the most profitable of the introduced species, as in south-eastern Australia, and it appears that agency personnel were largely trained in production forestry, an emphasis that must have coloured their own environmental attitudes.

Major conservation legislation for soils, water, and forests was passed in both countries in the early years of the Second World War, when the related promiscuous concept of the 'region' was finding favour. Of course it had been meat and drink to the 'synthesizing' geographers since the late nineteenth century, but, as Coombs describes, the emergence of 'ecological' issues, spanning the natural and social sciences, guaranteed a market for any number of organizational frameworks.[32] Regional approaches undoubtedly directed public and government attention to the whole gamut of 'decentralization' initiatives and to various aspects of 'selective' development—rural expansion in the Murray Valley, balanced industrial and agricultural progress in the Hunter Valley, and new settlement efforts in the Northern Territory and in the tropical zones in general. Holmes and his fellow geographers seized the chance to reinforce the 'planning' orientation connecting the human and physical branches, and continuing Russell's practical vision.[33]

But Coombs's reflections ring true: although the 'regional' concept became entrenched in the ruling Labor circles of postwar Australia, it was fundamentally flawed in its assumptions about 'self-contained' structures, and its proponents were deceived by idealizations of high degrees of self-sufficiency. The promise that the approach would harness the creative energies of communities was never fulfilled—on the contrary, it was very difficult to define the 'areal community of interest' that academic geographers and professional planners were noting. As Coombs explains, all such hopes and expectations naively ignored an internationalization of production and consumption that was rapidly destroying the cohesion of regional economies and community traditions.[34]

In the 1950s and 1960s ANZAAS meetings continued to encourage dialogue between academics and public servants on broad fields of national endeavour. It remained to be seen whether the use of ANZAAS as a focus could be maintained during the dramatic enlargement of higher education and the increasing diversification of perceived aims and requirements in the public sector.

STOP–GO FRONTIERS AND 'SUSTAINABLE' SOCIETIES

In Australia and New Zealand, as elsewhere, the postwar era began with earnest implementations of the regional approach in the management of the built and natural environments, and social interpretations continued to influence community appraisals of land and landscape. In different ways, Queensland, Western Australia and Tasmania consolidated earlier movements, accelerating pioneer settlement and mining development in the west and Queensland, and intensifying the 'hydro-economy' and local reliance on forest-based employment in the precarious economic structure of Tasmania. In the older settlement core of Australia's southeastern mainland, and to some extent in New Zealand, 'decentralization' programmes gave modest and much-needed support to country centres, and introduced cosmetic changes in our urban landscapes, without significantly deflecting a powerful centripetalism feeding the major towns and cities.

Remoteness became ever more ambiguous as Australia and New Zealand were drawn into global confrontations and annexed by transnational corporations and new cultural imperialisms. Both communities looked increasingly to Asia and the Pacific, and to their American protector. On the other hand, a new nationalism was expressed in both countries during the 1970s and 1980s, which was intimately connected to a growing appreciation of 'wilderness' ideals and the application of the 'limits to growth' to population control and the utilization of resources. In Australia, however, the very pace of social and economic change, combined with the restricted range of direct environmental experience available to the increasing numbers of immigrants and the native-born alike, was introducing anxious instabilities.

In another link with the past, Griffith Taylor was warmly welcomed in 1946 for a brief visit as a senior government consultant: the storm over his contribution to the debate on national planning had subsided, or so it seemed. One of Taylor's cruder regionalizations had distinguished between 'economic' and 'empty' Australia, largely on the basis of rainfall reliability, and that had won some general acceptance. But there was still a latent belief in the existence of an intermediary region, an undetermined but very wide marginal zone that might be made over to more useful purposes by the introduction of irrigation, new crop or livestock strains, and a more numerous and more enterprising population.

Irrigation, confidently extended between the wars, provided additional landscapes of hope to serve a white people's dreaming. Yet nothing came of J. J. C. Bradfield's audacious scheme for flooding the interior by channelling the northern rivers towards the Lake Eyre basin.[35] Sober analyses had repeatedly rejected large-scale irrigation on economic grounds since the late nineteenth century, and salinization and other warning signs had been detected in the earliest years, and again in the 1920s.[36] Fresh projects were successfully proposed by the Queensland and Western Australian governments during the period of *laissez-faire* under Robert Menzies between 1949 and 1966: state sovereignty in land matters had been retained under the federal Constitution, but Canberra was regularly approached for financial and other assistance. Except for the more demanding economic critiques, Queensland's new schemes for selected coastal rivers were initially given good reports; their marginal characteristics have been rudely exposed, however, in each recession. The Ord River project in Australia's far north-west was a financial blunder on an exaggerated (Bradfieldian) scale, reminiscent of the worst 'colonial speculations'. Its remote location resulted in extraordinary transport costs; the scheme was launched despite expert advice to the contrary—most researchers could not safely nominate a single economic crop—and the harrassed pioneers deposited tonnes of pesticide across the region in an ill-favoured attempt to establish minimum levels of productivity.

In the 1960s and early 1970s ANZAAS sessions debating the economic and ecological consequences of these projects received ample publicity. Davidson and Patterson locked horns over the over-arching wet–dry dilemma and the Ord scheme in particular, and others, examined the (strangely unforeseen or glossed-over) ecological consequences of an enormous artificial lake in a region that is so conveniently situated for large numbers of water birds from Asia.[37] Thus implications for human health and animal health were somewhat tardily added to the growing list of objections to the scheme. Social scientists suggested that the Menzies government had been currying favour with state authorities and with the northern electorates.[38] Our forebears had craved a great 'inland sea': political venalities and parochial ambitions had combined to give us a gigantic duck-pond. But the Ord project still limps on—an extravagant, interrogatory mistake, and therefore an undeniable opportunity in the 'land of the second chance'.

In the dry-farming sector, less notorious but far more extensive landscape changes included the clearing of millions of hectares of brigalow scrub, chiefly in Queensland, and the frontier development of wheat and sheep farming in the drier Western District of New South Wales and on the 'light lands' of Western Australia, including the Esperance sandplain. Federal loans and the continuing contribution of advanced agricultural science were common ingredients.[39] That was also true of a current of 'rationalization' coursing through the old farming frontiers created under closer settlement legislation. Associated with the aggressive restructuring of traditional manufacturing industries, in rural Australia the impact was harsh and pervasive, but dairy farmers bore the brunt. The more advanced, climatically favoured (and nationally more critical) New Zealand industry fared better, but did not entirely escape the effects of bankruptcy, forfeiture, and amalgamation. Rationalization in Australia produced complex local and regional disruptions in the rural communities of every state; overall, there was a severe contraction favouring the most temperate environments of Victoria and, to a lesser extent, Tasmania.[40]

From one perspective, the recent salutary blending of economic and environmental controls injected an overdue logic into Australian affairs. But frequent reminders of apocalyptic visitations—including bushfires, floods, and locust plagues, and the old enemy, drought—should have established the instructive image of a First World society in a Third World setting. No region of Australia can be classed as immune from major environmental hazards, and the interdisciplinary study of this central national theme has been well represented in ANZAAS and other mixed scientific bodies.[41] Too many Australians still pretend insulation, but there are interesting connections between an increasingly intrusive internationalization, cultural diversification, and heightened sensitivities to our own natural and social environments.

Recent imported notions of participatory democracy have been accompanied by a rising demand for accountability, which has challenged the competence of traditionally narrow forms of expertise in dealing with issues that are inescapably multifaceted and have implications extending beyond scientific discourse. Where social, aesthetic, and ecological considerations have come to the fore, Australian and New

Zealand versions of environmentalism have politicized local conservation activities, and wilderness has taken on international as well as national celebrity. Sharply focused research preferences have tended to isolate the scientific subculture everywhere. In ANZAAS itself, a well-meaning retention of very traditional disciplinary boundaries was increasingly scorned by specialists, and even that modest effort also failed to produce the desired brand of 'public science', despite anxious resort in the 1970s and 1980s to the declaration of strong congress 'themes', supported by interlocking symposia.

Modern environmentalism has clearly exerted almost as much influence in the transformation of social and political landscapes as in the management of the built and natural environments. It has also transformed the landscapes of science and its bold representation in ANZAAS meetings has done something to bolster the continuity of spirit and purpose. A transition may be discerned in the 1960s, when the universities were rapidly expanding. Many scientists still felt uncomfortable with philosophical ruminations and soul-searching statements of educational purposes and glibly repeated unfortunate claims for a supposed 'mastery over Nature'; but few of them were entirely oblivious to the crisis in confidence that had overtaken the public. The mood was made articulate in Thornton's paper at the Christchurch Congress in 1968.[42] Science had become embarrassingly entangled with issues of world power and was seen to be 'set in a desert where the struggle is hard, traversed by their generation with a heart-fear lest it quickly becomes a festival of death'. Further, 'the real step forward for "Science" lies in a new access to ethical insight'.[43]

ANZAAS's traditional utilitarian thrust was sustained in most of the congresses of that time, but a cautious balance was struck (or at least presented) between development and preservation interests. The *Australian Journal of Science* continued to publish contributions on the conditions for accelerated development in the north, and others examining optimum land use in the high rainfall zones of each country, the place of improved pastures in environmental change and economic progress, decentralization and regional development, and a Liversidge lecture on the potential role for chemists in the exploitation of mineralized beach sands.[44] At the same time, UNESCO's new conservation programmes were carefully noted in the journal, and several related ventures were reported for our part of the world, including discussions on the status of national parks and reserves, announcements of a proposal from the Flora and Fauna Committee of Australia's Academy of Science to establish a national biological survey, appeals for a National Museum in Canberra, and a characteristically reflective piece from the geomorphologist J. N. Jennings on the 'geological agency' of human society.[45] Editorialized propinquities were beginning to indicate the potential for rich alliances. Archaeology, for instance, was at last moved from the stones and bones category, and although the question of Aboriginal land rights was becoming highly contentious, there was good interdisciplinary progress towards a greater understanding of pre-European landscapes; similarly, the *terra nullius* assumptions of white Australia were trenchantly criticized, and conservationists identified lessons and alternatives in Aboriginal styles of stewardship.[46] The Australian Labor Party courted the conservationists; researchers were advised to get on with the job of controlling innovations instead of creating more of them; and scientists expressed horror at the gap between the wider community's behaviour in the environment and its understanding of proven sanctions—the situation signified a kind of cognitive dissonance, but some of the scientists saw it as dishonesty.[47]

The 1960s and early 1970s was a disputatious time, crowded with uncertainties. Resource conflicts erupted with disturbing frequency— over the proposed development of the Little Desert in Victoria; the flooding of Tasmania's Lake Pedder and contests over Lake Manapouri and the Clutha Valley in New Zealand; rival claims over the destruction of the Great Barrier Reef by people and starfish; and literally hundreds of other contests, ranging from city parks, freeways and neighbourhoods to national debates over pollution, energy consumption, and the impact of mining on outback landscapes.[48] Environmental degradation, including alarming descriptions of desertification, was the subject of several comprehensive official reports, and their tone returned us with a whimper to the 1930s.[49] And this dynamic situation reflected and incorporated the increased emphasis on 'environmental studies' under various titles in the schools, colleges, and universities. A proliferation of interdisciplinary science and social science programmes answered the challenge, and the geographers extended their commitment to that established constituency. While most mainstream scientists continued with business as usual, seemingly unaware of a coming crisis over declining public esteem, conservation continued to offer an invaluable meeting place for science, technology, politics, and the paying public. Once again, the contributions in the flagship journal of ANZAAS (*Search*, after 1970) conveniently document the ferment in, for example, papers on logging in the rainforests, uranium mining, energy planning and desertification, the kangaroo population, estuarine modelling, groundwater utilization and the management of the Murray–Darling system.[50]

The importation of American- and European-style 'ecotactics' in the 1970s promised revolutionary changes. Aesthetic and ecological considerations were given more prominent billing, and wilderness landscapes were invested with international as well as national values.[51] The overdue radicalization of the Australian Conservation Foundation made it a potent influence in several spheres and streamlined the co-ordination of activist strategies. Legislation for the registration and protection of significant parts of the National Estate followed.[52] Similarly, the introduction of the American notion of 'environmental impact statements' was warmly welcomed as a rational approach to even-handed arbitration in resource management disputes, but in the end it disappointed. After the introduction of the Federal Environmental Protection (Impact of Proposals) Act of 1974, the Environmental Impact Statement (EIS) concept was brought into all discussions in which the national government had a strong legal and financial interest. Export control and the monitoring of levels of foreign ownership lie within federal jurisdiction, and so the new legislation could be made to embrace all large mining, manufacturing, and transportation projects that involved overseas funding and were intended to serve overseas markets. The states themselves, however, still control most forms of land development, including the issuing of production leases, and they also have the power to seek impact studies.[53]

PERSPECTIVE

Perhaps the most salutary lessons of the postwar era are encountered in the intermediate spaces between the urban and outback or interior zones of Australia, and in the established farming districts of New Zealand. Our rural landscapes are being rationalized with some speed as we continue our course from provincial subordination to Britain to a far more perilous dependence upon transnational corporations and the vagaries of global trade competition. In Australia especially, those landscapes of hopes are also, perhaps, monuments to an economic impertinence, sometimes confirming our fretful estrangement from Nature. Identification with place remains a central task in both Australia and New Zealand, and in that consuming context scientists need not overstate their expert contributions; indisputably, quite as much has been achieved, in incremental experience, by the lay community, and by the creative efforts of a host of artists, writers, and poets. Russell's successors attempted to preserve the arts–sciences link, despite the rapid pace of cultural change and the institutionalization of narrow specializations, and their recent forays into 'science communication' recognize the breach between vision and practice.

The original idealism of the AAAS and later ANZAAS reminded local researchers of the dangerous gulf separating them from the wider community, and the unifying potentials of a secure coming-to-terms with difficult environments. Despite many disappointments, that idealism is still evident today, even though it has proven difficult enough to persuade researchers from different fields to come together. But the present cautionary pause may yet be turned to our mutual advantage. Shorn of yesterday's naive presumptions, we have been returned to Griffith Taylor's enlightened pessimism. Australians may face the larger challenge: scientific investigations repeatedly affirm that no region of Australia can be considered immune from major environmental hazards. Much of the scientists' imagery continues to suggest the precarious survival of a First World society in a Third World setting. Over most of the second century of white settlement that has offered an unusual national challenge, for scientists and non-scientists alike.

NOTES

1. The favoured example is the observation made by George Perkins Marsh in his *Man and Nature, or, Physical Geography as Modified by Human Action*, first published in New York by Scribner in 1864 (reprinted,

Cambridge, Massachusetts: Harvard University Press, 1964). The point is elaborated in J. M. Powell, *Environmental Management in Australia, 1788–1914. Guardians, Improvers and Profit: An Introductory Survey* (Melbourne: Oxford University Press, 1976)

2. S. P. Hays, *Conservation and the Gospel of Efficiency* (New York: Athenaeum, 1972); see also Powell, ibid.
3. Powell, op. cit., note 1.
4. D. W. Meinig, *On the Margins of the Good Earth: the South Australian Wheat Frontier, 1869–1884* (Chicago: Association of American Geographers, 1962); M. Williams, *The Making of the South Australian Landscape: A Study in the Historical Geography of Australia* (London: Academic Press, 1974).
5. R. L. Heathcote, *Back of Bourke. A Study of Land Appraisal and Settlement in Semi-Arid Australia* (Melbourne: Melbourne University Press, 1965).
6. The point is well made in Williams, op. cit., note 4, especially chapter 7, 'Changing the soil', 263–332.
7. H. C. Russell, 'President's Address', in *Report of the AAAS*, 1 (Sydney, 1888), 1–14, on 12–13.
8. ibid., 13–14. See also M. E. Hoare, 'Learned Societies in Australia: The Foundation Years in Victoria, 1850–1860', *Records of the Australian Academy of Science*, 1 (1967), 7–29, and 'Doctor John Henderson and the Van Diemen's Land Scientific Society', *Records of the Australian Academy of Science*, 1 (1968), 7–24. For the related activities of influential field naturalist groups, see S. Bardwell, *National Parks in Victoria, 1866–1956* (Unpublished PhD thesis, Monash University, 1974).
9. J. M. Powell, 'Exiled from the Garden. Von Mueller's Correspondence with Kew, 1871–81', *Victorian Historical Journal*, 48 (1977), 313–20, and 'A Baron under Siege: von Mueller and the Press in the 1870s', *Victorian Historical Journal*, 50 (1979), 18–35.
10. Powell, op. cit., note 1.
11. ibid.
12. J. M. Powell, 'Soldier Settlement in New Zealand, 1915–1923', *Australian Geographical Studies*, 9 (1971), 114–60, and 'The Debt of Honour: Soldier Settlement in the Dominions, 1915–1940', *Journal of Australian Studies*, 8 (1981), 64–87. I have essayed later developments in *An Historical Geography of Modern Australia: The Restive Fringe* (Cambridge: Cambridge University Press, in press).
13. For a brief guide see J. M. Powell, 'National Identity and the Gifted Immigrant: a note on T. Griffith Taylor, 1880–1963', *Journal of Intercultural Studies*, 2 (1981), 43–54. Taylor was a protégé of the pioneer geologist T. W. Edgeworth David (1858–1934); he was awarded an 1851 Exhibition to Cambridge for palaeontological research and subsequently led the western geological parties on Scott's ill-fated *Terra Nova* expedition to the Antarctic. On his return to Australia he took up an appointment as 'physiographer' to the infant Commonwealth Weather Service, and in 1920 returned to academia as foundation head of the University of Sydney's geography department.
14. ibid. See also E. J. Brady, *Australia Unlimited* (Sydney: George Robertson, 1918), and G. Taylor, *Australia: A Study in Warm Environments and Their Effect on British Settlement* (London: Methuen, 1940). His earlier writings on environmental constraints, the need for a national resources atlas, and the poor prospects for white settlement in the tropics were made from his public service base.

15. G. Taylor, 'Geography and Australian National Problems', *Report of the AAAS*, 16 (Wellington, 1924), 433–87.
16. ibid., 480.
17. G. Taylor, in the *Sydney Morning Herald*, 7 March 1925.
18. *Taylor Papers*, (National Library of Australia), Maiden to Taylor, 6 July 1924. Indeed, in the 1930s scientific colleagues in other fields still considered some of his comments dangerous, and campaigned against his sombre forecasts. An example is R. W. Cilento, 'Australia's Problems in the Tropics' (Medical Science and National Health), *Report of ANZAAS*, 21 (Sydney, 1932), 216–33.
19. *Sydney Morning Herald*, 5 July 1924. And so, in spirited defiance, 'The central portion and the north will be taken up in God's good time by British pioneers, and developed by them, for Australia is going to be for ever British, and whether red labour or yellow labour or green labour tries to hinder, it will be British sinew and British grit and British money that will win out in the end, as surely as it was British grit that won out in the beginning.'
20. J. M. Powell, 'Taylor, Stefansson and the Arid Centre. An Historic Encounter of "Environmentalism" and "Possibilism"', *Journal of the Royal Australian Historical Society*, 66 (1980), 163–83. As a late arrival in academia (at the age of 40), Taylor may be forgiven for unseemly haste in transforming his subject. He was neither the first nor the last to reject vernacular modes of learning in a new country of immigrants, but the charge of insensitivity blunted his timely campaign, marking him and his adopted field with an undeserved larrikin stigma.
21. J. M. Powell, 'James Macdonald Holmes, 1896–1966', *Geographers: Bibliographical Studies*, 7 (1983), 51–5. During the 1920s and 1930s, AAAS research committees made several recommendations on the need for a federal museum, the national problem of water conservation, the development of artesian water resources, and the demand for improved town planning.
22. J. Barrett (ed.), *Save Australia, A Plea for the Right Use of our Flora and Fauna* (Melbourne: Macmillan 1925). Contemporary ideas of a mastery over Nature may have been encouraged in the 1920s and 1930s by the astonishing success of the prickly pear eradication programme.
23. Williams, op. cit., note 4, 263–332.
24. J. A. Prescott, 'The Soils of Australia in Relation to Vegetation and Climate', *CSIR Bulletin*, 52 (1931); also his 'Soil Classification and Survey', *Report of the AAAS*, 18 (Perth, 1928), 724–8. Russian and American advances in the regional classification of soils and the continuing small-scale analyses in Britain were conscientiously monitored in Australia and New Zealand, cf., C. B. Wells and J. A. Prescott, 'The Origins and Early Development of Soil Science in Australia', in *Soils, An Australian Viewpoint* (Melbourne: CSIRO, 1983), 3–12, including a useful bibliography. Soil science was well established by the time Rothamsted's Sir John Russell presented his paper on 'Soil Conservation and Permanent Agriculture', *Report of ANZAAS*, 24 (Canberra, 1939), 260–8. For historical notes on New Zealand see K. B. Cumberland, *Landmarks* (Surry Hills: Readers Digest, 1981) and K. B. Cumberland and J. S. Whitelaw, *New Zealand* (Chicago: Aldine, 1970).
25. J. M. Holmes, *Soil Erosion in Australia and New Zealand* (Sydney: Angus & Robertson, 1946); K. B. Cumberland, *Soil Erosion in New Zealand; A Geographic Reconnaissance* (Wellington: Soil Conservation and Rivers Control Council, 1944); F. Ratcliffe, *Flying Fox and Drifting Sand. The*

Adventures of a Biologist in Australia (London: Chatto and Windus, 1938). Several ANZAAS Sections addressed the soils theme during the 1930s, for example E. Cheel (Botany), 'A Review of the Flora of the Arid and Semi-arid Regions of Australia', *Report of ANZAAS*, 23 (Auckland, 1937), 307–37. It was followed in the same volume by broader contextual statements, for example, L. T. Madigan (Geography and Oceanography), 'A Review of the Arid Regions of Australia and Their Economic Potentialities', 375–97 and Theodore Rigg's Liversidge Lecture, 'Soil Deficiencies in New Zealand', 401–22. The preferred disciplinary structure of ANZAAS was not well adapted to cope with such multifaceted themes, however; a few thorough theme-based congresses might have fulfilled a more useful function. More or less by default, the task was taken up by a handful of outspoken individuals.

26. G. H. Burvill (ed.), *Agriculture in Western Australia: 150 years of Development and Achievement, 1829–1979* (Perth: University of Western Australia Press, 1979), especially 157–72.
27. L. T. Carron, *A History of Forestry in Australia* (Canberra: Australian National University Press, 1985).
28. C. E. Lane-Poole, 'Forestry and Land Settlement', *Report of the AAAS*, 18 (Perth, 1928), 712–28, on 713. See also his 'Forestry as a profession', *Australian Quarterly*, 4 (1929), 991–6.
29. ibid., 713.
30. cf., Cumberland, op. cit., note 24. In contrast, the handful of Australia's specialists in historical geography have generally steered clear of the public history arena. And our historians continue to avoid environmental issues, despite such efforts as G. C. Bolton's *Spoils and Spoilers: Australians Make Their Environment* (Sydney: George Allen & Unwin, 1981). But the field still attracts the attention of gifted writers and popular modern versions of natural history: for example, Eric Rolls's *They All Ran Wild: The Story of Pests on the Land in Australia*. (Sydney: Angus & Robertson, 1969) and *A Million Wild Acres. 200 Years of Man and an Australian Forest* (Melbourne: Thomas Nelson, 1981); Vincent Serventy's many popular works; Judith Wright's *The Coral Battleground* (Melbourne: Thomas Nelson, 1977); and of course innumerable television documentaries, and the self-conscious but promising developments in science communication fostered by ANZAAS.
31. N. J. Ericksen, 'Sustained Yield Management of Indigenous Forests: Policies and Practices of the New Zealand Forest Service', *Proceedings, Institute of Australian Geographers Conference*, (Perth, 1987, in press); M. M. Roche, *Forest Policy in New Zealand. An Historical Geography* (Palmerston North, New Zealand: Dunmore Press, 1987).
32. H. C. Coombs, *Trial Balance* (Melbourne: Macmillan, 1981).
33. cf., J. M. Holmes, 'The Content of Geographical Study', *Report of ANZAAS*, 22 (Melbourne, 1935) 401–33, and *The Geographical Basis of Government: Specially Applied to New South Wales* (Sydney: Angus & Robertson, 1944); and several related efforts from the Sydney Department under Holmes's direction, noted in Powell, op. cit., note 12.
34. Coombs, op. cit., note 32, 64.
35. The most accessible reference may be J. J. C. Bradfield, 'Watering Inland Australia', *Rydges* (1 October 1941), 586–9, 606.
36. A good early example of research on salinization is A. J. Perkins, 'Statistical Data Relative to the Reclamation of a Salt-Impregnated Area in a River Murray Orchard', *Report of the AAAS*, 18 (Perth, 1926), 734–47. For the general background, see Powell, op. cit., note 1.

37. For example, R. A. Patterson, 'The Economic Justification of the Ord River Project', a widely circulated paper at the thirty-eighth ANZAAS congress (Hobart, 1965); B. R. Davidson, *Australia Wet or Dry? The Physical and Economic Limits to the Expansion of Irrigation* (Melbourne: Melbourne University Press, 1969); N. E. Stanley, 'Ord River Ecology', *Search*, 3 (1972), 7–12. Historians of the Ord project must start earlier, naturally: the context includes variously aggressive and opportunistic proclamations on the potentials and achievements of applied science over the preceding decades. Only a few examples have been cited in the present paper; others include two interesting presidential addresses: W. L. Waterhouse, 'Some Aspects of Plant Pathology', *Report of ANZAAS*, 24 (Canberra, 1939), 234–59 and I. Clunies-Ross (at that time CSIRO chairman), 'The Impact of Scientific Advances on Production in the Livestock Industries' (Kendall Oration), *Report of ANZAAS*, 30 (Canberra, 1954), 215–22.
38. S. Graham-Taylor, 'A Critical History of the Ord River Project', in B. R. Davidson and S. Graham-Taylor, *Lessons from the Ord* (St Leonards: Centre for Independent Studies, 1982), 25–55. Also G. McDonnell, 'The Ord Debate and Public Decision-Making: a View from the Terrace', *Australian Quarterly*, 38 (1966), 44–56, and K. J. Walker, 'The Politics of National Development: The Case of the Ord River Scheme', *Journal of Public Administration*, 32 (1973), 93–113.
39. Bureau of Agricultural Economics, *The Economics of Brigalow Land Development in the Fitzroy Basin, Queensland* (Canberra: AGPS, 1963); R. W. Johnson, *Ecology and Control of Brigalow in Queensland* (Brisbane: Queensland Department of Primary Industries, 1964); also P. J. Skerman, 'The Brigalow Country and its Importance to Queensland', *Journal of the Australian Institute of Agricultural Science*, 19 (1953), 167–76. For the west, see G. H. Burvill, 'The Development of Light Lands', op. cit., note 26.
40. G. R. Lewthwaite, 'Retreat, Reform and Regional Concentration: The Shifting Pattern of Australian Dairy-farming', *Queensland Geographical Journal*, 4 (1978), 1–20, and 'Product Diversification and Territorial Concentration: the Changing Status of Australia's Dairy States', *Queensland Geographical Journal*, 5 (1979), 21–42; G. Barrett and D. Gargett, 'Historical Review of Structural Adjustments on Australian Dairy Farms', Attachment E, 'Comparative Efficiency between Australian and New Zealand Dairy Industries and Implications for Trans-Tasman Trade', *Bureau of Agricultural Economics Occasional Paper no. 60* (Canberra: AGPS, 1981), 124–38.
41. Good overviews include R. L. Heathcote and B. G. Thom (eds), *Natural Hazards in Australia*, Proceedings of a Symposium sponsored by the Australian Academy of Science, the Institute of Australian Geographers and the Academy of Social Sciences in Australia (Canberra: Australian Academy of Science, 1975); J. Oliver (ed.), *Response to Disaster* (Townsville: James Cook University Centre for Disaster Studies, 1980); and G. Pickup and J. E. Minor, 'Assessment of Research and Practice in Australian Natural Hazards Management', *Northern Australian Research Bulletin*, 6, (Darwin, 1980).
42. H. Thornton, 'A Critique of Science', *Australian Journal of Science*, 31 (1969), 341–6.
43. ibid., 346.
44. All in the *Australian Journal of Science*: R. H. Greenwood, 'Will the North Remain Empty? An Analysis of Conditions for Accelerated Development',

26 (1964), 305–11; H. M. Caselberg *et al.*, 'Optimum Land Use in the High Rainfall Zones of Australia and New Zealand', 26 (1964), 379–85; C. M. Donald, 'The Progress of Australian Agriculture and the Role of Pastures in Environmental Change', (Farrer Memorial Oration), 27 (1965), 187–98; R. H. Greenwood, 'Decentralization and Regional Development in Western Australia', 30 (1967), 131–2; I. E. Newnham, 'The Chemical Challenge of the Beach Sands', 26 (1964), 320–34. The latter piece addressed a characteristic emphasis in one area of ANZAAS since the early postwar era. Cf., H. G. Raggatt, 'Depletion of Mineral Resources—a Challenge to Geology and Geophysics', *Report of ANZAAS*, 26 (Perth, 1947), 109–32, on 123: 'There is nothing like economic work to discipline the mind'. Similarly, the emphasis on national timber production had been clearly communicated in G. J. Rodgers's presidential address to the Agriculture and Forestry Section, 'Softwoods in Australian Forestry', *Report of ANZAAS*, 25 (Adelaide, 1946), 218–29.

45. The following appeared in the *Australian Journal of Science*: H. Daifuku, 'The Preservation of Monuments—a UNESCO Program', 26 (1964), 139–40; E. C. F. Bird, 'Approaches to Conservation', 28 (1966), 338–9; three papers in 30 (1968): J. G. Mosley, 'Trends in the Planning of Australia's National Parks and Reserves', 281–4, G. N. Bauer, 'The Role of Forestry in Conservation Management', 285–8, and R. G. Downes, 'Nature Reserves and National Parks in Relation to the Conservation of Man's Environment', 288–91; 'Australian Academy of Science Report on the Need for Study of the Australian Fauna and Flora', 28 (1966), 451–5; (Fauna and Flora Committee of the Australian Academy of Science), 'Proposal to establish a Biological Survey of Australia', 31, (1969), 377–82; J. N. Jennings 'Man as a Geological Agent', 28 (1965), 150–6 (a favoured theme with geographers, cf., Charles Lucas's presidential address to Section E during the Australian meeting of the British Association for the Advancement of Science in 1914, 'Man as a Geographical Agency', 426–39). See also F. Fenner (ed.), *A National System of Ecological Reserves in Australia* (Canberra: Australian Academy of Science, 1975).

46. The changes were signalled in various places, including the *Australian Journal of Science*. For example, D. J. Mulvaney, 'Australian Archaeology, 1929–1964: Problems and Policies', 27 (1964), 39–44; J. Cleland, 'Aborigines and Land Tenure', 28 (1965) 164–5; A. B. Pittock and J. Cleland, 'Aborigines and Land Tenure', 29 (1966), 20–1. Larger works in the 1970s included C. D. Rowley, *The Destruction of Aboriginal Society*, 3 vols (Canberra: Australian National University, 1970–1) and G. Blainey, *Triumph of the Nomads: A History of Ancient Australia* (Melbourne: Macmillan, 1975). It would be an injustice to omit the passionate appeals of earlier workers, for example, F. W. Jones, 'The Claims of the Australian Aborigine', *Report of the AAAS*, 18 (Perth, 1926), 497–519, had emphasized the need to make reparation 'for the filching of a whole vast continent from its real owners' (p. 497). Jones commented on the ethnocentric stupidities of the 'work and civilization panacea' and reminded the Anthropology Section of the high hopes entertained in the early years of Federation for more systematic and sensitive research programmes. Elkin, Kenyon, and others made similar pleas in the 1930s. The reference in this section of the text to editorialized initiatives is deliberate. The destructive fragmentation of scientific discourse was then well advanced, and from a historical viewpoint this is another case in which the convenient whole may be much more than the sum of the variously prominent and obscure parts. I mean that the most accessible published

evidence succeeded in suggesting a good response, organizational and spontaneous, to the ANZAAS tradition.

47. S. Encel, 'Science, Technology and Society', *Search*, 1 (1970), 12–17; E. G. Whitlam, 'A National Science Policy', *Search*, 1 (1970), 134–8 (admitting that 'urban man is diminished by any final severance of the links with nature and the countryside'); P. D. Dwyer, 'Beyond Biology: Alienation and Environment', *Search*, 2 (1971), 153–9.

48. Following the listed sequence, J. M. Powell, 'Action Analysis of Resource Conflicts: the Little Desert Dispute, 1963–72' in J. M. Powell (ed.), *The Making of Rural Australia* (Melbourne: Sorrett Press, 1974), 161–79; Australian Conservation Foundation, *Pedder Papers: Anatomy of a Decision* (Melbourne: Australian Conservation Foundation, 1972); T. O'Riordan, 'New Zealand Resource Management in the Seventies: A Review of Three Recent Conferences', *New Zealand Geographer*, 27 (1971), 197–210; R. G. Lister, 'The Clutha Valley: Multiple Use and the Problems of Choice', in A. G. Anderson (ed.), *The Land our Future, Essays on Land Use and Conservation in Honour of Kenneth Cumberland* (Auckland: Longman Paul, 1980), 175–91 (and other contributions in the same volume); see also R. J. Johnston (ed.) *Society and Environment in New Zealand* (Christchurch: Whitcombe and Tombs, 1974) (especially a note on the Lake Manapuri conflict, 198–9); in *Search*, 2 (1971) D. W. Connell, 'The Barrier Reef Conservation Issue', 188–92 and F. H. Talbot, 'Crown-of-Thorns Report', 192–3; A. Gilpin, *The Australian Environment: Twelve Controversial Issues* (Melbourne: Sun Books, 1980).

49. Examples include Research Directorate, Department of Environment, Housing and Community Development, *A Basis for Soil Conservation Policy in Australia, Report No. 1* (Canberra: AGPS, 1984); Department of Home Affairs and Environment, *National Conservation Strategy for Australia: Living Resource Conservation for Sustainable Development* (Canberra: AGPS, 1982); L. E. Woods, *Land Degradation in Australia* (Canberra: AGPS, 1984).

50. In *Search*: J. D. Ovington and R. J. Thistlethwaite, 'The Woodchip Industry: Environmental Effects of Cutting and Regeneration Practices', 7 (1976), 383–92; J. B. Kirkpatrick and D. M. J. S. Bowman, 'Clearfelling versus Selective Logging in Uneven-aged Eucalypt Forests', 13 (1982), 136–41; J. B. Kirkpatrick and K. J. M. Dickinson, 'Recent Destruction of Natural Vegetation in Tasmania', 13 (1982), 186–7; S. Myhra, 'Some Environmental Aspects of Uranium Mining and Milling in Australia', 9 (1978), 400–6; G. W. Butler, 'Land use and the Role of Research, with Particular Reference to Energy Planning in New Zealand', 12 (1981), 298–301; G. Caugley *et al.*, 'How many Kangaroos?', 14 (1983), 151–2; T. Beer, 'Australian Estuaries and Estuarine Modelling', 14 (1983), 136–40; C. W. Lyle, 'Groundwater Utilization in the Shepparton Region', 14 (1983–4), 326–9; P. Crabb, 'Whither the Murray: Politics and the Management of Australia's Water Resources', 15 (1984), 36–41.

51. J. B. Kirkpatrick and R. A. Haney, 'The Quantification of Developmental Wilderness Loss: the Case of Forestry in Tasmania', *Search*, 11 (1980), 331–5; K. McKenry, 'Value Analysis of Wilderness Areas', in D. C. Mercer (ed.), *Leisure and Recreation in Australia* (Melbourne: Sorrett Press, 1979), 209–21; M. Feller *et al.*, 'Wilderness in Victoria: An Inventory', *Monash Publications in Geography*, 21 (1979).

52. It fizzled out in the 1980s. A good summary of the earlier gains is offered in Australian Heritage Commission, *The National Estate in 1981* (Canberra: AGPS, 1982).

53. F. Talbot, 'Environmental Impact Assessment: Summary and Prospects', *Search*, 7 (1976), 273–9; R. H. Bradbury *et al.*, 'Prediction versus Explanation in Environmental Impact Assessment', *Search*, 14 (1983–4), 323–5; J. A. Barnes, *Who Should Know What? Social Science, Privacy and Ethics*, (Harmondsworth: Penguin, 1979); D. C. Mercer, 'Victoria's Land Conservation Council and the Alpine region', *Australian Geographical Studies*, 17 (1979), 107–30, and 'Freedom of Information and Geographical Practice', *Australian Geographical Studies*, 21 (1983), 8–32.

11

Developing Nature's Treasures: Agriculture and Mining in Australasia

Bruce Davidson

A nationwide body such as ANZAAS, consisting of research workers from all branches of learning, might be expected to perform a number of tasks. In the fields of applied sciences, such as agriculture, economics, veterinary science, geology, metallurgy, and mining engineering, these should include the identification of the major problems of the industries in which members of these disciplines were employed, the development of the methods needed to solve these problems, and the rapid recognition and recommendation of the adoption of the appropriate solutions by the industries concerned. In addition, one would expect such a group to examine and recommend the appropriate means of education and training workers for future research. In a country where government policy played a major part in the development of primary industry, any group of applied scientists might be expected to criticize existing government policies if they were detrimental to the nation's welfare and to suggest more appropriate policies.

The effectiveness of the Australasian Association for the Advancement of Science (AAAS) and later ANZAAS in serving Australasia's primary industries can be gauged by examining the degree to which the organization performed these tasks.

Its ability to do so should have been enhanced by the nature of its membership and the attitude of the farming and mining communities. Initially most of the papers concerned with agriculture and mining were delivered by academics, representatives from industry, and officers of the government departments carrying out research, extension, and regulatory work in agriculture and mining. Later, when organizations devoted purely to research, such as the CSIRO and the Bureau of Mineral Resources, were created by the Commonwealth government, they also played an active role in ANZAAS. There is no evidence that any one body dominated the Association and a great deal of evidence that both farmers and miners would readily support and adopt any profitable recommendation made by the Association.

The major difficulties were that the number of scientists involved was much smaller than in Europe and the United States, but the area of land they had to deal with was as large, and possibly even more diverse, than that of Europe or the United States. The climate, soils, and mineral lodes on which these industries were based were often very different from those encountered in the Northern Hemisphere. These differences limited the degree to which new discoveries in the Northern Hemisphere could be utilized in Australasia. Initially scientists suffered from having been mainly trained in an environment that was very different from the one they had to work in, as most were trained in the United Kingdom.

AGRICULTURE

Even in the 1980s the area of land capable of producing crops and highly productive pastures in Australasia, if expressed on a per capita basis, is twelve times as great as in western Europe and four times as great as in the United States.[1] Thus, agricultural land in Australasia was always cheaper than similar land in Europe, but agricultural labour was more expensive. In addition, such a vast expanse of agricultural land could quickly fill the limited domestic market available for any agricultural commodity.

The importance of the ratio of agricultural land to population was grasped as early as 1793 by John Macarthur, who realized that any agricultural industry in Australia would only succeed if it fulfilled four basic conditions, namely, that there should be a large export market for the commodity produced, and that it should be valuable enough in terms of its weight and volume, after meeting the cost of transport to distant markets, to leave a profit for the producer, and it must not deteriorate on the voyage. In addition, the commodity had to be produced using a small labour force if it were to compete with similar commodities produced in lands where wages were lower. The commodity most likely to succeed was one requiring large areas of land, as this was the only cheap resource in Australasia.[2] These rules are as valid today as in 1793.

Initially wool produced by sheep grazing native pastures was the only product that would meet the four conditions. After the gold rushes of the 1850s in Australia and 1860s in New Zealand, large numbers of people were settled on holdings of 160 to 260 hectares and a satisfactory living could not be made from wool production alone on such limited areas of land.[3] Between 1860 and 1888, however, Australasian farmers and applied engineers developed large cultivating and harvesting equipment, which, together with the railways constructed by the colonial governments, made it possible to produce wheat for export.[4] The development of refrigeration in 1879 enabled butter and meat to be exported.[5] In addition, a sugar industry using cheap Melanesian labour was established on the north coast of New South Wales and in coastal Queensland, and irrigation farms supplied with water from reservoirs constructed by the colonial government were established on the Murray River in Victoria. The growing demand for food in Britain, which had been created by the industrial revolution, provided a market for the produce of these new farming systems.

Stopping the Rot, 1888–1905

The introduction of a more intensive form of agriculture in south-eastern Australia quickly led to a number of problems that had not existed while the land had been used for sparse grazing. Wheat made a much heavier demand on soil nutrients than sparse grazing. In addition, the varieties of wheat used were not suited to Australia's climatic conditions, and both rust and smut emerged as major problems. Australian wheat yields declined from an average of 0.66 tonnes per hectare (10 bushels per acre) in the 1870s to 0.42 tonnes per hectare (6 bushels per acre) in the 1890s.[6] To make matters worse, the wild rabbit, which had been introduced in 1859, reached plague proportions at the end of the 1880s, at the very time when Australia was experiencing a series of severe droughts, which were to culminate in the great drought of 1899–1903.

Science began to play a part in overcoming these problems in 1882 when J. D. Custance, the principal of Roseworthy College in South Australia, discovered that one of the major reasons for decline in wheat yields was the lack of phosphorus.[7] The surface treatment recommended by Custance was not worth adopting, but this was overcome when the Correll brothers discovered in the York Peninsula that the same result could be obtained by sowing 55 kilograms per hectare of superphosphate with the wheat seed.[8] Eighteen years after Custance's discovery, 27 per cent of the wheat in South Australia and 12 per cent in Victoria was sown with superphosphate. Although the problem of soil fertility was frequently raised in papers given at the AAAS, the discoveries of Custance and the Correll brothers were not mentioned in any paper until 1900.[9]

A similar picture emerges from a perusal of the AAAS papers dealing with the deficiency of soil nitrogen. As early as 1890 R. L. Pudney of the New South Wales Department of Agriculture had pointed out that a long period of bare fallow before sowing allowed sufficient time for soil bacteria to break down organic matter and turn the nitrogen into nitrates that could be taken up by plants. Pudney stated that the system was used in Victoria and had been used for a number of years in South Australia.[10] It was described in the latter colony as early as 1876.[11] However, the practice was not mentioned in papers given to the AAAS until 1898.[12]

Of all the major problems facing the wheat industry at the end of the nineteenth century, it was only the development of disease-free varieties suitable for Australian conditions that attracted the attention of the

AAAS. At the very first meeting of the Association, a Rust Committee was set up and a report on the type of research required was published in 1891.[13] In 1895 the Association recognized that the breeding and selection of short season varieties of wheat, which were resistant to rusts, by William Farrer, were likely to overcome the problems of disease and produce a high-yielding variety.[14] In 1902 Farrer produced the variety Federation, which met most of the requirements of the Australian wheat industry. In the same year it was discovered that stinking smut could be prevented by treating the seed with copper sulphate or formalin.[15] The interest of the AAAS in plant breeding and plant pathology has persisted for many years.[16]

Although it was recognized that most of Australia's agricultural wealth came from the grazing industries, and their production had been greatly reduced by the droughts of the 1880s and 1890s, pasture improvement was scarcely mentioned in papers given at the AAAS congresses. It was noted that the native pastures had survived the droughts better than the introduced species in all areas except the coastal regions, but it was suggested that the dangers associated with introducing new species might outweight the benefits because of the risk of introducing weeds.[17] Members of the AAAS were apparently not told that in 1898 Amos Howard, a nurseryman in South Australia, had discovered the annual legume subterranean clover, which had been accidentally introduced from the Mediterranean region, and was attempting to sell the seed as a commercial product. Some Western Australian farmers introduced this plant into that state as early as 1902, and by 1910 some South Australian farmers reported that it was an excellent fodder plant.[18] While the plant did not become commercially important until the 1920s, when satisfactory methods of harvesting the seed and topdressing it with superphosphate were developed, it is strange that it passed unnoticed at the AAAS, but was sown by a number of farmers.

The only AAAS paper dealing with irrigation was limited to an insistence that all irrigable land should be rated at a level sufficient to pay for the reservoirs and distributory works built by the state.[19] The author failed to realize that the additional profits from irrigation were too low for farmers to pay such a rate.[20]

In spite of the multitude and magnitude of problems facing the scientific community in agriculture, the president of the Agricultural Section of the AAAS considered that the main aim of education in this field should be to produce a rural population trained in practical farming. This even applied to the education outlined for the agricultural colleges. University training was thought to be unnecessary in the 1890s.[21]

Although the Constitution of the Commonwealth of Australia was drawn up in the late 1890s, no paper given to the AAAS discussed the importance of the division of powers affecting agriculture between the Commonwealth and the states, although the retention by the states of the power to fix commodity prices, control Crown Lands and water supplies, and the transport system, on which agriculture depended, were to have important consequences in the future. Similarly, no papers discussed the effect of restricted immigration of the Melanesians, on which the sugar industry depended.

New Pastures and Healthy Animals, 1905–28

With the recovery from the drought of 1903, Australian agriculture entered a period of favourable prices, which continued until the mid-1920s. As far as farmers were concerned, these continued throughout the First World War, as special arrangements were made by the governments of the United Kingdom and Australia to purchase and store any commodities that could not be transported to the United Kingdom immediately. In these circumstances, farmers were able to introduce the new techniques that had been developed over the previous twenty years.

Although the area of wheat was expanded from 2.4 million hectares in 1905 to 6.5 million in 1929, and a far higher proportion of wheat was being grown in drier areas, average wheat yields in the 1920s were 0.72 tonnes per hectare (11 bushels per acre), or nearly double the 0.42 tonne (6 bushels per acre) average yields of the 1890s. This was the result of higher-yielding varieties and of a higher proportion of the crop being sown on fallowed land with superphosphate.[22]

In 1913 L. A. Musso of the New South Wales Department of Agriculture presented the AAAS with the results of his work on the trace elements. Plant analysis in England had revealed that plants contained significant amounts of magnesium, copper, and other minor elements. Additions of these elements to nutrient solutions in which plants were grown had varying effects. Musso noted the additions usually increased plant growth if only small quantities of these nutrients were added. He calculated that a 10 ton crop only removed 39 grams of magnesium, and 135 grams of copper per hectare.[23] This was the first suggestion in Australia that small quantities of some elements might be needed for some crops. Unfortunately this line of research was not followed up, and it was over twenty years before trace elements were used on field crops and pastures.

By 1927 the Western Australian farmer P. D. Forest had developed a practical means of harvesting subterranean clover seed and selected Dwalganup, an early variety, which would survive in regions with a growing season of only five months.[24] With these developments, the discovery of subterranean clover by Amos Howard in 1898 could at last be exploited. In the long term the plant was to lead to a doubling of Australia's livestock numbers and a revolution in wheat production. However, what was possibly the greatest technical innovation in Australian agriculture was acknowledged by the AAAS with a simple note in 1928 that pastures and legumes were being topdressed with superphosphate.[25] This was four years after the practice was being recommended to farmers in South Australia by the state's Department of Agriculture.[26]

The AAAS's record of participation in the first large-scale attempt at biological control of a weed is little better. Although prickly pear had been recognized as a serious problem in Queensland in 1880, it was not mentioned in any AAAS paper, or committee report, until 1913, by which time it covered 8 million hectares in Queensland and New South Wales.[27] A serious examination of how the problem might be overcome was presented to the AAAS in 1923, when the area affected had expanded to 18 million hectares. A number of methods of control were described, and it was rightly concluded that biological control was the only possible method.[28] By this time, however, the predators had been selected and in 1928 it was announced that *Cactoblastus cactorum* would eliminate the problem.[29]

A great deal of progress in the veterinary field was reported at AAAS meetings during this period. This included papers describing the narrow margins by which Australasia had escaped the introduction of foot and mouth disease, rabies, and rinderpest, and pointed out the need for laws enforcing the quarantine of imported animals and the destruction of infected livestock.[30] Other papers reported that some Australasian cattle were infected with mastitis and botulism, and black disease was identified in sheep.[31] The importance of the water snail as the intermediate host for liver fluke in Australia was described to the Association in 1923, as were better vaccines for pleuropneumonia which had been developed.[32] Veterinary science schools were established at the Universities of Melbourne in 1908 and Sydney in 1909 to ensure that an adequate number of well-qualifed veterinary surgeons were available.[33]

If the veterinarians accepted the need for tertiary education early in the twentieth century, the agriculturalists were still doubtful. This attitude probably arose because the well-established agricultural colleges were loath to lose their position as the leaders in the field of agricultural education. In a paper presented to the AAAS on this topic in 1913, H. Pye, the principal of Dookie Agricultural College, doubted that graduates would have sufficient knowledge of practical farming, and whether the multitude of knowledge needed could be provided by one faculty.[34] Even after agriculture faculties were established at the Universities of Melbourne and Sydney, doubts were expressed to the AAAS concerning employment opportunities for graduates and the ability of the community to afford them. It was thought that the establishment of chairs in agricultural science in Australian universities could only be justified by the need for additional research and not as a means of training agricultural scientists.[35]

It is unfortunate that there was no meeting of the AAAS between 1915 and 1921, for it was in this period that the disastrous decision was made to settle large numbers of inexperienced ex-servicemen on the land— many of them on small farms in marginal areas. It was assumed that the schemes would succeed because it was accepted that the high commodity prices prevailing during and immediately after the war would continue.[36] Whether any criticism of the scheme would have been made if the AAAS had met before its implementation is unknown, but perhaps some unease was felt. An attempt was made by the AAAS to establish the climatic limits of wheat production in 1921, but by this time it was too late.[37] By 1929 only 27 000 of the original 37 000 soldier settlers remained on their farms and a high proportion of the debts owed by many of the surviving settlers had to be written off.[38]

Science and the Great Depression, 1929–39

As early as 1928 J. A. Perkins, the director of agriculture in South Australia, described to the nineteenth ANZAAS Congress how narrow farmers' profit margins were and pointed out that declining commodity prices would eliminate profits on many farms. He called for a continuous examination of the economic position of different types of farming in all parts of Australia.[39] This was probably the first time that something like the rural surveys section of the existing Bureau of Agricultural Economics had been proposed in Australia.

As the Depression deepened unemployment increased. In a paper presented to the ANZAAS Congress in 1935, E. R. Hudson, of the New South Wales Department of Agriculture, pointed out that this problem could not be solved by settling the unemployed on the land. Australian agriculture was already producing more than could be sold and additional land settlement would lead to further increases in agricultural production. Hudson appears to have been aware that the assistance given to farmers by way of commodity bounties and two-price schemes during the Depression was not the most appropriate form of rural relief. These failed to distinguish between the inefficient and the efficient farmer. He told the ANZAAS Congress that such relief measures were inappropriate, as they perpetuated the problem by maintaining inefficient farmers on marginal land, and Hudson suggested that, if relief were given, it should be on an individual basis so that only the efficient farmer with a chance of survival was assisted.[40]

Having recognized many of the economic effects of the Depression on agriculture, it is strange that no published ANZAAS paper noted that the farmers' chief means of survival was by increasing production, and that this partly depended on the adoption of many of the techniques that the scientists had helped to find. The real value of agricultural production in Australia increased by 30 per cent between the three years ending in 1929 and the three years ending in 1939.[41] The Ottawa Agreement between the United Kingdom and the dominions in 1933, in which the latter were given preferential tariffs for primary produce imported by the United Kingdom in exchange for giving preferential tariffs on manufactures imported by them from the United Kingdom, provided a market for the additional produce.[42]

By the mid-1930s, a paper was at last given at ANZAAS recognizing that pastures of subterranean clover, Wimmera ryegrass, and *Phalaris*, topdressed with superphosphate, would carry two to three times as many livestock as native pasture in temperate regions of Australia with a growing season of five months or more, and that improved pastures could be established cheaply by surface sowing.[43] Between 1935 and 1939 the area of pasture topdressed in Australia increased from 1.6 million to 3.9 million hectares. All of the pasture improvement was, however, in the temperate regions. It was recognized that the productivity of pastures in tropical and subtropical areas could not be increased until a pasture legume could be found that would flourish in these regions.[44] The search for such a legume has continued until the present day.

In spite of the depressed economic conditions for agriculture, research carried out with trace elements in the 1930s was to have a significant effect in later years. It was during this period that it was discovered and reported to ANZAAS that minute quantities of copper, zinc, molybdenum, and manganese were essential for plant growth and that these were unavailable in some Australian soils.[45] It was also found that both copper and cobalt were essential in animal diets and were lacking in pastures in certain parts of Australasia.[46] Trace elements and the new pasture technology were to form the basis of the rapid expansion in Australia's agriculture after the Second World War.

In addition to the work with trace elements, papers given to the Veterinary Science Section of ANZAAS reported an effective vaccine for black disease and attempts to control the blowfly more effectively. Unsuccessful attempts were made to find a method of biological control, but the reduction in breech strike by using the surgical operation developed by the farmer J. H. W. Mules in South Australia in 1929 was described to ANZAAS in 1937.[47] By the end of the 1930s the chemists had realized that synthetic fibres could be a threat to Australia's wool industry and suggested to ANZAAS that an attempt be made to improve wool's competitiveness by reducing its shrinkability and prickliness.[48]

In 1935 the Australian Institute of Agricultural Science was established to provide a forum for scientists working in this field, but many of them still welcomed ANZAAS as a meeting ground where they could exchange ideas with colleagues working in allied fields such as chemistry and botany. In addition, in 1938 ANZAAS commenced publication of the *Australian Journal of Science*, in which papers given at ANZAAS, and other contributions, were published.[49]

Technology and Prosperity, 1940–86

A lack of capital rather than a lack of knowledge had prevented many farmers from introducing new technology in the 1930s. With the commencement of the Second World War, the price of agricultural commodities was guaranteed by the agreements between the governments of Australia and the United Kingdom.[50] The diversion of labour and materials to the armed services and munition factories, however, prevented expansion of agriculture during the war.

After the war it was planned to settle a large number of ex-servicemen on the land, and a careful survey was made by the specially appointed Rural Reconstruction Commission to determine the areas in which they were to be settled and the economic conditions that might prevail during the period of settlement.[51] This inquiry was the most extensive ever made into Australian agriculture, and it is surprising that no discussion of it was published in any paper given to ANZAAS. The Commission concluded that the United Kingdom would continue to be the major market for Australian agricultural commodities and that soldier settlement would only succeed if soldiers were settled in good rainfall areas and if the scheme were limited to men with adequate capital and farming experience. These conditions were mainly satisfied by training inexperienced ex-servicemen and by purchasing and subdividing the remaining large estates in good rainfall areas.

Contrary to expectations, the prices of rural produce continued to increase after the war, culminating in the Korean War boom of 1951. In addition, the rabbit was almost eliminated by the introduction of myxomatosis by CSIRO in the same year.[52] In these conditions soldier settlers were firmly established, and existing farmers invested in new technology. Between 1939 and 1970 the area of fertilized pastures in Australasia increased from 11 million to 25 million hectares. This included large areas where superphosphate had previously been ineffective because of a lack of trace elements.[53]

Part of the increase in the area of pasture sown was due to the development of aerial seeding in New Zealand in the 1940s. This practice, which was later adopted in Australia, made it possible to establish pasture in rugged country in which land-based vehicles could not be used. Grasslands research, particularly in breeding new grasses and the investigations into fertilizer usage, placed New Zealand research workers among the world's leaders in these fields.[54]

The productivity of pastures was further increased when trials in New Zealand in the 1950s and in Australia in the 1960s revealed that both rye grass and white clover pasture and subterranean clover could be grazed at higher stocking rates than had previously been thought possible.[55]

As a result of these changes Australasian sheep numbers increased from 143 million in 1939 to 240 million in 1970. Cattle numbers increased from 19 to 33 million during the same period. Leguminous pastures were also responsible for increasing wheat yields. In the late 1940s, farmers began to change from the old fallow–wheat rotation to subterranean clover or medic–wheat rotation without bare fallow. The large quantities of nitrogen fixed by legumes made the bare fallow superfluous.[56]

Between 1945 and 1979 the state constructed a number of large irrigation projects in southern Australia. Although none of these could be justified on economic grounds, they increased the area of irrigated land from 0.5 million to over 1.5 million hectares and led to the establishment of a large cotton industry and an expansion of rice growing.[57] Consideration of the economic aspects of these projects by ANZAAS was limited to one brief article, although this issue was widely debated elsewhere.[58]

Further work in veterinary science led to the development of vaccines for pulpy kidney and the introduction of more effective drenches to control fluke and stomach worms, and DDT, dieldrin, and BHC gave effective control against ticks in cattle, and lice and ked in sheep. In the late 1940s the organic herbicides 2,4,5,T and 2,4,D were introduced into Australia and these, and other similar chemicals, proved to be effective in selectively controlling weeds. They became an essential part of the minimum tillage techniques adopted by farmers in the 1970s as a means of reducing soil erosion and reducing the amount of highly priced machinery and fuel used in conventional cultivation. Later some sections of the public, alarmed by reports in the media, called for a cessation in the use of organic pesticides and herbicides. ANZAAS played an active part in attempting to dispel these fears.[59] P. Hall and B. Selinger, in a paper in *Search*, stressed the need for a permanent body of scientists to assess and inform the public of the effects of the new chemicals.[60]

The replacement of horses by tractors on most Australian farms in the 1940s and 1950s enabled farmers to sow more wheat. Large tractors and cultivating machinery were particularly important in expanding the wheat lands on the fertile black soils in northern New South Wales and in Queensland, as adequate weed control was impossible on these soils with the slower horse-drawn equipment.[61] As a result of these innovations the area of wheat expanded from 5.8 million hectares in 1939 to 10.3 million in 1970, and wheat production from 5.7 million tonnes to 11.0 million during the same period.

Increases in productivity arising from the new technology not only enabled farmers to retain real farm incomes when input prices increased faster than those of agricultural commodities after 1958, but they also increased the ability of farmers to change from lower to higher priced commodities. When wool prices declined in the late 1960s, the number of sheep decreased, and cattle increased in all of Australia's non-tropical grazing regions. When cattle prices declined in the 1970s, the area of wheat expanded and livestock numbers declined on wheat farms. The major factor limiting agricultural adjustment was the government's policy of protecting farmers from the market price messages by means of home-support price schemes and bounties. These enabled small, inefficient farmers to remain on the land long after they should have left it. It is strange that this matter was not discussed in the *Australian Journal of Science*.

Australian agricultural scientists also carried out extensive surveys of potential agricultural land in northern Australia, and after some years of experimental work C. S. Christian of the CSIRO concluded that peanuts, grain sorghum, and a number of fodder crops could be grown under dryland conditions in parts of the Northern Territory, and that the pasture legume Townsville Stylo could be established. It was also thought that rice, cotton, and safflower could be produced as irrigated crops on the Ord River.[62] This led to the recommendation that the Ord River Dam should be constructed.

Some scientists were aware that the approach to northern development, namely of building dams and finding suitable crops, was the wrong one. R. H. Greenwood pointed out in the presidential address to the Geography Section of ANZAAS that the commodities for which a market existed should be given first priority. He was also aware that the cost of northern development would be extremely high and that initially development would have to be heavily subsidized.[63] His advice was not heeded. Even using the experimental data that was all that was available at that time, it was possible to predict that farmers could not operate profitably in either region. The failure to properly evaluate these projects led to the fiasco of the Ord River Dam project.[64]

Many scientists showed an equal enthusiasm, after the large increase in oil prices in the 1970s, for producing ethyl alcohol as a fuel from plant material, although it was easily demonstrated that the same product could be produced more cheaply from coal.[65]

When the United Kingdom attempted to join the European Economic Community (EEC) in 1954 it was realized that the main market for all Australasian agricultural products except wool might be lost. New markets were found for all major products, with the exception of butter, in Japan, the United States, the Soviet Union, China, the Middle East, and South-East Asia. When Britain finally entered the EEC in 1973 it was no longer the major market for Australasia's agricultural produce. These provisions for change were timely, as the only concession made by the EEC to Australasia was to permit the United Kingdom to continue to import butter from New Zealand.

In the presidential address to ANZAAS in 1961, Sir Samuel Wadham of Melbourne University pointed out that because of the application of science to agriculture, the developed countries were producing more than they could consume and called for international agreements to fix commodity prices and limit production.[66] It is extremely doubtful if such agreements would have worked. Those that were attempted, to control wheat and sugar production, collapsed. In the meantime world prices were maintained by the United States limiting production and so holding world prices at a high level. With further increases in production, however, the EEC began to subsidize the export of surplus produce, and in 1986 the United States retaliated by adopting the same policy. As a result the world prices of wheat, sugar, and some other commodities rapidly declined. The question of overproduction, which Wadham had raised in 1961, was not discussed again at ANZAAS, although Australian agricultural economists had been calling for free trade in agricultural products for twenty years. The adoption of such a policy would have led to a decline in agricultural production in Europe and increased production in the United States, Canada, Australia, and New Zealand.

CONCLUSION

In spite of the initial slowness in recognizing important aspects of agriculture, such as pasture legumes and superphosphate, and the support of lines of development such as intensive agriculture in tropical Australia, which had little hope of success, ANZAAS played a part in enabling Australasian farmers to achieve a fivefold increase in agricultural production during the last century, through work in plant introduction, selection, and breeding, animal diseases, trace elements, myxomatosis, and pesticides and herbicides, all of which were discussed at ANZAAS meetings. This was achieved with a very limited development of new land and a declining work force. The farming community was less subsidized and had as high a standard of living as farmers in any other western country. The main products of Australasian farmers could be sold profitably in a free world market in competition with those of any other country.

AUSTRALASIA AND ITS MINES

To some extent the problems of developing large-scale mineral industries in Australasia were similar to those encountered in agriculture. Although it was not apparent at first, Australia had large reserves of minerals in terms of its population. As most of the output was exported, the prices received for minerals were completely dependent on the world market. Unlike agricultural products, no special arrangements were made with the United Kingdom for marketing minerals. The small Australasian work force demanded high wages, and where labour-saving devices could not be developed Australasia was at a disadvantage compared with its competitors.

The early Australasian goldfields were worked without the aid of science. Their main effect on Australasia was to increase its population and to provide the various colonies with enough capital to construct railways, roads, harbours, and other forms of infrastructure necessary for their development. The copper mines developed in the 1860s and 1870s in South Australia depended on skilled Cornish miners rather than on scientists.[67]

By the late 1860s, however, the need for professionally trained men was recognized by the mining industry. Schools of mines were established at Ballarat and Bendigo in the 1870s, and the teaching of geology and mineralogy commenced at the University of Sydney in 1866 and at the University of Melbourne at about the same time. Mining engineers were trained at both universities in the 1890s.[68]

Conquering the Ores

In 1861 mining contributed 16 per cent to Australia's gross domestic product (GDP), compared with the 21 per cent contributed by agriculture. However, as the more profitable gold and copper deposits were worked out, mining declined and by 1890 contributed less than 5 per cent to GDP, while agriculture's share had increased to 23 per cent.[69]

One of the first tasks undertaken by the AAAS was to attempt a census of Australian mineral deposits, including an assessment of their economic prospects. Although reports were only published for New Zealand, New South Wales, South Australia, and Queensland, the list is impressive.[70]

Mining revived as an industry in the 1880s and 1890s because of the discovery and working of large mineral deposits at Broken Hill, Kalgoorlie, Mount Morgan and in north-western Tasmania. All of these deposits had some features in common. All of the lodes were large and in isolated areas. With the exception of Kalgoorlie, more than one mineral was mined, and in general the minerals were difficult to separate from each other and from the rocks containing them. The geological formations encountered in Australasia were so large and so different from those in Europe that it was even suggested that a purely Australasian system of stratigraphy would have to be developed.[71] Techniques suitable for other countries could not be transplanted without modification to Australasia. A committee reporting to the AAAS in 1898 pointed out that satisfactory methods were not available to separate ores from the rock.[72] The isolation of the deposits placed a premium on techniques that required a minimum of inputs of large weight or volume, such as fuel for smelting, as transport costs could determine the life or death of a mine. Mount Bischoff's transport problems were reported in detail to the AAAS in 1892.[73]

This was a situation in which the skills of the geologist, metallurgist, and mining engineer were taxed to the full. As will be seen, however, most of the taxing was done in private and was not displayed before such open societies as the AAAS.

Large and often low-grade deposits could only be developed by large companies backed by adequate supplies of capital. Capital for such ventures was readily obtained from the United Kingdom, particularly in the 1890s, when a world-wide depression caused a cessation in government borrowing.[74] Large companies were not, however, prepared to hire geologists whose findings were difficult to hide and the benefits of which were readily available to others. It is for this reason that geologists played only a minor role in mining developments of this period. It was even difficult to find employment for all graduates in geology.[75] Excellent geological work was published by the AAAS on the nature of the coalfields in New South Wales and Queensland, but the full fruits of this research were not obtained until the twentieth century.[76] On the other hand, mining engineers and mineralogists had to be employed to develop new processes for smelting and separating one mineral from another and from the rocks that bore them. The processes developed were, however, carefully guarded from competitors. Often different firms were working along the same lines to solve the same problem on the same mining field without divulging their findings to each other.[77] Thus a full description of the smelting process at Broken Hill was given in a paper presented at the AAAS in 1898, stressing the need for a process to separate zinc from the tailings,[78] yet details of the flotation process that achieved this in the early 1900s were not published. The similar flotation processes developed separately by Charles Potter and Gillaume Delprat for different firms at Broken Hill, the skim process developed by Auguste de Bavay and the water froth process of F. J. Lyster were not mentioned in papers published by the AAAS.[79] It was not until 1946 that the theory of flotation was fully understood and described to the scientific world at ANZAAS.[80] Similarly, in 1896 Robert Sticht solved the problem of using the iron and sulphur in the copper ore as a fuel in the furnace, thus using less coke in smelting copper, and so greatly reducing transport costs at the isolated Mount Lyell mine. This was the first time in the world that the process had been used commercially, but it was not described before the AAAS until 1907.[81] Even when the developments of important processes were made public, such as John Sutherland's adoption of the filter press in 1897 and Ludwig Dhiel's discovery of the bromo cyanide process in 1899, both of which played a part in extracting gold from telluride on which much of Kalgoorlie's future depended, they were not mentioned in any paper published by the AAAS.[82].

Even if these developments were not formally discussed at the AAAS, they had a profound effect on the mining industry. The proportion of the nation's GDP contributed by the mining sector had increased from 5 per cent in 1890 to 11 per cent in 1900, and the industry employed 7 per cent of the nation's workforce.[83] Although the success of the industry was short-lived, it played an essential part in overcoming Australia's balance of payments problems in the 1890s.

In the Doldrums, 1920-45

By 1910 mining's contribution to GDP had declined to 6 per cent, in 1920 to 3 per cent, and to even lower values in the 1930s. The number of men employed in Australian mines in 1933 was less than half the number employed in 1901. The causes of the decline were many. Metal prices, with the exception of gold in the 1930s, were low. British investors were no longer interested. As early as 1908 they claimed to have lost $40 million in Australian mines. Between 1910 and 1930 most of the British shares in Broken Hill, Mount Lyell and Mount Morgan were sold to Australians who had little spare capital for further expansion.[84]

The only large mining development between 1910 and 1920 was the exploitation by Broken Hill Pty Ltd (BHP) of iron ore deposits in the Middleback Ranges of South Australia, which commenced in 1914. Iron ore was shipped to smelting works erected at Newcastle by BHP.[85] In addition, the brown coal fields of Gippsland were developed using modified German technology to provide electric power and briquettes in Victoria.[86]

Few new mineral discoveries were made during this period. The only large one, that at Mount Isa, was so isolated and its ores so difficult to work that it did not commence operating until 1931, and it only became profitable when copper prices rose in the Second World War.[87] In spite of the discovery of further rich iron ore deposits at Yampi Sound in north-western Australia, the Commonwealth government was convinced that the nation's iron ore deposits were too limited to allow exports.[88] The failure of the geologists to object to this decision before a proper survey was made is difficult to understand.

The Mining Boom

In his presidential address to the Geology Section of ANZAAS in 1947, H. G. Raggatt, director of the Bureau of Mineral Resources, pointed out that, excluding coal, 84 per cent of minerals produced in Australia came from fields discovered before 1900. He considered that the lack of new discoveries should be a challenge to geologists and that new methods should be adopted to find new ore lodes.[89] By this time, however, the search for new minerals had already been set in motion. In 1939 the CSIRO compiled a geological map of Australia under the direction of Sir Edgeworth David.[90] This was the first such map since 1908. In 1946 the Commonwealth government established the Bureau of Mineral Resources, and geologists were employed by this organization, the state governments, the CSIRO, and the large mining companies to search for minerals. Raggatt's confidence was justified. By 1958 he was able to report the finding of large deposits of copper at Mount Isa, bauxite at Gove and Weipa on the Gulf of Carpentaria, iron ore in the Northern Territory and north Queensland, ilmenite in the south-west of Western Australia, and uranium in several localities.[91] The next year deposits of iron ore were discovered in the Pilbara region in Western Australia and the government concluded that iron ore reserves were unlimited. Export restrictions were lifted in 1960, and the Hamersley Ranges were developed to export iron ore.[92]

In the meantime, the search for fossil fuels had not been neglected. The pioneering work of Edgeworth David and others formed a basis for new work. By 1949 J. A. Dalhunty, who had already made a study of the new methods in coal research, was calling for a study of all types of coal in Australia.[93] C. E. Marshall, in his presidential address to the Geology Section of ANZAAS in 1967, described new techniques that enabled the characteristics and quality of coal from different seams and parts of coal seams to be determined.[94]

Further exploration revealed that Australia's coal reserves were larger than had been thought and by the end of the 1960s substantial reserves of oil had been discovered in Bass Strait; natural gas was found in the same region and in the Cooper Basin in South Australia.[95] By 1970, 70 per cent of Australia's oil requirements were produced from domestic sources. Similar surveys in New Zealand led to the establishment of a large geothermal power supply.[96] In spite of these discoveries the limited Australian reserves of fossil fuels continued to worry scientists and this concern was increased when exports of coal to Japan commenced.[97] The pessimistic view of the Australian geologists is difficult to understand. It was not until 1978 that it was realized that coal reserves were increasing at a faster rate than they were being mined.[98]

Capital from Japan, Europe, the United States, and Canada enabled Australia's mineral reserves to be developed rapidly.[99] Between 1979 and 1984 mining's contribution to GDP exceeded 6.2 per cent and was only slightly less than the 6.4 per cent contributed by the rural industries. By 1986 Australia was the world's largest coal exporter, and the value of the returns obtained from coal exports were similar to those obtained from wool and wheat, each of which contributed 10 per cent to Australia's exports. Between 1953 and 1981 the number of workers employed in mining increased from 61 000 to 89 000, but this increase of 46 per cent was dwarfed by the sixfold increase in value added from mineral production.

CONCLUSION

In the late nineteenth and early twentieth centuries, the part played by the AAAS in the mining industry was limited, as the main work was carried out by mineralogists employed by large mining companies, who were not encouraged to reveal their newfound techniques to competitors. After the Second World War, when geological research was actively supported by the Australian state and federal governments it was possible for ANZAAS to play a larger part. The challenge made by H. G. Raggatt at the twenty-sixth meeting of ANZAAS, to go out and discover new mineral lodes, was taken up by the geologists and rapidly yielded results. It was because of these findings that Australia again became one of the world's major mineral producers.

NOTES

1. B. R. Davidson, *European Farming in Australia*, (Amsterdam: Elsevier, 1981), 3–21.
2. E. Macarthur Onslow, in S. Macarthur Onslow (ed.), *Some Early Records of the Macathurs of Camden* (Sydney: Rigby, 1973), 57–8.
3. Davidson, op. cit., note 1, 144–9.
4. F. Wheelhouse, *From Digging Stick to Rotary Hoe*, (Adelaide: Rigby, 1972), 14–100.
5. H. Williams, *Mechanical Refrigeration*, (New York: Pitman, 1941), 52–3.
6. C. M. Donald, 'Grass or Crop in the Land Use of Tomorrow', *Australian Journal of Science*, 25 (1963), 386–95.
7. J. D. Custance, *The Garden and Field*, 8 (1882–3), 118.
8. M. Williams, *The Making of the South Australian Landscape*, (London: Academic, 1974), 283–4.
9. W. Lowrie, 'That in Our Practice of Agriculture the Determining Influence of Climatic Conditions is not Sufficiently Recognized', *Report of the AAAS*, 8, (Melbourne, 1900), 124–32.
10. R. L. Pudney, 'Advantages of Bare Fallow', *Agricultural Gazette of New South Wales*, 1 (1890), 28–32.
11. *The Garden and Field*, 1 (1876), 56–7.
12. A. Liversidge, Presidential Address to AAAS, *Report of the AAAS*, 7 (Sydney, 1898), 47–9.
13. Rust in Wheat Committee, *Report of the AAAS*, 2 (Melbourne, 1890), xxi; *Report of the AAAS*, 3 (Christchurch, 1891), 547–50.
14. H. C. L. Anderson, Presidential Address to Section G, *Report of the AAAS*, 5 (Adelaide, 1893), 148–9; F. B. Guthrie, 'William J. Farrer, the Results of His Work', *Department of Agriculture NSW Sci. Bull.*, 22 (1922).
15. D. McAlpine, 'Experiments in Rust and Stinking Smut in Wheat during 1901', *Report of the AAAS*, 9 (Hobart, 1902), 610–11.
16. I. A. Watson and N. H. Luig, 'Asexual Intercrosses between Recombinants of *Puccinia graminis*', *Australian Journal of Science*, 25 (1962–3), 18–19.
17. F. Turner, 'Fodder Plants and Grasses of Australia', *Report of the AAAS*, 2 (Melbourne, 1890), 586–96.
18. Report of the Green Patch Farm Bureau; Report of the Mt. Gambier Farm Bureau; Report of the Narracoorte Farm Bureau, *South Australian Journal of Agriculture*, 14 (1910–11), 428, 642, 722.
19. W. W. Culcheth, 'Irrigation Works in Australia: How They May Be Made Remunerative', *Report of the AAAS*, 2 (Melbourne, 1890), 728–9.

20. Davidson, op. cit., note 1, 160–6.
21. Anderson, op. cit., note 14, 152–65.
22. W. Angus, 'The Relation of Science and this Section to the Further Development of Australian Agriculture', *Report of the AAAS*, 13 (Sydney, 1911), 515–17.
23. L. A. Musso, 'Catalysts and their Relation to Crops', *Report of the AAAS*, 14 (Melbourne, 1913), 667–71.
24. E. J. Underwood and J. S. Gladstones, 'Subterranean Clover and Other Legumes' in G. H. Burvill (ed.), *Agriculture in Western Australia, 1829–1979* (Perth: University of WA Press, 1979), 139–42.
25. R. H. Cambage, 'Presidential Address to the AAAS', *Report of the AAAS*, 19 (Hobart, 1928), 4–5.
26. W. J. Spafford, 'Subterranean Clover', *South Australian Journal of Agriculture*, 27 (1923–24), 636.
27. F. H. Campbell, 'The Destructive Distillation of Prickly Pear', *Report of the AAAS*, 14 (Melbourne, 1913), 104–7.
28. T. H. Johnston, 'The Australian Prickly-Pear Problem', *Report of the AAAS*, 16 (Wellington, 1923), 347–401; A. P. Dodd, *The Biological Campaign against the Prickly Pear* (Brisbane: Government Printer, 1940).
29. Cambage, op. cit., note 25, 6.
30. J. D. Stewart, 'Presidential Address to Veterinary Science', *Report of the AAAS*, 14 (Melbourne, 1913), 695–702; W. T. Kendall, 'Notes on the Early History of the Veterinary Profession in Victoria', ibid., 703–14; W. A. N. Robertson, 'Summary of Address on Rinderpest in Western Australia', *Report of the AAAS*, 17 (Adelaide, 1924), 710–11.
31. H. R. Seddon, 'Toxic Bulbar Paralysis in Horses and Cattle (Botulism and Parabotulism)', *Report of the AAAS*, 16 (Wellington, 1923), 822–8; H. A. Reid, 'Bovine Contagious Mastitis', ibid., 836–51; H. E. Albiston, 'Some Causes of Sheep Mortality in Victoria', *Report of the AAAS*, 17 (Adelaide, 1924), 703–5.
32. A. C. McKay, 'An Experimental Determination of the Intermediate Host or Hosts in Relation to the Incidence of the Parasite *Fasciola hepatica* in Sheep of New South Wales', *Report of the AAAS*, 18 (Perth, 1926), 794–805; G. C. Heslop, 'Studies in Bovine Pleuropneumonia', *Report of the AAAS*, 16 (Wellington, 1923), 829–36.
33. Stewart, op. cit., note 30; Kendall, op. cit., note 30.
34. H. Pye, 'The Importance of Agricultural Education to the Commonwealth', *Report of the AAAS*, 14 (Melbourne, 1913), 675–94.
35. A. J. Perkins, 'Agricultural Education in Australia', *Report of the AAAS*, 15 (Hobart, 1921), 244–57.
36. Rural Reconstruction Commission, *Second Report*, 'Settlement and Employment of Returned Men on the Land' (Canberra: AGPS, 1944), 1–8.
37. Griffith Taylor, 'Climatic Control of Wheat-Production in Australia Committee', *Report of the AAAS*, 16 (Wellington, 1923), 132–8.
38. Rural Reconstruction Commission, op. cit., note 36, 48.
39. A. J. Perkins, 'A Plea for Nation-Wide Research into the Economic Position of Our Various Rural Industries', *Report of the AAAS*, 19 (Hobart, 1928), 548–66.
40. E. R. Hudson, 'Rural Relief and Agricultural Extension', *Report of ANZAAS*, 22 (Melbourne, 1935), 313–14.
41. N. Butlin, *Australian Domestic Product and Foreign Exchange Borrowing, 1861–1938/39* (Cambridge: Cambridge University Press, 1962), 460–1.
42. Rural Reconstruction Commission, *Tenth Report*, 'Commercial Policy in Relation to Agriculture' (Canberra: AGPS, 1943), 48–50.

43. H. C. Trumble, 'The Relation of Pasture Development to Environmental Factors in South Australia', *Report of ANZAAS*, 22 (Melbourne, 1935), 315.
44. S. Marriot, 'Some Problems of Queensland Pastures', ibid., 316–17.
45. D. S. Riceman, C. M. Donald and C. S. Piper, 'Response of Copper on a South Australian Soil', *Journal of the Australian Institute of Agricultural Science*, 4 (1938), 41; C. S. Piper, 'Trace Elements in Soils and Plants, *Report of ANZAAS*, 26 (Perth, 1947), 82–108.
46. E. W. Lines, 'The Effect of the Ingestion of Minute Quantities of Cobalt by Sheep Affected with Coast Disease', *Journal of the CSIR*, 8 (1935), 117–19; H. R. Marston, 'Problems Associated with Coast Disease in South Australia', *Journal of the CSIR*, 8 (1935), 111–16; H. W. Bennetts, 'The Pathological Approach to Problems of Animal Disease' *Report of ANZAAS*, 25 (Adelaide, 1946), 231–44.
47. G. A. Gilruth, 'Recent Contributions to Veterinary Science by Australian and New Zealand Workers', *Report of ANZAAS*, 23 (Auckland, 1937), 289–306.
48. M. R. Freney, 'The Development of Artificial Fibres with Special Reference to Wool', *Australian Journal of Science*, 2 (1939), 72–5; A. R. Penfold, 'Recent Developments in Synthetic Fibres', ibid., 75–8.
49. *Australian Journal of Science*, 1 (1938), 1–2.
50. Rural Reconstruction Commission, op. cit., note 42, 53.
51. Rural Reconstruction Commission, *Reports 1–10* (Canberra: Government Printer, 1944–6).
52. F. N. Radcliffe and B. V. Fennessy, 'Mosquito-Borne Myxomatosis', *Australian Journal of Science*, 13 (1950–1), 103–5.
53. H. C. Trumble, 'The Ecological Relations of Pastures in South Australia', *Report of ANZAAS*, 27 (Hobart, 1949), 128–30.
54. J. D. Atkinson, *DSIR's First Fifty Years* (Wellington, NZ: Department of Scientific and Industrial Research, 1976); C. T. Bloomfield, *New Zealand Handbook of Historical Statistics* (Boston: Hall, 1984).
55. D. E. Walker, 'Meat Production per Acre', *Proc. of the Aust. Soc. of Animal Production*, 15 (1955), 51–6; W. M. Hamilton and K. J. Mitchell, 'Dairy Farming in Waipa Country', *Dairy Farming Annual* (Wellington: University of NZ, 1950); H. L. Davies, 'Pasture Utilization and Stocking Rate', *Annual Report* (Canberra: CSIRO Plant Division, 1969), 63.
56. Donald, op. cit., note 6, 392.
57. Davidson, op. cit., note 1, 367–75.
58. B. E. Butler, 'An Economic Study of Keepit Dam', *Search*, 2 (1971), 144–5; B. R. Davidson, *Australia Wet or Dry?* (Melbourne: Melbourne University Press, 1969), 161–266; W. F. Musgrave, 'The Management and Development of the River System' in H. J. Frith and G. Sawyer (eds), *The Murray Waters* (Sydney: Angus & Robertson, 1974), 301–419; K. O. Campbell, 'An Assessment of the Case for Irrigation in Australia' in *Water Resources Use and Management* (Melbourne: Melbourne University Press, 1964), 450–7.
59. M. T. Tanton, 'Some Problems Associated with the Use of Fertilizer and Pesticides in Agriculture', *Search*, 1 (1970), 341–6; Victorian Division of ANZAAS, 'Symposium on Pesticides', *Search*, 12 (1981), 427.
60. P. Hall and B. Selinger, 'Herbicide Statistics', *Search*, 13 (1982), 300–3.
61. C. J. McKeown, Large Scale Mechanised Agriculture in Queensland, (unpublished), *Report of ANZAAS*, 27 (Hobart, 1949), 130.
62. C. S. Christian, 'The Future Revolution in Agriculture in Northern Australia', *Australian Journal of Science*, 22 (1959–60), 138–47.

63. R. H. Greenwood, 'Will the North Remain Empty?', *Australian Journal of Science*, 26 (1963–4), 305–11.
64. B. R. Davidson, 'Economics of Irrigated Farming on the Ord River', *Farm Policy*, 3 (1963), 54–60.
65. G. A. Stewart, W. H. M. Rawlins, G. R. Quick, J. E. Begg and W. J. Peacock, 'Oilseeds as a Renewable Resource of Diesel Fuel', *Search*, 12 (1981), 107–15; J. R. McWilliam, 'Research and Development to Service a Sustainable Agriculture', *Search*, 12 (1981), 15–21.
66. Sir Samuel Wadham, 'The Winds of Agricultural Change', *Australian Journal of Science*, 24 (1961–2), 3–12.
67. G. Blainey, *The Rush that Never Ended* (Melbourne: Melbourne University Press, 1964), 116–27.
68. R. Tate, 'Inaugural Address to the AAAS', *Report of the AAAS*, 5 (Adelaide, 1893), 22, 26.
69. Butlin, op. cit., note 41, 12–13.
70. Report of the Committee No. 7, 'Mineral Census of Australasia', *Report of the AAAS*, 2 (Melbourne, 1890), 205–82.
71. Tate, op. cit., note 68, 35.
72. Report of Committee No. 14, 'The State and Progress of Chemical Science in Australasia, with Special Reference to Gold and Silver Appliances Used in the Colonies and Elsewhere', *Report of the AAAS*, 2 (Melbourne, 1890), 283–92.
73. H. W. F. Kayser, 'A Discussion of the Transport Faced at the Australian Mine, "Mount Bischoff"', *Report of the AAAS*, 4 (Hobart, 1892), 342–52.
74. Blainey, op. cit., note 67, 186–8.
75. Tate, op. cit., note 68, 38.
76. T. W. E. David, 'A Correlation of the Coal Fields of New South Wales', *Report of the AAAS*, 2 (Melbourne, 1890), 459–63.
77. Blainey, op. cit., note 67, 269–70.
78. G. H. Blakemore, 'Metallurgic Methods in Use at Broken Hill', *Report of the AAAS*, 7 (Sydney, 1898), 305–6.
79. T. H. Hoover, *Concentrating Ores by Flotation*, (London: The Mining Magazine, 3rd edn, 1916), 14–16; A. F. Taggart, *Handbook of Ore Dressing* (New York: Wiley, 1927), 878.
80. I. W. Wark, 'Australian Research on the Theory of Flotation', *Report of ANZAAS*, 25 (Adelaide, 1946), 23–51.
81. R. C. Sticht, 'Progress in Rapid Oxidation Processes as Applied to Copper-Smelting', *Report of the AAAS*, 11 (Adelaide, 1907), 57–130.
82. D. Clark, *Australian Mining and Metallurgy* (Melbourne: Critchley Parket, 1904), 29–37, 65–70.
83. Butlin, op. cit., note 41, 12–13.
84. Blainey, op. cit., note 67, 290.
85. Helen Hughes, *The Australian Mine and Steel Industry, 1948–1962* (Melbourne: Melbourne University Press, 1964), 155.
86. G. Serle, *John Monash: A Biography* (Melbourne: Melbourne University Press, 1983), 435–62.
87. Blainey, op. cit., note 67, 324–31.
88. Hughes, op. cit., note 85, 120–1.
89. H. G. Raggatt, 'Depletion of Mineral Resources—A Challenge to Geology and Geophysics', *Report of ANZAAS*, 26 (Perth, 1947), 109–33.
90. A. G. Maitland, 'The Geophysical Map of the Commonwealth of Australia', *Australian Journal of Science*, 1 (1938–9), 173–6.
91. H. G. Raggatt, 'Finding Ore: Some Australian Case Histories and Their

Bearing on Future Discovery', *Australian Journal of Science*, 21 (1958–9), 60–77.
92. Hughes, op. cit., note 85, 182–6.
93. J. A. Dalhunty, 'Recent Advances in Coal Research', Pt I, *Australian Journal of Science*, 8 (1945–6); Pt II, *Australian Journal of Science*, 9 (1946–7), 133–7; J. A. Dalhunty, 'Trends in Coal Research', *Australian Journal of Science*, 12 (1949), 98–100.
94. C. E. Marshall, 'Coal and the Commonwealth', *Australian Journal of Science*, 29 (1967), 248–63.
95. Australia, *Yearbook*, No. 58 (1972), 899–900.
96. J. D. Atkinson, *DSIR's First Fifty Years* (Wellington: Department of Scientific and Industrial Research, 1976), 92.
97. N. Y. Kirov, 'Fuels and Energy in Australia', *Search*, 2 (1971), 298–305.
98. J. R. Siemon, 'Three Curves for Coal Reserves', *Search*, 9 (1978), 70.
99. Australia, *Yearbook*, No. 54 (1968), 1059–61.

12

Professional Hygienists and the Health of the Nation

John Powles

To be able to cure a sick man is a power of the highest order, no knowledge seems greater or more desirable, but to lay down rules of guidance intelligible to the masses by which disease may be avoided is a yet higher attainment, and worthy of the name Sanitary Science.

So Joseph Bancroft, MD, of Brisbane told Section H (Sanitary Science and Hygiene) at the first Australasian Association for the Advancement of Science (AAAS) Congress in 1888.[1] Bancroft went on to note that, because 'the removal of a frightful ailment, and the restoration of a diseased person to health, seems more of the nature of a miracle', medical and surgical advancement was better able than sanitary science to claim a 'just reward' and therefore 'needs less help from Governments or learned societies striving for the advancement of science, which is the object of our meeting today'.

In the ensuing decades, the AAAS, like its British parent, helped public health professionals proclaim their achievements and promised future rewards in health to those who took their advice. In parallel with the Intercolonial Medical Congresses (later the Australasian Medical Congresses), the AAAS also provided an arena in which specific proposals to government could be developed and endorsed.

Since the 1880s, four overlapping periods can be identified in the public discussion of hygiene in Australia. The first, an eponymous 'sanitary' period, was concerned with the control of organic pollution and the more specific applications of bacteriology to the control of infections. The second, or what we can call a 'national' period, emerging in the first decade of the new century and, continuing until the late 1930s, was dominated by a set of ideas that can be associated with both 'national efficiency'[2] and 'Progressivism'.[3] The destiny of empire, race, and nation, the physical development of the citizens of the young nation, their fitness, and their discipline were central concerns. The third, a 'specialist' period, emerging after the Second World War, showed the effects of increasing scientific specialization. Discussion of the control of infections (tuberculosis and poliomyelitis, for example) became separated from such areas as nutrition and the control of chronic diseases and injury. Finally, over the last decade or so, the role of 'affluence' in the production of chronic disease has received systematic attention. This period might be referred to as one of 'post-affluent hygiene', in that it looks to reversing consumption trends hitherto associated with rising incomes—increased consumption of tobacco, alcohol, sugar, and animal fats.

This chapter will consider the role of the AAAS and later ANZAAS through these periods in the public discussion of hygiene. The second and longest period will receive the most attention because it is in many ways the most interesting.

SANITARY HYGIENE

The AAAS began life at a time of high interest in sanitary reform, and in microbiology and the more specific preventive measures that it heralded. The interval between 1880 and 1900 saw a peak of colonial legislative activity directed at controlling sanitary nuisances and limiting contagion. The smallpox epidemics of 1881–2 were the most traumatic of Sydney's encounters with infectious disease during the nineteenth century,[4] and they reawakened fears of imported infection. In 1884 representatives of all the colonies met in Sydney, at the invitation of the government of New South Wales, to consider the establishment of a federal system of quarantine. This Australasian Sanitary Conference was 'the first inter-state consultation on important health matters and formally introduced the idea of federation in health administration'.[5] A recommendation that all colonies should draft a uniform Federal Quarantine Act was not, however, implemented, either after the first meeting of colonial health authorities, or after a second in 1896; it had to await the arrival of the Commonwealth, which passed a Quarantine Act in 1908.[6]

Australia's cities and towns were late in installing sewerage systems, and mortality rates from typhoid were among the highest in the world.[7] There were few full-time public health officials,[8] but many members of the medical profession were enthusiastic about the prospects for preventing infectious diseases and made use both of the AAAS and of the Intercolonial Medical Congresses to proclaim this enthusiasm.

The Intercolonial Medical Congresses were initiated in 1887 by the South Australian branch of the British Medical Association, in response to the jubilee of Victoria's reign and the fiftieth anniversary of the founding of South Australia, both commemorated by the International Exhibition in Adelaide.[9]

When Joseph Bancroft presided over the Sanitary Science Section of the first AAAS Congress in Sydney, in 1888, he was thus continuing an experience of intercolonial discussion that had begun with his representation of Queensland at the Australasian Sanitary Congress of 1884. Over the subsequent decades, the kinds of issues raised for consideration in the Sanitary Science Section of the AAAS and later ANZAAS (and its successors) overlapped considerably with those considered at the Intercolonial Medical Congress (renamed the Australasian Medical Congress from 1905). Although the latter were, after 1920, formally and effectively to become the congresses of the British Medical Association in Australia, they began life with a much more public character. Colonial governments paid for their representatives to attend, and their proceedings were published by the colonial government printers. The respective roles of the AAAS and the Intercolonial Medical Congresses were thus not all that far apart.

What, then, was the early role of the AAAS in relation to health? The leaders of the profession would have welcomed it, not because it was the only public forum to which they had access, but because it was a valuable additional forum—and one that held promise of occurring every one or two years, in contrast to the three-year intervals that typically separated the medical congresses. An additional attraction would have been the range of other professional and intellectual leaders present, and the consequent opportunity for enlisting support for political action.

In this period the main target for such action was infectious disease. At the sixth Congress in Brisbane in 1895, recommendations from the Sanitary Science and Hygiene Section that the meeting endorse the compulsory notification of infectious diseases were at first dropped by the Recommendations Committee because they were 'contentious'.[10] Springthorpe protested on behalf of the Section, however, and amended recommendations in support of compulsory notification, federal quarantine, and the testing of livestock for tuberculosis were passed. At this time, the notification of virulent infections such as smallpox was already compulsory in most of the colonies. What the hygienists wanted was something more comprehensive, along the lines of the British Compulsory Notification Act of 1889.[11]

Joseph Bancroft of Brisbane and J. Ashburton Thomson of Sydney had been joined in the new Section H (Sanitary Science and Hygiene), at the first meeting, by two engineers speaking on sanitation, and by a W. E. Roth, BA, speaking on 'Theatre Hygiene'. The first congress also established a pattern of presenting papers concerned with health at other Sections: the first paper in Architecture and Engineering (Section J), for example, dealt with the deep drainage system of Adelaide.[12]

This pattern continued through the first decade of the new century. In 1909 J. H. L. Cumpston (ultimately to become first director-general of health of the Commonwealth) spoke to Section E (Social and Statistical Science) on pulmonary tuberculosis in Australia, and J. S. C. Elkington (director of health in Tasmania) made 'A Plea for the Australian Child Body' in Section J (Mental Science and Education).[13] Elkington played a significant role in the AAAS as its local secretary in Tasmania. The son of a Melbourne University history professor, he was a close friend of Norman Lindsay from the 1890s, and was married to Lindsay's sister-in-law. At the turn of the century, he and the Lindsays had been members of the Ishmaels, a Bohemian group that delighted 'to discuss everything from opera to Nietzsche'.[14] His concern that the fitness of the nation depended on the physical development of the young was one that assumed prominence during the second of our four periods.

THE NATIONAL PERIOD

Another protagonist of child development was Harvey Sutton. In 1909 the Victorian government appointed Sutton, Mary Booth, and Janet Greig as school medical inspectors.[15] Sutton was Victoria's first Rhodes Scholar, and while at Oxford he had run for Australia in the 1908 London Olympics. In 1921 he moved to the New South Wales School Medical Service, and in 1930 he became the first director of the School of Public Health and Tropical Medicine at the University of Sydney.[16] Sutton was to make extensive use of the AAAS platform. In 1911 he spoke to Section E (Social and Statistical Science) on 'The Importance of Nationality'. At Victorian school medical examinations, the birth-place of not only the child and its parents but also its grandparents was recorded. A complicated scoring system was used to classify children into five classes, ranging from foreign-born children to children with all grandparents Australian-born, in order

> to institute comparisons between these different types of children according to the degree of Australianship, so to speak, and so determine at the earliest possible moment what influence the environment of southern land and skies is having on our race. The responsibility of moulding and controlling the characters of that race falls on the shoulders of the present generation.[17]

Sutton's work in Victoria followed the establishment of the first state system of school medical inspections established in Tasmania by J. S. C. Elkington in 1907. In a paper to the AAAS in 1909 Elkington proclaimed that 'educationists are the builders of the ship of the nation, and upon the faithfulness and skill of their work will depend the safety and progress of the race'.[18]

The struggle for individual and national survival and the consequent need to strive for physical mastery—of the environment and of others—was a prominent theme of the national hygienists, one mirrored frequently by the AAAS congresses. Elkington (a boxer in his youth) wanted to see a 'broad chested, keen-sighted, hard-fisted race'.[19] Maintaining health would become a matter of duty. He looked forward to

> the day when the serious acceptance of a doctrine of national physical morality will cause preventable disease to be regarded as somebody's crime, and when the preservation and protection of health will occupy a place in the daily round of unquestioned duty to the State and to one's neighbours.[20]

In 1912 Elkington was transferred to Queensland to become commissioner of public health. There he was later to exert a strong influence on Raphael Cilento, another important figure of the 'national period'.[21]

In the years immediately before the First World War, Sutton and Elkington were far from alone in their concern with racial destiny. J. W. Springthorpe, a physician at the Melbourne Hospital and lecturer in hygiene at the University of Melbourne, had been president of the Sanitary Science and Hygiene Section at the Brisbane meeting in 1895, and in the following year he presided over the Intercolonial Medical Congress in Dunedin. He appears to have participated at least twice more in the AAAS after the Melbourne Congress of 1900. In 1914 he published *Therapeutics, Dietetics and Hygiene*,[22] Australia's first textbook in hygiene, which noted with implicit approval H. S. Chamberlain's judgement that 'Pure Race is the most important secret in all human history',[23] and gave expression to a sense of racial destiny of metaphysical intensity:

> The life of a whole people, also would be too short, if unity of race did not stamp it with a definite limited character, if the transcendent splendour of many sided and varying gifts were not concentrated by unity of stem which permits a gradual ripening, a gradual development in definite directions, and finally enables the most gifted individual to live for a superhuman purpose.[24]

Springthorpe does not shrink from the recognition that racial improvement depends on selective breeding and selective survival and goes on to cite five cardinal laws on which (according to Chamberlain) the development of a noble race depends. The young nation should be prepared to be tested: 'The struggle which eliminates or destroys the weak heals the strong... the storm of war has always raged around the childhood of great races.'[25]

Artificial selection—'the careful elimination of everything that is of indifferent quality'—would be necessary. Springthorpe's text elaborates on the means by which the production of a physically fit and robust youth may be pursued. Exercise programmes, open-air classrooms, and medical inspection of school children to identify remediable defects were the recognized means to physical improvement. But was the challenge being met?

> The warm sun, the bright skies, the rich lands, the absence of indigenous disease, the easy living conditions of Australia are capable of producing such a race as the world has never before seen. All the knowledge and experience of the ages is at our command to improve our children—and they degenerate.[26]

Under the Commonwealth Defence Act of 1909, physical training was made obligatory for all 'accepted [male] scholars between the ages of 12 and 18'. Springthorpe thought it 'a unique experiment in the military organisation of the Empire'.[27] In Springthorpe's view, the physical development of females was also to be supervised. His text gives a series of exercises—including one to 'strengthen a girl where she needs it most'. He quotes approvingly from a Dr Sergent, director of the Harvard University Gymnasia, who noted

a wonderful advance since 1890. Stooping shoulders are disappearing, the back is better developed, the gait and general physique have notably improved and American women of the middle and upper classes have approached the primeval days when women were so like men in form that it was well nigh impossible to tell them apart.[28]

Harvey Sutton, in his long association with the AAAS and later ANZAAS returned to this question of child development. In his presidential address in 1923 to the Sanitary Science and Hygiene Section (relettered from H to I in 1921), he spoke on 'Recent Progress in Child Hygiene'.[29] In 1930 he was president of the Section again (now called Medical Science and National Health) and gave his address on 'The Child as the Test of Progress'. He continued to be especially concerned with 'the progress or degeneration of our race in its new Australian environment' and could, on the latter occasion, look back at marked and sustained declines in mortality from tuberculosis and in infant mortality.

With considerable insight, he attributed the decline in tuberculosis to fundamental economic and social changes:

The remarkable fall in the death rate from tuberculosis began even before the discovery of the tubercle bacillus by Koch, and its curve of descent corresponds to the curve of the rise in value of wages...Public health is purchaseable. The gordian knot of poverty and poor food, ignorance and infection, can be cut by a sword of gold. Food comes before Education, and Education precedes Health.

The reduction of infant mortality was the greatest triumph of preventative medicine. Yet the conclusion could not be avoided that:

the basal control has been the improvement of general education which, culminating in 'The Compulsory Education Act of 1871' (New South Wales, 1880), has bred up a generation of more or less educated mothers and fathers in a more enlightened community and eliminated the illiterate and diffused scientific ideas.[30]

In the year that J.H.L. Cumpston was appointed as Australia's first director-general of health (1921), he also served as president of Section I. He had spoken at three earlier meetings and was to be president of the Section again in 1928. Cumpston was born in Melbourne in 1880 to a Methodist warehouseman and pioneer kindergarten teacher. He graduated well in 1902 and developed an interest in the new possibilities for preventing disease. He travelled via east Asia to Europe and secured a Diploma of Public Health at London University before returning to the employ of the West Australian Central Board of Health in 1907. He moved to the Quarantine Service in Melbourne in 1911 and became head of the federal Quarantine Service from 1913.[31] During the war years, Cumpston worked towards the formation of a Commonwealth

Department of Health. He was inspired by the role of science as saviour and by 'the more mystical ideas as to the War exalting mankind to sacrifice and altruism'.[32] In a 'War Lecture' at Melbourne University in 1915, he noted that:

> The stirring of a nation's soul under a common danger is one of the most wonderful effects of the war, and is, I firmly believe, destined to be one of the most profound and far-reaching effects of the war upon the public health ... [Returning troops would have the war-taught knowledge that] ... things can be accomplished by resolute men, activated by a common impulse, and determined to achieve any objective which represents an advance on their previous position.[33]

At the Australasian Medical Congress in Brisbane in 1920 Cumpston spoke as president of the Public Health Section on 'The New Preventive Medicine'. He referred to himself and fellow-thinkers as 'we, who aspire to use this opportunity, who dream of leading this young nation of ours to a paradise of physical perfection'.[34]

An important external influence at this time was the Rockefeller Foundation, through its International Health Board. Its regional director, V. G. Heiser, visited Australia in 1916 and again in 1921, when he planned with Cumpston and Elkington to force the federal government's hand in establishing a Commonwealth Health Department. Heiser met W. M. Hughes several times in January 1921, and promised that, if a department were created, the International Health Board would supply for at least one year experts in industrial hygiene, sanitary engineering, and tropical health, while arranging that Australians should be trained in the United States to take over these jobs.[35] The federal ministry decided to accept the offer and to create a new department.[36] Between 1922 and 1924 four Australians were trained on Rockefeller grants in the United States.[37] But subsequent development of the Department was not to fulfil early expectations, as it was drastically cut back in the early depression years.

Raphael Cilento was another leading representative of the national period of Australian hygiene. He was to take up, at the AAAS and elsewhere, an issue that had been vigorously debated since before Federation: the capacity of 'Whites' to successfully exploit the tropics. In 1910 the Australian Institute of Tropical Medicine was established in Townsville—'the Commonwealth's most significant support so far of science and learning'[38]—and in 1922 Cilento became its director. A graduate of the University of Adelaide and the London School of Tropical Medicine, he was a passionate believer in Australia's right-cum-duty to settle its tropics. In 1924, at the age of 30, Cilento became director of public health in New Guinea, and from 1928 he took over Elkington's position as director of the Commonwealth Health Department's Division of Tropical Hygiene. In 1934 he became director-general of Health and Medical Services for Queensland. From 1937 he occupied Australia's first chair of Social Medicine (actually 'Social and Tropical Medicine') at the University of Queensland, concurrently with his official position. Both ended in 1945, when he moved into the international arena as director of the United Nations' refugee and resettlement agency in the British zone of occupied Germany.³⁹

For Cilento, hygiene was above all a weapon to be deployed in the contest between nations. He spoke on 'Preventive Medicine and Hygiene in Tropical Areas' at the AAAS meeting in Wellington in 1923, and served as Section president in 1932. On the latter occasion, speaking on 'Australia's Problems in the Tropics', he observed that

Tropical and sub-tropical products play a part so increasingly great in modern commercial and industrial life that it is not surprising to find all the great trading nations of the world in active competition for those raw materials pre-eminently available in tropical areas.

To those who heard, he concluded that the successful colonization of tropical Australia: 'the great problem before this generation—the problem, indeed, which may make that our "place in the sun", or may, perhaps, mean no son of ours in that place—is essentially a problem of applied public health.'⁴⁰

Cilento was active as Queensland's representative on national bodies. In 1926 a Federal Health Council had been established on the recommendation of the Royal Commission on Health. The ANZAAS general council at the twenty-second Congress, in Melbourne in 1935, resolved on the recommendation of Section I that 'a Medical Research Council on similar lines to that in England should be formed to control medical research in Australia'.⁴¹ The National Health and Medical Research Council (NH&MRC) was formed the following year, incorporating the public health activities of the Federal Health Council.

In 1944 Cilento published *The Health of the Nation*, 'a key document of post-war reconstruction literature'.⁴² In it, he reiterated his conviction that 'the first "real" problem of positive health is quantitative: it is to ensure our actual survival as a race; the second is qualitative: it is to make up in quality what we lack in quantity'.⁴³ Survival was at risk because 'the aging civilisation of Europe with its offshoots in America, Australia, and elsewhere is pressed on all its borders by the increasing hordes of Mongol and Mongoloid races'.⁴⁴ He deplored urbanization, the economic penalties of parenthood, and contraception, for their lowering effect on the birth rate. As for qualitative improvement, the NH&MRC had noted at its fifth meeting in 1938 that: 'In the constant struggle for economic survival, progress is determined, other resources being equal, by the relative proportions of the fit and unfit.' The 'unfit and the untrained' would represent a handicap 'should the increasingly intense economic competition of today end in war'.⁴⁵

The Council, pressed by Cilento, recommended the establishment of a National Council for Physical Fitness. Cilento saw the whole nation as needing the intensive training that was currently being given to commando units in the armed forces: 'The Australian of the present and the next generation has to make for himself and his descendants the deliberate physical and psychological choice between the status of the "commando" and the status of the coolie.'⁴⁶

For Sutton, Elkington, Springthorpe, Cumpston and Cilento, the AAAS provided a platform for expounding ideas that achieved a remarkable prominence in Australian hygiene between Federation and the Second World War. Concerns with the biological destiny of the young nation and with the physical vigour of its members were not the only ones to be considered in public fora such as the AAAS and later ANZAAS through those decades, but many of the more detailed considerations of topics ranging from vitamins and rat fleas to tuberculosis took place within the shadow of nationalist ideology. With the exception of Spingthorpe, our exemplars of national hygiene were all salaried public health officials. From such positions they were well used to addressing both the political élite and the public at large. ANZAAS provided another means of doing so. At ANZAAS, more than in their official reports, they could give vent to the ideas and sentiments that animated them. But how should these ideas be characterized and their resonance within Australia assessed? The influences of 'progressivism' and 'national efficiency' were particularly important in giving them shape and direction.

Michael Roe sees Australian progressivism as a social and political movement derived from European modernism, via Rooseveltian progressivism. He argues that progressivism was an important influence on Australian politics in the post-Federation period. The concerns of the progressives spread widely, but health, closely linked to their ideas of vitality, efficiency, purity and virtue, was among the most important of them.[47] Progressivism was as notable for its political style and application (exemplified in the career of Theodore Roosevelt) as for its internal consistency. Its sense of vitalism and purpose was in tension with a belief in rationality; elitism in conflict with democracy; and authoritarianism with liberalism. Progressivism was essentially 'bourgeois': 'The expert and elite groups which it esteemed so highly comprised educated, more or less successful Anglo-Saxon Protestants [and one could add "males"] ... It was a movement of the established and possessing classes, seeking to save capitalist society from its excesses'[48] — excesses that had been manifest in the 'boom' of the 1880s and subsequent collapse of the 1890s.

Beginning in the 1890s, and stimulated by British failures in the Boer War, there emerged a complementary ideology, equally influential in the Australian public health movement. G. R. Searle has described the ideology of 'national efficiency' 'as an attempt to discredit the habits, beliefs and institutions that put the British at a handicap in their competition with foreigners and to commend instead a social organisation that more closely followed the German model'.[49]

To a belief in the state as a creative and moral agency were added aspirations to improve the national physique, to reform the machinery of government, to reform the education system towards the production of first-rate experts (rather than amateurs), to effect a closer union between government and science, to operate government in a more 'business-like' manner, to transcend the party system, and, as an efficient nation and race, to triumph in the international struggle for existence—leaving lesser nations to fall by the wayside.[50]

The ideas of national efficiency were taken up widely across the British political spectrum, from conservative military leaders on the right, to Fabian socialists on the left. Under the emergency conditions of the First World War, a strong government was able to force through many of the proposals of the 'efficiency group', which had been politically impossible in times of peace. During the 1930s, similar ideas were to come to prominence again under the guise of 'technocracy'; again, they crossed the political spectrum.[51]

Australians, oriented towards Britain, absorbed a good deal of this 'efficiency' ideology. For example, in the first decade of the century, official Australian reaction to the decline in the birth rate borrowed substantially, though not exclusively, from contemporary British debates.[52] In the same period, the related issue of the wastage of infant life was also discussed in terms typical of the British debate on national efficiency.[53]

While some aspects of the thinking of the 'national hygienists' remain to be clarified, several observations can be made. First, they occupied a significant number of strategically important posts. Second, the ideology they promulgated was largely uncontested. (Certainly there is very little evidence of ideological debate at the AAAS or ANZAAS meetings!) Third, in Australia, doctors appear to have been the leading proponents of ideas (notably eugenics) that in other countries found their strongest support outside the medical profession.[54] Finally, belief in the importance of 'national efficiency' continued to feature prominently in public health literature well after it had been supplanted in other discourse on public affairs.

THE SECOND WORLD WAR AND AFTER: SPECIALIST HYGIENE

In the decade or so following the Second World War, various issues of public hygiene were taken up at ANZAAS but the content and emphases within Section I followed a checkered course. In 1949 the name of Section I changed to Medical Science, National Health and Physiology; in 1951 there was apparently no Section; in 1952 Microbiology replaced Physiology; and in 1955 the name changed to Microbiology, Epidemiology and Preventive Medicine. The broad drift was to increasingly specialized discussions of microbiology, with nutrition as a minor theme. Most speakers came from research and academic environments and by the late 1950s were addressing topics as specialized as soil and marine microbiology and the molecular basis of the immune response. Presidential addresses more often reflected developments of the basic sciences. Thus in 1949 Macfarlane Burnet served as Section president and gave an address on 'The Nature of Immunity',[55] and in 1952 E. V. Keogh (director of TB Services in Victoria) spoke on 'Virulence and Infectivity'.[56]

If direct discussion of the application of recently aquired knowledge to the improvement of human health found less place in the body of the congress, there was still room for it in the public lectures. At the thirty-first meeting in Melbourne in 1955, Macfarlane Burnet gave a public lecture on 'Preventive medicine—past and future'.[57] Two years later, as president of Section I, he gave his presidential address on the more inward-looking topic of 'Biology and Medicine'.[58]

With the continuing retreat of fatal infection, there were fewer dragons left of the kind that established public health practitioners had been taught to slay. Recognition of the public health aspects of emergent epidemics of chronic disease was slow in coming. Nutritional topics were addressed in Section I, and for a while in the mid-1950s they had a formally designated subsection. Then, in 1959, Nutrition moved across to join Physiology and Biochemistry in Section N. In neither location did the nutritional causes of coronary heart disease occupy centre stage, and yet that disease was far and away the leading cause of premature death and was well on its way to its historic peak in incidence, which came around 1967–8. It is not that ANZAAS or the Australian scientific community in general were lagging behind international opinion on this issue. The pleasurable aspects of postwar affluence were immediately enjoyed, here and elsewhere; the harmful effects took longer to come to terms with.

THE LAST FIFTEEN YEARS: POST-AFFLUENT HYGIENE

The publication, in October 1986, of the three volume report of the Better Health Commission[59] marked the culmination of a reconstruction of hygiene over the preceding two decades. Over that period had developed, in Australia and in many other western countries, a systematic understanding of the ways in which postwar affluence was producing a characteristic burden of chronic disease and injury. Death rates had risen from causes such as heart attack, lung cancer, suicide, cirrhosis of the liver, and car smashes. During the 1960s, these rising causes of mortality had been sufficient to bring the secular improvement of male life expectancy to a halt.

The characterization of the new hygiene as 'post-affluent' is based on its intent to reverse many of the health-damaging consumption and activity patterns that come with the modern increase in incomes: rising consumption of sugar, animal fat, alcohol, and tobacco, and decreased activity consequent on automation. In many of its central notions, the 'new hygiene' harks back to classical hygiene: it is a guide to citizens on how to live in order to preserve health (in contrast to the 'state' orientation of the 'national hygiene' period); it is centrally concerned with diet and physical activity (in contrast with the concentration on cleanliness in the 'sanitary' period); and it rests on the supposition that luxury is potentially corrupting.[60]

The professional sources of the 'new hygiene' have been epidemiologists specializing in chronic disease, and those who have sought to put this new knowledge to work—dietitians, health educators, and professionals who have become active in voluntary bodies such as the National Heart Foundation and the Anti-Cancer Councils. Public health officials in state and federal governments have not played an active role: none of the Better Health Commissioners were from such backgrounds (though one was a state director of health promotion).[61]

The prime constituency of the 'new hygiene' is the vastly expanded professional middle class, whose health and fitness awareness and knowledge is continually fed by the mass communication media. Such a world has little need for the periodic pooling of expertise and enthusiasm that was the foundation of the AAAS and ANZAAS, and ANZAAS congresses have in recent years contributed little to the development of Australian attitudes toward hygiene and public health.

ANZAAS AND THE NATION'S HEALTH: A RETROSPECT

The contribution of the AAAS and later ANZAAS to the formulation and articulation of ideas, and to public and political programmes of action has inevitably changed with the times. During the first two periods it (along with the Intercolonial Medical Congresses) undoubtedly played a significant and positive role. But the AAAS and ANZAAS congresses were more obvious places for discussing public, as distinct from professional, interests. This became even more noticeable during the 'national' phase, which saw strong ideas propounded by individuals occupying positions of bureaucratic strength. They may not have needed ANZAAS but, being there, it provided them with a platform where they could speak at least partially freed of the conventional restraints of office.[62]

Since the Second World War, the role of ANZAAS has become progressively more marginal to the public consideration of what needs to be done to protect and further improve health. Microbiologists and others who participated in the 'specialized' period seemed largely to be addressing an assumed popular interest in their science, rather than making programmatic statements for the further improvement of health. Their professional reference groups were the evolving scientific specialties. When we come to the emergence of 'post-affluent hygiene', the contribution of ANZAAS appears to have been particularly slight.

The relation of ANZAAS to the political processes bearing on health appears to have varied similarly. By the time of the second Congress, in Melbourne in 1890, two research committees were reporting on practical health topics: number 6 on The Construction and Hygienic Requirements of Places of Amusement in Sydney, and number 9 on Town Sanitation. Ashburton Thomson was on Committee 6 and about half of the members of Committee 9 were medical.[63]

The support of ANZAAS in 1935 for the establishment of so central an institution in the development of Australian public health as the

National Health and Medical Research Council has been noted. The 1949 Congress at Hobart recommended that chairs of child health be established in each of the Australian medical schools.[64] Little policy advice appears to have come forward since.

Finally, what of health itself? When the AAAS began, life expectancy in Australia and New Zealand compared very favourably with that in the nations of northern Europe. In 1901 Australia ranked third behind New Zealand and Norway. Our life expectancy was about three years greater than the unweighted mean of New Zealand, England and Wales, the United States, France, Netherlands, Denmark, Norway, and Sweden.[65] By mid-century, this group of comparison countries had gradually overtaken Australia. We ranked seventh out of nine for male life expectancy and fifth out of ten for female. Through the 1950s and 1960s, we fell further behind, especially in the case of males. By the late 1960s, we were two years behind the mean in the case of males and about one year behind for females. Since around 1970, Australia's improvement has been more rapid than the average of the comparison countries, so that by the early 1980s we were set to overtake them as a group.[66]

The explanation of mortality trends remains largely elusive. There are still great uncertainties about the reasons for the dramatic improvements evident in Australia during the last two decades. Perhaps the greatest irony, however, is that in the period from Federation to mid-century, while our national hygienists were proclaiming loudly a gospel of national invigoration, and aiming for nothing less than the physical perfection of the race, we were being progressively overtaken, in longevity, by many of the older states of Europe.

NOTES

This chapter draws inspiration from the work of others, especially Michael Roe and Graeme Davison. The chapter would never have reached its current state without Roy MacLeod's careful scrutiny and advice and his unremitting encouragement. Thanks are due to Jane McGlashan for conscientious research assistance and to Kate Brittlebank and Joan Roberts for typing.

1. *Report of the AAAS*, 1 (Sydney, 1888), 494.
2. G. R. Searle, *The Quest for National Efficiency* (Oxford: Basil Blackwell, 1971).
3. M. Roe, *Nine Australian Progressives: Vitalism in Bourgeois Social Thought, 1890–1960* (St Lucia: University of Queensland Press, 1984).
4. A. J. C. Mayne, *Fever, Squalor and Vice: Sanitation and Social Policy in Victorian Sydney* (St Lucia: University of Queensland Press, 1982), 187.
5. E. Ford, 'The Bancroft Memorial Lecture: The Life and Influence of Joseph Bancroft', *Medical Journal of Australia*, 1 (1961), 163.
6. D. Gordon, *Health, Sickness and Society: Theoretical Concepts in Social and Preventive Medicine.* (Brisbane: University of Queensland Press, 1976), 806.

7. Connections to Melbourne's first trunk sewer (to Werribee) were not made till 1897. J. W. Springthorpe, *Therapeutics, Dietetics and Hygiene* (Melbourne: James Little, 1914), 1, 270. D. Gordon, op. cit., note 6, 641.
8. J. Ashburton Thomson had been appointed temporary medical officer to the Central Board of Health in Sydney in 1884. D. Gordon, op. cit., note 6, 349.
9. *Transactions of the Intercolonial Medical Congress of Australasia, First Session* (Adelaide, 1887), 1.
10. *Report of the AAAS*, 6 (Brisbane, 1895), xx.
11. The British Act permitted local government to implement compulsory notification and most had done so by 1895. A. Newsholme, *The Elements of Vital Statistics in their Bearing on Social and Public Health Problems*, new edition (London: George Allen & Unwin, 1923); W. Love, 'Report [on] Compulsory Notification of Infectious Diseases', *Report of the AAAS*, 6 (Brisbane, 1895), 829–35.
12. *Report of the AAAS*, 1 (Sydney, 1888), ix–x.
13. J. S. C. Elkington, 'A Plea for the Australian Child Body', *Report of the AAAS*, 11 (Brisbane, 1909), 774–9.
14. M. Roe, op. cit., note 3, 90.
15. Greig was to become a member of the 1926 Royal Commission on Health.
16. Anon., 'Obituary: Harvey Sutton', *Medical Journal of Australia* (1964), 496–99; Roe, op. cit., note 3, 140.
17. H. Sutton, 'The Importance of Nationality', *Report of the AAAS*, 13 (Sydney, 1911), 508–10.
18. J. S. C. Elkington. op. cit., note 13, 774–9.
19. ibid.
20. As cited in Roe, op. cit., note 3, 111.
21. From 1921 to 1928 Elkington was head of the Commonwealth Health Department's Division of Tropical Hygiene.
22. J. W. Springthorpe, op. cit., note 7.
23. ibid., 27.
24. ibid.
25. ibid.
26. ibid., 128.
27. ibid., 99.
28. ibid., 108. Sergent's specifications for the ideal female figure is given in a series of nineteen measurements extending in detail down to the girth of the wrist; ibid., 97.
29. H. Sutton, 'Presidential Address Section I: Recent Progress in Child Hygiene', *Report of the AAAS*, 16 (Wellington, 1923), 645–65.
30. H. Sutton, 'Presidential Address Section I: Modern Development of Public Health—The Child as the Test of Progress', *Report of ANZAAS*, 20 (Brisbane, 1930), 335–64. His interpretation anticipates the current opinion of the World Health Organization that maternal literacy is perhaps the most important influence on infant mortality, World Health Organization, *World Health Statistics Annual* (Geneva: WHO, 1985), 8.
31. Roe, op. cit., note 3, ch 4.
32. ibid., 124.
33. ibid., 125.
34. J. H. L. Cumpston, 'Presidential Address in Public Health and State Medicine: The New Preventive Medicine', *Australasian Medical Congress: Transactions of the Eleventh Session...Brisbane...1920* (Brisbane: Government Printer, 1921), 77–85.
35. Roe, op. cit., note 3, 132.

36. Anon., 'A Federal Ministry of Health [editorial]', *Medical Journal of Australia* (1921), 133–5.
37. Roe, op. cit., note 3, 132–3.
38. ibid., 105.
39. ibid., 135–6. Anon., 'Sir Raphael W. Cilento' in D. Lu (ed.), *Notable Australians: The Pictorial Who's Who* (Sydney: Paul Hamlyn, 1978), 259.
40. R. W. Cilento, 'Australia's Problems in the Tropics', *Report of ANZAAS*, 21 (Sydney, 1932), 216–33.
41. *Report of ANZAAS*, 22 (Melbourne, 1935), xxix.
42. R. Cilento, *Blueprint for the Health of a Nation* (Sydney: Scotow Press, 1944); Roe, op. cit., note 3, 146.
43. ibid., 92.
44. ibid., 93.
45. ibid., 110.
46. ibid., 113.
47. Roe, op. cit., note 3, 12.
48. Roe, however, is perhaps more successful in demonstrating that several of our leading hygienists exemplified progressive concerns than he is in showing that these ideas were transmitted from American progressive sources.
49. G. R. Searle, op. cit., note 2, 54.
50. ibid., ch. 3.
51. The Webbs' admiration for Soviet Russia in 1932 did not rest on sympathy with Marxism as a doctrine but rather on their vision of an élite-directed society with 'an elaborately planned network of more than a thousand research laboratories'. There were also scientists, such as J. D. Bernal, P. M. S. Blackett and J. B. S. Haldane, who 'joined the cult of Soviet Russia while at the same time making sweeping claims for science as a solvent of contemporary economic and political difficulties', Searle, op. cit., note 2, 262.
52. N. Hicks, '*This Sin and Scandal*': *Australia's Population Debate 1891–1911* (Canberra: Australian National University Press, 1978).
53. M. Lewis, 'The Problem of Maternal Mortality and Population Growth in Australia, 1880–1939: Some Ideological Considerations', *Community Health Studies*, 4 (1980), 104–10; G. Davison, 'The City-bred Child and Urban Reform in Melbourne 1900–1940' in P. Williams, *Social Process and the City* (Sydney: George Allen & Unwin, 1983), 143–74.
54. Doctors were apparently less central to the development of eugenics in the United States and Britain. S. Garton, 'Sir Charles Mackellar: Psychiatry, Eugenics and Child Welfare in New South Wales, 1900–1914', *Historical Studies*, 22 (1986), 21–35; 23 (footnote). In Britain, at a time of widespread concern over 'physical deterioration' following the reverses of the Boer War, important elements of the medical profession remained sceptical. When asked for their opinion on why so many military recruits were being rejected on grounds of physical disability the Royal College of Physicians did not find itself in possession 'of any evidence which satisfies it that there is any physical degeneration of the urban population generally', *British Medical Journal* (1903), 1338. The *British Medical Journal* ran a series of articles on physical degeneration, emphasizing environmental causes and environmental solutions, ibid., 1338–41, 1430–1, 1471–4, 1555–7, 1614–15, 1652; ibid. (1904), 45–7, 86–88, 140–2, 197–9, 272–3.) It concluded editorially that 'if the State requires the children of the very poor to work, they should be so fed as to enable them to learn without injury'. The *Journal* therefore recommended the provision of cooked dinners at school 'at a small cost to parents', ibid., 37.

55. F. M. Burnet, 'Presidential Address: The Nature of Immunity', *Report of ANZAAS*, 27 (Hobart, 1949), 106–12.
56. E. V. Keogh, 'Presidential Address: Virulence and Infectivity', *Report of ANZAAS*, 29 (Sydney, 1952), 211–16.
57. F. M. Burnet, 'Public Lecture: Preventive Medicine—Past and Future', *Report of ANZAAS*, 31 (Melbourne, 1955), 13–19.
58. F. M. Burnet, 'Biology and Medicine', *Report of ANZAAS*, 32 (Dunedin 1957), 9–16.
59. Australia, Better Health Commission, *Looking Forward to Better Health. Volume 1: Final Report. Volume 2: The Taskforces and Working Groups: Reports to the Better Health Commission. Volume 3: The Workshops and Consultations: Reports to the Better Health Commission* (Canberra: Australian Government Publishing Service, 1986).
60. H. E. Sigerist, *Landmarks in the History of Hygiene* (London: Oxford University Press, 1956); J. Sekora, *Luxury* (Maryland: Johns Hopkins University Press, 1977).
61. The 'Better Health' Commissioners were: Derek Llewellyn-Jones (chairman), associate professor of obstetrics and gynaecology, University of Sydney; Lisa Currey (resigned September, 1985), Olympic sportswoman; Richard Charlesworth, Member of Parliament, Olympic sportsman; Anne Deveson, director, Australian Film and Television School; Dawn Fraser, Olympic sportswoman; Robert Gradwell, former assistant secretary, ACTU; Janet Irwin, (appointed October, 1985), director, University of Queensland Health Service; Tony Kearney, former registrar, University of Tasmania, former chairman, Hobart District Hospitals Board; Suzanne Kellaway, television personality; Stephen Leeder, professor of community and geriatric medicine, University of Sydney, Australian president of ANSERCH; Tony McMichael, head of the Epidemiology Research Unit, CSIRO Division of Human Nutrition; Rod Muir, executive director, 2MMM-FM radio, Sydney; Lynda Stephens, health consultant, Department of Health, Victoria.
62. The preventive medicine sections of the Australian Medical Congresses during this period had a more pragmatic orientation, being largely concerned with the practical detail of controlling infectious disease.
63. *Report of the AAAS*, 2, (Melbourne, 1890), 356, 693.
64. *Report of ANZAAS*, 27 (Hobart, 1949), xxiv. The Commonwealth had established an Institute of Child Health at Sydney University the previous year, D. Gordon, op. cit., note 6, 806.
65. J. Powles, International Life Expectancy Trends Since the 1890s (unpublished tabulation) (Melbourne: Department of Social and Preventive Medicine, Monash Medical School, Prahran, 1986).
66. Many of them had experienced a period of relative stagnation through the 1970s (analogous to Australia's experience in the 1960s).

13

Social Responsibility of Science: The Social Mirror of Science

Ron Johnston

Earlier chapters have examined the ways in which science has contributed to the shaping of Australasia over the past one hundred years through the development and application of new technologies in the agricultural, mining, and manufacturing sectors. The sciences have also been shown to have played a significant cultural role in the development of independent, national identities.

There is, however, a third, political dimension, through which the sciences have both shaped and reflected national and international political agendas. For the place of science in human affairs is neither natural, nor necessarily pre-eminent. Its influence and authority have varied, not only with the nature and quality of the knowledge produced, but also with its apparent relationship to what are perceived as the most important elements of the social climate. Ben-David has argued that the emergence of modern science can be linked to the developing professionalism associated with the gradual fashioning of a social *role* for the scientist.[1]

The interests, in the case of seventeenth century Europe, lay in establishing the basis of a 'changing, pluralistic, and future-oriented society'.[2] Thus, it was a scientific movement, which saw in modern science both a symbol and model for the route to truth, mastery over nature and a transformed social order, that provided the legitimation for the institutionalization of science. Such legitimation, however, was not achieved once and for all. There exists a continual process of negotiation between science and the social values of the community on which it relies. At times the relationship has been strong and supportive, at other times weak, hostile, or tinged with doubt.

Institutionalization of the social role of science in Australia has taken many forms. In this essay I argue that a recurring vehicle for expressions of the appropriate social role of the scientist has been provided by the notion of 'social responsibility'. In particular, it is at times when the social role of science is uncertain, in the process of redefinition, or under attack, that the strongest demands or claims for social responsibility are made. This is reflected not in the presence or absence of social responsibility rhetoric, rather, we shall see that when the social role of science is unchallenged, social responsibility is presented in the form of demands from science on other social institutions. Conversely, when the social role of science is a matter of public debate, social responsibility arguments are primarily designed to defend and justify the scientific enterprise. In this sense, the content of social responsibility 'claims' provide a mirror of the social status of science.

Various attempts have been made to provide an authoritative definition of the social responsibility of science.[3] At least seven distinct usages can be identified. The first four usages share, in most respects, an essentially Baconian view of modern science: if properly practised, science should give rise, automatically, to social benefits and progress. The special responsibility of the scientist, then, is to ensure that science is not harmed or impeded by the forces of darkness, and that the wisdom of science is made available directly to those in power. These usages reflect an essentially positive, or optimistic view, of the social role of science:

1. the responsibility of scientists to promote the model and norms of science as an exemplar of knowledge and human conduct, and to seek to counter and overcome ignorance and superstition;
2. the responsibility to protect science, and scientists, from the attacks of those who would deny their truth and seek power for their own purposes;
3. the responsibility of scientists, because of their expert knowledge, to advise governments and politicians on courses of action and the implication of scientific advances; and
4. the responsibility to educate the public about the methods and outcomes of science, so that they will understand its value, and support and acknowledge it more readily.

In contrast, another three common usages of social responsibility have at their root a cautious, if not negative, view of science. These concern:

5. the responsibility for the application of scientific knowledge and its consequences;
6. the responsibility for the content of scientific knowledge, and in particular its congruence (or lack of it) with human and social needs; this incorporates a responsibility to be aware of the needs of society; and
7. the responsibility for the nature and values of the scientific enterprise itself.

The wide varieties in meaning reflect different perceptions of the social role of science and derivative scientistic models, by, and between scientists and non-scientists, at different times. Moreover, the chosen interpretation of the social role of science is undoubtedly a reflection, at least in part, of political purposes. In other words, it may be in the interests of those seeking either to maintain or oppose the authority of science or those seeking to draw support for their cause from the practice of science to attempt to establish a particular view of the social responsibility of science.

This array of views of the social responsibility of science, and the social role of science it implies, is well represented in the development and application of science in Australasia. In this essay, the focus will be on the period from 1938 to 1988, corresponding to the second fifty years of ANZAAS. This emphasis has been chosen not only to limit the scope of the essay, but also because the more evidently political nature of the scientific enterprise over that period, along with its greater scale, has made the issue of the social role of science, as reflected through social responsibility claims, a matter of much wider concern and debate. In addition, the activities of ANZAAS have reflected, and on occasion contributed to, the debates about the appropriate social role of science.

SOCIAL RESPONSIBILITY AS SCIENTISM

The development in the 1930s of a strong concern about the issue of social responsibility among scientists in Britain and Australia, was led by what became labelled as the Social Relations of Science movement. This was spearheaded by a group of influential British scientists, including J. D. Bernal, J. B. S. Haldane, P. M. S. Blackett, Joseph Needham, and Julian Huxley, when the breakdown of capitalism in the Great Depression, the rise of Fascism, and the emergence of the Soviet Union as a powerful socialist nation created the climate for radical new thinking.

The essence of their view was Baconian: science, properly organized and directed, was capable of solving all the problems of humankind. Moreover, in the light of the evident failure of economics and politics, science provided the basis for the most effective organization and conduct of human affairs:

Science puts into our hands the means of satisfying our material needs. It gives us also the ideas which will enable us to understand, to co-ordinate, and to satisfy our needs in the social sphere. Beyond this science has something as important though less definite to offer: a reasonable hope in the unexplored possibilities of the future, an inspiration which is slowly but surely becoming the dominant force of modern thought and action.[4]

These views had strong adherents in Australia. That Australian scientists should follow the lead of their British counterparts is not surprising. At this time, the primary reference point and measuring stick for Australian scientists was British science. The Australian scientific community was still small, and the majority had received postgraduate education at either Oxford or Cambridge. Indeed, a significant number had come under the direct influence of Bernal himself.[5]

One line that was argued particularly strongly in Australia was the special responsibility of the scientist as an influence on public policy. As

Professor O.U. Vonwiller, professor of physics at the University of Sydney, argued in 1938:

The scientist today is gradually realising that his duty goes beyond the acquisition of knowledge for its own sake or commercialising such knowledge for private or public gain. He must insist on being heard when policies are formulated and when methods of administration are being discussed, but first he must make a critical examination of the influence of scientific endeavour on society: he must ascertain those faults of past and present methods which have led to the present unsatisfactory position; he must determine the part which he can play in effecting reform in those methods. The scientist should be able to insist, not necessarily that he take a part in the forming and administration of national policy, but that those charged with that work be educated to understand the possibilities of science and of the application of scientific methods in other fields of human endeavour.[6]

An informal discussion of the issue of 'science and society' modelled on the British Association meeting in 1936 was held at the January 1939 meeting of ANZAAS in Canberra. Sir David Rivett, chairman of the Council for Scientific and Industrial Research (CSIR), argued that the justification for the pursuit of knowledge rested on insistence upon two facts:

1. that the potential applications leading to social good were greater in number and significance than those leading to ill; and
2. that if the pursuit of the good of society through science were pressed forward with vigour and enthusiasm, the incentive towards an application for evil ends would largely vanish.[7]

Professor Vonwiller repeated his view that those in power should be trained to apply the essentials of scientific method in the attack on the problems of society, but also warned of potentially serious effects of the application of scientific knowledge:

The most important influence of development in science on society was the change it had produced in the quality of mankind. Through the advances of medicine and connected sciences everyone enjoyed a protection from disease and hardship which might introduce a lack of resistant power which would later prove costly. The present generation was flabbier in body, mind and soul than its predecessors. Flabbiness of the body was illustrated by the inability of present-day cricketers to stand up to the strenuousness of tours as adequately as could the players of 30 years ago.[8]

Clearly, the driving concerns of the time are not so different from those of today.

Subsequently, a meeting was held in Sydney, chaired by Professor Eric Ashby, and attended by 110 people. Given that at this time the majority of Australian scientists were government employees, working in the CSIR, and that the number of university scientists was still small, this attendance reflected a high level of interest. At this meeting the

Australian Association of Scientific Workers (AASW) was established[9] to 'secure the wider application of science and the scientific method for the welfare of society, and to promote the interests of science'.[10] The twin goals of 'extended and conscious application of science for the welfare of the community', and 'the protection of the status of the scientific worker'[11] went hand-in-hand, for if science were to solve all human problems, the welfare and freedom of the scientist was paramount.

To the question of whether the scientist has 'a special social responsibility different from and perhaps greater than that of the non-scientist citizen',[12] the answer was a resounding yes, 'by virtue of their work and knowledge'.[13] How then to discharge this responsibility? Here a degree of equivocation entered, as the superior position and responsibility of the scientist led logically to a dictatorship of scientists. The conclusion was to call for greater education for everyone in the methods and principles of science, and for an 'extension of scientific methods to the problems of social organisation'.[14]

Wartime saw the full-scale application of the principles of science and the energies of scientists to the human problems presented by war. The AASW played an important role in Australia in a variety of projects.[15] After the Second World War various attempts were made to regain the drive of the social relations of science movement. UNESCO, for example, sought to promote and sponsor studies on the social relations of science.[16] The experience of war, however, had introduced two new elements. The first arose from the construction and use of the atomic bomb. The second, connected to the first, was the establishment of a pattern of secrecy in the conduct of science.

THE ATOMIC EFFECT

It was the pattern of secrecy established in wartime conditions that first raised new problems about the social role of science after the war. Most scientists expected the secrecy requirements of wartime to be rapidly dismantled. However, the emergence of the cold war with the Soviet Union, and spy revelations in Britain and America created an environment of deep suspicion, used by demagogues—in the United States, Senator Joseph McCarthy, and in Australia, W. C. Wentworth and Joseph Abbott—to promote an anti-Communist witch-hunt. In this atmosphere, calls for the abolition of secrecy in science became interpreted as a threat to national security.

The concerns of the time are reflected in a letter to the *Australian Journal of Science* in 1957:

Holders of research studentships financed by the Commonwealth through the CSIRO must agree not to 'engage publicly in party political controversy, whether by speaking, broadcasting, writing of letters to the press, or by publishing books, articles or leaflets' (CSIRO Head Office Circular). This restriction operates no matter what subject is being studied. Science students are deterred from taking an active interest in public affairs by the knowledge that a security check may have to be passed in order to obtain employment. When the grounds for deeming a person to be a 'security risk' are not defined by the Government, many will play safe by avoiding controversy.[17]

These circumstances, more than anything else, prompted the demise of the AASW and the decline of the 'social relations in science' movement. To a significant extent, as the result of the magnitude of their contribution to the war effort, science and scientists had become too important as a national resource to be allowed to return to their own devices. There was pressure from the government for scientists, particularly in the CSIR, to be treated like other public servants, and their capabilities exploited in what was seen as the national interest. Their special expertise and their well-publicized commitment to internationalism required that they be controlled all the more effectively, rather than allowed special authority in government decision making.

This view did not, however, carry through to scientists involved in the British atomic weapons tests held in Australia between 1952 and 1963. Great confidence was placed in the advice of the Australian Weapons Testing Safety Committee (AWTSC) consisting of five scientists, including Professors L. H. Martin, E. W. Titterton and J. P. Baxter. While there was considerable public concern over fallout, and regular reports were made by Titterton and his colleagues in the *Australian Journal of Science*, there was little involvement of other scientists. One exception was Hedley Marston, chief of the Division of Animal Biochemistry and General Nutrition, who prepared a report on the uptake of radioactive iodine into the thyroid of grazing animals. The report became the subject of some controversy and of a protracted, though largely private, difference of opinions between Marston and the Safety Committee, running into 1958. Claims were advanced by Marston that the Committee has acted to suppress his findings.[18]

The 1950s, however, were largely marked by close relations between the leaders of the scientific community and government. Under the patronage of Robert Menzies, science flourished and expanded. The Australian Academy of Science was established with a royal charter; the Murray Report led to a substantial increase in funding to the universities; and the Australian National University's Institute of Advanced Studies, with responsibilities for research and postgraduate education only, expanded greatly. The CSIRO, under the guidance of its Minister, R. G. Casey, also prospered mightily.

Through this conjunction of interests, the institution of science grew at a great rate, and Australia's élite scientists played their role in supporting and reinforcing an ordered, Anglo-centric view of the world. There were costs for science, however, as their expansion was traded for a steadily increasing government influence over the direction of research.[19]

One issue over which the Australian government and at least a significant section of the Australian community took rather different views was the proliferation of nuclear weapons. Through the Pugwash movement, a group of international scientists sought to put their prestige behind a move to control nuclear weapons and argued that scientists had a special responsibility to resist the use of their own research for harmful purposes.

The initiative for this movement came from a letter written by Bertrand Russell and Albert Einstein in 1955, calling for scientists to assess the perils to humanity that had arisen as a result of the development of weapons of mass destruction. With the support of Cyrus Eaton, a wealthy Canadian industrialist, a meeting of twenty scientists drawn from ten nations and widely representative of different political and economic opinions was held at his birthplace, Pugwash, in July 1957.

Three points were on the agenda: the hazards arising from the use of atomic energy in peace and war, the problems of the control of weapons, and the social responsibility of scientists. The statement issued at the end of the meeting included an assertion of 'our common conviction':

That we should do all in our power to prevent war and to assist in establishing a permanent and universal peace. This we can do by contributing to the task of public enlightenment concerning the great dilemma of our times, and by serving to the full extent of our opportunities in the formation of national policies.[20]

This statement reveals a partial resurrection of the claims of the social relations of science movement, though directed specifically towards the issue of nuclear weapons control. This group of eminent scientists included Professor Marcus Oliphant who, on his return to Australia, distributed the Pugwash documents widely. A Melbourne Pugwash Committee was formed to circulate the Pugwash statement, and fifty scientists registered as signatories. Subsequently Pugwash groups were formed in Melbourne, Sydney, Adelaide, and Canberra; their major activity was dissemination and discussion of the reports from the annual International Pugwash meetings. Study groups were formed, public meetings called, and a newsletter published over the period until 1968. Following an initial burst of support, however, enthusiasm gradually declined.[21]

The dominant form and content of arguments about responsibility of science changed significantly between the 1930s and the 1950s in Australia. Debates overseas were of major influence, but local war-related experiences also contributed. There persisted a strong belief that 'Scientists are, because of their special knowledge, well equipped for early awareness of the danger and the promise arising from scientific discoveries. Hence, they have a special competence and a special responsibility in relation to the most pressing problems of our times.'[22] This special expertise was, however, no longer extended to an unqualified justification and promotion of science itself: 'We believe it to be a responsibility of scientists in all countries to contribute to the education of the peoples by spreading among them a wide understanding of the dangers and potentialities offered by the unprecedented growth of science.'[23] Science was now part of the problem, rather than the solution.

A NEW SOCIAL CLIMATE FOR SCIENCE

The mass blackout of the eastern United States in 1965 signalled for Barry Commoner the end of the 'age of innocent faith in science and technology.'[24] This can be linked to the emergence of three broad areas of social concern during the 1960s. The first of those was the state of the environment. Triggered initially by Rachael Carson's damning account of the effect of pesticides on native animals in *Silent Spring*, there swept across the United States, Europe, and subsequently Australia, a broadly based concern over the effects of pollution and over-population. Barry Commoner's *Science and Survival* and Paul Erlich's *The Population Bomb* were extremely influential, and ecology became the watchword, the fashionable discipline, and the basis of many attempts at a new ideology.[25] The response in Australia, as elsewhere, took two forms. There was a general reaction against science and technology, particularly as generators of pollution, and a call for a reorientation of science to objectives likely to be more beneficial to society. Second, biologists were able to claim a special knowledge and responsibility to advise.[26]

The battle between the images of science and technology—'despoliator' versus 'saviour'—was complicated by a second broad social movement, aroused by the Vietnam War. The passions for and against this war spilled into almost all elements of Australian society, including science. In particular, for many opponents of the war, science was seen as an agent of death and an oppressor of technologically disadvantaged nations. At the same time, the failure of extraordinary levels of technology to achieve decisive victories revealed a critical flaw in the argument that control of science led inevitably to superiority. While these two views might be inconsistent, together or apart they revealed serious doubts about the traditional values underlying the pursuit of science. The social role of science became a matter for anxiety, questioning, and potentially major revision.

The universities provided a major source of this revision. Academics were prominent at rallies and demonstrations, contributed to the membership of protest groups, and were largely responsible for importing and promoting the American 'teach-in' forum.[27] Scientists were well represented in this process. In 1966, a full-page anti-Vietnam newspaper declaration included a high proportion of scientific and medical academics.[28] Even more prominent was a statement in the *Australian Journal of Science*, in 1967, signed by 677 Australian scientists, including 61 professors, and also staff of CSIRO. Mindful of the special responsibilities of science and scientists everywhere, they registered our 'deep concern and revulsion' at the country's involvement in this war,[29] and argued for the scaling down of military activities by all parties, an expression of willingness to negotiate from all belligerents, and the channeling of funds from the war into solving the 'tremendous biological and sociological problems now facing mankind, whatever their politics or persuasion'.[30]

With the environmental crisis, exacerbated by the Vietnam conflict, a third area of concern struck at the foundations of the institution of science itself. This became known as the 'anti-science movement'. Criticisms from a variety of standpoints coalesced to provide the basis for a widespread rejection of not only the fruits, but also the values, of western science. Commoner, along with many others, saw scientists as sorcerer's apprentices, with insufficient knowledge to control the effects of their actions.[34] For others, the alleged 'neutrality' of science, and its *lack* of response to social demands was the major concern.[32] Herbert Marcuse[33] combined attacks on American involvement in Vietnam with criticisms of the political status of science. He argued that, far from the Baconian ideal of science as freeing humanity from material want, science had been captured by sectional capitalistic interests, particularly those of the industrial-military establishment. As a consequence, science had become an instrument in the domination of 'man over man'. The attack was continued by Theodore Roszak, who condemned science as 'a bewilderingly perverse effort to demonstrate that nothing, *absolutely nothing* is particularly special, unique or marvellous, but can be lowered to the status of mechanized routine'.[34]

This 'counter culture' was transported to Australia, but there was little or no contribution here to its intellectual development. Some believed it discouraged student interest in science, though the decline in participation rates in science education may have followed more from the general anti-intellectualism of the day. However, in response to the emergence of these three broad areas of social concern, there were calls overseas for the complete separation of science from politics,[35] and even dire predictions of an end to the scientific enterprise itself.[36] In Australia, the response of some scientists was to establish organizations to promote the concept and practice of social responsibility in science.

THE SOCIAL RESPONSIBILITY OF SCIENCE INSTITUTIONALIZED

A variety of organizations emerged to meet this challenge. In the United States these included the Union of Concerned Scientists, formed at MIT in 1969 after a 'research strike', which prepared public reports on such issues as anti-ballistic missile systems, and chemical and biological weapons, and Barry Commoner's Institute for Public Information. More radical was the Scientists and Engineers for Social and Political Action (SESPA), formed in 1969 following the refusal of the American Physical Society to take a stance on the use of physics in Vietnam. In Britain, the British Society for Social Responsibility in Science (BSSRS) was formed, also in 1969.[37] Eventual dissatisfaction with the 'political' drift of the British society and its failure to address what many senior scientists considered the key issue—public education—led to the establishment of the Council for Science and Society in 1973.

These developments in the United Kingdom and the United States and the increase in political consciousness caused by the Vietnam War produced a variety of institutional responses in Australasia. Within ANZAAS the *Australian Journal of Science* was replaced by a new journal, *Search*, revealingly sub-titled 'Science, Technology and Society', and given the mandate of publishing 'articles which deal with the social and economic consequences of advances in science and technology'.[38]

The contents of *Search* over the following ten years provide a fair reflection of the wider issues that concerned scientists. The essential conflicts are captured in a Bruce Petty cartoon drawn for the Brisbane Congress in 1971. The major issues addressed included science education, conservation and the environment, the crown-of-thorns starfish attack on the Barrier Reef, the supply of energy, the Concorde Supersonic Transport and its potential effects, implications of biological advances, population control, the relation of science to industry, science and the press, technology and unemployment, science in government, food supply, and health care.

It is more difficult to assess the effectiveness of ANZAAS in its role as a vehicle for expressing the social responsibilities of science. Through *Search* and the congresses, it assumed a significant level of communication with at least sections of the public. But these remained essentially passive vehicles. More activist approaches were adopted by the organizations established specifically in response to the new social climate for science in the 1960s: the Society for Social Responsibility in Science (SSRS) groups in Australia and the New Zealand Association of Scientists.

The Society for Social Responsibility in Science

The establishment of a set of SSRS groups in Australia can be traced largely to the actions of Professor Charles Birch[39] of the University of Sydney, who returned from an overseas trip impressed by developments in the United States. He called a meeting of academics in July 1969, which led to the formation of the Social Responsibility in Science–Sydney group. Subsequent initiatives at the August 1969 ANZAAS Congress and by mail led to the establishment of two other strong groups, in Melbourne and Canberra, and a small one in Rockhampton.[40]

Both the Sydney and Melbourne groups soon attracted 250–300 members; they peaked in the period 1970–2, but declined rapidly afterwards. They informed the public and stimulated social awareness through lectures and public meetings on Concorde, population control, intermediate technology, chemical and biological warfare, energy, and genetics.

The ACT group, formed in March 1970, continued actively into the mid-1980s.[41] Their aims, as set out in the constitution, are remarkably similar to those of ANZAAS, and also those of the British Society for Responsibility in Science.[42] These aims were to be pursued by public meetings, study groups, dissemination of findings, and discussions between scientific experts and decision makers. The ACT membership grew to 150, most of whom were researchers in the biological sciences employed by CSIRO. Probably as a consequence SSRS–ACT operated essentially as a community action group with a focus on environmental issues. Thus, its early activities included lobbying against the Telecom Tower proposed for Black Mountain, establishing mechanisms for waste recycling in Canberra, studying pollution in the waterways around Canberra, and opposing the Molonglo Parkway. In addition, a schools programme entitled INSPECT (Inquiry into the State of Pollution and Environmental Conservation by Thoughtful people), designed to encourage awareness of environmental problems, was established in Canberra and was rapidly taken up elsewhere. A sign of their standing in the environmental field came with an invitation to attend the United Nations Conference on the Human Environment in 1972.

The most obvious achievement of SSRS–ACT was the success of their campaign, over the period 1973–6 to persuade the Commonwealth government to withdraw its financial support from the mass chest X-ray campaign. During this period, it also addressed nuclear power and the effects of nuclear weapons testing, environmental health issues, and science and energy policy. Some of these activities led to clashes with the establishment, particularly in the medical and health fields. Perhaps as a result, by the end of the 1970s, the emphasis had shifted from political activism to the provision of information. This, together with further changes in the social climate, and an inability to recruit new members led to the effective, if not formal, demise of the organization.

The history of this particular phase of the social responsibility of science reveals a pattern commonly experienced in Australia, in both political and cultural movements. Their origin is elsewhere, in Europe or the United States. The ideas are transported, frequently by visiting academics, but if they find a fertile environment in Australia, their flourishing will produce rapid mutation to correspond with the special needs, interests, and opportunities of the local climate, but its foreign origin can also mean that it never effectively takes root, and quickly withers and dies.

When measured against the sheer size as well as the political and economic basis of the forces shaping science and technology, it was perhaps naive to expect too much from the approach. The present structure of Australian science is probably such that it prevents any movement towards greater social responsibility having much effect on the actions, and beliefs, of Australian scientists.[43]

The effect of the groups on both the scientific community and the public was limited. SSRS provided the organizational context for some scientists to work out their views of an appropriate social role of science. It was, however, unable to disseminate this view into the mainstream of the scientific enterprise, and as the social role changed, so the support for its particular perspective on the social responsibility of science weakened.

The New Zealand Association of Scientists

The institutionalization of social responsibility in New Zealand took a different form, with one organization, the New Zealand Association of Scientists (NZAS), in many ways performing the combined functions of AASW, ANZAAS, and SSRS in Australia.

The Association was formed in the wartime conditions of 1941 with the aim of securing the widest application of science and of the scientific method for the welfare of society and to protect the interests of scientific workers.[44] Its initial aims were a mixture of promotion, social responsibility, and employment protection. After the war, secrecy became a major issue in New Zealand. Cold-war politics led to a severing of connections with the World Federation of Scientific Workers, because of alleged Marxist bias, and the adoption by the New Zealand Association of a determinedly non-political stance. In the 1960s and early 1970s, the Association's major preoccupation was with salaries. It was not until the mid-1970s that its focus broadened to include such topical issues as the 'Limits to Growth' debate, women in science, and the New Zealand fishing industry.

The commitment to a non-political stance led to a fairly cautious approach to social responsibility:

> The Association remained strictly imparital over the issue of nuclear power, as it has in general over the issues in science. However, during the past year [1978] it has taken a stand on genetic engineering, not because it is opposed to such work ... but because it believes that the public has a right to be fully informed about possible risks.[45]

It organized a conference on Social Responsibility in Science in 1979,[46] but thereafter interest in these issues and the membership of the Association declined.

The prominent position of the Department of Scientific and Industrial Research in New Zealand has undoubtedly played a significant role in the country's economic, social, and cultural development. But the fact that the majority of scientists are employed by government has restricted the development of an independent and more critical conception of social responsibility.

The Australian Academy of Science

The response of the Australian Academy of Science to the changing social climate for science was inevitably different from that of younger and more radical scientists, showing a greater concern with the protection of the science enterprise itself. The first organized activity of the Academy specifically dealing with social responsibility was a symposium on 'Science, Technology and Society' in conjunction with its annual meeting on 3 May 1968.[47] For the Academy, at least as reflected in this Symposium, the changing social climate did not raise problems for the values, organization, and practice of science. Rather, it was only a matter of linking research more effectively to economic and social objectives.

In 1972 the Academy published two reports on environmental issues: the use of DDT and the atmospheric effects of supersonic aircraft. The approach revealed the Academy's self-image as a purveyor of definitive knowledge:

> Knowing that important matters of public interest were imminent in this country relating to supersonic transport, knowing that the public had been assailed by propaganda from both sides and might well have become confused, the Academy instituted an enquiry by a group of uncommitted scientists to review the various dangers which have been suggested, and put the risk of climatological impact into perspective.[48]

However, the Academy's findings—that the hazards of DDT and SST were not great—met with considerable opposition, something for which the Academy did not appear to be prepared. Its report on Concorde was even described as an exercise in 'social irresponsibility'.[49] The working group that produced the supersonic transport report comprised technical experts in a number of fields, including atmospheric physics. It collected and analysed data, consulted with experts, and arrived at conclusions. The group's opponents disagreed not only with the interpretation of technical data, but also with the values inherent in the analysis, and the research procedure itself, claiming it was designed to manufacture a consensus that the limited state of knowledge did not justify.

The supersonic transport report provided an occasion for a clash within the scientific community between two distinct sets of assumptions about the social role of science. For the Academy team, the issue was a matter for objective determination; for the critics, science and politics were inextricably mixed. That the debate was conducted in public, and over issues of some community interest, served only to diminish the authority of the first view of science as the source of superior wisdom, and to reinforce the claims of the second. The lesson apparently drawn by the Academy of Science was that it should avoid, or at least handle far more circumspectly, potentially controversial issues.

There was, however, another move within the Academy, or more properly among some of its Fellows, to address the new social responsibility of science. This was led by Sir Otto Frankel, who, in delivering the second John Edwin Falk Memorial Lecture in Canberra on 3 March 1972, on conservation of crop genetic resources, emphasized the need for proper debate through a Science and Society forum where 'dissent and constructive criticism are given the orderly freedoms of institutionalization'.[50]

The establishment of the Science and Society Forum was not announced until August 1973, the major delay being a result of difficulties in obtaining appropriate membership of the steering committee, particularly suitable 'younger scientists'. Among the first projects selected for study were the value of environmental impact statements, water and its uses (in particular the feasibility of towing icebergs), the 'limits to growth' concept, and the value of nature conservation.

The first forum was held on 9 November 1974, under the title Science and Society in Australia. The topics to be discussed were, Major Issues Confronting Science and Society in Australia, Medical Science—Are Our Priorities Right?, and The Wired City—Science and Urban Life. Sir Macfarlane Burnet chaired the meeting, and speakers were to include the Ministers for Science and for Health, John Gorton, (the former Prime Minister), Sir Otto Frankel, and Professor Gus Nossal. Despite this prestigious guest list, the proceedings could hardly have been less auspicious.

First, the opening of the meeting was interrupted by what the press described as a 'wild brawl'. Thirty students attempted to break up the meeting, accusing the first speaker, Professor Roger Russell, (vice-chancellor and professor of psycho-biology at Flinders University) of being an agent of the United States Department of Defence. Second, Mr Bill Morrison, Minister for Science, used the occasion to make a long-awaited announcement of the establishment of the Australian Science and Technology Council. Third, the Minister for Health, Dr Everingham, announced a plan to introduce a diploma of community nursing to ease the shortage of general practitioners. Under these circumstances the principal objective of providing a Forum for discussion of issues of scientific and social relevance was totally undermined.

This first meeting sounded the deathknell for the Science and Society Forum. After some discussion, it was dissolved in July 1975. Various committees met sporadically to consider the matter until 1979, but there was little pressure for action. It appeared that the experiences of physical violence and of being used by Ministers persuaded the Fellows to withdraw into their dome. With the passage of time, the pressure for visible expressions or affirmations of social responsibility declined.

SOCIAL RESPONSIBILITY IN THE 1980s: AND STILL THE WHEEL TURNS

The period from 1978 to 1983 witnessed a remarkable turnaround in Australian public and political attitudes to science and technology. Indeed, the very success of scientists in promoting the value of their research to economic progress had the effect of tying the public image of science very close to that of technology.

By 1978 there was growing concern with the effects of technology on employment. ANZAAS had perhaps been rather percipient on this issue, organizing a symposium on the consequences of automation in 1968,[51] when there was interest in the United States, but little in Australia. By the late 1970s, however, computers and other modern technology were widely presented as 'job killers', and a threat to the Australian economy and social order.[52] The growing body of literature on the beneficial role of technology in the advanced economies, a strengthening Department of Science and Technology, and an awareness of the Silicon Valley, and other similar, spectacular commercial and technological breakthroughs, however, began a shift in the perception of the role of technology, and of science, in national economic performance. The election of a Labor government in 1983, and the appointment of a Minister of Science and Technology, Barry Jones, who had a strong commitment to the place of science in a modern society, established technology on the political and economic agenda in a previously unheard-of way.

The growing difficulties of the Australian economy, and the dramatic decline of the Australian dollar in 1986 completed the transformation. Technology was widely accepted as a central, and much neglected element in economic performance. In addition, its very importance had the effect of placing great pressure on the research system to deliver new technology more effectively and rapidly to the market-place. Whereas in 1975 the Science Taskforce of the Royal Commission on Australian Government Administration had been able to argue strenuously against a place for the planning of science,[53] by 1985, scientists were largely united in seeking ways to bend their research efforts more effectively to the advantage of the national economy.[54]

Throughout this period, ANZAAS, through its congresses and *Search*, continued to provide a vehicle for the expression of concerns about science, and the relating of science to social issues.

A number of issues raised wider concerns. Thus, a new society, Scientists Against Nuclear Arms (SANA), was founded in 1982, with the main aims of halting and reversing the arms race in nuclear, biological, and chemical weapons. Concern about the safety of recombinant DNA research led to the establishment of a committee to oversee and report on 'genetic engineering research', a responsibility later taken over by the Commonwealth Recombinant DNA Advisory Committee. Much of the work of this Committee has been oriented towards establishing appropriate safeguards so that recombinant DNA research and the commercialization of its results can proceed without hindrance.

A new focus of social responsibility pressure is the powerful animal rights lobby group concerned about the experimental use of animals. This movement has been a source of growing pressure on scientists to show greater concern for their animal subjects and to take into account large moral issues.

CONCLUSION

The perception of social responsibility held by scientists in Australia, as elsewhere, is shaped by the wider social context. The values of the scientific enterprise itself are rational, progressive, and optimistic. When such values are questioned, there is pressure to define and reassess the direction of research and its application. In this sense, the social responsibilities attributed to science mirror the social standing of science.

In Australasia, the colonization and development of a hostile land required a deep commitment to progress and belief in a better future.

The scientific spirit was admirably suited to providing the moral reinforcement for such a task, and it flourished in the pioneering eras of Australia and New Zealand. War, urbanization and affluence have, however, combined at various times to cast doubt on these scientistic values, when pursued in isolation. In these circumstances, since the war the scientific community, frequently in a very unreflective way, has found itself called upon to defend science, and to justify its contributions to society.

The centenary of ANZAAS is being celebrated at a time when the Australian economy is under great pressure, and the country is faced with the very real possibility of a substantial decline in living standards. Under these conditions, it is likely that the values of science should again reassert themselves. The achievement 'of unlimited economic expansion and improved human welfare' is no longer regarded as automatic. But that these benefits cannot be achieved without science is widely accepted. In its new guise, science is again at centre stage.

NOTES

1. J. Ben-David, *The Scientist's Role in Society* (Englewood Cliffs, New Jersey: Prentice Hall, 1971), 169–70.
2. ibid., 170.
3. For example, M. Brown, *The Social Responsibility of Science*, (New York: Free Press, 1971).
4. J. D. Bernal, *The Social Function of Science* (London: Routledge and Kegan Paul, 1939), 408.
5. See Sir Rutherford Robertson's account of the influence of Bernal and the ideas of the Social Relations of Science movement in 'Scientific Advice and Government Decision Making', *Leon Peres Memorial Seminar* (University of Melbourne, September 1985, mimeo).
6. O. U. Vonwiller, 'The Social Relations of Science', *Australian Journal of Science*, 1 (1938), 30.
7. Reported in 'Science and Society', *Aust. J. Sci.*, 1 (1939), 116–17.
8. ibid., 117.
9. For a detailed history, see Jean Moran, Scientists in the Political and Public Arena: a Social-Intellectual History of the Australian Association of Scientific Workers, 1939–49 (unpublished MPhil thesis, Griffith University, 1983).
10. O. U. Vonwiller, quoted in R. Howard, The Involvement of Scientists with the Social Relations of Science in Australia: the Nature of Social Responsibility in Science (unpublished Master of Science and Society thesis, University of New South Wales, 1978.)
11. Reported in 'Science and Society', *Aust. J. Sci.*, 2 (1939), 15–16.
12. Editorial, 'Science and Social Reconstruction', *Aust. J. Sci.*, 5 (1962), 2.
13. ibid., 2.
14. ibid., 4.
15. Moran, op. cit., note 9.
16. Reported in 'International Council of Scientific Unions', *Aust. J. Sci.*, 9 (1946), 71.

17. K. P. Barley, 'Security Checks the Scientist', *Aust. J. Sci.*, 19 (1957), 203–4.
18. J. L. Symonds, *A History of British Atomic Tests in Australia* (Canberra: Commonwealth Department of Resources and Energy, 1985), 433.
19. R. Johnston and J. Buckley, 'Post War Science and Politics in Australia' in R. Home (ed.), *Bicentennial History of Australian Science* (Canberra: Academy of Science, in press).
20. Reported in 'A Statement Issued by Scientists Meeting at Pugwash, Nova Scotia', *Aust. J. Sci.*, 20 (1958), 175.
21. Howard, op. cit., note 10, 36–7.
22. 'The Third Pugwash Statement and Questionnaire, *Aust. J. Sci.*, 21 (1959), 56.
23. ibid.
24. B. Commoner, *Science and Survival* (London: Ballantine, 1965), 3.
25. For example, J. Monod, *Chance and Necessity* (New York: Knopf, 1971).
26. For example, R. O. Slatyer, 'Man's Use of the Environment—The Need for Ecological Guidelines', *Aust. J. Sci.*, 32 (1969), 146; the language of this quotation demonstrates clearly the precedence then of environmentalism over sexism.
27. H. S. Albinski, *Politics and Foreign Policy in Australia* (Durham, North Carolina: Duke University Press, 1970), 145–6.
28. *Melbourne Herald* (5 October 1966).
29. 'Statement on the War in Vietnam by Australian Scientists', *Aust. J. Sci.*, 30 (1967), xlvi–xlvii.
30. ibid.
31. Commoner, op. cit., note 24, 73.
32. S. Rose and H. Rose, 'The Myth of the Neutrality of Science' in J. Fuller (ed.), *The Social Impact of Modern Biology* (London: Routledge and Kegan Paul, 1971), 223.
33. H. Marcuse, *One Dimensional Man* (London: Routledge and Kegan Paul, 1967).
34. T. Roszak, *The Making of a Counter-Culture* (London: Faber, 1971), 229.
35. J. Bronowski, 'The Disestablishment of Science' in Fuller op. cit., note 32, 233–46.
36. G. S. Stent, *The Coming of the Golden Age: A View of the End of Progress* (Garden City, NY: Natural History Press, 1969).
37. Howard, op. cit., note 10, 27.
38. Editorial, 'The Art of the Possible', *Search*, 1 (1970), 1.
39. C. Birch, 'Social Responsibility in Science', *Search*, 3 (1972), 110.
40. Their development is reviewed in Howard, op. cit., note 10 and in D. Biggins, 'Social Responsibility in Science, *Social Alternative*, 3 (1978), 54–60.
41. A detailed history is available: E. Baker, A Study in the Social Relations of Science in Australia; an Anatomy of the Society for Social Responsibility in Science (ACT) (unpublished Master of Science and Society thesis, University of New South Wales, 1980).
42. 1. To draw the attention of natural and social scientists to the social consequences and implications of scientific developments and to stimulate in scientists a sense of their social responsibilities, both as individuals and as members of groups.
    2. To draw the attention of the public to the social consequences and implications of scientific developments.
    3. To draw the attention of decision makers to the social consequences

and implications of scientific developments (SSRS–ACT constitution, 1970).
43. R. Johnston, 'Structural Silence in the Conduct of Science', in W. Green (ed.), *Focus on Social Responsibility in Science* (Christchurch: New Zealand Association of Scientists, 1979), 169.
44. *New Zealand Science Review*, 1 (1942), 10.
45. F. B. Shortland, 'A Short History of the New Zealand Association of Scientists', *New Zealand Science Review*, 36 (1979), 10.
46. Green, op. cit., note 43.
47. 'Science, Technology and Society', *Aust. J. Sci.*, 31 (1968), 203–20.
48. C. H. B. Priestley, 'Reply to Four Criticisms of the Academy Report on Atmospheric Effects of Supersonic Aircraft', *Search*, 3 (1972), 298–9.
49. K. A. W. Crook, 'The Academy Report on 'Concorde': An Exercise in Social Irresponsibility', *Search*, 3 (1972), 290.
50. O. H. Frankel, 'Genetic Conservation—a Parable of the Scientist's Social Responsibility', *Search*, 3 (1972), 198.
51. G. W. Ford (ed.), *Automation: Threat or Promise? Impact and Implications in Australia* (Sydney: Law Book Co., 1969).
52. Committee for Investigation of Technological Change in Australia, *Technological Change in Australia*, 4 volumes (Canberra: AGPS, 1980).
53. The argument is outlined in J. R. Philip, 'Towards Diversity and Adaptability: An Australian View of Governmentally Supported Science', *Minerva*, 16 (1978), 397–415.
54. See, for example, National Science and Technology Analysis Group, *Science and Technology in Australia: A Review of Government Support* (Canberra: Australian Defence Force Academy, 1986).

14

The Political Economy of Technology

Ted Wheelwright and Greg Crough

This chapter attempts to relate the development and use of technology since the First World War to the changing structure of the Australian economy, its relation to the world economy, and the resultant configurations of political and socio-economic forces, including the role of ANZAAS in the discussion. In the discussion of technology in its widest context, the AAAS has from the 1880s played an important role, in airing economic strategies, and in shaping and consolidating public attitudes towards the uses of technology in Australia. In many respects, however, that role has been conservative, even custodial. Today, however, ANZAAS has a responsibility to extend progressive discussion and to point towards the use of research in ways that will be appropriate to the best interests of the people of Australia and New Zealand.

In this chapter, technology is taken to mean the 'systematic application of collective human rationality to the solution of problems by asserting control over nature and over human processes of all kinds'.[1] Technology is to be seen as one special kind of human skill, the 'know-how' derived from scientific knowledge and incorporated in some object, process or activity. So defined, technology affects development in several ways: it can be a major source of new wealth creation, it can be an instrument allowing its owners to exercise social control in various forms, and it can affect modes of decision making and patterns of alienation.

It is becoming increasingly clear that as socio-economic activity becomes more complex and further removed from the simple utilization of natural resources, the humanly created resources of technology comes to be more important than natural resources.[2] This will be particularly important for the future development of Australia, given its historical reliance on the production and export of primary commodities. Therefore, the question of what kinds of science and technology should be fostered and how this should be done need to receive a much greater prominence in public debate than it has in the past.

SOCIAL FORCES SHAPING TECHNOLOGY

Traditional economic analysis is not very helpful in tackling the wide range of problems confronting the world. Such analysis cannot handle the long run; it cannot explain the fact of technological changes, nor cope with the interdisciplinary nature of socio-economic change. Economists tend to 'gloss over important issues of political economy through which technological changes may be viewed as an instrinsic part of the bureaucratic and political operations of the modern industrial economy'.[3] In addition, as Freeman has noted, the competition of 'new products, new processes and new systems is far more important than traditional price competition with equal technology and invariant conditions of production... the implications of placing technical innovation at the centre of the stage are very great indeed because it brings down the whole elaborate house of cards upon which orthodox economic theory depends'.[4]

A wider approach is essential, one that recognizes that 'our technology of production is in some sense the *result* of our social relations. Industrial innovation is a product of a historically specific activity carried out by social groups for particular purposes.'[5] These social relations include not only those between capitalist and worker, but also those between groups of workers themselves, and they can alter the balance in the choice of technology, particularly to secure managerial power and weaken trade unions.[6] In this connection, it is important to note that technology is usually defined as a male activity, that it can be a source of male power, and that there exists a new feminist analysis that is providing a new perspective on the social shaping of technology.[7]

Social relations, including those relating to the extent of monopoly or competition, can affect the framework of economic calculations relating to technological change. These presuppose a structure of costs, but they are affected by the way society is organized, which may have changed by the time a technological innovation comes onto the market, often a decade or so later. Property rights play an important role in a system where the social relations of production are based on private property. Intellectual property becomes a marketable commodity through the device of patenting, which bestows a temporary monopoly on its owners.

Class, culture, and education also fundamentally shape technology through their effect on social relations and the value systems of society. Veblen as early as 1915 showed how differences in the origin and nature of business classes in the United States, Britain, Japan, and Germany affected the *modus operandi* of the economic system, especially in relation to technology and innovation. As industry matured and developed into larger and larger units, businessmen began to lose touch with technology and came to be 'wholly absorbed in banking, underwriting, insurance, activities that have nothing to do with material equipment or technology or the tangible performance of industry'.[8] Stretton, writing in Australia in the 1980s, pointed out that among the hundred richest Australians 'luck, inheritance and predatory activities figure largely, wholesome productivity rather less...very few have invented anything or done much technical research and development; scarcely any have made any significant use of Australian science'.[9]

The interaction of these factors is complex and is sometimes best revealed by comparisons between countries. For example, a comparative study of agricultural development in Australia and Argentina has shown how the interaction of differing patterns of land ownership and the productivity of the land fostered or hindered the search for new agricultural technologies during particular periods.[10] Similar differences could be highlighted in a comparative study of the industrial development experiences of Australia, Canada, or New Zealand.[11]

THE ROLE OF THE STATE

The social relations of production are not the only forces shaping technology. It is arguable that state power, especially through the sponsoring of military technology, has been at least as influential in promoting technological innovation as other social forces. In Australia's case, the role of the state has been critical, although the technology sponsored was not primarily of a military nature, but in many cases essential for survival in a hostile physical environment. The development of the telecommunications system is a particularly interesting example, and the history of communications generally provides a good illustration of the critical role state enterprises, such as Telecom, play in Australia's economic and social development.[12]

The 'socialization' of the costs of agricultural research has been an important manifestation of the importance of the state. Schedvin has pointed out that one important consequence of a wool-based economy was that livestock production absorbed a disproportionate share of the nation's expenditure on scientific research and development.[13] At Federation, the newly developed state Departments of Agriculture concentrated their research on crops and viticulture since pastoralists were not their clients. In the inter-war period, research for the pastoral industry assumed a large part of the total effort, and for a considerable period after the Second World War research on wool fibre, sheep biology, and rabbits completely dominated the CSIRO's research expenditures.

The results have to a large extent been successful, although Australia continues to export its wool largely in an unprocessed form. While government support for such research allowed smaller farmers and graziers access to the latest technology at no direct cost, this situation may now be changing as the large corporations that dominate agricultural inputs increasingly privatize research, and federal government funding of the CSIRO continues to be reduced.

On the other hand, government finance for research undertaken for manufacturing industry in Australia was virtually non-existent until relatively recently. Despite substantial manufacturing expansion after the Second World War, there was hardly any relationship between that expansion and Australian-based industrial innovation. The dominance of many of the important sections of the manufacturing industry by foreign-controlled companies strongly reinforced this tendency. Many manufacturers sought tariff protection against imports rather than investing in new technology and equipment.

A further factor that needs to be taken into account in discussing the role of the state in Australia is the federal system of government. In many respects, the fact that political power is divided between the states and the federal government makes it more difficult for the country to develop an integrated national strategy for industrial and technological development. The experience of the postwar period shows how companies, particularly but not exclusively foreign companies, were able to play off the states against each other for the attraction of investment and employment opportunities. It is not surprising that many of these investments were approved with few conditions attached to them, and only rarely did such conditions relate to technology.

THE ROLE OF WARS

State funding for research, even for the rural sector, was begun in earnest because of the two world wars. For all practical purposes, the CSIRO began life as a delayed offspring of the First World War, and the 'imperial connection'.[14] It was modelled directly on the British Department of Scientific and Industrial Research, which was established during the First World War in recognition of Germany's technological superiority. In Australia, where the needs were economic rather than military, efforts were at first concentrated on rural commodities and with the incentive of the imperial preference the focus of the research was on agricultural products and processes.[15]

It was really not until 1936 that the CSIR began secondary industry research, and this was a result of government inquiries about establishing an engines research laboratory, which led to the appointment of a committee to survey the research needs to the whole of secondary industry. The Wimperis Report, presented in 1937, defined the chief areas of scientific activity needed to foster manufacturing production as scientific testing, standardization, research, and the wider provision of technical information to industry. As a result, an Aeronautics Research Laboratory was established in Melbourne in 1939, and a chair of Aeronautics was established at the University of Sydney in 1940.

Sir David Rivett, then the president of ANZAAS, feared that expenditure on research for secondary industry might endanger rural research, although extra funds were granted to the CSIR in 1938 for secondary industry research. The chairman of CSIR, Sir George Julius, addressed groups of leading manufacturers, and commercial and engineering groups, advising them about the existing problems. 'Many of them began to realise that whereas hitherto they had thought that there was little need for research and that all they needed for manufacturing processes was to obtain plans and designs from overseas to copy them, they now began to see research as an essential adjunct to manufacture.'[16] By the end of the Second World War, however, defence and industrial problems took a more prominent place in CSIR's activities, although it did not directly engage in defence or atomic research, which was left to the Weapons Research Establishment and the Atomic Energy Authority.[17]

More generally, the tremendous impetus given to technological development in secondary industry before, during, and after the Second World War is documented in Mellor's classic study *The Role of Science and Industry*.[18] It is clear that the record of Australia's wartime industry was remarkable, particularly in areas such as munitions, drugs, tropic proofing, radar, aeronautics, and engineering, and these developments profoundly affected the course of science and technology in subsequent years. A greater appreciation of the role and influence of institutions producing scientists and undertaking relevant research on industrial development was fostered during this period and was followed by an unprecedented expansion of the facilities for such training and research. Government funding of such research was stronger than ever before, and the 1950s saw the beginnings of a major expansion of the tertiary education system. In the immediate postwar period the Australian National University was founded, with strong medical and physical science research schools, and the New South Wales University of Technology was established.

The great postwar expansion of secondary industry in Australia was almost certainly based on the variety and complexity of manufacture and in technology generally that was telescoped into the war years, but would otherwise have taken much longer. A significant proportion of the country's resources had gone into a wide variety of buildings and equipment designed to produce the sinews of war, not just in the capital cities but also in rural areas. There were over 4000 wartime factories, and in due course many were sold or leased for peacetime production. Some were the basis of entirely new industries, such as special types of machine tools, scientific equipment, and new chemicals.

Of special importance was the establishment and expansion of the aluminium and automobile industries. In the case of the automobile industry, the crucial issue was the perceived need to create employment opportunities in the engineering and metal trades, which had expanded rapidly in wartime; it was thought they would severely contract if alternative uses were not found for these resources. The Secondary Industries Commission believed that the capacity to create an integrated motor vehicle industry existed, and that wartime experience showed that large overseas firms, who had handled large and difficult contracts, were in the strongest position to undertake it. As Schedvin has noted, 'in the event the government was persuaded that foreign technology could only be introduced to Australia if no significant conditions were imposed. This was tantamount to rejection of the efficiency argument in favour of the familiar trio: employment creation, industrialisation and population growth.'[19] As a result, GMH succeeded in having most of its own terms accepted as government policy. Undoubtedly, in later years, the establishment of an automotive industry based on American technology proved to be the pivot on which the expansion of secondary industry turned. By the mid-1980s, the international and domestic restructuring of the automobile industry had become one of the most important issues of industry and social policy facing governments in many countries, including Australia.

THE DISSIPATION OF THE EFFORT IN THE POSTWAR PERIOD

As Australia enters the latter part of the 1980s, however, it is becoming clearer that the indigenous manufacturing industry capability so laboriously built up over half a century was dissipated in the postwar period, and that the country is now dangerously dependent on the export of a relatively smaller number of primary commodities. With continuing reductions in protection, many parts of Australia's fragmented industrial structure are now feeling the full effects of international competition.

As numerous commentators have observed, Australia has a pitiful record in industrial research and development, and only three comparable OECD countries (New Zealand, Iceland, and Portugal) have a worse record for private sector expenditure on research and development. The proportion of gross domestic product (GDP) spent on business research and development was 0.21 per cent in Australia, 0.47 per cent in Denmark, 0.6 per cent in Canada, 1.26 per cent in Sweden, 1.35 per cent in the United States, and 1.5 per cent in Japan.

The factors behind such a poor performance are complex. Nevile has identified some of these as the poor investment climate, lack of financial incentives (although the 150 per cent taxation concessions introduced in 1986 should help), inadequate assistance for small or new businesses, the lack of an entrepreneurial ethos, managerial attitudes towards industrial relations, high levels of protection, and foreign ownership.[20] He noted that at least 40 per cent of foreign controlled companies undertaking research and development in Australia have controls on such expenditure by the parent company, that some are prevented from undertaking local research and development, and that others are restricted in the types of products they can produce and their level of exports.

Over the last decade or so, a significant literature has emerged showing that, by any criteria, Australia is now a technologically dependent country. Over 90 per cent of all patents lodged have been foreign; over 90 per cent of royalty and technical payments have gone overseas; most of the research and development carried out in large companies is by foreign-controlled ones, and the net cost of the 'technological balance of payments' was of the order of $A100 million a year (60 per cent of which goes to the United States), although the ratio of receipts to payments has been improving slightly in recent years.[21] As long ago as 1975, the Committee to Advise on Policies for Manufacturing Industry observed that 'Australia's technology is, in the main, derived from abroad and reflects the resource endowment and market size of its originators. Technology developed abroad is not always appropriate to Australian conditions and comparative advantage ...there is a feeling that the work done here concentrates on adapting basic work done abroad, limiting expansion of the skill base and opportunities for Australians below its potential.'[22]

It is, therefore, not surprising that the concept of technological sovereignty has developed in Australia. Grant has defined this as 'the capability and the freedom to select, to generate or acquire and to apply, build upon and exploit commercially, technology needed for industrial innovation'.[23] He argues that Australia's past failure to take the sovereignty factor into account has far-reaching implications for future industry and technology strategies. The federal political system in Australia has not, however, facilitated industry and technology strategies based on *national* sovereignty.

Grant claims that Australia is almost unique in its failure to recognize the critical role technological sovereignty has to play in achieving growth in manufacturing production and exports. The technology-intensive industries that have been developed in Australia are in many respects ill-equipped to compete in international markets because of their technological subservience. Although the exigencies of the Second World War led to the development of indigenous capability in many areas of secondary industry, much of it was lost in the postwar period by the foreign acquisition of local companies, the growth of licensed technology from abroad over which there is little local control, and failure of local industry to retain 'technological competence'. As a result, Australia's technology-intensive industries are now largely foreign-controlled and highly concentrated; by the end of the 1970s at least 77 per cent of declared payments for imported technology were being made by intra-company transfers.[24] A very select and dwindling group of Australian subsidiaries of transnational corporations made local innovations that were exploited internationally, the most prominent of which was probably ICI.

Noting that proprietary technology is less common in mining and agriculture than in manufacturing, Grant observes that technological sovereignty has particular significance for the service sector, which now accounts for approximately 70 per cent of GDP and employment in Australia. The service sector is the base of the high value-added exports, such as engineering and consulting services, computer software and its application, and financial skills. Although Australia built up a strong base of professional consultants in mining, exploration, engineering, including the Snowy Mountains Electricity Commission, in the immediate postwar period, this base was allowed to dissipate. The service sector is one of the fastest growing components of world trade, and Australia does not appear to be well-placed to benefit from these developments.

MECHANISMS OF TECHNOLOGY TRANSFER

By the early 1970s six countries—the United States, the Soviet Union, West Germany, Japan, France and the United Kingdom—accounted for about 85 per cent of the world's research and development expenditure of $US96 billion, and employed about 70 per cent of its scientists and engineers. By the beginning of this decade the expenditure reached $US150 billion, and the number of scientists and engineers involved was three million. At least one-quarter was for designated military programmes, and if the military-backed space research is added the figure would be about one-third.[25] Most of the private research is financed by transnational corporations, some of which spend more on research and development than entire countries.

There is a large and growing international trade in technology, which is increasingly becoming the main source of the power and prosperity of nations. In so far as it can be measured, international trade in technology was estimated to be at least $US15 billion by 1980. Few countries have a positive 'technological balance of payments', although the United States is the major exception. The technological lead of the United States in many fields is, however, fast disappearing, and large transnational corporations are no longer the exclusive channels of transfer of technology they once were.[26]

At least until the early 1980s it could be said that virtually no consideration was given to the problems associated with the international transfer of technology in Australia. Perhaps the first serious manifestation of increasing concern was its adoption as a theme of a workshop held in 1979 by the Australian Academy of Science. The general conclusion was that while the Australian development model of the liberal import of technology had generally been successful in producing rapid and balanced industrial development, disadvantages were beginning to emerge more clearly as import replacement ceased to be the driving force of development.[27]

More recently, the Australian Science and Technology Council (ASTEC), in its 1986 report on *Mechanisms for Technology Transfer into Australia*, found that 'Australia's dependence on overseas technology is, of necessity, high and very little can or should be done about this situation'. Its principal conclusions were that overseas technology is readily accessible and that effective transfer mechanisms exist, but the flow is inadequate because demand is limited, as many firms either do not make enough effort to tap into the international technology market, or do not have sufficient in-house research and engineering capacity to make use of it. This was reflected in a steady decline in capital expenditure and a slow rate of innovation over more than two decades.

A number of authors, however, have taken issue with some of ASTEC's recent reports. In particular, Johnston has criticized ASTEC for 'the conservatism and narrowness of its bureaucracy', and for its seeming inability to fully consider the social and economic costs of certain technological developments.[28] In many cases, the main beneficiaries of such developments were the researchers themselves, Australia's international scientific standing, the rural sector, and foreign companies and economies.

It is also revealing that in the many reports and studies prepared in recent years on the subject of technological development and technology transfer in Australia, very little mention is ever made of the voluminous studies made by the United Nations Conference on Trade and Development in Geneva, and its attempts to reach agreement on a Code of Conduct on the Transfer of Technology, which almost succeeded in 1985. The facts that these international deliberations seem to have been largely ignored and that Australia has played an insignificant role in relation to them on the world stage indicate a disturbing degree of complacency.

Part of the problem is that, in many respects, Australia is identified as a 'developed' country, with needs and values similar to the other industrialized capitalist countries, whereas it is in fact an 'in-between' country: a country with a relatively high standard of living based on the production and export of mainly unprocessed food and raw materials. Technologically speaking it is, in many respects, an under-developed country.

Action on the transfer of technology need not wait for a code of conduct, and many other countries have taken diverse initiatives. Technology transfer agencies have been established in at least fifteen developing countries to attempt to ensure that suitable technology is acquired under reasonable terms and conditions.

The role of the patent system in a country such as Australia needs to be critically examined. A study commissioned for the ASTEC report on *Patents, Innovation and Competition in Australia*, for example, concluded that the economic effects of the patent system to the innovative process in Australia are small and extremely subtle. Patents are not a significant determinant of research and development activity, but they do have some importance for small inventors. Patent information is a relatively unimportant source of research and development or technological information for most sectors, except large foreign-based transnational firms. The direct costs of the patent system can be quite high, and restrictive practices often occur. The study suggested that there is justification for reducing the negative effects of patents by stricter examination, by reducing the length of the term and scope of patent monopolies, and by dealing with restrictive practices. Reform is not easy because 'those who would lose by it are concentrated, powerful and active defenders of their interests'.[29]

It is also important for Australia, given its geographical location, to be aware of developments taking place in the Asian–Pacific region. In particular, it should be noted that Japan has now become the prominent supplier of technology to the Asian–Pacific countries. The costs of imported technology to some countries have been growing five- or sixfold over the last decade, and in many instances technology sales now provide steadier and quicker returns then those from invested capital. A number of countries in South-East Asia, including Singapore, South Korea, and Taiwan Province, have been able to achieve transfer of technology on favourable terms from both the United States and Japan, and have experienced rapid export-oriented industrialization.

A NATIONAL TECHNOLOGY STRATEGY?

There has been considerable discussion in recent years about the need for a national technology strategy, a debate that has been considerably assisted by the efforts of the energetic Minister of Science, Barry Jones, whose book *Sleepers, Wake!* has been widely read and discussed.[30] The process of developing such a strategy by wide-ranging discussion began in September 1983, with the National Technology Conference and the preparation of a draft Strategy statement. One problem with the draft is that it basically sees government economic strategy in terms of assistance to support technological innovation in line with industry's perception of priorities—according to 'market forces'.[31]

The discussion in the scientific and technological community about the strategy is often reflected in the pages of *Search*. Reports of the National Meeting of Concern on Science and Technology in April 1985 clearly indicate that the science and technology community believed it had failed to have any significant impact on government policy and had only poor communication with industrial management. Accordingly, a new organization was established, the Federation of Australian Scientific and Technological Societies, in an attempt to overcome some of these problems.[32]

More recently, in November 1986, a National Science and Technology Forum was organized by the Australian Academy of Science, the Australian Academy of Technological Sciences, the Federation of Australian Scientific and Technological Societies, and the Institution of Engineers. This took as its major premises that:
1. Australia is at the economic cross-roads and its future economic direction will depend critically upon the success or failure of its national policies on science and technology;
2. fundamental structural changes are taking place in Australia's relations with the global economy; and
3. a new dialogue with government needs to be created, and scientists need further self-education in the realities of science and technology policy

Of all the numerous reports on science and technology in Australia, the 1984 report of the OECD is probably the best, and it has certainly influenced the federal government's thinking on the development of such a national strategy.³³ It begins by noting that, at the time of the OECD's previous report in 1974, Australia had virtually no policy for technology. By 1984 there had been established an extensive collection of science and technology policy instruments at federal and state level. The challenge for the 1980s was to make these new policies and instruments work to stimulate innovation. The challenges were international, and each principal theme identified in Australia was common to almost all OECD countries.

The previous decade was seen as a period of disappointment, and even deterioration. The Report noted, for example, that in the period 1969–81 industry funded research and development had fallen from 0.48 per cent of GDP to 0.21 per cent. Total research and development had fallen from 1.34 per cent to 1.01 per cent of GDP. The comparable current figure for all OECD medium performers was 1.64 per cent of GDP, of which 0.93 per cent was funded by industry. It is possible that technology may have been less important to Australian industry than it was in the early 1970s. Another indicator was the small and declining proportion of tertiary education students enrolled in engineering and technology subjects—8 per cent. Funds for university research had not increased in real terms.

The diversity of technology is particularly important. The value of technology can only be understood by considering the various sectors of the economy and their requirements. Clearly policies for technology and growth must be appropriate to a country's historical and geographical strengths, not to some 'international norm' of industrial structure. A sector-by-sector approach is recommended, with some emphasis on the public sector as the major user of new technology, and the major employer of technically skilled people. The OECD report emphasized the need to promote the understanding of the relationship between technology, science, and economic well-being, as there seemed to be less agreement on these issues in Australia than in other industrialized countries.

There are, however, limits to the effect of science and technology policies, because the condition of a country's science and technology effort is determined by a complex of cultural, social and psychological factors. Hence there is a tension between what is expected and what can realistically be achieved. The Australian attitude to technology was seen to be almost one of alienation, which may be due in part to the fact that most of the techniques used in industry are imported from overseas, and that there is a tradition of importing technical and professional workers. Technical procedures are often seen as being imposed from outside the office or factory, with minimal consultation, and access to technology and interest in technical questions seem to be mainly male functions. This attitude continues to lead to the consistent undervaluation of national technological achievements and possibilities.

Similar debates have been taking place in other countries, including New Zealand. In a paper delivered to the ANZAAS Congress in 1984, for example, Ellis and Hunt point out that science planning is not new in New Zealand, but the May 1984 plan was the first step in developing

TABLE 14.1

Total Commonwealth government expenditure on research and development by socio-economic objective (constant (1985–6) prices)

Objective Category	78–9*	79–80*	80–1*	81–2	($m) 82–3	83–4	84–5	85–6	Proj. 86–7 Preliminary
National security									
Defence	146.05	144.30	147.06	133.82	134.62	135.40	141.34	144.2	143.8
Economic development									
Agriculture	128.01	157.71	166.89	161.67	154.89	129.59	134.87	154.1	153.7
Other primary industry	22.89	24.41	22.15	22.80	23.37	29.62	26.63	29.9	29.4
Mining	28.23	22.58	26.03	30.11	24.09	24.93	23.88	22.9	23.1
Manufacturing	124.67	133.11	153.37	114.19	138.54	129.22	136.97	170.9	162.4
Construction	14.77	12.89	11.39	9.97	9.81	9.21	10.95	14.3	15.5
Energy	53.07	556.92	63.39	68.88	82.08	75.80	67.73	59.1	47.6
Transport	8.72	8.32	6.84	4.58	4.19	4.00	3.64	4.0	3.5
Communications	50.40	47.80	57.02	53.01	55.85	52.33	53.23	59.3	63.9
Other economic services	23.28	25.00	26.58	22.56	21.37	25.59	32.45	21.7	29.1
Sub-total	454.0	488.8	533.7	487.8	514.2	480.3	490.3	536.2	528.3
Community welfare									
Urban & Regional Planning	2.65	3.01	2.11	2.31	3.46	2.91	2.73	0.3	0.3
Environment	30.04	33.71	30.69	32.53	43.37	35.63	32.83	29.3	30.0
Health	49.05	50.86	52.94	58.90	66.18	74.76	83.56	88.3	91.4
Education**	6.67	5.52	6.05	5.07	3.05	3.36	4.14	4.4	4.1
Welfare	1.66	2.01	2.60	2.64	3.20	1.44	2.05	2.7	3.1
Other community services O/seas development assistance	22.79	20.84	22.39	26.85	36.87	36.92	40.21	43.9	42.3
Other commercial services	4.29	4.33	5.04	4.39	4.49	6.29	9.80	7.0	5.5
Sub-total	117.1	120.3	121.8	132.7	160.6	161.3	175.3	175.8	176.7
Advancement of knowledge									
Earth, ocean and atmosphere	51.27	54.87	63.67	60.74	73.72	77.30	67.60	52.6	51.7
General advancement of knowledge	171.73	167.36	165.71	166.23	182.21	204.14	192.51	216.5	221.5
Sub-total	223.0	222.2	229.4	227.0	255.9	281.4	260.1	269.1	273.2
Total	940.2	975.6	1031.9	981.3	1065.4	1058.4	1067.1	1125.3	1122.0

Note: *Before 1981–82, there were differing superannuation arrangements applying to certain statutory authorities (especially CSIRO and AAED) with the effect that superannuation payments were not included in the expenditure for such authorities. In this table notional increases have been applied to data for 1978–9, 1979–80 and 1980–1 for trend comparison purposes. (See Science and Technology Statement 1984–5 for data without such notional increases.)

**Research and development funded by the Minister for Education for the purpose of producing qualified researchers or for supporting normal academic activities has been included in 'General advancement of knowledge'. Only research mainly directed towards education processes or education administration has been included in the 'Education' objective.

Source: *Science and Technology in the Budget*, (Canberra: National Science & Technology Analysis Group, November 1986).

comprehensive national priorities. Political concern that scientific and technological effort should be relevant to the goals of economic development was growing, and hence growing support for the incorporation of technology policy in overall government planning.[34]

Any technology strategy adopted by Australia will, to some extent, impinge on New Zealand and vice versa, especially following the implementation of the Closer Economic Relations Agreement. For example, financial incentives for research and development in Australia will attract some New Zealand companies to establish branch plant facilities there. As both countries have to compete in the Asian–Pacific region, there is much to be said for Australasian co-operation. The two countries should look to promoting the most efficient use of their limited research and development resources, particularly in areas where both countries already have strong research programmes underway, such as oceanography, agricultural waste utilization, environmental monitoring, and plant and animal diseases. In general, however, despite the existence of ANZAAS, Australian–New Zealand academic relations tend to be weak and need strengthening.[35]

While it is becoming clearer that countries need to undertake some form of science and technology planning, governments in both Australia and New Zealand have, in recent years, deregulated large sectors of economic activity, eschewed national planning, and exposed their economies even more to the 'market forces' of world capitalism. How national planning of technology is to take place in such circumstances is far from clear. Nor is it clear what purpose it is to serve—other than to make parts of the economy more 'competitive'.

CONCLUSION

The development of technology and attitudes towards it reflect Australia's origins as a colonial economy, having for at least its first hundred years large export sectors dominated by livestock and grain production. Its class of pastoral and agrarian capitalists was strong, while the indigenous industrial capitalists were relatively weaker. Complacent attitudes towards technology and technical education were inherited from the imperial power.

Technology was only fostered on any significant scale when the empire was threatened in the First World War, through the creation of a publicly funded research organization catering mainly for the rural sector. Technology in the industrial sector was stimulated to some extent by that war, particularly through the creation of a munitions industry, but not until the Second World War was there a substantial manufacturing industry with a significant indigenous technological capability. Hence the most important factor fostering this was the role of the state in relation to two world wars, and to a large extent the Depression of the 1930s. Consequently by far the greatest part of investment in research and development was in the public sector.

In the 1950s this capability began to be dissipated, as Australia industrialized further to meet the needs of a growing domestic market, boosted by immigration, on the basis of substantial uncontrolled foreign investment by transnational corporations. By the early 1980s it was clear that Australia had lost most of whatever technological sovereignty it had been able to achieve and was dependent in key areas on the transfer of technology from these corporations. Little attempt was made to monitor this transfer to ensure that the benefit to Australia was maximized, nor to make some foreign investment conditional on suitable research and development expenditure domestically. Australia played virtually no role in the international controversy over these issues.

The first serious attempts to develop a national strategy on science and technology were made in the early 1980s, but they were hampered by the dogmatism of economists who regarded 'market forces' as sacrosanct, public servants oriented more towards bureaucratic regulation than national development, politicians (particularly state politicians) subservient to foreign capital, the elitism of organizations of scientists and technologists, and archaic political processes.

Nevertheless some progress was made, particularly in private expenditure on research and development, which doubled between 1982 and 1985, and was expected to equal government-funded research in universities and the CSIRO in 1987. This was largely as a result of the 150 per cent tax deduction, which by subsidizing research and development in effect made Australia one of the cheapest places in the world to undertake research. Promising areas include scientific instruments, especially for medical use, biotechnology, advanced ceramics, new aluminium alloys, and optical fibres.[36]

It is no coincidence that these attempts to develop a national strategy came at a time when the economy was in serious difficulty, its export base having been eroded by irreversible changes in the structure of the world economy, some of them caused by changes due to technology that significantly reduced Australia's comparative advantage in the production of traditional export items. These advantages stemmed from the possession of vast quantities of easily extractable natural resources. Technology inevitably produces substitutes, as in the case of synthetic fibres for wool, or corn syrup for sugar; or it reduces the demand by fostering more economical usage, enabling steel to be made with less iron ore and lower quality coal; or it replaces natural materials by those that are scientifically designed specifically to suit new products and processes, inverting the traditional relation by which products were designed to suit the raw materials available.

The very foundations of Australian prosperity are under threat; in 1986-7 its current account deficits, its gross foreign debt, and the servicing costs were at levels unprecedented since the Great Depression of the 1930s. Unfortunately, for many policy makers technology has come to be regarded as a kind of saviour, a 'cargo cult', that could arrest and turn around the result of decades of misguided and mismanaged 'development'. This is far too great a burden for it to bear: it can make an important contribution, but does not ultimately offer a solution.

In all of these deliberations ANZAAS has been virtually bypassed: it has had no effective role, probably because for some time it has had no organic connection with an increasingly pecuniary culture that is now dominated by paper entrepreneurs, takeover merchants, tax avoiders, and others of that sort who can make far more money from such activities than by investing in technology, even at the new subsidized rates.

There has also been little discussion of the central issues raised some years ago by H. C. Coombs:

> It is necessary, I believe, to call into question the basic assumptions on which industrial society rests—that increasing quantities of commodities of themselves add to human welfare, that the indirect costs of the division of labour, specialisation and the domination of human affairs by the market can be safely ignored...and whether we must acquiesce in growing concentrations of power which, from our experience, we know will be abused.[37]

There seems to have been minimal discussion of the issue of who benefits from the so-called planning of technology. Stretton has noted that new wealth is not created at the point technology is introduced (which halves the workforce); this will happen 'only if you can re-employ the disused half of your labour or other resources in further production'; Stretton questioned whether market forces can adequately perform this task. If this does not happen, the result is not any great increase in wealth but a radical redistribution.[38]

These are some of the central issues of the political economy of technology in Australia that have not been faced or resolved by all the conferences, reports, and plans thus far. Until they are the success of these strategies is problematic, as Barry Jones has suggested:

> This could be a period for using technology creatively to expand human capacity and enlarge human happiness at both the individual and collective levels. The responsiveness of our political systems is so slow that I see little chance of it happening.[39]

NOTES

The authors would like to thank Wayne Pash and Cristine Cifuentes for their research assistance for parts of the paper.

1. D. Goulet, *The Uncertain Promise: Value Conflicts in Technology Transfer* (New York: IDOC, 1977), 6.
2. See the special issue on 'Materials for Economic Growth' in *Scientific American*, 255 (October 1986).
3. N. Clark, *The Political Economy of Science and Technology* (Oxford: Blackwell, 1985), 229–30, 137.
4. C. Freeman, in the foreword to E. Arnold, *Competition and Technological Change in the Television Industry* (London: Macmillan, 1985), xiii.

5. D. MacKenzie, and J. Wajcman (eds), *The Social Shaping of Technology* (Milton Keynes: Open University Press, 1985), 70.
6. For examples from the mining industry, see K. Tsokhas, *Beyond Dependence: Companies, Labour Processes and Australian Mining* (Melbourne: Oxford University Press, 1986).
7. MacKenzie and Wajcman, op. cit., note 5, 10–11.
8. See J. P. Diggins, *The Bard of Savagery, Thorstein Veblen and Modern Social Theory* (New York: Seaberry Press, 1978).
9. H. Stretton, 'The Quality of Leading Australians' in S. R. Graubard, *Australia: the Daedalus Symposium* (Sydney: Angus & Robertson, 1985), 201.
10. R. E. Gerardi, *Australia, Argentina and World Capitalism: A Comparative Analysis, 1840–1945* (Sydney: Transnational Corporations Research Project, 1985).
11. See for example W. Armstrong, J. Bradbury, 'Industrialisation and Class Structure in Australia, Canada and Argentina: 1870 to 1980' in E. L. Wheelwright and K. Buckley (eds), *Essays in the Political Economy of Australian Capitalism*, vol. 5 (Sydney: ANZ Book Company, 1983).
12. See A. Moyal, *Clear Across Australia: A History of Telecommunications* (Melbourne: Nelson, 1984).
13. C. B. Schedvin, *The Australian Economy on the Hinge of History*, Second Henry George Memorial Lecture, Macquarie University, Sydney (October 1986).
14. G. Currie and J. Graham, *The Origins of CSIRO* (Melbourne: CSIRO, 1966), 11.
15. R. Johnston, 'Political, Administrative and Policy Aspects of Science and Technology' in A. T. A. Healy (ed), *Science and Technology for What Purpose?* (Canberra: Australian Academy of Science, 1979), 98.
16. Sir George Currie and J. Graham, 'CSIR, 1926–39', *Public Administration*, XXXIII (1974), 240–1.
17. D. M. Lamberton, *Science, Technology and the Australian Economy* (Sydney: Tudor Press, 1973), 21.
18. D. P. Mellor, *The Role of Science and Industry*, vol. V (Civil Series of Australia in the War of 1939–45 (Canberra: Australian War Memorial, 1958).
19. C. B. Schedvin, 'The Formation of Post-War Motor Vehicle Policy in Australia', *Monash Papers in Economic History*, 2 (Melbourne: Monash University Economic History Department, 1976), 2.
20. J. Nevile, 'The Disaster of Private Sector Research and Development in Australia', *Science, Technology and the Economy*, Occasional Paper no. 11 (Sydney: University of New South Wales, 1986).
21. See, for example, Report of the Senate Standing Committee on Science and the Environment, *Industrial Research and Development in Australia* (Canberra: Australian Government Publishing Service, 1979); Report of the Trade Development Council, *Export Franchise Restrictions* (Canberra: Department of Trade, July 1983); Report of the Industrial Property Advisory Committee to the Minister for Science and Technology, *Patents, Innovation and Competition in Australia* (Canberra: Australian Government Publishing Service, 1984); *Science and Technology Statement 1985–86* (Canberra: Australian Government Publishing Service, 1985).
22. Committee to Advise on Policies for Manufacturing Industry (Jackson Committee), *Policies for Development of Manufacturing Industry*, vol. 1 (Canberra: Australian Government Publishing Service, 1975), 98–9.
23. P. Grant, 'Technological Sovereignty: Forgotten Factor in the 'High-Tech' Razzamatazz', *Prometheus*, 1 (December 1983), 240.

24. S. Murray-Smith, 'Technical Education in Australia: A Historical Sketch' in E. L. Wheelwright (ed), *Higher Education in Australia* (Melbourne: Cheshire, 1965), 175.
25. C. Norman, *Knowledge and Power: The Global Research and Development Budget*, Worldwatch Paper no. 31 (Washington: Worldwatch Institute, 1979).
26. See R. Vernon, *Exploring the Global Economy* (Boston: Harvard College, 1985), ch. 2.
27. Healy, op. cit., note 15.
28. R. Johnston, quoted in A. Moyal, 'Science Research in Australia: Who Benefits', *Search*, 14 (7–8) (August–September 1983), 174.
29. T. D. Mandeville, D. M. Lamberton, and E. J. Bishop, *Economic Effects of the Australian Patent System* (Canberra: Australian Government Publishing Service, 29 August 1984), 79–80.
30. B. O. Jones, *Sleepers, Wake! Technology and the Future of Work* (Melbourne: Oxford University Press, 1982).
31. 'Science in Government: The Draft Technology Strategy', *Search*, 15 (5–6) (June–July 1984), 141.
32. F. Smith, 'National Meeting of Concern on Science and Technology', *Search*, 16 (5–6) (June–July 1985), 131.
33. Organization for Economic Cooperation and Development, *Review of National Science and Technology Policy: Australia* (Paris: OECD, 1985).
34. A. J. Ellis and D. M. Hunt, 'Technology Planning: The New Zealand Viewpoint', *Search*, 15 (7–8) (August–September 1984), 217–8.
35. D. West, 'Australian/New Zealand Academic Relations Weak', *Search*, 15 (9–10) (October–November 1984), 253.
36. P. Roberts, *Australian Financial Review* (31 December 1986), 10.
37. H. C. Coombs, 'Science and Technology—For What Purpose?', in Healy, op. cit., note 15, 35–6.
38. H. Stretton, 'Conference Luncheon Address', in National Technology Conference, *Proceedings and Report of September 1983 Conference* (Canberra: Australian Government Publishing Service, 1984), 206.
39. B. Jones, 'Policies for Promoting Promising Technologies and Industries', *Search*, 17 (1–2) (January–February 1986), 6.

15

Technology, Employment and Post-Industrial Society

Sol Encel

Despite its central concern with the 'advancement of science', ANZAAS has a significant history of concern also with the social, economic, and political implications of science and technology. In the last twenty years, particularly, there has been a high level of interest in the effects of technological change on work and employment, which is also the major theme of this paper.

Much of today's thinking about science, technology, and employment stems from ideas developed in the 1930s under the influence of the world depression and the experience of prolonged mass unemployment. One of the first people to use the term 'technological unemployment' was J. M. Keynes in an essay written in 1930, where he defined it as 'unemployment due to our discovery of means of economising the use of labour outrunning the pace at which we can find new uses for labour'.[1] Not long afterwards, the subject was given an airing at the twenty-first ANZAAS Congress, held in Sydney in 1932. Professor T. H. Laby, undoubtedly the most distinguished physicist in Australia at the time, opened a general discussion on 'Science and the Depression' with an address that still after more than half a century, commands attention for its breadth and foresight.[2]

ANZAAS, he observed, was a particularly appropriate venue to discuss these questions because it represented both the natural and the social sciences. Economists, however, were at fault because they had neglected the role of labour-saving machinery in producing unemployment. Nevertheless, he did not lay the principal blame for unemployment on technological progress. Rather, the problem lay in an economic system that was not able to avail itself of greatly increased productivity and in the rise of economic nationalism, which would only 'defeat the triumphs of science in improving transport, facilities for travel, and communications between nations'. Nationalistic rivalries, which had led to the First World War, had also led to the use of scientific discoveries to effect destruction on the greatest scale the world had ever known.

Only because of this conflict between the ideals of nationalism and science does it seem possible to account for the failure of scientific progress and discovery to prevent the present impoverishment of the world... Man has not yet learnt to put science to its best uses: he has always first used a discovery for a destructive end, and he has learnt little of the spirit which has actuated the greatest scientists which at its best has been a desire to serve all mankind.

Laby's emphasis on the link between mechanization, productivity, and employment, his attack on economic nationalism, and his apprehension about the destructive potentialities of scientific discovery have a strong contemporary flavour. An updated version of his address, apart from drawing on a wealth of literature that did not exist in 1932, would also include the more recent concern about environmental problems that is absent from his speech. Those of us who have made efforts to advance the study of science, technology, and society as part of the school curriculum will pay special attention to his peroration:

It is evident that the progress of science has created more social and political problems than mankind has been able to solve. Scientific discovery and its application shows no sign of diminishing, but rather of increasing, and social and political problems thereby created will increase... The effect of science on the work, wealth and happiness of man, and the relation of nations, is a subject very much in need of study... Should not our secondary schools and universities consider seriously whether the educational system is training the people and their leaders in the best way possible for them to solve the very difficult and complex problems of the modern world?

Another physicist, Professor Oscar Vonwiller, picked up this theme again in the first issue of the *Australian Journal of Science* in 1938. Scientists should be concerned, he wrote, about the misuse of science. 'The scientist sees, as the result of invention based on scientific research, revolution in methods of manufacture, without adequate economic adjustment, followed by widespread distress as the increased efficiency causes unemployment far in excess of employment added by the development of new luxuries and necessities.'[3]

The pessimism of the 1930s was overtaken by the prospect of apparently endless expansion following the end of the Second World War. The role of science and technology in stimulating this expansion was expressed in such exuberant statements as Vannevar Bush's report to President Franklin Roosevelt, *Science—the Endless Frontier*. It was a long time before the problems generated by rapid technological change and its effects on employment and working life began to receive serious attention. It was not until 1957 that a paper dealing with the effects of automation was delivered at an ANZAAS Congress in Dunedin.[4] The significant change came in the 1960s, partly as a result of the rapid expansion of the social sciences in the new universities. By 1967 the New South Wales division was able to reflect the new mood by organizing a full-scale symposium, whose proceedings were later published in book form.[5] The contributors stressed that automation was already occurring at a rapid rate, but its specific local consequences had received very little systematic inquiry. A further collection, entitled *Redundancy* and consisting of papers originally published in *Search* dealing with the 'post-industrial challenge', appeared in 1973.[6] Paraphrasing the title of the well-known best-seller by Alvin Toffler, James Davenport, the editor of *Search*, entitled his preface 'Present Shock'.

The real shock, however, was yet to come, with the reappearance of significant levels of unemployment, which rose steadily from 1973 onwards. This was reflected in the title of a further conference organized by the New South Wales division of ANZAAS in 1979 under the title of 'Automation and Unemployment'.[7] Opening the conference, the New South Wales Minister for Industrial Relations, Technology and Energy, P. D. Hills, expressed his concern that there were no reliable figures about the number of people who had lost their jobs because of technological change; he declared that his department would be paying close attention to the collection of such information. The same theme recurred in the addresses of a number of other speakers. A. W. Goldsworthy, a leading figure in the Australian Computer Society, called for a survey of the impact of computing and related technologies on employment and the structure of the workforce.[8] Guest speaker at the conference was Professor Christopher Freeman, director of Sussex University's Science Policy Research Unit (SPRU). Since its establishment in 1965, the Unit has been a major centre of research on the social effects of scientific and technological change, and since the mid-1970s it has been the source of numerous publications dealing with the employment implications of industrial innovation. In his address Professor Freeman paid particular attention to the effects of developments in microelectronics. The intelligent use of microelectronics could make possible the elimination of monotonous, boring, and dangerous jobs, but there was great danger of a 'creeping growth' of unemployment to much more serious levels than those already encountered.[9]

Not all writers on the subject are pessimistic. One of the best-known expressions of optimism concerning the growth of science-based technology is to be found in the concept of the 'post-industrial society', associated particularly with the writings of the American sociologist Daniel Bell. This concept has had a considerable vogue in Australia, so we may look briefly at the nature of the idea and the criticisms that have been levelled at it.

THE CONCEPT OF THE POST-INDUSTRIAL SOCIETY

Since Daniel Bell coined the term 'post-industrial society' in the 1960s, its constant use as a metaphor to denote a fundamental change in western capitalist society has reduced its status to that of a cliché. The term has also stimulated the proliferation of similar metaphors, including post-modern, post-service, post-capitalist, and post-bourgeois, as well as a vast critical literature. A sign of its gradual fall from grace is the growing frequency of allusions to the 'so-called post-industrial society'.

Richard Badham has recently suggested that the concept represents only a minor extension of the idea of industrial society, which dates back to St-Simon's book *L'Industrie* of 1812. Badham identifies five characteristics as constituting post-industrial society:

(a) a transition from the 'industrial' to the 'service' economy
(b) substantial growth of white-collar, professional and technical occupations
(c) the rise of information technology as the dominant branch of technological activity
(d) the key role of scientific and theoretical knowledge
(e) growth of concern with leisure and the 'quality of life'.[10]

These five characteristics do not represent a decisive break with the past, and most of them were foreseen by nineteenth century writers, including Henri de St-Simon, Max Weber, and Emile Durkheim. This is not wholly true as regards information technology, whose spectacular development was not predictable until the advent of electronics and quantum mechanics. Hence the concept of the 'information society' has emerged as a subset of the broader theory of post-industrial society. We shall return to this point later.

Heilbroner has also disputed the novelty of the post-industrial concept, maintaining that the 'post-industrial future' should be regarded as a stage of capitalist development and not as a step beyond capitalism. It will, he remarks, 'manifest many of the structural attributes of industrial capitalism, including concentrated economic power and wealth, a highly unequal distribution of pre- (and probably post-) tax income, and macro-malfunctions and misallocations of resources that arise from the predominance of the market as the principal allocatory mechanism'.[11]

Bell's picture of the post-industrial society was essentially romantic, reflecting the kind of optimism about the beneficial effects of science and technology shown by earlier writers such as H. G. Wells, J. D. Bernal, and Vannevar Bush. The 'axial principle' of post-industrial society, according to Bell, is the 'centrality of theoretical knowledge', and its roots lie in the 'inexorable influence of science on productive methods.'[12] As knowledge becomes the prime resource, so the professionally trained expert will replace the entrepreneur and the financier. In Bell's own words:

If the dominant figures of the past 100 years have been the entrepreneur, the businessman, and the industrial executive, the 'new men' are the scientists, the mathematicians, the economists, and the engineers of the new computer technology. And the dominant institutions of the new society—in the sense that they will provide the most creative challenges and enlist the richest talents—will be the intellectual institutions. The leadership of the new society will rest, not with businessmen or corporations as we know them (for a good deal of production will have been routinized) but with the research corporation, the industrial laboratories, the experimental stations, and the universities.[13]

Those are fine words, and when they were written they may have sounded plausible. Few people would now endorse that kind of optimism, even if they take a generally cheerful view of the future. As one of Bell's American critics remarked sardonically, the post-industrial society 'was a period of 2 or 3 years in the mid-sixties when G.N.P. [gross national product], social policy programmes, social research and universities were flourishing. Things have certainly changed.'[14]

The theoretical inadequacy of the views put forward by Bell and like-minded writers such as Clark Kerr and Herman Kahn has generated a wide range of criticism, which I have examined elsewhere.[15] Some aspects of this inadequacy may be mentioned briefly. First, the evidence for a shift of employment from manufacturing to 'services' is based on a simplistic use of statistics. Heilbroner warns against the notion of a 'massive emigration from industrial work'.[16] Gershuny has demonstrated in detail that Bell's impressive array of statistical tables dealing with occupational and educational changes does not support his concept of the 'service economy'.[17] Second, the original view of post-industrial society makes virtually no mention of the problems of unemployment caused partly by technological change and partly by long-term structural trends in the international economy. Again, the post-industrial theorists claim that technological change has become the major moving force in contemporary society, causing the obsolescence of traditional class differences. The evidence for this contention is, to put it mildly, highly disputable. As Barry Jones comments, technological change is not a neutral force, but carries with it the potentialities of intensifying the worst aspects of a competitive society, of widening the gap between rich and poor, and of leading to a technocratic authoritarianism.[18]

Jones's criticism echoes the earlier remarks of J. H. Goldthorpe, who describes the underlying philosophy of Kerr and Bell as 'technocratic historicism', which assumes that western-style political and economic systems are the norm for the future development of other societies. This new style of historicism, like the earlier versions attacked by Popper, is based on a conception of historical inevitability that rules out a variety of theoretical and normative alternatives.[19]

The theory of post-industrial society also ignores the role of international conflict and the arms race in generating a large part of the scientific and technological activity that this theory ascribes to industrial development. The 'military-industrial complex' identified in President Eisenhower's farewell speech in 1961 figures little, if at all, in the writings of the post-industrial theorists. The aerospace industry, a prime example of the new technologically advanced sectors, would virtually cease to exist if it were not underwritten by governments. Many of the major advances in electronics, solid-state physics, nuclear engineering, and computing have depended on the military budget.

Finally, a theory that takes industrial development as the central process of change cannot account for the social and political history of the Third World, whose inevitable 'modernization' is a corollary of theories about industrial and post-industrial society. Modernization has indeed taken place in the traditional societies of the Third World. But, like other social changes, it has evoked strong reactions and violent conflicts. Instead of the implementation of development plans by modernizing élites, many Third World countries are governed by military dictatorships whose prime interest is to stay in power and enhance their privileges. Attempts at industrialization have worsened the economic situation of many 'developing countries' and generated the North–South problem, a new metaphor that underlines the fact that world poverty is growing rather than decreasing.

Most spectacular of all is the emergence of Islamic fundamentalism. Whereas the emergence of military dictatorships and other corrupt, repressive regimes underlines the incidental difficulties of modernization, fundamentalist Islam poses an ideological challenge of the most basic kind to the very idea of modernization itself. Theories based on the assumption that the scope of rational thought will steadily increase have no answer to such phenomena as deliberate martyrdom on the part of Shi-ite terrorists, emulating their predecessors, the original 'assassins'of the eleventh and twelfth centuries. A recent analysis of the Iranian revolution of 1979 and its galvanic effects on Islamic society in the Middle East argues that the violence and fanaticism of militant fundamentalism can only be understood as a deliberate rejection of the values of western culture, which is seen as a force destructive of Islamic values and independence.[20]

Partly because of the ambiguities and inadequacies inherent in the concept of post-industrial society and partly because of the spectacular expansion of information technology, there has been a shift of attention to the 'information economy' and the 'information society'. This is clear in the work of the Japanese writer Y. Masuda, who first attracted attention with his *Plan for an Information Society*, published in 1971. In a more recent book, Masuda rejects 'post-industrial' as an excessively vague term in favour of the 'information society', which will be characterized by the fact that 'the production of information values and not material values will be the driving force'.[21] This society, which he also calls 'Computopia', will move away from centralized government and class hierarchies to a system of participatory democracy, horizontally functional, maintaining social order by autonomous and complementary functions of a voluntary civil society. M. Marien, in a critique of 'Computopia', notes that it is but one of many labels generated by the advance of information technology, none of which has lasted very long. He identifies eight areas in which the growth of information technology is likely to have ambiguous or conflicting consequences: work, commerce, health, entertainment, education, politics, intergroup relations, and family life.[22] The outcome in these and other areas will be decided not simply by the technology, but also by economic, social, political, and cultural factors—the very opposite of technological determinism. The exploitation of information technology in order to promote industrial competitiveness and to improve the quality of life is of the greatest importance, but to subsume it under rhetorical notions, such as the post-industrial society or the information society, is to place these issues in a false perspective.

TECHNOLOGICAL CHANGE—PROBLEMS AND POSSIBILITIES

Without accepting the excessive optimism of theories about post-industrial society, we can recognize that the nature of capitalist–industrial society has been profoundly affected by the rapid and widespread technological changes of the past fifty years. These effects are more accurately summed up in the term 'neo-capitalism'.[23] As Galbraith has pointed out in detail, many characteristics of the large modern business corporation are dependent upon the exigencies of contemporary technology.[24] Viewed from this perspective, technological change manifests a spectrum of problems as well as possibilities. In the remainder of this chapter, I shall briefly examine some of the issues that have received particular attention in this country, including technological unemployment, the quality of working life, the decline of the manufacturing sector, the special role of information technology, and the problems associated with education and training.

Technology and Job Destruction

The impact technology on jobs has been examined by a variety of groups, official and unofficial, in the past ten years. The major union in the Australian metal industry, the Amalgamated Metal Workers Union (AMWU), examined the subject at its national conference in 1980, and a report is attached to the conference proceedings. The report asserts that 65 000 jobs had been lost in the metal industry as a result of technological change and consequent rationalization. In a report to the conference, the assistant national secretary of the Union, Laurie Carmichael, identified seven effects of technology on employment and working life:

1. Large-scale job displacement—as high as ten jobs displaced for every one created
2. Virtual elimination of employment for process and transfer operations in mass production
3. Reduction of skilled employment required in the use of machine tools
4. Displacement of maintenance tasks through computerized 'self-diagnosis'
5. Demarcation disputes, e.g. in relation to responsibility for programming robots
6. Multi-skilling requirements; and,
7. Rapid change in education and training.[25]

These concerns are fairly typical of the period from 1978 to 1983, when fears of large-scale technological unemployment were at a high level. A long-running industrial dispute over the automation of telephone exchanges[26] led to the establishment of the Committee of Inquiry into Technological Change in Australia (CITCA), which reported in 1980. Although its report (generally known, after its chairman, as the Myers Report) was largely ignored by the Commonwealth government of the day, its four volumes provide a useful summary of available research and speculation about the effects of technological change in industry. Broadly speaking, CITCA concluded that the progress of technological change is likely to reduce job opportunities, that 'jobless growth' is more likely than widespread reductions in numbers of employees, and that new technologies will increase the number of unskilled tasks.[27] CITCA also pointed out that general estimates of the rate and effects of change are virtually impossible because of inherent differences, not only between technologies, but also between industrial situations.

Policy recommendations in the Myers Report were of a very general character. The report emphasized the need for industry to use the high level of expertise in the Australian community to develop new products and processes that would provide employment opportunities, and to keep up with international technological developments. People affected by technological change should be properly informed and consulted, and a 'safety net' provided to assist people to adapt to change. A watching brief on technological change should be maintained by establishing a standing committee attached to the Australian Science and Technology Council (ASTEC), as well as a national consultative committee on occupational safety and health.

The last recommendations were among the very few to elicit positive action on the part of the federal government. Otherwise the Report remained more or less of a dead letter until after the change of government in 1983—the effects of which are described below. In the meantime, however, the Report generated a large volume of commentary, most of it critical. One exception was the response from the computer industry, which welcomed the report for its stress on keeping up with new technology and its balanced approach to social issues.[28] The level of economic analysis was widely criticized, on the grounds that the Report said nothing about the role of large international corporations in the Australian economy, and very little about the relationships between technological change and wider structural changes in the economy.[29] The Report was also criticized for its relative lack of attention to the effects of technological change on employment.[30] Finally, several critics commented unfavourably on the lack of attention to education and training in the Report.[31]

There is, in fact, some material on the employment effects of technology in the fourth volume of the Myers Report, which contains the results of a number of commissioned studies dealing with the impact of technological change on employment in a number of settings: the introduction of word processors, computer-aided manufacture, new printing technology in the newspaper industry, and the computerization of the Totalisator Agency Board (TAB) in South Australia. In Sydney, the Fairfax newspaper group introduced computerized typesetting and 'cold metal' printing, with a reduction of production staff from 659 in

1977 to 227 in 1980. Similar reductions took place in other newspaper establishments in Sydney, Brisbane and Melbourne.[32] The Printing and Kindred Industries Union (PKIU) made a submission to the Senate Committee on Science, Technology and the Environment in 1985, which reflects the total impact of new printing technology in New South Wales, where the number of jobs in the printing and allied industries fell by more than 2500 between 1973 and 1979. There were losses in twenty-four out of thirty-eight job categories, especially among compositors, letterpress printing machinists, cardboard box and carton makers, stationery printers, and rotary printing machinists. The membership of the union, which reached a peak of 59 000 in 1974, had fallen to 48 000 in 1985.[33]

Clerical employment has been similarly affected, for example, in two important areas of clerical work, banking and insurance. The AMP Society, Australia's largest life insurance company, presented a submission to CITCA that recorded the employment effects of the introduction, since 1974, of centralized computing services linked with visual display terminals at local branches. In 1974 AMP employed a total of 4042 persons under industrial awards (i.e. junior grades of clerical/administrative staff); by 1979 this had fallen to 3387. The most dramatic decline was at the lowest levels. In the same five year period, the number of Grade 1 employees fell by 35 per cent, and of Grade 2 by 32 per cent. Much of this reduction was achieved by 'natural wastage', but the long-term effect has been a shrinkage of opportunities for young people, especially school leavers without specialized skills. Large insurance companies (and banks) used to have annual recruitment drives in both city and country areas. This is no longer the case, and incentives such as the 'living aways from home allowance' that AMP used to offer to young people in country areas have ceased. With the advent of computerized systems linked by telecommunications, insurance companies have centralized their functions, so that 90 per cent of AMP staff now work in the state or national headquarters offices.[34]

The case of AMP is repeated, more or less, in other life offices, and the effects have been intensified by a series of mergers and amalgamations. In its submission, the AMP Society noted the 'rapidly falling demand for the conventional typist and filing clerk'. Another large life office, MLC Assurance, shed 9 per cent of its clerical staff between 1967 and 1978. The T&G Assurance Co. (since amalgamated with another large life office) reduced its staff by 11 per cent between 1973 and 1979 and shed 47 per cent of its typists and stenographers between 1973 and 1977. Another spectacular instance of the effects of technological rationalization occurred in 1978, when Ampol Petroleum Ltd established a centralized and computerized national accounting system which resulted in the disappearance of 154 clerical positions and the retrenchment of 116 clerical staff.[35]

In general, the evidence regarding clerical employment conforms to the proposition stated by Freeman, that 'of all activities information processing, storage and dissemination are probably the most vulnerable to the impact of the micro-processor and the micro-computer'.[36]

There is much less information about the employment effects of technological change in other sectors. In one case—waterfront employment—a detailed submission was made to CITCA by the Waterside Workers Federation that documents the spectacular drop in the number of jobs since the 1950s as containerization, mechanical handling, and computerized accounting systems were introduced. In 1955-6, the average number of registered waterside workers was over 25 500; ten years later it had fallen to 18 500; and in 1977-8 it dropped to 10 000. During the same twenty year period, productivity per man rose from 0.64 tonnes handled per work hour to 4.4, almost a sixfold increase. For individual ports, the fall was even more dramatic. At the sugar port of Mackay Queensland, the number of men on the main register fell from 401 to 148 in the three years from 1955 to 1958. Other Queensland ports north of Brisbane were also run down as mechanization affected coal, metallic ores, and chemicals. Inter-union problems, resulting from these sharp reductions in membership, have subsequently led to a string of industrial disputes.[37]

Apart from the claims made by the Metal Workers Union (quoted above), there has been little detailed research on the employment effects of technological change in manufacturing industry. One exception is a study carried out by the author in the chemical industry, centred on the chlorine and caustic soda plant operated by ICI (Australia) Ltd at Botany, an industrial suburb of Sydney. This plant was established during the Second World War, using the old-established mercury cathode process for the electrolysis of salt. The operation of each electrolytic cell required a team of thirteen men, whose main job was the maintenance of the graphite anode in each cell. Since 1960 the graphite anodes have been replaced by titanium electrodes, which require much less maintenance, and the teams of thirteen have been phased out, so that the workforce has shrunk dramatically. The next stage will be the computerization of the control system and the establishment of a centralized management structure for the various factories on the site, with further staff reductions and the introduction of a small maintenance staff of 'super-technicians' plus their assistants. This will involve the disappearance of the present skilled jobs of instrument fitter and mechanical fitter.[38]

Experience in Australia is closely comparable to that in other countries that are members of the OECD. Freeman and Soete have reviewed studies undertaken in the United States, Canada, Japan, the Federal Republic of Germany, France, and the United Kingdom whose conclusions are similar whether they are based on national employment statistics or on case studies. A Canadian paper identifies three main features common to all the countries examined:

(a) A decline in demand for unskilled workers in both manufacturing and clerical occupations
(b) A parallel growth of demand for highly qualified technical personnel, especially computer-trained staff and engineers in both manufacturing and services
(c) A need for a minimum level of 'computer literacy' in a wide variety of occupations.[39]

These three tendencies have all been observed in Australia. The computer industry, for example, is chronically short of skilled personnel, and something like 25 per cent of the higher levels of skill—systems analysts, programmers, designers, and engineers—are recruited from immigrants.[40]

Technology and Economic Growth

The simultaneous occurrence of technological unemployment and manpower shortages illustrates the complexity of the relationships between technological change and the availability of jobs. Since the upsurge of concern between 1978 and 1983 mentioned above, the argument that technological change 'destroys' jobs and generates massive 'deskilling' has become less popular. Perhaps the most comprehensive review of the topic is to be found in a report published by the Technological Change Committee of ASTEC, established as a result of the Myers Report. The report concludes that it is 'the combination of technological change with comparatively low levels of economic activity and with other structural adjustment problems, rather than technological change alone, that can lead to employment problems when new technology is introduced'.[41] This conclusion is shared by a wide range of economists. Tom Mandeville and Stuart Macdonald, who have conducted a number of studies in the area, note very skimpy evidence on which predictions about a 'holocaust of jobs' were based in the late 1970s.[42] Dixon, reviewing Barry Jones's best-selling book *Sleepers, Wake!*, points out that Jones's stress on Australia's technological backwardness ignores the fact that this is not the sole or even the major cause of our economic problems.[43]

Barry Jones's ideas have been influential in stimulating the notion that these economic problems — the decline of manufacturing industry, the continuation of unemployment, and inflation — can be overcome by recognizing that we live in a 'post-industrial' or 'post-service' society, in which prosperity can be generated by encouraging high-technology, 'sunrise' industries. Apart from the fact that high-technology industry is capital-intensive rather than labour-intensive, so that its capacity to create jobs is strictly limited, the extent to which the Australian economy is capable of generating and consolidating a high-technology sector is open to considerable doubt.

Australia belongs to a group of countries, which also includes Canada, Argentina, and New Zealand, whose economic and social progress has depended heavily on the export of primary and extractive products, as well as a continuous inflow of overseas investment. Following the arguments of the economic historian Immanuel Wallerstein, the general label of 'dominion capitalist' societies has been suggested for this group of countries, all of them products of European colonization. Economic growth in these countries led to high living standards and to political autonomy, but their economies have never achieved a high standard of diversification and have remained dependent on foreign capital. Their economic problems, aggravated by the world recession since 1970, underline the tension between a sophisticated urban way of life and an economic base that depends for its strength essentially on commodity exports.[44] If the rest of the affluent, capitalist–industrial world is moving into some kind of 'post-industrial' phase, the implications for Australia may be increased economic, technological, and political dependence—a 'client state' as the economist Ted Wheelwright has mordantly described the situation.[45]

STRATEGIES FOR THE FUTURE

The concerns articulated by Wheelwright were expressed even more forcibly by the Australian Treasurer Paul Keating, in several speeches during 1986, when he warned of the danger that Australia might become a 'banana republic'. This prospect was a major influence on the policies of the Hawke government from the time it came to power in 1983. The aim of recovering a degree of technological sovereignty was promoted especially by Barry Jones, who was given the portfolio of Science and Technology in the new ministry. A national summit conference on technology was held in 1983, largely promoted by Jones, from which emerged a 'national technology strategy' (the implementation of which was later transferred to Jones's senior colleague, Senator John Button, whose portfolio of Industry and Commerce was enlarged to include technology).

One of the consequences of the easy prosperity and the prolonged conservative dominance of national politics that marked Australian society for the twenty years preceding the 'oil shock' of 1973 was a lack of concern with policy development in science and technology. The conservative coalition governments that held national office unbrokenly from 1949 to 1972 resisted the urgings of a variety of critics within the academic, business, and research communities. ANZAAS was an important vehicle for this kind of criticism. In 1974 the national council of ANZAAS appointed a commission to prepare a report on science and technology policy, which laid out a detailed agenda for government action in a variety of fields.[46] Unfortunately, the Whitlam Labor government, elected in 1972, continued to treat the subject as a matter of secondary importance, and this situation did not improve under its successor, the Fraser administration (1975–83).[47] Under the Hawke government, there was a significant change, including a growing realization of the narrow social base for science and technology in Australia and their lack of relationship to industries other than agriculture.

Senator Button himself remarked, in an address given in 1985, that it was essential to develop a 'culture where science and technology is [sic] seen by practitioners and users alike as an integral and vital part of industry and the economy'. But he also emphasized that science and technology were not a panacea for deep-rooted economic problems.[48] A similar view has been expressed by Christopher Freeman:

> The institutional framework for future economic recovery is only now being shaped. Most of our institutions and ideologies are still geared to the old post-war technological paradigm. Only through social and political debate and conflict shall we determine how we re-shape our institutions and our way of life

to match the potential of the new technology and to humanise its innumerable potential applications. The new patterns of employment which will emerge should be of a kind which encourage great variety in hours of work and in continuing education and training, but which ensure to everyone who is seeking paid employment the opportunities to work in socially useful activity.[49]

Recognition of the need for this combination of expansionary economic policies together with the exploitation of new technologies is evident in numerous statements made by government leaders since 1983. In an address to a regional OECD conference in 1986, the Prime Minister, R. J. Hawke, acknowledged that Australia could not compete with countries that had lower costs, and that competitiveness must be bolstered by enhancing the non-price attributes of the goods and services that Australia produces. Innovation, he declared, enhances competitiveness in areas such as product quality, lower rates of defects, and better product performance.[50]

Similar views have been expressed by the Minister for Science, Barry Jones, who has also argued for a rationalization of the funding arrangements for research. Selectivity and concentration, he maintained in an address to a national scientific conference in November 1986, would greatly improve the ability of research and development to provide a focus for applications-oriented research and industrial support. To this end, he proposed the establishment of a National Research Council with a more positive role than the present scatter of agencies for supporting academic research, of which he estimated there were no less than twenty-five.[51] It is ironic that a similar suggestion was made as far back as 1964 by a committee of inquiry into tertiary education.[52]

Although government policy statements indicate a realization of the need to co-ordinate science and technology with general economic objectives, the tendency of governments in the past decade to retreat from objectives such as full employment and to cut back on the activities of the public sector poses a basic contradiction. As Freeman writes, economic improvement will only come about through a renewal of commitment to full employment in a new social context that takes account both of changes in technology and changes in society.[53] It is on these lines that strategies for the future should be based.

NOTES

1. J. M. Keynes, 'Economic Possibilities for Our Grandchildren', in *Essays in Persuasion* (London: Macmillan, 1931), 364.
2. T. H. Laby, 'Science and the Depression', *Report of ANZAAS*, 21 (Sydney, 1932), 432–5.
3. O. U. Vonwiller, 'The Social Relations of Science', *Aust. J. Sci.*, 1 (1938), 30–2.
4. A. Denning, 'Automation—What Is It and What of the Future', *Aust. J. Sci.*, 19 (1957), 116.
5. G. W. Ford (ed.), *Automation: Threat or Promise?* (Sydney: Law Book Co., 1969).

6. G. W. Ford (ed.), *Redundancy* (Sydney: ANZAAS, 1973).
7. ANZAAS (NSW Division) (ed.), *Automation and Unemployment: An ANZAAS Symposium* (Sydney: Law Book Co., 1979).
8. ibid., 31.
9. ibid., 112–13.
10. Richard Badham, *Theories of Industrial Society* (Beckenham: Croom Helm, 1986), 72–3.
11. Robert L. Heilbroner, 'Economic Problems of a Post-Industrial Society', *Dissent*, 20 (2) (1973), 170.
12. Daniel Bell, *The Coming of Post-Industrial Society* (New York: Basic Books, 1973), 14.
13. Daniel Bell, 'Notes on the Post-Industrial Society', *The Public Interest*, 1 (1) (1967), 27.
14. S. M. Miller, 'Notes on Neo-Capitalism', *Theory and Society*, 2 (1)(1975), 27.
15. S. Encel, 'The Post-Industrial Society and the Corporate State', *Australia and New Zealand Journal of Sociology*, 15 (2) (1979), 37–44.
16. Heilbroner, op. cit., note 11, 164.
17. J. L. Gershuny, *After Industrial Society?* (London: Macmillan, 1978).
18. Barry Jones, *Sleepers, Wake! Technology and the Future of Work* (Melbourne: Oxford University Press, 1982), 254–6.
19. J. H. Goldthorpe, 'Theories of Industrial Society', *European Journal of Sociology*, 12 (1971), 263–88.
20. R. Wright, *Sacred Rage: the Wrath of Militant Islam* (New York: Simon and Schuster, 1985).
21. Y. Masuda, 'Computopia' in Tom Forester (ed.), *The Information Technology Revolution* (London: Blackwell, 1985), 620–35.
22. M. Marien, 'Some Questions for the Information Society', in ibid., 648–60.
23. S. Encel, 'Capitalism, the Middle Classes and the Welfare State', in E. L. Wheelwright and Ken Buckley (eds), *Essays on the Political Economy of Australian Capitalism*, (5 vols) vol. 2 (Sydney: ANZ Book Co., 1977), 148–68.
24. J. K. Galbraith, *The New Industrial State* (Harmondsworth: Pelican, 1973), ch. 2.
25. Laurie Carmichael, 'The Process of Technological Change: A Union Perspective' in Russell Lansbury and Edward Davis (eds), *Technology, Work and Industrial Relations* (Melbourne: Longman Cheshire, 1985), 29.
26. Stephen Deery, 'Trade Unions. Technological Change and Redundancy Protection in Australia', in ibid., 203–4.
27. *Report of the Committee of Inquiry into Technological Change in Australia* (4 vols) vol. 1 (Canberra: AGPS, 1980), 8.
28. 'Myers Report Welcomed by the Computer Industry', *Modern Office and Data Management* (October 1980), 18–19.
29. For example, J. Selby Smith, 'The Report of the Committee of Inquiry into Technological Change: A Review', *Australian Quarterly*, 53 (2) (1981), 177–86; P. C. Stubbs, 'Technological Change in Australia: A Review of the Myers Report', *Economic Record*, 57 (158) (1981), 224–31; M. Carter, 'Technological Change in Australia: A Review of the Myers Report', *Discussion Paper No. 20*, Centre for Economic Policy Research (Canberra: Australian National University, 1981).
30. Ian Reinecke, 'Inside the Myers Report—1', *Australian Financial Review* (5 August 1980), 14–15; S. Encel and Susan Walpole, 'Technological

Change—Some Case Histories', *Technology and Society Papers*, no. 2 (Sydney: University of New South Wales, 1981).
31. B. W. Smith, 'Technological Change—What the Real Questions Are', in National Information Technology Committee, *Impact of Information Technology, 1981* (Canberra: AGPS, 1981), 49–59.
32. Report, op. cit., note 27, 4, 143–87.
33. Personal communication from Carmel Robinson, Industrial Democracy Project Co-ordinator, PKIU.
34. Encel and Walpole, op. cit., note 30.
35. Patti Burke, 'Clerical Work and Technological Change', *Technology and Society Papers*, no. 1 (Sydney: University of New South Wales, 1981), 24.
36. Christopher Freeman and Ray Curnow, Technological Change and Employment—A Review of Post-War Research (unpublished paper, Science Policy Research Unit, University of Sussex, 1978).
37. S. Encel, 'Working Life' in S. Encel and L. Bryson (eds), *Australian Society*, 4th ed (Melbourne: Longman Cheshire, 1985), 75.
38. Encel and Walpole. op. cit., note 30, 8–14.
39. Christopher Freeman and Luc Soete, *Information Technology and Employment* (Brussels: IBM-Europe, 1981), 70–1.
40. S. Encel, 'Information Technology and Occupations' (London: Technical Change Centre, 1984); A. E. Daniel, S. Encel, M. Markus, 'Work in the Computer Industry', *Search*, 17 (5–6) (1986), 126–31.
41. ASTEC, *Technological Change and Employment* (Canberra: AGPS, 1983), 70.
42. T. Mandeville and S. Macdonald, 'Reflections on the Technological Change Debate in Australia' in *Readings on Technology and Change* (Monash University: Community Research Action Centre, 1985), 14–20.
43. P. B. Dixon, 'Review of Barry Jones, *Sleepers, Wake!*, in R. Castle *et al.* (eds), *Work, Leisure and Technology* (Melbourne: Longman Cheshire, 1986), 14–15.
44. S. Encel, 'Metropolitan Societies and Dominion Societies' in S. N. Eisenstadt (ed.), *Patterns of Modernity*, vol. 1 (London: Frances Pinter, 1987).
45. E. L. Wheelwright, *Australia—the Client State* (Sydney: Transnational Corporations Research Project, University of Sydney, 1981).
46. Report of ANZAAS Science Policy Commission, *Search*, 5 (11–12) (1974), 557–88.
47. S. Encel, 'Pushing the Barrow Uphill' in S. Encel, P. Wilenski and B. Schaffer (eds), *Decisions* (Melbourne: Longman Cheshire, 1981), 21–40.
48. Senator Button, *Science, Technology and the Economy*, Occasional Papers, no. 11 (Sydney: University of New South Wales, 1986).
49. Freeman and Soete, op. cit., note 39, 148.
50. R. J. Hawke, Address to OECD Regional Science Conference, Canberra, 11 November 1986.
51. Barry Jones, Address to National Forum on Science and Technology in the Budget, Academy of Science, Canberra, 6 November 1986.
52. *Report of the Committee of Inquiry into Tertiary Education* (Canberra: AGPS, 1964), 1, 53.
53. Freeman and Soete, op. cit., note 39.

APPENDICES

APPENDICES

APPENDIX 1
Officers of the Australasian Association for the Advancement of Science (afterwards ANZAAS)

Year	Place of meeting	President	General treasurer	Permanent honorary secretary	Honorary editor (created in 1954)	Chairman (created in 1970)
1888	Sydney	H. C. Russell	Sir Edward Strickland	Prof. A. Liversidge (until 1907)		
1889						
1890	Melbourne	Baron von Mueller	H. C. Russell (until 1904)			
1891	Christchurch	Sir James Hector				
1892	Hobart	Sir Robert G. G. Hamilton				
1893	Adelaide	Prof. Ralph Tate				
1894						
1895	Brisbane	The Hon. A. C. Gregory				
1896						
1897						
1898	Sydney	Prof. A. Liversidge				
1899						
1900	Melbourne	R. L. J. Ellery				
1901						
1902	Hobart	Captain F. W. Hutton				
1903						
1904	Dunedin	Prof. T. W. Edgeworth David	D. Carment (until 1934)			
1905						
1906						
1907	Adelaide	A. W. Howitt				
1908						

361

Year	Place of meeting	President	General treasurer	Permanent honorary secretary	Honorary editor (created in 1954)	Chairman (created in 1970)
1909	Brisbane	Prof. W. H. Bragg		J. H. Maiden (until 1922)		
1910						
1911	Sydney	Prof. Orme Masson				
1912			Prof. H. G. Chapman (1912–13)			
1913	Melbourne	Prof. T. W. Edgeworth David				
1914						
1915						
1916						
1917						
1918						
1919						
1920						
1921	Hobart (held in Melbourne)	Prof. Sir W. Baldwin Spencer				
1922				E. C. Andrews (until 1926)		
1923	Wellington	Sir George H. Knibbs				
1924	Adelaide	Lieutenant-General Sir John Monash				
1925						
1926	Perth	Prof. Edward H. Rennie		1926 — Honorary general secretary A. B. Walkom (until 1947)		
1927						
1928	Hobart	R. H. Cambage				
1929						
1930	Brisbane	E. C. Andrews				
1931	Sydney	Sir Hubert Murray				
1933			G. A. Waterhouse (until 1946)			

1: Officers of ANZAAS

Year	City					
1935	Melbourne	Sir Douglas Mawson				
1936	Auckland	Sir David Rivett				
1937						
1938						
1939	Canberra	Prof. Ernest Scott				
1940						
1941						
1942						
1943						
1944						
1945						
1946	Adelaide	Prof. P. Marshall	Prof. J. R. A. McMillan (until 1952)			
1947	Perth	A. E. V. Richardson		Prof. N. A. Burges (until 1952)		
1948						
1949	Hobart	A. B. Walkom				
1950						
1951	Brisbane	Sir Kerr Grant				
1952	Sydney	Sir Douglas B. Copland			Prof. J. L. Still (until 1962)	
1954	Canberra	Sir Theodore Rigg				
1955	Melbourne	Prof. R. v. d. R. Woolley			Prof. J. R. A. McMillan (until 1970)	
1956						
1957	Dunedin	Prof. Sir Macfarlane Burnet				
1958	Adelaide	Prof. M. L. E. Oliphant				R. L. Aston
1959	Perth	H. C. Coombs				J. B. Thornton
1960						Dr. K. W. Knox (until 1970)
1961	Brisbane	Emeritus Prof. Sir Samuel Wadham				
1962	Sydney	Prof. N. S. Bayliss				K. W. Knox
1964	Canberra	Sir Frederick White				N. A. Esserman
1965	Hobart	Prof. R. N. Robertson				F. J. Lehany (until 1973)

Year	Place of meeting	President	General treasurer	Permanent honorary secretary	Honorary editor (created in 1954)	Chairman (created in 1970)
1967	Melbourne	Sir Fred Schonell				
1968	Christchurch	Prof. Sir John Crawford				
1969	Adelaide	Dr Charles A Fleming				
1970	Port Moresby	Prof. S. Warren Carey		N. C. Manning	J. B. Davenport (until 1977)	Sir Frederick White (until 1977)
1971	Brisbane	Prof. Gustav J. V. Nossal		C. W. Davis (until 1977)		
1972	Sydney	Dr R. G. Ward				
1973	Perth	Prof. Eric J. Underwood	Dr G. de V. Gipps (until 1977)			
1974						
1975	Canberra	The Hon. Mr Justice J. H. Wootten				F. J. Lehany
1976	Hobart	Prof. W. D. Barrie	R. H. Scott (until 1981)			
1977	Melbourne	Dr Lloyd Evans		J. H. Elliott	R. Strahan (until 1984)	J. B. Davenport (until 1984)
1978						
1979	Auckland	Dr K. L. Sutherland	N. Taylor	R. Strahan (Acting)		
1980	Adelaide	Prof. Sir Geoffrey Badger	N. Taylor	Dr Diana Temple (until 1986)		
1981	Brisbane	Dr Graham W. Butler				
1982	Sydney	Sir Zelman Cowen				
1983	Perth	Prof. R. O. Slayter				
1984	Canberra	Sir Gustav J. V. Nossal				
1985	Monash University	Sir Edmund Hillary		Dr D. Bartels	Ann M. Moyal	Prof. I. G. Ross (until 1986)
1987	Palmerston North	Sir David Beattie			B. J. Walby	A. Morgan
1987	Townsville	Sir Bruce Watson				

APPENDIX 2
Sections of the AAAS (ANZAAS)

At its first meeting in 1888, the subject divisions of the Australasian Association for the Advancement of Science were arranged as follows:

A Astronomy, Mathematics, Physics, and Mechanics
B Chemistry and Mineralogy
C Geology and Palaeontology
D Biology
E Geography
F Economic and Social Science, and Statistics
G Anthropology
H Sanitary Science and Hygiene
I Literature and Fine Arts
J Architecture and Engineering

By Christchurch in 1891, Mineralogy had migrated from Chemistry (Section B) to Geology (C); Anthropology (G) had been married with Ethnology and renamed F; while F (Economics and Social Science and Statistics) lost Statistics and acquired Agriculture. Sadly, Section I, Literature and Fine Arts, ceased by 1891, replaced alphabetically (if not in sensibility) by Mental Science and Education (Section J) in 1893.

In 1893 the order was rearranged, so that Ethnology and Anthropology was lettered F instead of G; Agriculture was added to Economic Science and designated G; and Engineering and Architecture came together under H.

In March 1888, even before the inaugural meeting, the College of Pharmacy had asked for a Section, but Pharmacy resided with Section H (Sanitary Science and Hygiene) until 1911, when it became a subsection under B (Chemistry). Also in 1911 History joined Geography in a new Section (E), and Agriculture separated from Economic Science and was made a Section (K) in its own right. Veterinary Science, arriving as a subsection of Agriculture in 1911, became a full Section (L) in 1913.

In 1921 the Sections were as follows:

A Astronomy, Mathematics, and Physics
B Chemistry
C Geology and Mineralogy
D Biology (renamed Zoology in 1924)
E Geography
F Ethnology
G Social and Statistical Science
H Architecture and Engineering
I Sanitary Science and Hygiene
J Mental Science and Education
K Agriculture (created in 1913)
L Veterinary Science (first introduced in 1913 as a subsection to K)

It was proposed at Brisbane in 1909 that Biology be divided into Zoology and Botany, but it was decided to hold the Section together, with Section presidents chosen alternatively from the two areas. This compromise lasted until 1923, when Section D (Biology) was renamed Zoology, and Botany was constituted a separate Section (M).

In 1923 Section K (Agriculture) was designated Agriculture and Forestry. Section N (Physiology and Experimental Biology) was added in 1926. In 1928 Section I became Sanitary Science and National Health, and Section O (Pharmaceutical Science) made its first appearance.

In 1928 History was separated from Geography, becoming Section E, while Geography and Oceanography formed Section P in 1930.

By the twentieth Congress in Brisbane in 1930, the Sections had grown and had altered significantly:

A   Mathematics, Physics and Astronomy
B   Chemistry
C   Geology
D   Zoology
E   History
F   Anthropology
G   Economics, Statistics, and Social Science
H   Engineering and Architecture
I   Medical Science and National Health
J   Education, Psychology, and Philosophy
K   Agriculture and Forestry
L   Veterinary Science
M   Botany
N   Physiology and Experimental Biology
O   Pharmaceutical Science
P   Geography and Oceanography

The advances and specialization within medical science were reflected in frequent additions and changes to the medical sections. In 1949 Section I was renamed Medical Science, National Health and Physiology; in 1952, Microbiology replaced Physiology, which moved to Section N. In 1954 Microbiology was added to Section I, and Biochemistry to Physiology in N. In 1955 Section I changed again, to become Microbiology, Epidemiology and Preventive Medicine.

In 1958 a sub-committee appointed to report on ANZAAS Sections commented:

After every ANZAAS meeting within our experience, and particularly since World War II, it has happened that some scientific body in the city where the meeting was held has felt that its own—sometimes fairly specialised—interests were not adequately catered for by the existing arrangement of the Sections. (ML MSS 1613 add-on 344, *ANZAAS General Committee Minutes*, 20 August 1958, 66)

Despite a decision on that occasion not to alter the Sections for these reasons, the requests continued. In 1965 the general committee recommended that the constitution be altered to make the Sections more flexible. The question was taken up by the recently created Planning Committee, and it is in a 1967 report of that Committee that the present numbering scheme first appeared: Sections are numbered 1 to 8, 11 to 17 and 21 to 26. At later congresses, Sections were created for some of the numbers not included in this scheme (ML MSS 1613 add-on 344, *Report Minutes*, 1965–69).

The Sections at Adelaide in 1969 were:

1   Physics
2   Chemistry
3   Geology
4   Architecture and Town Planning
5   Engineering
6   Pharmaceutical Sciences
7   Optometry (first appeared in A in 1949)
8   Mathematical Sciences
11 Zoology
12 Botany
13 Agriculture and Forestry
14 Microbiology and Immunology
15 Physiology and Nutrition
16 Veterinary Science
17 Biochemistry
18 Plant Pathology
21 Geographical Sciences
22 Education
23 Psychology
24 Economics
25 Anthropology
26 History
31 Junior ANZAAS

Between 1970 and 1984 (the fifty-fourth Congress at Canberra) there were no major changes to this scheme, but many new subjects were added. These included:

27   Sociology (1972)
18   Food Science and Nutrition (replaced Plant Pathology)
25A   Archaeology (1976)
25B   Linguistics (1976)
27A   (later 29) Criminology and Forensic Science (1976)
28   Industrial Relations (1976)
32   (later 33) Communication in Science (1976)
34   Sports Science (1979 only)
35   Oenology (1979 only)
36   Musicology (1979)
37   Trace Element Research (1979 only)
37   History, Philosophy, and Sociology of Science (1980)
38   Health Education (1980 only)
40   Environmental Studies (1980)
41   Computer Science (1981)
42   Data Processing (1981: in 1982 joined 41)
42   Law (1982)
43   Robotics (1982)
44   Women's Studies (1982)
45   Social Welfare (1983)

In 1985 (at the fifty-fifth Congress at Monash University in Melbourne), a much greater change was introduced as part of an attempt to make ANZAAS even more flexible and of wider appeal. The forty specialized Sections were replaced with just nine Areas—broad subject groupings—as follows:

Area 1 Major Symposia and Other Special Events
Area 2 Mathematical and Physical Sciences
        (Sections 1, 2, 8)
Area 3 Engineering and Applied Sciences
        (Sections 3, 5, 41, 45)
Area 4 Biological Sciences
        (Sections 11, 12, 13, 15, 16, 17, 23, 25A)
Area 5 Medical Sciences
        (Sections 6, 7, 14, 18, 38)
Area 6 Environmental Sciences
        (Sections 4, 21, 40, 25A)
Area 7 Education and Communication Sciences
        (Sections 22, 25B, 33, 34, 36, 37, 44)
Area 8 Social Sciences
        (Sections 24, 25, 27, 28, 29, 42, 45)
Area 9 Youth ANZAAS
        (Section 31)

In 1987 (at the fifty-sixth Congress in Palmerston North, New Zealand) a series of Interest Groups were established for programming purposes. The slight changes for 1987 were as follows:

Group 2 Technological and Biochemical Sciences
Group 3 Biological Sciences
Group 4 Health Sciences
Group 5 Social Sciences
Group 6 Geographical and Forestry Sciences
Group 7 Community Sciences
Group 8 Education Sciences
Group 9 Youth ANZAAS

APPENDIX 3
Rules and Constitutional Changes

During its first century, the AAAS/ANZAAS has introduced several changes in its constitution. These have in many ways modified the Rules of the British Association for the Advancement of Science, with which it began. Most were to affect the internal government of the Association and followed from its desire to encourage new subjects; to maintain a balance between 'federal' and 'state' interests; and to serve the interests of administrative efficiency. After 1921 the Association and its rules were increasingly bound up with the Australian National Research Council (ANRC). The constitutional history is confusing, however, and not least because of the use of similar expressions to mean quite different things. What follows is an attempt to mark and explain the more significant changes that have occurred.

The ten rules adopted from the British Association were as follows:

1. All persons who signify their intention of attending the first Meeting shall be entitled to become original Members of the Association, upon agreeing to conform to the Rules.
2. The Officers, Members of Council, Fellows, and Members of Literary and Philosophical Societies publishing Transactions or Journals in the British Empire, shall be entitled in like manner to become Members of the Association, Annual Subscribers, or Associates for the year, subject to the payment of the prescribed subscription, and the approval of a General Committee.
3. All Members who have paid their subscriptions (£1 per annum) shall be entitled to receive the Publications of the Association.
4. The Association shall meet for one week or longer. The place of meeting shall be appointed by the General Committee two years in advance.
5. There shall be a General Council, having the supreme control, to be composed of Delegates from the different Colonies or Colonial Scientific Societies. The number of Delegates from each Society or Colony shall be proportionate to the number of Members from the particular Colony or Society—subscribing or otherwise—taking part in the proceedings (i.e., after the preliminary meetings). Each Colony or Society shall be allowed to nominate a Delegate for each one hundred of its Members.
6. There shall be a General Committee consisting of Members of the Council, President, Vice-Presidents and Secretaries of Sections, Contributors of Papers to the Association, and such others as may be elected.
7. A Local Committee shall be appointed at the place of meeting to make arrangements for the reception and entertainment of the visitors and to make preparations for the Business of the General Meetings.
8. Sectional Committees shall be appointed for the following Subjects:—
   Section A—Astronomy, Mathematics, Physics and Mechanics.
   Section B—Chemistry and Mineralogy.
   Section C—Geology and Palaeontology.
   Section D—Biology.
   Section E—Geography.
   Section F—Economic and Social Science, and Statistics.
   Section G—Anthropology.
   Section H—Sanitary Science and Hygiene.
   Section I—Literature and Fine Arts.
   Section J—Architecture and Engineering.
9. Ladies are eligible for Membership.

10. The rights and privileges of Membership shall be in the main similar to those afforded by the British Association, subject to revision and alteration after the first meeting of the Australasian Association for the Advancement of Science.

The first decade saw several important alterations. At the outset, the rules vested supreme control in the hands of a general council, consisting of delegates from the various colonial scientific societies, in a proportion of 1 to 100 members. (What defined an acceptable society was decided on a case basis.) This assembly was to act through a general committee, a wider parliament, consisting of the council, the presidents, vice-presidents, and secretaries of the Sections, and all those who read papers. The congresses were to be organized by local committees. By the second meeting in Melbourne, however, it was clear that the ten 'commandments' borrowed from the British Association were too broad in some respects, and too limiting in others.

In particular, the role of the various Australasian societies Liversidge initially approached was unclear. Certainly Liversidge depended upon their membership, their goodwill, and their premises, but he had also to distance himself from their local trappings. The Royal Societies of New South Wales, Queensland, and Victoria were accused of elitism and aloofness; others were identified with local, particular, and not always mutually compatible, interests. Moreover, even if only twenty-eight of the thirty-four societies that agreed to join the enterprise actually sent delegates, the membership of the remaining twelve could neither be neglected nor sacrificed to the interests of the participating twenty-eight by an Association that spoke to all Australasia. It was important to make the Association both embrace and transcend local societies, without creating a federation of them: thus, a federation of science, without a federation of societies.

A committee was appointed at Melbourne in 1890 to consider a revision of the rules. This was reported at Christchurch in 1891 where a new set of forty-four rules was adopted, which were later ratified at the Hobart Congress in 1982. By this time the Association had generated sufficient past and present office bearers to formalize structures based upon them. These changes altered the government of the Association in three significant ways. First, the general committee was abolished, and in its place came a new council, consisting of the current and past presidents, officers, and officers of the Sections, together with the authors of papers. Second, five vice-presidents were to be elected from among the ex-presidents; provision was made for one or more general secretaries and local secretaries, appointed by local committees (members of council resident in the colony), and later confirmed by council. As A. B. Walkom remarked in 1962, 'hardly a meeting passed without proposals being put forward for change in one or more rules' (*A Goodly Heritage* (Sydney: ANZAAS, 1962) 17).

The changes of 1891 had the twin effect of loosening the Association from any possibility of control by colonial societies and removing any possibility of acting in competition with them. Societies continued to send delegates to council, but in 1913 the council decided to restrict this right to societies that undertook and published research (Minutes, second meeting). In practice, the Association consulted the societies, but it gave notice that, 'as a matter of general policy', as the Pharmaceutical Society found to its cost, 'it did not recognise the extension of other societies lest claims for special representation should become too numerous' ('Pharmacy and the Australasian Association for the Advancement of Science', *The Chemist and Druggist of Australasia* (February 1895), 18).

The constitution provided, in principle, for a considerable number of people to join in the government of the Association, through the general council. But the Association's possible leadership, drawing upon a population of less than 100 government and academic scientists, was inevitably small. Even among this small number, the actual running of the Association was unavoidably left to those whose creature it was. Thus the openness of the AAAS was in fact, a fairly closed circle, particularly during its first quarter century. This paradox persisted into the 1920s—an unmanageably large general council, counterpoised by a few senior men.

In 1892, greater definition was given to the committee structure, with the appointment of local committees (hereafter, members of the council resident in the colony where congress was held), sectional committees (nominated by the local committees in which ensuing congresses were to be held), a recommendations committee (which would be appointed by the council at its first meeting each year), and a publications committee (similarly by election each year). Section presidents would be elected from among men living outside the colony where a given congress would be held but there was no limit to the number of times the same men could hold office. Overall, these changes had the effect of strengthening the hands (and responsibilities) of the principal officers.

By 1924, at Adelaide, it was recognized that the general council was far too large to execute the affairs of the Association. It could only meet at congresses and in the inter-congress period the president, general secretary, and general treasurer were all-powerful. There was, however, a strong belief that the state societies were being under-represented. So in 1926, at the Perth Congress, a new constitution was adopted. This expanded the council by adding delegates from what were called associated societies and institutions, and by creating a smaller executive committee of the general council, consisting of the president, immediate past president, the general secretary, the general treasurer and the local secretaries. This committee was charged with managing the general affairs of the Association between congresses.

At the Auckland Congress in 1937, as a result of the merger with the ANRC, a class of 'Fellows' was added to the classes of membership. The initial ANZAAS Fellowship consisted of the past and present presidents, general treasurers, and general secretaries of ANZAAS and ANRC; the presidents and past presidents of the Sections of ANZAAS; all present full members of ANRC not included in the foregoing; and not more than twenty other Fellows co-opted for their special qualifications. The Fellows of ANZAAS, thus defined, became the new ANRC. The ANRC was charged with publishing a journal, and the *Australian Journal of Science* was born. In 1939 detailed procedures for the election of Fellows were adopted.

At the Dunedin Congress in 1957, for reasons that are today unclear, a new general committee, distinct from the general council (which continued) was established (or rather re-established). This larger body included the officers of the previous executive committee of the general council with the addition of the chairman of the publications committee, the editor of publications, the chairman and organizing secretary of the next congress, and representatives from each Section. It did not, however, include the state (or affiliated) societies. Thus the Association moved again towards a large and unwieldy body to manage its affairs.

The constitution arising from the Crawford Committee of Review in 1968–9, recommended radical changes in the Association's management structure, which were adopted by the general committee at the Adelaide Congress in 1969. First, the general committee of 1957 was abolished, and the council, shorn of any decision-making function, was renamed the convocation. The convocation met during congresses, and could make recommendations to a new council (the equivalent of the former general committee). Second, the Association was formally incorporated under the ACT Associations Incorporation Ordinance in

1970. Third, the presidency was filled by the council upon the recommendation of the organising committee of congress, and retained as an honorific post. Next, chairman of council was appointed for five congress terms, in order to establish continuity of administration. The new council consisted of the office bearers, representatives of each state and territory divisions where they existed; and New Zealand, with three members appointed by the council to spread its range of representation. It was still too large a body to meet regularly, so a smaller management committee, consisting of the office bearers and three divisional representatives, rotating every few years among the various divisions, was established. The divisions comprised the members of ANZAAS resident in each state or territory. These gained their own constitutions, and were thus able to operate financially independently of the controlling body. In cases where no formal division was established, the local Royal Society provided the representatives for council. Fellows were abolished but reintroduced in 1980 at the fiftieth Congress in Adelaide, in order to honour distinguished service to the Association and the scientific estate. Next, local scientific societies were given a place as affiliated societies, defined as bodies, not involved in 'business', with which ANZAAS corresponded and exchanged journals. 'Institutional' members were societies and other bodies that fell outside this definition.

Fully democratic procedures were not put in place, however, until the implementation of a new constitution in 1982. This followed on a review of the Association by a committee of past presidents and was influenced by the wish of successive councils to 'democratize' the appointment of council members. The convocation was abolished and a general meeting of all members was held during congresses. The office bearers were elected for fixed terms by a vote of all the members. The only officers who were not democratically elected were the president, who was appointed by council, on the recommendation of the congress organizing committee, as before, and the treasurer and editor of *Search*, who were appointed by council.

We gratefully thank Rupert Best, James Davenport, and Peter Lever-Naylor for the information on which this appendix is based.

APPENDIX 4
Local Secretaries and Divisional Representatives of the AAAS (ANZAAS)

New South Wales

1890–1908	Prof. Archibald Liversidge
1909–1921	Joseph H. Maiden
1922–1925	Ernest C. Andrews
1926–1951	Dr A. B. Walkom
1952–1953	J. M. Vincent
1954–1961	Prof. J. R. A. McMillan
1961–1963	Emeritus Professor A. P. Elkin
1964–1968	Prof. J. R. A. McMillan
1969–1970	James B. Davenport
1971–1974	Dr G. de V. Gipps
1975	J. H. Elliott
1976–1981	R. E. Cooke-Yarborough
1982	P. Keam
1983–1984	Ronald Strahan
1985–1986	A. M. Cooke
1986—	Dr D. J. O'Connor

South Australia

1890	Prof. W. H. Bragg
1891–1892	F. Wright
1893–1906	Prof. E. H. Rennie
1907	W. Howchin
1908	J. P. V. Madsen
1909–1920	W. Howchin
1921–1922	Prof. J. W. Wilton Carey
1923–1925	L. Keigh Ward
1926–1963	Dr R. S. Burdon
1964	Dr R. S. Burdon and Dr Rupert J. Best
1965–1976	Dr Rupert J. Best
1977–1979	Dr P. S. Delin
1980–1984	Dr Brian Daily
1985–1986	Dr R. Southcott
1986—	Dr J. T. Wiskich

Victoria

1980–1891	Prof. W. Baldwin Spencer
1892	A. H. Lucas
1893–1906	E. F. J. Love
1907–1915	T. S. Hall
1919	A. C. D. Rivett
1920–1921	Dr Georgina Sweet
1922–1951	F. R. Pitt
1952–1958	N. H. Olver
1959–1963	J. F. H. Wright
1964–1966	Dr R. B. Johns
1967–1968	W. W. Fee
1969–1971	Dr R. B. Johns
1972–1975	Dr P. G. Thorne
1976–1979	Rev. Alan Scott
1980–1983	W. E. Purnell
1984	Rev. Alan Scott
1985—	A. Grossbard

Western Australia

1909–1910	E. A. Mann
1911–1929	A. Gibb Maitland
1923–1926	Prof. N. T. M. Wilsmore (joint secretary)
1930–1945	Prof. E. de C. Clarke
1946–1958	Prof. A. D. Ross
1959–1960	D. E. White
1961–1963	Prof. N. F. Stanley
1964–1969	D. W. Oxnam
1970–1976	Prof. Martyn J. Webb
1977–1981	S. T. Waddell
1982–1984	Dr G. Chandler
1985–1986	Prof. B. Kakulas
1986–1987	vacant

Queensland

1890–1921	Dr John Shirley
1922–1926	C. T. White
1927–1956	Dr D. A. Herbert
1957–1963	Prof. Mansergh Shaw
1964–1967	Prof. R. H. Greenwood
1968–1972	Prof. O. E. Budtz-Olson
1973–1979	Noel V. Verney
1980–1981	Dr R. Gardiner
1982–1984	Prof. J. M. Thomson
1985—	vacant

Tasmania

1890–1908	A. Morton
1909–1910	Dr John Elkington
1911–1912	Robert Hall
1913–1921	Prof. T. Thomson Flynn
1922–1933	Clive E. Lord
1934–1945	Dr J. Pearson
1946–1951	Dr H. D. Gordon and Prof. S. Warren
1952–1957	Prof. S. Warren Carey
1958–1974	K. D. Nicolls
1975–1981	Dr P. W. Smith
1986—	Dr. P. G. Quilty

Australian Capital Territory

1931–1945	Dr G. A. Currie
1946–1958	Dr C. Barnard
1959–1963	W. Hartley
1964	Prof. L. D. Pryor
1965–1972	A. F. Gurnett-Smith
1973–1978	H. R. Webb
1979–1983	Peter Judge
1984–1985	R. Scott
1986—	P. J. Judge

Northern Territory

1975–1985	Dr Colin Jack-Hinton
1986—	vacant

New Zealand

1890–1891	Prof. A. P. W. Thomas
1892–1893	A. de Bathe Brandon
1894–1897	Prof. T. J. Parker
1898–1900	Captain F. W. Hutton
1901–1908	George M. Thomson
1909–1934	Prof. C. Coleridge Farr
1935–1954	Gilbert Archey
1955–1960	Prof. Gordon J. Williams
1961–1964	Prof. L. R. Richardson
1965–1971	Prof. J. F. Duncan
1972	Prof. R. Langer
1973–1974	Dr R. W. Willett
1975–1978	Sir Malcolm Burns

1979–1981   Dr R. K. Dell
1982   Dr E. Bollard
1983–1984   Prof. N. Curtis
1985—   Dr. R. D. Batt

Papua New Guinea

1958–1960   Dr G. A. V. Stanley
1961–1974   Dr R. F. R. Scragg
1975–1982   Karol Kisokau
1983—   Vacant

APPENDIX 5
Attendance at AAAS (ANZAAS) Congresses

Congress number	Year	Place	Attendance
1	1888	Sydney	850
2	1890	Melbourne	1162
3	1891	Christchurch	550
4	1892	Hobart	600
5	1893	Adelaide	488
6	1895	Brisbane	524
7	1898	Sydney	685
8	1900	Melbourne	693
9	1902	Hobart	550
10	1904	Dunedin	689
11	1907	Adelaide	335
12	1909	Brisbane	647
13	1911	Sydney	820
14	1913	Melbourne	626
15	1921	Hobart (in lieu of Melbourne)	950
16	1923	Wellington	670
17	1924	Adelaide	850
18	1926	Perth	1045
19	1928	Hobart	970
20	1930	Brisbane	780
21	1932	Sydney	1200
22	1935	Melbourne	1410
23	1937	Auckland	1296
24	1939	Canberra	1200
25	1946	Adelaide	1400
26	1947	Perth	1450
27	1949	Hobart	1650
28	1951	Brisbane	N/A
29	1952	Sydney	2620
30	1954	Canberra	1987
31	1955	Melbourne	3500
32	1957	Dunedin	2070
33	1958	Adelaide	2525
34	1959	Perth	2000
35	1961	Brisbane	2800
36	1962	Sydney	3600
37	1963	Canberra	3860
38	1965	Hobart	N/A
39	1967	Melbourne	N/A
40	1968	Christchurch	2760

Congress number	Year	Place	Attendance
41	1969	Adelaide	N/A
42	1970	Port Moresby	N/A
43	1971	Brisbane	N/A
44	1972	Sydney	2170
45	1974	Perth	3600
46	1975	Canberra	2200
47	1976	Hobart	2750
48	1977	Melbourne	2800
49	1979	Auckland	3500
50	1980	Adelaide	2020
51	1981	Brisbane	2930
52	1982	Sydney	2816
53	1983	Perth	3800
54	1984	Canberra	2500
55	1985	Monash University	3500
56	1987	Palmerston North	1750
57	1987	Townsville	1300

These figures are approximate. In each year, the figure includes associates and accompanying spouses. Where no figure is supplied, no records are available.

Sources: Between 1888 and 1902, tables given in congress reports; between 1904 and 1949, lists of members in congress reports (these lists also include members outside Australia and New Zealand); from 1952, the reports of the honorary general secretary, appearing in the *Australian Journal of Science*. Since 1971 these reports have been held at the ANZAAS office in Sydney.

APPENDIX 6
AAAS/ANZAAS Medallists and Lecturers

ANZAAS MEDAL

The ANZAAS Medal is awarded for services to the advancement of science in Australia and New Zealand; 'science' is taken to include all subjects in the various Sections of ANZAAS. The first medal was awarded at the Hobart Congress in 1965 to Professor J. R. A. McMillan, dean of the faculty of agriculture at the University of Sydney (and honorary general secretary of ANZAAS) for his contributions to agricultural science.

1965	Professor J. R. A. McMillan
1967	L. A. Bull
1968	Professor R. N. Robertson
1969	E. H. Derrick
1970	A. B. Walkom
1971	Sir John Crawford
1972	C. A. Fleming
1973	Sir Ian Wark
1975	Sir Frederick White
1976	Professor E. J. Underwood
1977	H. C. Coombs
1979	Sir Mark Oliphant
1980	Professor F. Fenner
1981	Sir Geoffrey Badger
1982	Sir Gustav Nossal
1983	Emeritus Professor Dorothy Hill
1984	J. P. Wild
1985	Professor Mollie Holman
1987	Emeritus Professor Robert Hanbury Brown

MUELLER MEMORIAL MEDAL

At the Hobart meeting of the general council held on 8 January 1902, it was decided to accept the offer of the committee of the Baron von Mueller National Memorial Fund to hand over money collected by it to the AAAS for the purpose of founding a medal. The recipients of the medal, chosen by a select committee, were to be involved in 'work having special reference to Australasia'. The first medal was awarded to A. W. Howitt, a distinguished ethnologist and geologist, at the Dunedin Congress in 1904

1904	A. W. Howitt
1097	J. P. Hill
1909	Professor T. W. Edgeworth David

1911	Robert Etheridge
1913	Walter Howchin
1921	R. T. Baker and C. Chilton
1923	J. H. Maiden
1924	A. Gibb Maitland
1926	F. Wood-Jones
1928	Leonard Cockayne
1930	Sir Douglas Mawson
1932	J. M. Black
1935	R. J. Tillyard
1937	E. W. Skeats
1939	T. Harvey Johnston
1946	E. C. Andrews
	C. T. White
1949	W. J. Dakin
1951	W. N. Benson
1952	Heber Longman
1954	J. A. Prescott
1955	L. B. Bull
1957	Emeritus Professor A. P. Elkin
1958	Hedley R. Marston
1959	William R. Browne
1961	Ian M. Mackerras
1962	Sir Macfarlane Burnet
1964	Frank J. Fenner
1965	M. J. D. White
1967	Professor Dorothy Hill
1968	Norman H. Taylor
1969	J. C. Beaglehole
1970	R. N. Robertson
1971	Professor W. E. H. Stanner
1972	D. F. Waterhouse
1973	R. J. Moir
1975	A. E. Ringwood
1976	L. D. Pryor
1977	A. K. McIntyre
1979	W. V. Macfarlane
1980	J. Waring
1981	J. F. Sprent
1982	Isobel Bennett
1983	Dr L. J. Webb
1984	Dr L. A. S. Johnson
1985	Dr R. Woodall
1986	H. B. S. Womersley

THE LIVERSIDGE LECTURE

The 'founding father' of ANZAAS, Professor Archibald Liversidge, died in September 1927. He bequeathed a sum of money to the Association for a research lecture in chemistry to be delivered at each ANZAAS congress.

1930	Professor N. T. M. Wilsmore
1932	Dr A. C. D. Rivett

1935 Sir D. O. Masson
1937 Dr T. Rigg
1939 Professor H. G. Denham
1946 Dr A. Albert
1947 Associate Professor E. Heymann
1949 A. R. Penfold
1951 Professor D. P. Mellor
1952 Dr J. R. Price
1954 Professor A. E. Alexander
1955 Dr F. B. Shorland
1957 Professor R. H. Stokes
1958 Professor G. M. Badger
1959 Dr F. P. Dwyer
1961 Professor R. D. Brown
1962 Dr A. H. Cole
1964 Dr I. E. Newnham
1965 Professor S. J. Angyal
1967 Professor J. M. Swan
1968 A. T. Wilson
1969 W. I. Whitton
1970 Dr K. L. Sutherland
1971 Professor D. R. Stranks
1972 Professor N. S. Bayliss
1973 Dr A. J. Parker
1975 R. W. Rickards
1976 Professor L. E. Lyons
1977 Dr D. H. Solomon
1979 Professor D. A. Buckingham
1980 Dr L. N. Mander
1981 Dr M. Gerlock
1982 Professor L. F. Phillips
1983 Professor A. M. Sargeson
1984 Professor J. S. Anderson
1985 Professor T. W. Healy
1987 Dr A. Brodie

THE GIBLIN LECTURE

This lecture was instituted in 1958 in honour of Professor L. F. Giblin, an economist and statistician of great renown. The lecturer was to be chosen for his contributions to economics, public administration, and government.

1958 Sir Douglas Copland
1964 Sir Ronald Walker
1965 W. B. Reddaway (Cambridge)
1967 Sir Leslie Melville
1968 A. R. Low (New Zealand)
1969 Dr H. C. Coombs
1970 Sixto K. Roxas
1871 G. A. Rattigan
1972 T. W. Swan
1973 Professor J. Parkin
1975 Professor P. Karmel

1976 Sir Roland Wilson
1977 Hon. J. E. Isaacs
1979 A. J. Culyer
1980 Austin Holmes
1981 Professor N. Park
1982 Professor F. Gruen
1983 Professor R. H. Snape
1984 Professor G. Brennan
1985 Professor Richard Blandy
1987 Professor A. Lloyd

A GUIDE TO SOURCES

Members of ANZAAS and students of its history have much material on which to base further enquiry. An exhaustive bibliography would be difficult to compile, but it is possible to suggest in outline the range of these sources. In general terms, they can be divided into four distinct areas:

1. the Association's administrative records and congress reports;
2. the Association's publications and periodicals;
3. personal papers and other sources; and
4. historical and contemporary accounts and commentaries.

ADMINISTRATIVE RECORDS OF AAAS/(ANZAAS)

In its original constitution, the Association was governed by a general council, composed of delegates from the different colonies, and scientific societies in those colonies. There was also a general committee consisting of members of the council, presidents and other Section officers, and contributors of papers to the Sections. However, new rules were confirmed at Hobart in 1892, which gave the general council the same membership as the general committee. As the committee thus practically duplicated the council, it was done away with. The general council continued in this much larger form until 1970, when on the recommendations of the Crawford Committee, it was renamed convocation, and a new smaller council of about twenty members was created (The Crawford Report was published in *Search*, 1 (1970), 3–5). At the Melbourne Congress in 1955 the general committee was reinstated and made responsible for the control and management of the Association's affairs.

Not surprisingly, given the peripatetic character of the ANZAAS congresses, many administrative records were not returned to head office in Sydney. Today, the bulk of the ANZAAS archives will be found in the Mitchell Library, within the State Library of New South Wales; with a few remaining in the libraries of other capital cities where the congresses were held. The records in the Mitchell Library were transferred from head office in a series of donations between 1914 and 1979. These records are complete in some areas but meagre in others. Archives dating from later than 1950 are as yet uncatalogued. A reasonably complete listing of records that are catalogued is available on application to the Mitchell Librarian. Throughout this guide, numbers in parentheses after record titles refer to the manuscript (ML MSS) numbers of the Mitchell Library, unless otherwise stated. Outside Sydney, libraries and archives in the other capital cities have collections of newspaper clippings, ANZAAS publications, local committee records and personal papers of ANZAAS members.

The Mitchell holds volumes of council minutes from 1888 to 1911 (ML MSS 988/1, 988/7, 988/5). It should be noted that as the general council and the general committee met only during congresses, the minutes held are only for congress years. There are no council minutes for 1913 or 1921, but from 1923, the second congress after the First World War, the minutes continue uninterrupted until 1952 (ML MSS 988/6). Minutes between 1952 and 1972 have been deposited, but are as yet uncatalogued (ML MSS 1613 add-on 344, add-on 1121). Researchers will find the general committee's minutes with those of the council for the years 1888–91 (ML MSS 988/1). The second general committee held its first meeting on 19 August 1955, and its minutes for 1955 onwards are an excellent source for the major concerns of ANZAAS (are kept at our ML MSS 1613 add-on 344, add-on 1121).

In the years between congresses, most administrative work fell to the local committees in each colony or state, and to the permanent honorary secretary (or honorary general secretary as the position was later styled), based in Sydney. However, there are few records of the different local committees, no distinct records for the honorary secretary, and very little correspondence surviving from the years before 1950. Registers of 'letters received' between 1892 and 1905 do exist, and occasionally other letters will be found in the general files. Some correspondence between Archibald Liversidge and the Linnean Society, and the circular letters sent delegates to the first meeting of the Association, have been preserved (ML MSS 1613, 988/1).

The AAAS and ANZAAS had many committees for special purposes. The research and investigation committees are far too numerous to list here, but their proceedings can be found in the congress reports and in the minutes of the recommendations committee. Unfortunately, no original records of the early research committees survive. However the records of some of the more recent short-term committees, such as Steering Committees and Symposiums on Road Safety (1966–71), Science Policy (1967–74) and the aged (1968–69), do remain (ML MSS 1613 add-on 1121 boxes 3K 21206–21212).

The publications committee, responsible for producing the congress reports, the *Australian Journal of Science*, and *Search*, is represented by minutes for the years 1888 (ML MSS 988/1) and 1954–68 (ML MSS 1613 add-on 344). Correspondence, in alphabetical order, survives for 1958–70 (ML MSS 1613 add-on 1121). General files from 1952 onwards contain miscellaneous records of articles published, journal exchanges, orders, reviews, and circulation figures (ML MSS 1613 add-on 1121).

When Section meetings at congresses passed resolutions concerning research, government initiatives, and changes to Sections or to the Association, these were examined by a recommendations committee, which undertook to decide which should be placed before the general council for action. The recommendations committee held its first meeting in January 1891. Minutes are held for the years 1891 to 1911 only (ML MSS 988/17).

In 1930 it was found necessary to create an executive committee of the general council to manage the general affairs of the Association. The Mitchell holds executive committee minutes only for the years 1967–8 (ML MSS 1613 add-on 1121) and correspondence for 1952–72 (ML MSS 1613 add-on 1121). The finance committee's minutes for 1969–70 (ML MSS 1613 add-on 1121) and correspondence for 1968–72 (ML MSS 1613 add-on 1121) has also been kept.

Information on the finances of the AAAS and ANZAAS until 1954 can be found in the congress reports, which give accounts of receipts and expenditure for the previous financial year. There are also ledgers although their sequence is not complete; these cover the period 1933–55 (ML MSS 1613 add-on 168), and there are also account books, cash books, ledgers, and postage books for the years 1955–72 (ML MSS 1613 add-on 1121).

AAAS/(ANZAAS) PUBLICATIONS

The first publications of ANZAAS were its congress reports, beginning with the Sydney meeting in 1888. (The first volume is often found mistakenly labelled '1887' on library shelves.) The reports are often the best starting point for research. This is especially true of the earlier reports, which chronicle the sequence of officers of special committees, and give abbreviated accounts of the proceedings of the council and committees, as well as the papers presented to the Sections, and (until 1952) membership lists. A complete set of these reports printed in separate volumes (1888–1954) can be found in the Mitchell Library, in the State Libraries in Sydney, Melbourne, Brisbane and Adelaide, and in the National Library in Canberra.

Congress proceedings were usually covered in great detail by the local press, especially in the early years. In fact, the first set of volumes of minutes of the Section meetings, covering the period 1888–1935, consists largely of newspaper cuttings (ML MSS 988/9–16, Sections B to J only). Researchers will find that newspapers often reported proceedings almost verbatim and are often more informative than official minutes. Their accuracy can be measured against the minutes, or by using the volumes of newscuttings, annotated by the Section secretaries, held in the Mitchell for the years 1888–1909 (ML MSS 988/89–100).

In 1914 the British Association for the Advancement of Science met in Sydney, an event anticipated for almost thirty years. There are records of the preparations made in Australia, including the minutes of the general committee (ML MSS 988/18), the business committee (ML MSS 988/19), the executive committee (988/20), and the reception and hospitality committee (ML MSS 988/21). The Australasian Association also produced a *Handbook for New South Wales* (Sydney: BAAS, 1914) for British visitors, containing chapters on the state's history, culture, legislation, natural history etc. In Britain, the British Association for the Advancement of Science holds files on Australasian matters; not only on the 1914 meeting, but also on scientific research, organizations and expeditions in Australia, New Zealand, and the Pacific. There are five boxes of material, covering the period 1914–1952. The Association records are housed in the Bodleian Library, Oxford.

There were no congresses between 1914 and 1921, but during this period two progress reports were produced, in 1918 and 1920. Copies are available in the Mitchell (ML MSS 1613). They were written by the honorary secretaries, and reported on correspondence, finances, and the activities of the research committees during this period.

The organization of ANZAAS congresses in a host capital was conducted by small groups known variously as local councils or local committees. These committees also managed the affairs of the Association in each colony or state between congresses. Only the records of the local committee in Hobart, 1926–7 (ML MSS 988/18), and in Sydney, 1908–11 (ML MSS 988/3, 988/4), are held at the Mitchell. However, some local committee minutes have been preserved in other capitals. The Barr Smith Library in the University of Adelaide has a small number of publications relating to the 1893, 1907, and 1924 Congresses. More material is available in the South Australian Collection of the State Library in Adelaide, including programmes, handbooks, and newspapers, and in the Archives of the small library there is a minute book of the local committee 1888–93, papers relating to the 1969 Congress, and records and programmes of the 1907 and 1908 Congresses.

In Brisbane the John Oxley Library has congress reports, handbooks for 1895 and 1930, and newspapers clippings. The State Library of Tasmania has programmes and circulars for the Hobart meetings and handbooks on Tasmania for the 1892, 1902, 1928, and 1949 Congresses, as well as various publications of congresses in other cities, and folders of miscellaneous material concerning ANZAAS for 1925–30, and the Anthropological Section in 1928.

The Library of the University of Western Australia has no original documents, but holds reports for 1930–58, and the handbooks for the 1926 and 1947 meetings in Perth. The Western Australian Branch of the Australian Archives holds a Department of External Affairs life on the Congress in Perth in 1973.

A number of libraries in New Zealand have records concerning congresses held there. There are some records of the second Congress held in Christchurch in 1968, in the University of Canterbury's Library including reports from organising committees and Sections. The City of Dunedin Public Library has reports and programmes for the meetings held there in 1904 and 1957, as well as newspaper cuttings. The University of Auckland Library holds press cuttings and minutes of the publicity committee from the 1979 Congress, and two handbooks from the 1937 Congress. The Auckland Public Library holds many more publications from the 1979 Congress. The records of the Department of Scientific Industrial Research, in the National Archives of the Department of Internal Affairs, include several files referring to ANZAAS. Of particular interest are files (SIR 39/12) on overseas scientific organizations and ANZAAS, 1921–58, and of the Board of Science and Art, which mentions ANZAAS congresses (1A 2/48/4).

For the 1970 Congress held in Port Moresby, there are organizational records in the Archives of the University of Papua New Guinea, deposited by the then secretary, Professor Donald Drover. There is also some secondary material in the Papua New Guinea Collection of the National Library.

There are also approximately thirty boxes of 'Congress Material' for the years 1959 to 1977 in the Mitchell (ML MSS 1613 add-on 1121). These contain correspondence concerning arrangements for congresses, and minutes and agendas for council and committees on presidential addresses, etc. Those looking for pictorial records of congresses will find a small selection, mostly of the earlier period, in the Mitchell (PIC ACC 332). There were also often pictorial supplements in the illustrated papers of the period.

ANZAAS has produced two periodicals, the *Australian Journal of Science*, and *Search*. A scientific journal with the name *Australian Journal of Science* was proposed by Archibald Liversidge as early as 1905, but this did not eventuate. The journal with the same name, which first appeared in 1938, was published by the Australian National Research Council (ANRC) until 1954 (vols 1 to 16, 1938 to 1953/4), when ANZAAS itself took over publication. When the *Journal of Australian Science* ceased publication in 1970 it was replaced by *Search* (vol. 1 no. 1, July 1970).

From the Melbourne Congress of 1955 to the Adelaide congress of 1958, reports were published as supplements to the *Australian Journal of Science* (vols 18 to 20). However, from the Perth Congress of 1959 to the present, only presidential addresses and the titles of papers were published, interspersed with other articles. When, in 1970, the *Australian Journal of Science* was replaced by *Search* this practice continued. *Search* still publishes the principal presidential address, memorial lectures and sectional presidential addresses, and some papers submitted by authors, but researchers will find more complete information in the continuing series of microfiche copies of ANZAAS papers and programmes compiled by the University of New South Wales. These microfiche sets have copies of all papers submitted by authors to the University from all congresses since 1970. This project includes author, subject, and keyword indexes. The only other index to ANZAAS papers was compiled sixty years ago by George M. Thomson, *Index to Volumes 1 to XVI: 1888 to 1923* (Sydney: ANZAAS, 1925).

Apart from reports and journals, the Association also produced handbooks and programmes for each congress. Handbooks for the years 1888 to 1951 are held in Mitchell (ML MSS 988/66–88). An almost complete set of both handbooks and programmes is held in the ANZAAS head office in Sydney. The University of New South Wales has microfiche copies of the programmes from 1970 onwards. Many local libraries and archives include copies in their collections of ANZAAS material. The handbooks include the following:

1888   William M. Hamlet (ed.), *Handbook of Sydney, for the Use of the Members of the Australasian Association for the Advancement of Science* (Sydney: Turner and Henderson, 1888).

1890   W. Baldwin Spencer (ed.), *Handbook of Melbourne, for the Use of the Members of the Australasian Association for the Advancement of Science* (Melbourne: Spectator, 1890).

1891   F. W. Hutton (ed.), *Handbook of Christchurch, for the Use of the Members of the Australasian Association for the Advancement of Science* (Christchurch: Government Printer, Lyttelton Times Office, 1891).

1892   Alexander Morton (ed.), *Handbook for the Use of Members of Australasian Association for the Advancement of Science* (Hobart: Government Printer, 1892).

1898   W. M. Hamlet (ed.), *Handbook of Sydney and the County of Cumberland* (Melbourne: G. Robertson & Co., 1898).

1898   W. M. Hamlet (ed.), *Handbook of Sydney and the County of Cumberland, for the Use of the Members of the AAAS* (Sydney: George Robertson, 1898).

1900   W. Baldwin Spencer (ed.), *Handbook of Melbourne, for the Use of Members of the Australasian Association for the Advancement of Science* (Melbourne: Ford & Son, 1900).

1900   H. Hill, *The Extension of University and Science Work in New Zealand* (New Zealand: Transactions of the New Zealand Institute, 1900), 33, 395–406.

1902   L. Rodway (ed.), *Handbook of the Australasian Association for the Advancement of Science* (Hobart: Government Printer, 1902).

1904   A. Bathgate (ed.), *Dunedin and its Neighbourhoood* (Dunedin: Otago Daily Times, 1904).

1911   J. W. Joynt, 'University Development in New Zealand', *British Empire Review*, 5 (1911), 257–62.

1914   G. H. Knibbs (ed.), *Federal Handbook prepared in Connection with the Eighty-Fourth Meeting of the British Association for the Advancement of Science* (Canberra: Commonwealth Government Printer, 1914).

1914   W. S. Dun (ed.), *Handbook for New South Wales* (Sydney: AAAS/ Edward Lee, 1914).

1914   A. M. Laughton (ed.), *Handbook for Victoria* (Melbourne: AAAS, 1914).

1914   AAAS, *Excursions Arranged for Members of [the] British Association for the Advancement of Science* (Sydney: Government Printer, 1914).

1921   G. Sweet and A. C. D. Rivett (eds), *Hobart Meeting, held in Melbourne, January 1921* (Hobart: Government Printer, 1921).

1924   Bessie Threadgill, *et al.* (eds), *Handbook for the Use of Members of the [Australasian] Association [for the Advancement of Science]* (Adelaide: Government Printer, 1924).

1926   A. Gibb Maitland (ed.), *Science in Western Australia* (Perth: Government Printer, 1926).

1927 L. F. Giblin et al. (eds), *Handbook to Tasmania* (Hobart: Government Printer, 1927).
1928 L. F. Giblin, A. N. Lewis and Clive Lord (eds), *Handbook to Tasmania prepared for the Members of the Australasian Association for the Advancement of Science* (Hobart: Government Printer, 1928).
1930 W. H. Bryan, H. A. Longman and J. F. F. Reid (eds), *Handbook for Queensland, prepared for Members of the Australasian Association for the Advancement of Science* (Brisbane: Government Printer, 1930).
1932 G. A. Waterhouse (ed.), *Handbook for New South Wales, prepared for the Members of the Australian and New Zealand Association for the Advancement of Science* (Sydney: Government Printer, 1932).
1935 H. J. Green (ed.), *Handbook for Victoria* (Melbourne: Government Printer, 1935).
1937 R. A. Falla (ed.), *Handbook for New Zealand, prepared for Members of the Australian and New Zealand Association for the Advancement of Science* (Wellington: Government Printer, 1937).
1937 John Barr, et al. (eds), *Auckland* (Auckland: ANZAAS, 1937).
1939 Kenneth Binns (ed.), *Handbook for Canberra, prepared for the Members of the ANZAAS* (Canberra: Commonwealth Government Printer, 1939).
1946 C. Fenner (ed.), *Handbook of South Australia* (Adelaide: Government Printer, 1946).
1947 J. B. Campbell (ed.), *Handbook of Western Australia* (Perth: Government Printer, 1949).
1948 L. Bastings (ed.), *Directory of New Zealand Scientists* (Wellington: Harry H. Tombs, 1948).
1949 L. Cerutty (ed.), *Handbook for Tasmania prepared for the Members of the ANZAAS* (Hobart: Government Printer, 1949).
1951 George Mack (ed.), *Handbook of Queensland* (Brisbane: ANZAAS, 1951).
1954 H. L. White (ed.), *Canberra: A Nation's Capital* (Sydney: ANZAAS, 1954).
1955 G. W. Leeper (ed.), *Introducing Victoria* (Melbourne: University of Melbourne Press, 1955).
1957 F. R. Callaghan (ed.), *Science in New Zealand* (Wellington: Reed, 1957).
1958 R. J. Best (ed.), *Introducing South Australia* (Adelaide: ANZAAS, 1958).
1961 W. H. Bryan et al. (eds), *Introducing Queensland: A Handbook prepared for ANZAAS* (Brisbane: Government Printer, 1961).
1965 J. L. Davies (ed.), *Atlas of Tasmania* (Hobart: Government Lands and Survey Dept, 1965).
1969 G. W. Ford (ed.), *Automation: Threat or Promise? Impact and Implications in Australia* (Sydney: Law Book Co., 1969).
1969 M. Williams (ed.), *South Australia from the Air* (Melbourne: Melbourne University Press, 1969).
1969 B. Daily, *Geological Excursions Handbook* (Adelaide: ANZAAS, 1969).
1970 R. G. Ward and A. M. Lea David (eds), *An Atlas of Papua and New Guinea* (Port Moresby: University of New Guinea, 1970).
1971 P. Wood *Brisbane and its River* (Brisbane: University of Queensland, 1971).
1971 G. Playford, *Geological Excursions Handbook* (Brisbane: ANZAAS, 1971).
1977 G. Parkville, *Urbanisation* (Centre for Environmental Studies, University of Melbourne: ANZAAS, 1977).
1979 P. J. Brook *Natural History of Auckland: An Introduction* (Auckland: Auckland War Memorial Museum Handbook, 1979).

1979   ANZAAS, 49th Congress, *A Vanishing Heritage—The Problem of Endangered Species and their Habitats* (Wellington: Nature Conervation Council, 1979).

In recent years ANZAAS has also encouraged the organization of symposia on particular themes, frequently held in conjunction with its congresses and as a function of the active New South Wales division. We can find no exhaustive list and no single complete collection, but the extensive range of these reports merits an indicative list:

1968   ANZAAS (NSW Division), *Papers Based on the Theme of The Symposium entitled the Planning and Management of Australia's Natural Resources* (The Natural Resources of Australia Tables, Armidale: University of New England, 1968), (vii), 334.
1971   ANZAAS (N.S.W.) Division, *Noise Pollution* (Sydney: ANZAAS, 1971).
1972   G. D. McColl, *Nuclear Power: Economic Considerations* (Sydney: ANZAAS and University of New South Wales, 1972).
1972   J. A. Sinden, *The Natural Resources of Australia: Prospects and Problems for Development* (Sydney: ANZAAS, 1972).
1973   G. W. Ford (ed.), *Redundancy* (Sydney: ANZAAS, 1973).
1973   J. E. Glover, and G. Playford (eds), *Mesozoic and Cainozoic Palynology: Essays in Honour of Isabel Cookson* (Canberra: Geological Society of Australia, 1973).
1973   G. A. Kidd, *Professional Manpower in Australia: Prospects, Problems and Policies* (Perth: ANZAAS, 1973).
1973   E. F. Kunz, *Australian Professional Attitudes and the Immigrant Professional Section 27, Sociology* (Perth: ANZAAS, 1973).
1974   M. Dawson (ed.), *Families: Australian Studies of Changing Relationships within the Family and Between the Family & Society* (Sydney: ANZAAS, 1974).
1974   P. F. Gloss (ed.), *Medicine, Technology, Society and Health Care* (Sydney: ANZAAS, 1974).
1974   R. L. Whitemore (ed.), *Minerals: the Future of Australia's Mineral Industry* (Sydney: Wymond Morell Press, 1974).
1975   ANZAAS *Papers Presented by the National Capital Development Commission* (Canberra: AGPS, 1975).
1976   H. Gelber (ed.), *International Politics and World Energy* (Hobart: ANZAAS, Dept of Political Science, University of Tasmania, 1976).
1976   R. M. Gifford, *Energy in Agriculture* (Sydney: ANZAAS, 1976).
1976   J. M. Gilbert (ed.), *Woodchip Symposium Papers* (Hobart: Tasmanian Forestry Commission, 1976).
1977   A. Nelson Johnston *Relevance of Education to Future Changes in Agriculture and Forestry* (Melbourne: Victorian Dept. of Agriculture, 1977).
1977   N. D. McGlashan, *Studies in Australian Morality* (Hobart: University of Tasmania, Board of Environmental Studies, 1977).
1977   R. Monroe and N. S. Stevens (eds), *The Border Ranges: a Land Use Conflict in Regional Perspective* (Brisbane: The Royal Society of Queensland, 1977).
1978   G. Seddon (ed.), *Urbanisation: Papers delivered at the 48th Congress* (Melbourne: Centre for Environmental Studies, University of Melbourne, 1978).
1979   ANZAAS (NSW Division) (ed.), *Automation and Unemployment: An ANZAAS Symposium* (Sydney: Law Book Co., 1979).

1979 Mari Davis (ed.), *Directions for the Future—Communication Papers* (Auckland: ANZAAS, 1979).
1979 J. R. McKinlay and K. L. Jones (ed.), *Archaeological Resource Management in Australia and Oceania* (Wellington: New Zealand Historic Places Trust, 1979).
1980 Wolfgang Kasper, et al., *Australia at the Crossroads* (Sydney: HBJ, 1980).
1980 S. Sax (ed.), *The Aged in Australian Society* (Sydney: Angus & Robertson, 1980).
1980 J. W. Zillman, et al., *Energy, Climate and the Future* (Melbourne: ANZAAS, 1980).
1981 A. G. Palthe and T. M. Parkes, *Messaging the Message—PR and Media Campaign* (Canberra: CSIRO and ANZAAS, 1981).
1982 J. C. Aldred, et al., *Tasmania's Economic Future—Prospects and Possibilities* (Hobart: ANZAAS, Tasmanian Division, 1982).
1982 J. M. Bennett, et al., *Venture Capital and Technological Innovation in Australia* (Sydney: ANZAAS, New South Wales Division, 1982).
1982 M. Roe et al., *Tasmania's Economic Future—Prospects and Possibilities* (Hobart: ANZAAS, Tasmanian Division, 1982).
1983 J. M. Bennett and G. C. Lowenthal (eds), *Manpower Planning and Industrial Development in Uncertain Times* (Sydney: ANZAAS, New South Wales Division, 1983).
1983 Mari Davis (ed.), *Australia's Industrial Future: Communication* (Selected papers from Section 33 delivered to the 52nd Congress, 1982) (Melbourne: Transknowledge Associates, 1983).
1983 Alec Lazenby, et al., *Science, Technology and Employment* (Hobart: ANZAAS, Tasmanian Division, 1983).
1984 Michael Booth and Colin London (eds), *Western Australia: It's Land, Its Future* (Perth: ANZAAS, Western Australian Division, 1984).
1985 R. B. McKern and G. C. Lowenthal (eds), *Limits to Prediction* (Sydney: Australian Professional Publications, 1985).

PERSONAL PAPERS AND OTHER SOURCES

The best guide to other sources is still Ann Mozley Moyal, *A Guide to the Manuscript Sources of Australian Science* (Canberra: Australian Academy of Science with Australian National University Press, 1966). The papers of Professor Archibald Liversidge (1846–1927) can be found in the Sydney University Archives; they include three boxes of material on the Association, newscuttings, programmes, and circulars. There are often references to AAAS affairs in his correspondence and photographs of the proceedings. Michael E. Hoare's, 'Turnbull Library Manuscript Holdings in the History of Australian Science: A Review', *Turnbull Library Record,* 9 (1976), 7–19 is a valuable introduction to an invaluable collection. Accounts may also be found of Dorothy Mabel Laurie, who accompanied number of scientists, including T. Griffith Taylor, L. A. Cotton and C. A. Sussmilch, to the 1923 Congress in Wellington. Her typescript, with photographs and newspaper cuttings, is in the Mitchell Library (ML MSS 4245).

Historians will also find abundant comment in the newspapers and periodicals of the day, especially at the time of the Association's foundation, Most of the Sydney papers featured articles on the birth and growth of the Australasian

Association for the Advancement of Science in 1888, including the *Bulletin* (1880+), the *Australian Town and Country Journal* (1870–1919), the *Sydney Mail* (1860–1938), the *Nation* (1887–90), and *The Echo* (1875–93), as well as the *Sydney Morning Herald* and the *Daily Telegraph*. They continued to do so in the years from 1888 to *c.* 1910. Other colonial newspapers worth scanning are the *Brisbane Courier* (1846–1933); *The Queenslander* (1866–1939); the *South Australian Register* (1839–1900); the *Argus* (1846–1957); the *West Australian* (1833+); the *Australasian* (1864–1946); the *Otago Daily Times* (1861+); the *New Zealand Times* (1854–1927); and the *Canberra Times* (1920+). Some of the illustrated periodicals, such as the *Illustrated Sydney News* (1853–94) and the Hobart *Critic* (1905–1924), reported on the social side of the congresses. Detailed reporting in these local papers was usually restricted to years in which a congress was being held in the same city. Professional periodicals such as the *Chemist and Druggist of Australasia* (1886–1934) and the *Building Engineering and Mining Journal* (1884–1905) also reported on the Association, as did the influential British science journal *Nature*.

From 1920 the newspapers tended to exchange verbatim reporting of proceedings for more enquiring articles on the issues discussed at the congresses. The major daily papers, such as the *Sydney Morning Herald*, the *Daily Telegraph*, and the *Argus* continued to report them, but when researching individual congresses, it is best to make use of the local papers mentioned.

HISTORICAL AND CONTEMPORARY DOCUMENTS

The history of the AAAS and ANZAAS has, until recently, attracted few historians. Today, it has become necessary to set this history into the wider context of Australasian intellectual endeavour. The history of Australasian science has received considerable encouragement and visibility through the *Historical Records of Australian Science*, published by the Australian Academy of Science, and scholars treat L. Carlson's regular bibliographies in that journal as sources of first resort. The references in the foregoing chapters will, of course, include much that is relevant to this subject. For a complement to these, and to highlight other (particularly, more recent) sources relevant to the interpretation of the AAAS and its history, the following list may also prove useful.

PRIMARY SOURCES

Bailey, F. M., 'Concise History of Australian Botany', *Proc. Roy. Soc. of Qld*, 8 (1890–1), xvii–xiii.

Cameron, A. M., 'Australian Meteorology', *Sydney University Magazine*, 1 (1878).

David, T. W. Edgeworth, 'The Aims and Ideals of Australasian Science' *Report of the AAAS*, 10 (Dunedin, 1904), 1–43.

——, 'University Science Teaching' in *Record of the Jubilee Celebrations of the University of Sydney* (Sydney: William Brooks & Co. Ltd, 1903), 93–121.

Dayton, W. T., *Catalogue of the Scientific Serial Literature in the Following Libraries in Sydney, N. S. W.: The Australian Museum, Free Public Library, Linnean Society of N. S. W., Observatory, Parliamentary Library, Royal Geographical Society of N. S. W., Royal Society of N. S. W., Technological*

Museum and the University of Sydney (Sydney: Government Printer, 1889).

Hutton, F., and Drummond, J., *The Animals of New Zealand* (Christchurch and Wellington: Whitcombe and Tombs, 1904).

Lauterer, J., 'Progress of Science in Australia', *Proc. Roy. Soc. of Qld*, 12 (1897), viii–xix.

Lindsay, W. Lauder, 'The Place and Power of Natural History in Colonisation with Special Reference to Otago, New Zealand', *Edinburgh New Philosophical Journal*, 17 (Jan–June 1863), 125–46.

Liversidge, Archibald, 'The Australian Association for the Advancement of Science', *Nature*, 82 (1909–10), 264–6.

MacCallum, M. W., 'The Federation of the Universities', *Hermes* (Jubilee Issue, 1902), 51–58.

——, 'University Development', *Australian Journal of Education*, 4 (12) (1907), 6–8.

Macdonald, A. C., 'The Utility of, and Necessity for, a Geographical Society' *Proc. of the Geog. Soc. of Australasia (N. S. W.) and Victoria*, 1 (1883–4), 133–6.

Maiden, J. H., 'A Century of Botanic Endeavour in South Australia', *Report of the AAAS*, 11 (Adelaide, 1907), 158–99.

Maiden, J. H., 'Portraits of Scientific Men of New South Wales', *Proc. Roy. Soc. of NSW*, XLVI (1912), 17–20.

——, 'Records of Queensland Botanists', *Report of the AAAS*, 12 (Brisbane, 1909), 373–84.

——, 'Records of Tasmanian Botanists', *Papers and Proc. Roy. Soc. of Tas.* (1908), 101–17.

——, 'Records of Western Australian Botanists', *Journal of Western Australian Natural History Society* (1909).

Newman, A. K., 'A New Zealand Association of Science', *NZ Journ. of Sci.*, 1 (1883), 145–50.

Park, James, *The Cyanide Process of Gold Extraction* (Auckland: Chamtaloupe & Cooper, 1897).

Robertson, J. Steele, 'Natural Science in Australia', *Centennial Magazine*, 2 (1889–90), 523–7.

Russell, H. C., 'President's Address', *Report of the AAAS*, 1 (Sydney, 1888), 1–19.

Stephens, William J., 'Biological Science: A Necessary Factor in University Work', *Sydney University Review*, 4 (December 1882), 393–407.

von Haast, J., 'On the Progress of Geology: An opening address delivered to the students of Canterbury College on March 28th 1883', *NZ Journ. of Sci.*, 1 (1883) 395–406.

SECONDARY SOURCES

Abbie, A. A., 'The History of Biology in Australia', *Aust. Journal of Science*, 17 (1954), 1–9.

Allen, J. A., 'Government/Industry Collaboration in Industrial Science', *Search*, 6 (1975), 410–14.

——, 'Scientific Innovation and Industrial Prosperity', *Proc. Royal Aust. Chem. Inst.*, 30 (1963), 377–400.

Anderson, J. R., 'Science Policy for Australia: A Scientific View', Paper read at

the Australian Industrial Research Group Symposium: Science Policy for Australia, Canberra, 10–11 February 1972, *Search*, 3 (1972), 209–12.
Antill, J. M., 'Role of the Expert Witness', *Journal Institution Engineers Aust.*, 48 (July/August 1976), 33–4.
ANZAAS, 'Report of the ANZAAS General Committee on Bringing Science to the Public', *Aust. Journal of Science*, 24 (1961), 27.
——, *National Goals and Research Needs*, Report no. 9 (Canberra, February 1974).
Australian Academy of Science, 'Science and Society Forum', *Science and Society in Australia*, Public Forum, University of Melbourne, 9 November 1974.
ANU, *Science in Australia* (Melbourne: Cheshire, 1951).
ASTEC, *Basic Research and National Objectives* (Canberra: AGPS, 1981).
——, *Future Directions for CSIRO* (Canberra: AGPS, 1985).
Atkinson, J. D., *DSIR'S First Fifty Years* (Wellington: Dept of Scientific and Industrial Research, 1976).
Australian Academy of Science, *From Stump-Jump Plough to Interscan* (Canberra: Australian Academy of Science, 1977).
Badger, Sir Geoffrey, 'The Role of Government in Australian Science', *Australian Physicist*, 17 (10) (November, 1980), 157–60.
Baldock, C. V. and Lally, J., *Sociology in Australia and New Zealand* (Westport: Greenwood Press, 1974).
Bambrick, Susan, 'First Commonwealth Statistician, Sir George Knibbs', *Proc. Roy. Soc. of NSW*, 102 (1969), 127–35.
Barrett, J. (ed.), *Save Australia. A Plea for the Right Use of our Flora and Fauna* (Melbourne: Macmillan, 1925).
Basalla, George, Coleman, William and Kargon, Robert H. (eds), *Victorian Science: A Self-Portrait for the Presidential Addresses to the British Association for the Advancement of Science* (New York: Doubleday & Co., 1970).
Bastings, L. (ed.), *Dictionary of New Zealand Scientists* (Wellington: Harry H. Tombs, 1948).
Bastow, S. H., 'CSIRO and the Universities', *Vestes*, 4 (1961), 21–7.
Baxter, J. P., 'Industrial Research and Development in Australia', *Symposium on the Industrial Development of Australia* (1960), 33–42.
Bolton, G. C., *Spoils and Spoilers: Australians Make Their Environment* (Sydney: George Allen & Unwin, 1981).
Borchardt, D. H. (ed.), *Some Sources for the History of Australian Science*, Historical Bibliography Monograph, no. 12 (Sydney: University of New South Wales, History Project, 1982).
Brady, E. J., *Australia Unlimited* (Sydney: George Robertson, 1918).
Branagan, D. F., 'The Geological Society of Australasia, 1885–1907', *Journ. of Geol. Soc. of Aust.*, 23 (2) (1976), 169–82.
——, 'J. W. Gregory, Traveller in the Dead Heart', *Historical Records of Australian Science*, 6 (1) (1984), 71–84.
——, 'Putting Geology on the Map', *Historical Records of Australian Science*, 5 (2) (1981), 30–57.
——, 'Words, Actions, People: 150 Years of Scientific Societies in Australia', *Proc. Roy. Soc. of NSW*, 103 (1971), 123–41.
—— 'The *Challenger* Expedition and Australian Science', *Proc. Roy. Soc. of Edinburgh*, 73 (10) (1971–2), 85–95.
—— (ed.), *Rocks-Fossils-Profs: Geological Sciences in the University of Sydney, 1866–1973* (Sydney: Science Press, 1973).
Branagan, D. F. and Holland, Graham (eds), *Ever Reaping Something New: a*

Science Centenary (Sydney: University of Sydney, Science Centenary Committee, 1985).

Branagan, D. F. and Townley, K. A., 'The Geological Sciences in Australia—A Brief Historical Review', *Earth Science Reviews*, 12 (1976), 323–46.

Branagan, D. F. and Vallance, T. C., 'The Geographical Society of Australasia, 1885–1905', *Journ. of Geol. Soc. of Aust.*, 14 (1967), 249–51.

Brown, A. W., 'The Economic Benefit to Australia from Atomic Absorption Spectroscopy', *Economic Record*, 45 (1969), 158–80. Comment by Converse, A. O., 'The Economic Benefit to Australia from A. S. S.', *Economic Record*, 46 (1970), 261–2. Reply by Brown, A. W., *Economic Record*, 46 (1970), 263.

——, 'The Role of Government in the Encouragement of Research in Industry', *Public Administration*, 29 (1970), 339–55.

Bryan, W. H., 'Samuel Stutchbury and Some of Those who Followed Him', *Queensland Government Mining Journal*. LV (1954), 641–6.

Buckley-Moran, Jean, 'Australian Scientists and the Cold War' in B. Martin *et al.*, *Intellectual Suppression* (Sydney: Angus & Robertson, 1986).

Burntee, R. I., The Life and Work of James Hector with Special Reference to the Hector Collection (unpublished MA Hons (History) thesis, University of Dunedin, New Zealand, 1936).

Butcher, B. W., 'Science and the Imperial Vision: The Imperial Geophysical Experimental Survey, 1928–1930', *Historical Records of Australian Science*, 6 (1) (1984), 31–43.

Cairns, D., *Scientific Institutions in New Zealand, 1949* (Christchurch: C.S.W., 1949).

Callaghan, F. R., (ed.), *Science in New Zealand*, (Wellington: A. H. & A. W. Reed, 1957).

Cambie, R. C. and Davis, B. R., *A Century of Chemistry at the University of Auckland* (Auckland: Percival Publishing Co., 1982).

Campbell, D., Culture and the Colonial City: A Study in Ideas, Attitudes and Institutions: Sydney, 1870–1890 (unpublished PhD, thesis, University of NSW, 1982).

Caroe, G. M., *William Henry Bragg, 1862–1942* (Cambridge: Cambridge University Press, 1978).

Carr, D. J. and S. G. M. (eds), *Plants and Man in Australia* (Sydney: Academic Press, 1981).

Chapman, H. G., 'A History of Science in New South Wales' in R. H. Cambage (ed.), *Handbook for New South Wales* (Sydney: British Association for the Advancement of Science, 1914).

Cheney, B. J., 'Is Science Dehumanising Scientists?', *Science Review*, 18 (Melbourne, 1963), 15–18.

Clark, Betty, 'Men, Minerals and Museums: A Century (and more) of Mineral History in New South Wales. Parts 1–3', *Australian Lapidary Museum*, 11 (November 1974), 4–15; (December 1974), 27–34; (January 1975), 25–30.

Clarke, N., *The Political Economy of Science and Technology* (Oxford: Blackwell, 1985).

Cockayne, L., *The Vegetation of New Zealand* (Leipzig: Verlag von Wilhelm Engelmann, 1928).

Cockburn, S. and Ellyard, D., *Oliphant: The Life and Times of Sir Mark Oliphant* (Adelaide: Axiom Books, 1981).

Connell, R. W., 'The Colonial Mentality in Social Science', *Search*, 15 (1984), 100–11.

Corbett, D. W. P. *et al.*, 'Geology', in *Ideas and Endeavours—The Natural Sciences in South Australia* (Adelaide: Royal Society of SA, 1986), 29–67.

Craig, D. P., 'Physical Chemistry at the University of Sydney', *Aust. Journal of Science*, 16 (1954), 138-9.
Crawford, Sir John, 'Report by the Chairman of the Committee of Review', *Search*, 1 (1970), 3-5.
Crawford, R. H., '*A Bit of a Rebel*'. *The Life and Work of George Arnold Wood* (Sydney: Sydney University Press, 1975).
Crough, Greg and Wheelwright, Ted, *Australia: A Client State* (Ringwood: Penguin, 1982).
Currie, George and John Graham, 'C. A. Julius and Research for Secondary Industry (CSIR)', *Records of the Aust. Academy of Science*, 2 (1970), 10-28.
——, 'CSIR, 1926-1939', *Public Administration*, 33 (September 1974), 230-52.
——, 'Growth of Scientific Research in Australia: The Council for Scientific and Industrial Research and the Empire Marketing Board', *Rec. of Aust. Academy of Science*, 1 (3) (1968), 25-35.
——, *The Origins of CSIRO: Science and the Commonwealth Government, 1901-1926* (Melbourne: Melbourne University Press, 1966).
David, M. E., *Professor David. The Life of Sir Edgeworth David* (London: Edward Arnold & Co., 1937).
Davidson, B. R., *European Farming in Australia* (Amsterdam: Elsevier, 1981).
Davidson, F. G., *The Industrialisation of Australia*, 3rd edn (Melbourne: Melbourne University Press, 1962).
Denham, H. G., 'Modern Development in the Industrial World', *Report of the AAAS*, 16 (Wellington, 1923), 1-46.
Denmead, A. K., 'The Geological Survey of Queensland is 100 Years Old', *Queensland Government Mining Journal*, 69 (1968), 145-8.
Dick, Ian, 'The History of Scientific Endeavour in New Zealand', *New Zealand Scientific Review*, 9 (September 1951), 139-43.
Dixon, J. K. and Callaghan, F. R. (eds), *Science in New Zealand* (Wellington: A. H. and A. W. Reed, 1957).
Donath, E. J., *William Farrer* (Oxford: Oxford University Press, 1962).
Dugan, K. G., 'The Zoological Exploration of the Australian Region and its Impact on Biological Theory' in N. Reingold and M. Rothenberg (eds), *Scientific Colonialism: A Cross Cultural Comparison* (Washington: Smithsonian Institution Press, 1987), 79-100.
Dwyer, P. D., 'Beyond Biology: Alienation and Environment', *Search*, 2 (1971), 153-9.
Elkin, A. P. (ed.), 'ANZAAS: A History', *Aust. Journal of Science*, 25 (1962), 2-4.
——, 'The Australian National Research Council', *Aust. Journal of Science*, 16 (1954), 203-11.
——, *A Century of Scientific Progress: The Centenary Volume of the Royal Society of New South Wales* (Sydney: Royal Society of NSW, 1968).
——, 'Centenary Oration: The Challenge to Science, 1866; The Challenge of Science, 1966', *Journ. and Proc. Roy. Soc. of NSW*, 100 (1966), 105-18.
——, 'The Development of Scientific Knowledge of the Aborigines' in H. Sheils (ed.), *Australian Aboriginal Studies* (Melbourne: Oxford University Press, 1963), 3-28.
——, (ed.), *A Goodly Heritage: ANZAAS Jubilee Science in New South Wales* (Sydney: ANZAAS, 1962).
Encel, S., 'The Post-Industrial Society and the Corporate State', *Aust. and New Zealand Journ. of Sociology*, 15 (1979), 37.
——, *Science, Technology and Public Policy: An Australian Perspective* (Sydney: Pergamon, 1979).

——, 'Science, Technology and Society', *Search*, 1 (1970), 12–17.
——, 'Science, Discovery and Innovation: An Australian Case Study', *Int. Social Science Journ.*, 22 (1970), 42–53.
——, 'Science, Technology and the Future', *Search*, 5 (1974), 387–93.
——, 'The Support of Science without Science Policy in Australia', *Minerva*, 9 (1971), 349–60.
Esplin, Y., 'TT takes Two', unpublished paper presented to *ANZAAS*, 52 (Sydney, 1982).
Eutican, A. R., *Forest Education and Training in New Zealand* (Wellington: NZ Forest Service, 1957).
Evans, L. T., 'The Divorce of Science: Presidential Address, 48th ANZAAS Congress', *Search*, 8 (1977), 403–10.
Farrer, K. T. H., *A Settlement Amply Supplied: Food Technology in Nineteenth Century Australia* (Melbourne: Melbourne University Press, 1980).
Fell, H. B., *The First Century of New Zealand Ecology, 1769–1868* (Wellington: Department of Zoology, Victoria University College, 1953).
Fenner, F., 'Scientific Societies in Australia. The Australian Academy of Science', *Proc. Royal Aust. Chem. Inst.*, 27 (1960), 289–94.
Fenner, F., and Rees, A. L. G. (eds), *The First Twenty Five Years: The Australian Academy of Science, 1954–1979* (Canberra: The Academy, 1980).
Fenwick, G., *Romance of the Flora of New Zealand* (Dunedin: Otago Daily Times, 1922).
Fleming, C. A., 'The Contribution of New Zealand Geoscientists to the Development of Scientific Institutions', *Journal of the History of the Earth Sciences Society*, 5 (1986), 3–11.
——, 'J. A. Thompson's Proposals for Reform of the New Zealand Institute in 1917—A Chapter in the History of the Royal Society of New Zealand', *Trans. Roy. Soc. of NZ*, 2 (8) (17 April, 1969), 129–33.
——, 'The Royal Society of New Zealand—A Century of Scientific Endeavour', *Trans. Roy. Soc. of NZ*, 2 (6) (1968), 99–114.
——, 'Science, Settlers and Scholars. Centennial History of the Royal Society of New Zealand', *Roy. Soc. of NZ Bulletin*, 25 (Wellington: 1987).
Fleming, D., 'Science in Australia, Canada and the United States: Some Comparative Remarks', *Proc. 10th International Congress of the History of Science*, 1 (1962), 179–96.
Flood, Joe, 'The Advent of Strategic Management in CSIRO: A History of Change', *Prometheus*, 2 (1) (1984), 38–72.
Fowler, R. T. 'A History of Chemical Engineering Education in Australia', *Chemical Engineering in Australia*, 5 (1980–1), 40–8.
Francis, John, 'Sir Joseph Banks, Architect of Science and Empire', *Proc. Roy. Soc. of Qld*, 83 (1972), 1–19.
Frankel, O. H., 'The Social Responsibility of Agricultural Science', *Aust. Journal of Science*, 25 (1963), 301–7.
Froggatt, Walter W., 'The Curators and Botanists of the Botanical Gardens, Sydney', *Journ. Royal Aust. Historical Society*, 18 (1932), 101–33.
Gardner, W. J., *Colonial Cap and Gown. Studies in the Mid-Victorian Universities of Australasia* (Christchurch: University of Canterbury, 1979).
Gellatly, F. M., 'Science and Industry; Establishment of Commonwealth Institute: How the University can Assist', *Hermes*, 24 (June 1918), 18–22.
Gibbs, W. J., *The Origins of Australian Meteorology* (Canberra: AGPS, 1975).
Gilbert, Lionel A., 'Plants and Parsons in Nineteenth Century New South Wales', *Historical Records of Australian Science*, 5 (3) (1982), 17–32.
——, Plants, Politics and Personalities in Colonial New South Wales' in D. J. and S. G. M. Carr (eds), *People and Plants in Australia* (Sydney: Academic Press, 1981), 220–58.

——, 'Plants, Politics and Personalities in Nineteenth Century New South Wales', *Journ. Royal Aust. Historical Society*, 56, part 1 (March, 1970), 15–35.
Grainger, Elena, *The Remarkable Reverend Clarke* (Melbourne: Oxford University Press, 1982).
Graubard, Stephen, R., *Australia: The Daedalus Symposium* (Sydney: Angus & Robertson, 1985).
Green, Ken, 'Research Funding in Australia: A View from the North', *Prometheus*, 4 (1) (1986), 68–92.
Gurnett-Smith, A. F., 'Interface Between Science and Society' *Aust. Journal of Science*, 32 (1969), 143–6.
Hays, S. P., *Conservation and the Gospel of Efficiency* (New York: Athenaeum, 1972).
Healy, A. T. A. (ed.), *Science and Technology for what Purpose? An Australian Perspective* (Canberra: Australian Academy of Science, 1979).
Heathcote, R. L. and Thom, B. G. (eds), *Natural Hazards in Australia*, Proceedings of a Symposium sponsored by the Australian Academy of Science, the Institute of Australian Geographers and the Academy of Social Sciences in Australia (Canberra: Australian Academy of Science, 1975).
Herbert, D. A., 'A Story of Queensland's Scientific Achievement, 1859–1959', *Proc. Roy. Soc. of Qld, LXXI* (1959), 1–15.
Hill, Stephen and Johnston, Ron, *Future Tense? Technology in Australia* (St Lucia: University of Queensland Press, 1983).
Hind, Robert J., 'The Internal Colonial Concept', *Comparative Studies in Society and History*, 26 (1984), 543–68.
Hoare, Michael E., ' 'All Things are Queer and Opposite': Scientific Societies in Tasmania in the 1840's', *Isis*, 60 (1969), 198–209.
——, *Beyond the 'Filial Piety': Science History in New Zealand. A Critical Review of the State of the Art* (Melbourne: The Hawthorn Press, 1977).
——, 'The Board of Science and Art, 1913–1930: A Precursor to the DSIR' in M. E. Hoare and L. G. Bell (eds), *In Search of New Zealand's Scientific Heritage, Roy. Soc. of NZ Bulletin*, 21 (Wellington: 1984), 25–48.
——, 'Botany and Society in Eastern Australia' in D. J. and S. G. M. Carr (eds), *People and Plants in Australia* (Sydney: Academic Press, 1981), 183–219.
——, 'The Challenge of Science Accepted in New South Wales', *Records of the Aust. Academy of Science*, 1 (4) (1969), 32–7.
——, 'Doctor John Henderson and the Van Diemen's Land Scientific Society', *Records of the Aust. Academy of Science*, 1 (3) (1968), 7–24.
——, 'The History of Australian Science: Prospect and Retrospect', *Aust. Assn for the History and Philosophy of Science Newsletter*, no. 5 (1974), 21–36.
——, 'The Intercolonial Science Movement in Australasia, 1870–1890', *Records of the Aust. Academy of Science*, 3 (2) (1976), 7–28.
——, 'Learned Societies in Australia: The Foundation Years in Victoria, 1850–1860', *Records of the Aust. Academy of Science*, 1 (2) (1969), 7–29.
——, 'Light in our Past: Australian Science in Retrospect', *Search*, 6 (1975), 285–90.
——, *Reform in New Zealand Science, 1880–1926*, Third Cook Lecture (Melbourne: Hawthorn Press, 1977).
——, 'The Relationship between Government and Science in Australia and New Zealand: The Comparative Experience', *Jour. Roy. Soc. of NZ*, 6 (3) (1976), 381–94.
——, Science and Scientific Associations in Eastern Australia, 1820–1890 (unpublished PhD thesis, Australian National University, 1974).
——, 'Some Primary Sources for the History of Scientific Societies in Australia in the Nineteenth Century', *Records of the Aust. Academy of Science*, 1 (4) (1969), 71–6.

——, 'Some Recent Writings in the History of Australian Science: A Critical Review', *Aust. Assn for the History, Philosophy and Social Studies of Science Newsletter*, 7 (1975–6), 1–25.

Hoare, M. E. and Bell, L. G. (eds), *In Search of New Zealand's Scientific Heritage*, Roy. Soc. of NZ Bulletin, 21 (Wellington: 1984).

Home, R. W., 'First Physicist of Australia: Richard Threlfall at the University of Sydney, 1886–1898', *Historical Records of Australian Science*, 6 (3) (1986), 331–56.

——, 'Origins of the Australian Physics Community', *Historical Studies*, 29 (1982–3), 383–400.

——, 'The Problem of Intellectual Isolation in Scientific Life: W. H. Bragg and the Australian Scientific Community, 1886–1909', *Historical Records of Australian Science*, 6 (1984), 19–30.

Howard, K., 'Science Reshapes Australia', *Walkabout*, 29 (1963), 30–4.

Howarth, O. J. R., *The British Association for the Advancement of Science: A Retrospect, 1831–1931* (London: The Association, 1931).

Howchin, W., et al., 'The Growth of Scientific Knowledge' in *Centenary History of South Australia* (Adelaide: Royal Geographical Society of Australasia, 1936).

Inkster, Ian, 'Scientific Enterprise and the Colonial "Model": Observations on Australian Experience in a Historical Context', *Social Studies of Science*, 15 (1985), 677–704.

Jenkins, C. F. H., *The Noah's Ark Syndrome (One Hundred Years of Acclimiatization and Zoo Development in Australia)* (Perth: Zoological Gardens Board, WA, 1977).

Jenkinson, S. H., *New Zealanders and Science* (Wellington: Department of Internal Affairs, New Zealand, 1940).

Johns, R. K., *History and Role of Government Geological Surveys in Australia* (Adelaide: Government Printer, 1940).

——, (ed.), *History and Role of Government Geological Surveys in Australia* (Adelaide: Government Printer, 1976).

Johnson, M. M., *The Botanical Explorers of New Zealand* (Wellington: Reed, 1950).

Johnston, R. J. (ed.), *Society and Environment in New Zealand* (Christchurch: Whitcombe and Tombs, 1974).

Johnston, R., 'Structural Silence in the Conduct of Science' in W. Green (ed.), *Focus on Social Responsibility in Science* (Christchurch: New Zealand Association of Scientists, 1979).

Jones, Barry, *Sleepers, Wake! Technology and the Future of Work* (Melbourne: Oxford University Press, 1982).

Joseph, Richard, 'Recent Trends in Australian Government Policies for Technological Innovation', *Prometheus*, 2 (1) (1984), 93–111.

Kargon, Robert H. (ed.), *The Maturing of American Science: A Portrait of Science in Public Life Drawn from the Presidential Addresses of the American Association for the Advancement of Science, 1920–1970* (Washington, D. C.: American Association for the Advancement of Science, 1974).

Knibbs, Sir George, 'Science and its Service to Men', *Report of ANZAAS*, 24 (Canberra, 1939), 1–16.

Kohlstedt, Sally Gregory, *The Formation of the American Scientific Community: the American Association for the Advancement of Science, 1848–60* (Urbana: University of Illinois Press, 1976).

——, 'Savants and Professionals: The American Association for the Advancement of Science, 1848–1860' in Alexandra Oleson and Sanborn C. Brown (eds), *The Pursuit of Knowledge in the Early American Republic* (Baltimore and London: Johns Hopkins University Press, 1976), 299–325.

Kynaston, Edward, *A Man on Edge: Life of Baron Sir Ferdinand von Mueller* (Ringwood: Allen Lane, 1981).

Lamberton, D. M., *Science, Technology and the Australian Economy* (Sydney: Tudor Press, 1973).

Lassack, E. V. and McCarthy, T., *Australian Medicinal Plants* (Sydney: Methuen, 1983).

Le Fevre, R. J. W., 'The Establishment of Chemistry within Australian Science—Contributions from New South Wales' in A. P. Elkin (ed.), *Century of Scientific Progress* (Sydney: Royal Society of NSW, 1966), 332–78.

Lewis, M. J., 'The Royal Society of Australia: An Attempt to Establish a National Academy of Science', *Records of the Aust. Academy of Science*, 4 (1) (1978), 51–62.

Love, Rosaleen, 'Science and Government in Australia, 1905–14: Geoffrey Duffield and the Foundation of the Commonwealth Solar Observatory', *Historical Records of Australian Science*, 6 (2) (1985), 171–88.

McCall, G. (ed.), *Anthropology in Australia: Essays to Honour 50 Years of 'Mankind'* (Sydney: Royal Anthropological Society, 1982).

Macdonald, Stuart *et al.*, *The Trouble with Technology—Explorations in the Technology of Social Change* (London: Frances Pinter, 1983).

McKay, Andrew, *Surprise and Enterprise, Fifty Years of Science for Australia: Commonwealth Scientific and Industrial Research Organisation* (Melbourne: CSIRO, 1976).

MacKenzie, D. and Wajcman, J. (eds), *The Social Shaping of Technology* (Milton Keynes: Open University Press, 1985).

MacLeod, Roy, 'On Visiting the Moving Metropolis: Reflections on the Architecture of Imperial Science', *Historical Records of Australian Science*, 5 (3) (1982), 1–6.

MacLeod, Roy M. and Andrews, E. K., 'Scientific Careers of 1851 Exhibition Scholars', *Nature*, 218 (1968), 1011–16.

MacLeod, Roy M. and Collins, Peter (eds), *The Parliament of Science: The British Association for the Advancement of Science 1831–1981* (London: Science Reviews Ltd, 1981).

Maiden, J. H., 'A Contribution to a History of the Royal Society of New South Wales', *Journ. and Proc. Roy. Soc. of NSW*, 52 (1918), 215–361.

Mandeville, Thomas and Macdonald, Stuart, 'Technological Change and Employment in the Information Economy: The Example of Queensland', *Prometheus*, 3 (1) (1985), 71–85.

Martin, W. M. E., *Forestry in New Zealand: Statement prepared for the British Empire Forestry Conference* (London: July 1920).

Martyn, D. F., 'Personal Notes on the Early Days of Our Academy', *Records of Aust. Academy of Science*, 1 (December 1967), 53–72.

Maxwell, W. Bridges., 'Science and Politics', *Search*, 1 (1970), 145–6.

Mellor, D. P., *The Role of Science and Industry; Australia in the War of 1939–45* (Canberra: Australian War Memorial, 1958).

Middleton, B. S., 'Science Policy Developments in Australia', *Search*, 3 (1972), 62–6.

Milligan, R. D., *Physiological Approach to Biology: Presidential Address to the Canterbury Philosophical Institute* (Christchurch, 1930).

Moran, Jean, 'Rhetoric and Representation in Australian Science in the 1940s and 1980s', *Prometheus*, 1 (2) (1983), 271–302.

——, Scientists in the Political and Public Arena: a Social-Intellectual History of the Australian Association of Scientific Workers, 1939–49 (unpublished MPhil thesis, Griffith University, 1983).

Morrell, Jack and Thackray, Arnold, *Gentlement of Science: Early Years of the British Association for the Advancement of Science* (Oxford: Oxford University Press, 1981).

Moyal, Ann, *A Bright and Savage Land: Science in Colonial Australia* (Sydney: William Collins, 1986).

——, *Clear Across Australia: A History of Telecommunications* (Melbourne: Nelson, 1984).

——, 'Collectors and Illustrators: Women Botanists in the Nineteenth Century' in D. J. and S. A. M. Carr (eds), *People and Plants in Australia* (Sydney: Academic Press, 1981), 333–56.

Moyal, Ann Mozley, 'The Australian Academy of Science: The Anatomy of a Scientific Elite. Part I (History and Sociology)', *Search*, 11 (1980), 231–8; Part II (Relations with Government), *Search*, 11 (1980), 281–8.

——, 'The Making of the Federal Government's Science Policy' in R. Lucy (ed.), *The Pieces of Politics* (Sydney: Macmillan, 1975).

——, 'Medical Research in Australia: A Historical Perspective', *Search*, 12 (1981), 302–09.

——, Science and the Press in Australia', *Search*, 4 (1973), 133–8.

——, 'Scientific and Technological Change, 1939–1988' in 'Australia, 1939–1988', *Bicentennial History Bulletin*, 3 (April 1981), 58–67.

——, *Scientists in Nineteenth Century Australia: A Documentary History* (Sydney: Cassell Australia, 1976).

——, 'Sir Richard Owen and his Influence on Australian Zoological and Palaeontological Science', *Records of the Aust. Academy of Science*, 3 (1975), 41–56.

Mozley, Ann, 'ANZAAS and the Public Communication of Science', *Search*, 5 (1974), 589–94.

——, 'Checklist of Publications on the History of Australian Science', *Aust. Journal of Science*, 25, (1962), 206–14; 27, (1964), 8–15.

——, 'Evolution and the Climate of Opinion in Australia, 1840–1876', *Victorian Studies*, 10 (June 1967), 411–30.

——, 'The Foundations of the Geological Survey of New South Wales', *Journ. and Proc. Roy. Soc. of NSW*, 98 (1965), 91–100.

——, *A Guide to the Manuscript Records of Australian Science*, (Canberra: Australian Academy of Science in association with Australian National Univerity Press, 1966).

——, 'The History of Australian Science', *Historical Studies of Australia and Zealand'*, 11 (1963), 258–9.

——, 'Richard Daintree, First Government Geologist of Northern Queensland', *Queensland Heritage*, 1, part 2 (1965), 11–16.

Mulvaney, D. J., 'Anthropology in Victoria 100 Years Ago', *Proc. Roy. Soc. of Vic.*, 73 (15 February 1961), 47–50.

——, 'Patron and Client: The Web of Intellectual Kinship in Australian Anthropology' in N. Reingold and Marc Rothenberg (eds), *Scientific Colonialism* (Washington: Smithsonian Institution, 1987.)

Mulvaney, D. J. and J. H. Calaby, *'So Much That Is New': Baldwin Spencer, 1860–1929, A Biography* (Melbourne: Melbourne University Press, 1985).

Nadel, G., *Australia's Colonial Culture: Ideas, Men and Institutions in Mid-Nineteenth Century Eastern Australia* (Melbourne: Cheshire, 1957).

Newland, Elizabeth, 'Forgotten Early Australian Journals of Science and Their Editors', *Journ. Royal Aust. Historical Society*, 72 (1986), 59–68.

OECD, *Reviews of National Science Policies: Australia* (Paris: Organization for Economic Cooperation of Development, 1985).
O'Leary, J. T. and Schaffer, R. H., *Scientific and Engineering Manpower in New Zealand Industry* (Wellington: DSIR, 1958).
Oliphant, Sir Mark, 'Science and Humanity', *Aust. Journal of Science*, 32 (1970), 377–82.
Osborne, W. A., *William Sutherland: A Biography* (Melbourne: Lothian Book Publishing Co., 1920).
Parton, Hugh, *The University of New Zealand* (Auckland: Auckland University Press, 1979).
Passmore, J. A., 'The Revolt Against Science', *Search*, 3 (1972), 415–22.
Philip, J. R. and Conlon, T. J. *Science and the Polity: Ideals, Illusions and Realities*, Silver Jubilee Symposium, 1 (Canberra: Australian Academy of Science, 1980).
Powell, J. M., *Environmental Management in Australia, 1788–1914. Guardians, Improvers and Profit: An Introductory Survey* (Melbourne: Oxford University Press, 1976).
——, 'Exiled from the Garden. Von Mueller's Correspondence with Kew, 1871–81', *Victorian Historical Journal*, 48 (1977), 313–20.
—— (ed.), *Making of Rural Australia* (Melbourne: Sorrett Press, 1974).
—— (ed.), 'National Identity and Gifted Immigrant: a Note on T. Griffith Taylor, 1880–1963', *Journal of Intercultural Studies*, 2 (1981), 43–54.
——, *An Historical Geography of Modern Australia: The Restive Fringe* (Cambridge: Cambridge University Press, in press).
——, Taylor, Stefansson and the Arid Centre. An Historic Encounter of 'Environmentalism' and 'Possibilism', *Journ. Royal Aust. Historical Society*, 66 (1980), 163–83.
Poynter, J. R., *Russell Grimwade* (Melbourne: Melbourne University Press, 1967).
Prescott, R. T. M., 'The Royal Society of Victoria from Then, 1854 to Now, 1959', *Proc. Roy. Soc. of Vic.*, 73 (15 February, 1961), 1–40.
Price, J. R., 'CSIRO: Fifty Years of Research—Looking to the Future', *Nature*, 261 (1976), 631–2.
Priestley, R. E., 'Sir Edgeworth David', *Australian Quarterly*, 10 (June 1938), 34–9.
Radford, Joan T., *The Chemistry Department of the University of Melbourne. Its Contribution to Australian Science, 1854–1959* (Melbourne: Hawthorn Press, 1978).
——, 'Chemistry in Nineteenth-Century New Zealand', *Chemistry in New Zealand*, 49 (1984), 35–7, 60–2, 125–9.
Read, John, 'The Study of Science Abroad', *Hermes*, 32 (Trinity 1926), 5–7.
Reingold, N. and Rothenberg, M. (eds), *Scientific Colonialism: A Cross-Cultural Comparison* (Washington: Smithsonian Institution Press, 1987).
Rivett, Sir David, 'The Scientific Estate', *Report of ANZAAS*, 33 (Auckland, 1937), 1–13.
Rivett, Rohan, *David Rivett: Fighter for Australian Science* (Melbourne: Dominion Press, 1972).
Robertson, Peter, 'Coming of Age: The British Association in Australia, 1914', *Australian Physicist*, 17 (1980), 23–7.
Robertson, R. N., 'Scientists and Government in Australia', *Impact of Science in Society*, 22 (1972), 187–96.
Ronayne, Jarlath, 'Further Thoughts on Diversity and Adaptability in Australian Science Policy', *Minerva*, 17 (1979) 444–58.

——, *Science in Government* (Melbourne: Edward Arnold, 1984).
——, 'Uneasy Alliance: Science and Politics in Australia', *Search*, 7 (1976), 85–9.
Room, T. C., 'The Royal Society and Australasia', *Etruscan* (June 1960), 4–8.
Ross, A. D., 'The Origin and Meaning of ANZAAS', *Royal Perth Hosptial Journal*, 11 (September 1959), 165–7.
Russell, Archer, *William James Farrer* (Melbourne: F. W. Cheshire, 1949).
Schedvin, Boris, 'The Culture of CSIRO', *Australian Cultural History*, 2 (1982–3), 76–89.
Schedvin, C. B., *The Australian Economy on the Hinge of History*, Second Henry George Memorial Lecture, Macquarie University, Sydney (October, 1986).
——, 'Environment, Economy and Australian Biology, 1890–1939', *Historical Studies*, 21 (1984), 17–26.
——, 'Science in Australasia', *Nature*, 316 (18 July 1985), 185–208.
Scott, Ernest, 'The History of Australian Science', *Report of ANZAAS*, 24 (Canberra, 1939), 1–16.
Seddon, George, 'Eurocentrism and Australian Science: Some Examples', *Search*, 12 (1981–2), 446–50.
Shann, Edward, *An Economic History of Australia* (Cambridge: Cambridge University Press, 1930).
Sinden, A. J. (ed.), *The Natural Resources of Australia. Prospects and Problems for Development* (Sydney: Angus & Robertson, in association with ANZAAS, 1972).
Spicer, B. M., 'Physics at the University of Melbourne', *Australian Physicist*, 17 (1980), 113–16.
Spurling, T. H., 'William Sutherland: Australia's First Physical Chemist', *Proc. Royal Aust. Chem. Inst.*, 41 (1974), 313–14.
Summers, H. S., 'The Teachers of Geology in Australian Universities', *Jour. and Proc. Roy. Soc. of NSW*, 81 (1947), 122–46.
Taylor, Griffith, *Journeyman Taylor: The Education of a Scientist*, abridged and edited by Alasdair Alpin MacGregor (London: Robert Hale Limited, 1958).
Thomson, G. M., *The Naturalisation of Animals and Plants in New Zealand* (Cambridge: Cambridge University Press, 1922).
——, *Wild Life in New Zealand—Part I: Mammalia* (Wellington: Govt Printer, 1921).
——, *Wild Life in New Zealand—Part II: Introduced Birds and Fishes* (Wellington: Govt. Printer, 1926).
Tisdell, Clem, 'International Scientific Cooperation, Technology Transfer and Aid: ASEAN Countries, Australia and New Zealand', *Prometheus*, 4 (1) (1986), 111–27.
Todd, Alexander, *A Time to Remember* (Cambridge: Cambridge University Press, 1983).
Tomlin, S. C., 'William Henry Bragg, 1862–1942', *Australian Physicist*, 13 (1976), 76–99.
Turney, C. (ed.), *Pioneers of Australian Education*, 2 vols (Sydney: Sydney University Press, 1969, 1972).
Vallance, T. G. and Branagan, D. F., 'New South Wales Geology—Its Origin and Growth' in *Royal Society of NSW. A Century of Scientific Progress* (Sydney: Royal Society of NSW, 1968), 265–79.
——, 'Origins of Australian Geology', *Proc. Linnean Society of NSW*, 100 (1975), 13–43.
——, 'The Start of Government Science in Australia: A. W. H. Humphrey, His Majesty's Mineralogist in New South Wales, 1803–1812', *Proc. Linnean Society of NSW*, 105 (1981), 107–46.

Vonwiller, O. U., 'The Social Relations of Science', *Aust. Journal of Science*, 1 (1938), 30–2.
Walby, B. J., 'Australian Journals of Scientific Research', *Nature*, 261 (1976), 661–2.
Walkom, A. B., 'A Short History of the Association', in Kenneth Binns (ed.), *ANZAAS Handbook for Canberra* (Canberra: Commonwealth Government Printer, 1938), 7–10.
——, *The Linnean Society of New South Wales: Historical Notes of its First Fifty Years* (Sydney: Australian Medical Publishing Company, 1925).
Walsh, A., 'Atomic Absorption Spectroscopy. Stagnant or Pregnant', *Analytical Chemistry*, 46 (1974).
——, 'Atomic Absorption Spectroscopy', *Australian Physicist*, 4 (November 1967), 185–9.
——, 'Invention and Innovation', *Search*, 4 (1963), 69–74.
Ward, R. G., 'The Case for Government Support of Industrial R and D', *Search*, 6 (1975), 407–9.
——, 'Science and Industry', *Search*, 3 (1972), 371–5.
——, 'The Role and Function of Science in the Modern Community', *Public Administration*, 27 (1968), 99–112.
Watts, W. W., 'Progress of the Geological Survey', *Discovery*, 13 (1932), 152–6.
Webster, E. M., *Whirlwinds in the Plain: Ludwig Leichhardt, Friends, Foes and History* (Melbourne: Melbourne University Press, 1980).
Wells, C. B. and Prescott, J. A., 'The Origins and Early Development of Soil Science in Australia', in *Soils, An Australian Viewpoint* (Melbourne: CSIRO, 1983), 3–12.
Wheeler, Edward, 'The First Hobart Congress' *Search*, 7 (1976), 202–4.
——, 'The First Melbourne Congress', *Search*, 8 (1977), 275–82.
——, 'The First New Zealand Congress', *Search*, 9 (1978), 452–5.
——, 'The First Adelaide Congress', *Search*, 11 (1980), 152–6.
——, 'The First Brisbane Congress', *Search*, 12 (1981), 123–7.
——, 'The First ANZAAS Congress', *Search*, 13 (1982), 82–6.
——, 'The First Perth Congress', *Search*, 14 (1983), 96–8.
——, 'The First Canberra Congress', *Search*, 15 (1984), 104–7.
Wheelhouse, F., *From Digging Stick to Rotary Hoe* (Adelaide: Rigby, 1972).
Wheelwright, E. L., and Crough, Greg *Australia—The Client State* (Sydney: Transnational Corporations Research Project, University of Sydney, 1981).
White, F., 'Administrative Problems in the Development of Science and Research', *Public Administration*, 27 (Sydney, 1968), 113–40.
——, 'CSIR to CSIRO; The Events of 1948–49', *Public Administration*, 34 (December 1975), 281–93.
——, 'A Personal Account of the Historical Development of CSIRO', *Nature*, (1976), 261.
Whitlam, E. G., 'A National Science Policy', *Search*, 1 (1970) 134–8.
Williams, M., *The Making of the South Australian Landscape* (London: Academic Press, 1974).
Williams, P. P. (ed.), *Chemistry in a Young Country* (Christchurch: New Zealand Institute of Chemistry, 1981).
Williams, R. D., 'Research Personnel in Industry—Cabbages or Kings?', *Journ. and Proc. of the Aust. Chemical Institute*, 13 (1946), 108–115.
Wilsmore, N. T. M., 'Chemical Research and the State', *Report of ANZAAS*, 20 (Brisbane, 1930), 546–69.
Wise, Tigger, *The Self-Made Anthropologist. A Life of A. P. Elkin* (Sydney: George Allen & Unwin, 1985).

NOTES ON CONTRIBUTORS

David Branagan studied at the University of Sydney where he is currently associate professor in geology. He has worked in many parts of Australia, the United States, Canada, and United Kingdom and has published papers on many aspects of geology. A major research interest is the history of Australian geology.

Greg Crough is a research fellow with the Transnational Corporations Research Project at the University of Sydney, where he has been employed for ten years. His other appointments have been as senior economic adviser to the Deputy Prime Minister, Lionel Bowen, and the United Nations Conference on Trade and Development in Geneva. His books include *Foreign Investment and Transnational Corporations in Australia: An Annotated Bibliography* (Sydney: Alternative Publishing Cooperative, 1977); *Transnational Banking and the World Economy* (Sydney: Transnational Cooperative Research Project, 1979); *Money Work and Social Responsibility: The Australian Financial System* (ed.) (Sydney: Transnational Cooperative Research Project, 1980); *Australia and World Capitalism* (With Ted Wheelwright) (Ringwood: Penguin, 1980); and *Australia: A Client State* (with Ted Wheelwright) (Ringwood: Penguin, 1982).

James Davenport, a graduate of Queensland and Cambridge Universities, spent most of his professional career as a research biochemist in the Division of Food Research, CSIRO. He was a Broodbank Fellow of the University of Cambridge and science liaison officer for the Reserve Bank of Australia administering an agricultural research fund. He founded *Search*, the journal of ANZAAS, and was its first honorary editor, and he has been chairman of the ANZAAS council. He is a Fellow of ANZAAS, and has been a member of the Visual Arts Board of the Australia Council.

Bruce Davidson is a senior lecturer in agricultural economics at the University of Sydney. After graduating from the University of Melbourne, he undertook postgraduate work at the University of London, examining land use problems in Britain. He later established, and was head of, the Farm Management Research Unit in Kenya. On returning to Australia he joined CSIRO and carried out research into the economic prospects of establishing intensive agriculture in tropical Australia. In recent years most of his research has been devoted to the economics of resource usage in Australian agriculture and to studying the economic history of Australian farming. His best known publications are *The Northern Myth* (Melbourne: Melbourne University Press,

1965); *Australia Wet or Dry* (Melbourne: Melbourne University Press, 1969); and *European Farming in Australia* (Amsterdam: Elsevier, 1981).

Sol Encel, professor of sociology at the University of New South Wales since 1966, abandoned a science course at the University of Melbourne to enlist in the RAAF during the Second World War, but retained a keen interest in the natural sciences. In 1960 he was a visiting fellow at the Science and Public Policy Programme of Harvard University, and has been working in the field of science, technology, and society ever since. He has also been a visiting fellow at the Science Policy Research Unit, University of Sussex, on several occasions. His publications include *Science, Technology and Public Policy: An Australian Perspective* (Sydney: Pergamon, 1979) and numerous papers on science policy, technological change, and post-industrial society. In 1975 he was a foundation member of the Australian Science and Technology Council (ASTEC). He is currently engaged on a study of the data processing industry in Australia.

Linden Gillbank studied science at the University of Melbourne, where she gained her PhD on the biochemistry of a plant hormone. She has taught various aspects of biology, mainly botanical and biochemical, at several tertiary institutions—the University of Melbourne, Melbourne State College (now Melbourne College of Advanced Education), and a teachers college in Malaysia. She is currently investigating postwar CSIRO biological research, while maintaining a general interest in nineteenth and twentieth century Australian biology.

R. W. Home has been professor of history and philosophy of science at the University of Melbourne since 1975. He has written extensively on the history of classical physics, especially the sciences of electricity and magnetism. More recently he has turned to the history of Australian science. He has been editor of *Historical Records of Australian Science* since January 1984 and is preparing a history of the Australian physics community to 1945.

Ron Johnston is a director of the Centre for Technology and Social Change and foundation professor of science and technology at the University of Wollongong. He was educated at the New South Wales, Manchester, and Northwestern universities, and University College, London. He has worked extensively on issues of science and technology, primarily from sociological, economic, and political angles over a period of eighteen years in Europe, North America, Asia, and Australia. His current research interests include policy making for science and technology, and the social and economic implications of technological change, and the history of postwar Australian science and technology.

Roy MacLeod is professor of history at the University of Sydney, where he teaches, among other things, the history of science and technology. He was educated at Harvard and Cambridge, and has held appointments at Cambridge, Sussex and London. He has written many articles and edited several books in social history and in the history of science, including (with P. Collins) *The Parliament of Science: Essays in Honour of the British Association for the Advancement of Science* (London: Science Reviews Limited, 1981).

John Mulvaney, CMG, graduated in history at the University of Melbourne, and in prehistoric archaeology at Cambridge. He taught ancient world history at Melbourne, before moving to a research post in Australian prehistory at the Australian National University in 1965. He was appointed foundation professor of prehistory in the Faculty of Arts at ANU in 1971, retiring in 1985. He is co-author of a biography of Sir Baldwin Spencer and of studies of A. W. Howitt. He has served terms as chairman of the Australian Institute of Aboriginal Studies and as an Australian Heritage Commissioner.

J. M. Powell arrived in Australia in 1964 as an assisted immigrant; he has been reader in geography at Monash University since 1977. Formerly president of the Institute of Australian Geographers and editor of *Australian Geographical Studies*, he acted as president of the Geographical Sciences Section at Adelaide's jubilee ANZAAS Congress and has contributed several papers on geographical works of Griffith Taylor and other earlier ANZAAS participants. His major publications include *The Public Lands of Australia Felix* (Melbourne: Oxford University Press, 1970), *Environmental Management in Australia, 1788–1914* (Melbourne: Oxford University Press, 1976), and *Mirrors of the New World* (Canberra: Australian National University Press, 1978). He has recently completed *An Historical Geography of Modern Australia: The Restive Fringe* (Cambridge University Press, in press).

John Powles studied at the University of Sydney Medical School, the School of Sociology at the University of New South Wales and the University of Sussex, before taking a post at Monash Medical School in 1975. He has published on the limitations of modern medicine and on socio-economic inequalities in health. He is currently engaged on a study of the effects on health of migration from the Greek island of Levkada to Melbourne.

Ian Rae is an associate professor of chemistry at Monash University in Melbourne where he was first appointed as research fellow in 1967. A graduate of Footscray Technical College, Melbourne University, and the Australian National University, he is an organic chemist, interested in the use of nuclear magnetic resonance techniques for structure determination. He has served for a number of years as chief examiner for chemistry in Victoria, and publishes a regular column, 'Letter from Monash', in *Chemistry in Australia*. His recent work includes a history of wood distillation in Australia and a review of the work of Alexander Borodin, the Russian chemist–composer.

Alison Turtle is a graduate of the University of Sydney, with BA Honours in both history and psychology, and a Master's degree in psychology. She now works in the history of psychology within the department of psychology at the University of Sydney and has just co-edited a book on the development of the discipline in Asian and South Pacific countries. She is at present working on a history of psychology in Australia.

T. G. Vallance, associate professor in petrology at the University of Sydney, is a graduate of that institution and has held visiting appointments at Berkeley, Cambridge, and Geneva. His research publications are mainly in the fields of metamorphic petrology and Australian geological discovery. He is at present vice-president of the International Commission on the History of Geological Sciences.

Ted Wheelwright has been studying and teaching political economy for forty years. A graduate of the University of St Andrews, in Scotland, he has worked in universities and research institutions in five continents. He has specialized in the changing structure of world capitalism through the operations of transnational corporations, with special reference to the impact on Australia. Recently retired from the University of Sydney his most recent book (co-edited with Ken Buckley) is *Communication and the Media in Australia* (Sydney: George Allen & Unwin, 1986). He is currently working on a history of capitalism and the common people in Australia over the last 200 years.

Index

Sections are indexed under general subject headings, for example, biology, meteorology. For changes in Section names and labels see pages 365-8.

AAAS: administration, 50-1; constitution, 45-6, 369-72; effectiveness, 151, 272-3; Federation and, 140-2; finances, 52, 63; function of, 7-8; leadership, 53-4; membership, 43-4, 54; moves to form, 19, 31-5; popularity of, 51-2; reform of, 59, 63; research committees, 42, 43, 45, 49; venues, 51; *see also* ANZAAS; congresses
Abbott, Joseph, 312
Academy of Social Sciences, 84, 240
Academy of Technological Sciences, 90, 92
Adair, J.F., 150
agriculture, 64, 106-7, 114, 273-82
Ainsworth, John, 207
Albert, Adrien, 179, 381
Alexander, A.E., 381
Allan, F.M., 176
Allan, Harry, 110
Anderson, Francis, 57, 224, 225, 232, 239
Anderson, H.C.L., 51
Anderson, J.S., 381
Andrew, Henry M., 150, 169
Andrews, Ernest C., 54, 59, 63, 138, 362, 373, 380
Andrews, Peter, 186
Anet, Frank, 185
Angyal, S.J., 381
Antarctic research, 41, 42, 45, 56, 62
anthropology, 49, 62, 196-216
anthropometry, 226, 227-31
ANZAAS: aims, 64-5, 182, 186; changing role, 84-5, 141-3, 162-3, 186, 214-16, 238-42; effectiveness of, 10-11, 272-3, 317, 340; created, 63; 'Fellows', 62; function, 8, 9, 78, 84-5, 89, 272-3; Medal, 379; political involvement in, 90; public role for, 9, 10; reform of, 9-11, 89; rules, 369-72; *see also* AAAS; congresses
archaeology, 209, 213-14, 215, 262
Archey, Gilbert, 375
Archibald, Jules François, 2
Armstrong, H.E., 174
Ashby, Eric, 311
Aston, R.L., 363
astronomy, 8, 23, 134, 147, 148-9, 151, 153, 161-2

Australasian Association for the Advancement of Science *see* AAAS
Australian Academy of Science (AAS), 9, 62, 80, 83-4, 86, 87, 115, 116, 262, 313, 319-21
Australian and New Zealand Association for the Advancement of Science *see* ANZAAS
Australian Association of Scientific Workers (AASW), 312
Australian Council of Educational Research (ACER), 235, 236-7, 238, 240, 242
Australian Institute of Agricultural Science, 279
Australian Journal of Science, 10, 62, 79, 85-6, 89, 179, 262, 279, 281, 312, 313, 315, 344
Australian National Research Council (ANRC), 9, 61-2, 64, 77, 78, 79, 83-4, 154-5, 156, 177, 179, 206-7, 208, 209, 210-12, 240
Australian Science and Technology Council (ASTEC), 87-8, 142, 333-4, 350, 353

Badger, Sir Geoffrey M., 88, 179, 364, 379, 381
Badham, Richard, 346
Baker, R.T., 104, 172, 380
Balfour, F.M., 29
Bancroft, Joseph, 292, 294
Baracchi, Pietro, 134-5, 153
Barnard, C., 375
Barrett, James, 255
Barrie, W.D., 364
Bartels, D., 364
Basalla, George, 6
Batchelard, P.M., 236
Bates, Daisy, 202, 203, 204, 254
Bateson, William, 57
Batt, R.D., 376
Bavay, Auguste de, 284
Baxter, J.P., 313
Bayliss, Noel, 179, 363, 381
Beaglehole, J.C., 380
Beeby, Clarence, 232
Beattie, Sir David, 364
Bell, Daniel, 345, 346-7

408

Ben-David, J., 308
Bennett, George, 33, 34, 41, 53, 100
Bennett, Isobel, 380
Benson, W.N., 380
Bentham, G., 100
Bernal, J.D., 310, 346
Berndt, Catherine, 212
Berndt, Ronald, 212
Best, Elsdon, 204
Best, Rupert J., 373
Bickerton, A.W., 150, 167, 170
biology, 42, 99–117
Birch, A.J., 91
Birch, Arthur, 185
Birch, Charles, 317
Black, J.G., 42, 170
Black, J.M., 380
Blackett, P.M.S., 310
Blackwell, E.C., 234
Blainey, Geoffrey, 2, 3
Blandy, Richard, 382
Board, Peter, 223, 235
Boas, I.H., 178
Boldrewood, Rolf, 47
Bolger, P., 202
Bollard, E., 376
Booth, Mary, 232, 295
Bosisto, Joseph, 168
botany, 100, 103–4, 115–16, 251, 262
Bowden, Philip, 159
Bowen, E.G., 161
Bowman, Isaiah, 254
Bradfield, J.J.C., 260
Bragg, William H., 1, 41, 46, 47, 54, 55, 150, 151, 152, 173, 362, 373
Brandon, A. de Bathe, 375
Brennan, G., 382
Briggs, Edna, 157
Briggs, G.H., 157
Briggs, L.H., 184
British Association for the Advancement of Science (BAAS): model, 8, 19, 24, 29–31, 34; visit of, 7, 29, 31–2, 40, 56–7, 105, 137, 174, 204–6
Brodie, A., 381
Brown, F.D., 150, 170
Brown, R.D., 179, 381
Brown, Robert Hanbury, 379
Brown, T.A., 47
Browne, William R., 139, 380
Bruce, Stanley Melbourne, 58, 177
Bryce, James, 60
Buck, Peter, 207
Buckingham, A. David, 185, 381
Budtz-Olson, O.E., 374
Bull, L.A., 379
Bull, L.B., 380
Bullen, K.E., 162
Burdon, R.S., 373
Burges, N.A., 363

Burhop, E.H.S., 159
Burnet, Sir Macfarlane, 1, 301, 302, 321, 363, 380
Burns, Sir Malcolm, 375
Burrows, George, 179
Bush, Vannevar, 344, 346
Butler, Graham W., 364
Button, John, 91, 354

Caldwell, W.H., 28–9
Callister, C.P., 177
Cambage, R.H., 54, 61, 362
Cameron, R.G., 233
Campbell, A.J., 105
Campbell, Alan, 42
Campbell, W.D., 207
Cannon, Michael, 2, 21
Capell, Arthur, 209
Carey, J.W. Wilton, 373
Carey, S. Warren, 364, 375
Carment, David, 50, 361
Carmichael, Laurie, 349
Carnegie Corporation, 212, 213, 236
Carroll, Alan, 199, 202, 226
Carslaw, Horatio S., 54, 153
Carson, Rachel, 89, 315
Carstensz, William, 197
Casey, R.G., 76, 313
Chamberlain, H.S., 296
Chandler, G., 374
Chapman, Henry G., 54, 61, 362
Chapman, R.W., 150
chemistry, 40, 41, 42, 45, 46–7, 49, 55, 60, 166–86
Chifley, J.B., 77
child development, 226–34, 295–7
Chilton, C., 380
Chinnery, E.W.P., 217
Christian, C.S., 281
Cilento, Raphael, 296, 298–9, 300
Clark, Donald, 171
Clark, C.M.H (Manning), 2
Clarke, E. de C., 374
Clarke, W.B., 22, 25, 26
Clarke, Sir William, 43, 51
Cleland, J.B., 210
Clemens, Samuel, 45
Cockayne, Leonard, 110, 380
Coghlan, T.A., 229–30, 231
cold war, effects of, 77, 78, 80–1, 312–13
Cole, A.H., 381
Coleman, Peter, 2
Collins, David, 198
Commoner, Barry, 315, 316
Commonwealth Scientific and Industrial Research Organization (CSIRO), 77–8, 91–2, 93, 112, 113, 114, 115, 141–2, 181, 183, 184, 272–3, 280, 281, 286, 313, 315, 328, 329, 339; see also

Council for Scientific and Industrial Research (CSIR)
Compton, Carl T., 79
Conder, W.J., 33
congresses, AAAS/ANZAAS: 50, 377–8; **1888**, 31–5, 40–2, 105, 133, 135, 136, 140, 147–51, 171, 196, 197, 199, 292; **1890**, 43–4, 105, 133, 134, 136, 138, 151, 152, 171, 224, 251, 303; **1891**, 44–5, 46, 51, 105, 136, 137, 138, 275; **1892**, 46–7, 51, 105, 136, 171, 284; **1893**, 47, 51, 171, 200, 224, 227, 241; **1895**, 47–8, 227, 275, 294, 296; **1898**, 48–9, 53, 134, 152, 274, 283, 284; **1900**, 54, 140, 171, 241, 274, 297; **1902**, 135, 138, 229; **1904**, 53, 151; **1907**, 54, 103, 171, 204, 284; **1909**, 51, 139, 152, 294, 295; **1911**, 51, 54, 137, 153, 224, 227, 299, 295; **1913**, 105, 140, 204, 231, 276, 277; **1921**, 59, 106, 111, 138, 154, 175–6, 177, 204, 234, 277, 297; **1923**, 60, 63, 138, 139, 204, 234, 253, 277, 297, 299; **1924**, 59, 178, 234; **1926**, 60, 210, 224, 257; **1928**, 234, 241, 276, 277, 278, 297; **1930**, 208, 235, 297; **1932**, 65, 111, 135, 137, 178, 228, 299, 343; **1935**, 65, 111, 137, 139, 278; **1937**, 62, 64, 279; **1939**, 1, 66–7, 111, 134, 139, 140, 178, 179, 311; **1946**, 78, 138, 139, 182, 224, 284; **1947**, 182, 285; **1949**, 182, 301; **1951**, 141, 162, 182, 301; **1952**, 162, 301; **1954**, 9; **1955**, 162, 183, 301, 302; **1957**, 302, 344; **1958**, 10, 114, 286; **1959**, 302; **1961**, 282; **1962**, 86; **1965**, 183; **1967**, 286; **1968**, 224; **1969**, 224; **1971**, 115, 317; **1972**, 10, 224; **1973**, 10, 90; **1975**, 116; **1985**, 10, 186; **1987**, 10, 186
Connell, R.W., 1
Considen, Dennis, 167
Cooke, A.M., 373
Cooke-Yardborough, R.E., 373
Coombs, H.C., 77, 81, 259, 340, 363, 379, 381
Copland, Sir Douglas B., 82, 363, 381
Cornforth, John, 185
Correll brothers, 274
Cotton, L.A., 135
Council for Scientific and Industrial Research (CSIR), 9, 58, 61–2, 64, 76–8, 107–9, 111, 112, 158–9, 161, 178, 181, 256, 329–30; *see also* Commonwealth Scientific and Industrial Research Organization (CSIRO)
Cowen, Sir Zelman, 364
Cox, J.C., 41
Cox, S. Herbert, 33
Craig, David, 185
Crawford, Sir, John, 10, 89, 364, 379
Crawford, R.M., 2, 8, 60, 64

Creswell, D.A., 171
CSIR *see* Council for Scientific and Industrial Research
CSIRO *see* Commonwealth Scientific and Industrial Research Organization
Culyer, A.J., 382
Cumberland, K.B., 256
Cuming, James, 168
Cumming, A.C., 175
Cumpston, J.H.L., 294, 297–8, 300
Cunningham, K.S., 232, 233, 235, 237, 240
Curr, E.M., 198
Curran, J.M., 136
Currie, G.A., 375
Curtin, John, 240
Curtis, N., 376
Custance, J.D., 274

Daily, Brian, 373
Daintree, Richard, 136
Dakin, W.J., 380
Dale, Henry, 172
Dalhunty, J.A., 286
Darwin, Charles, 199
Davenport, James B., 345, 364, 373
Davey, Constance, 233
David, T.W. Edgeworth: and AAAS, 43, 50, 59, 63, 136, 138; AAAS president, 54, 173, 361, 362; and ANRC, 61, 206, 207; and British AAS visit, 204, 205; geological work, 137, 141, 285, 286, 379; on Liversidge, 48; on science, 55; wartime service, 58, 154
Davidson, B.R., 261
Davies, William, 178
Davis, C.W., 364
Davison, Graeme, 3
Dawson, James, 198
Deakin, Alfred, 56, 57
Dedman, J.J., 77
Delin, P.S., 373
Dell, R.K., 376
Delprat, Gillaume, 284
Dendy, Arthur, 102
Denham, H.G., 178, 180, 381
Denison, Sir William, 20
Department of Scientific and Industrial Research (DSIR), 58, 99, 107–8, 111, 112, 113
Derrick, E.H., 379
Dhiel, Ludwig, 284
Dixon, P.B., 353
Dixon, S., 105
Doherty, W., 171
Duffield, Geoffrey, 152
Dun, W.S., 137
Duncan, J.F., 375
Dunk, W.E., 77

Durkheim, Emile, 202, 346
Dwyer, Francis, 179, 381

Earl, J.C., 175, 178
Easterfield, T.H., 170, 171, 172, 173
Eaton, Cyrus, 314
Eddington, A.S., 57, 154
education, 48, 155–6, 173, 222–34
Edwards, A.B., 139
efficiency, national, 300–1
Einstein, Albert, 314
Elkin, A.P., 62, 197, 204, 206, 209, 210, 211–12, 215–16, 238, 239, 373, 380
Elkington, J.S.C., 294–5, 296, 298, 300, 375
Ella, S., 200
Ellery, Robert L.J., 1, 23, 25, 27, 28, 42, 50, 54, 147, 189–9, 153, 361
Elliott, J.H., 364, 373
Elliott brothers, 168
Ellis, A.J., 336
Embree, E.R., 208
Engel, Frederick, 199
environmental issues, 48, 49, 103, 104–5, 109–11, 114–16, 251–64, 273–4, 315–16, 318, 320
Erlich, Paul, 315
Esserman, N.A., 363
Esson, Louis, 2
Etheridge, Robert, 135, 137, 204, 205, 380
Evans, Lloyd, 364
Evans, W.P., 167
Everett, Arthur, 136
Everingham, Douglas, 321
evolutionary theories, 101–5, 198–9
Ewart, Professor, 106
Eyre, E.J., 198

Farr, C. Coleridge, 134–5, 375
Farrall, Lindsay, 3
Farrer, William, 106, 275
Faulding, Francis, 168
Fawsitt, Charles, 55
Fee, W.W., 374
Felton, Alfred, 168
Fenner, Frank J., 379, 380
Firth, Raymond, 209, 210, 212
Fisher, Andrew, 56, 227, 231
Fison, Lorimer, 199, 201, 203, 216
Fitzpatrick, Kathleen, 3
Fleming, Charles A., 364, 379
Fleming, Donald, 6
Fletcher, J.J., 32
Flynn, T. Thompson, 375
Forest, P.D., 276
Forrest, John, 41, 42
Fortune, Reo, 209
Fowler, H.L., 235
Frankel, Sir Otto, 320, 321

Fraser, Malcolm, 87, 88, 354
Frazer, Sir James, 199, 202, 207
Freeman, Christopher, 327, 345, 351, 352, 354, 355
Fry, H.K., 211

Galbraith, J.K., 74, 349
Galton, Sir Francis, 225, 228, 230, 232
Gardiner, R., 374
Garrett, Thomas, 32
geography, 8, 41, 45, 46
Geological Society of Australia, 141–2
geology, 8, 41, 42, 45, 47, 130–42, 284, 285–6
geophysics, 134–5
Gepp, H.W., 178
Gerlock, M., 381
Gershuny, J.L., 347
Gibbons, F.B., 150
Giblin Lectures, 381–2
Gibson, A.J., 108
Giffen, Robert, 46
Gill, E.D., 213
Gillen, F.J., 199, 201, 202, 203, 216
Gillespie, Roselyn, 226
Gipps, G. de V., 373
Glaessner, M.F., 138
Goldsworthy, A.W., 345
Goldthorpe, J.H., 347
Golson, Jack, 213, 214
Gordon, H.D., 375
Gorton, John, 81, 321
Gould, John, 100
Graham, Dr, 229
Grant, Sir Kerr, 173, 178, 363
Grant, P., 332
Grattan, C. Hartley, 67
Gray, George, 134
Graydon, N.A., 148
Greenwood, R.H., 281, 374
Gregory, A.C., 361
Gregory, J.W., 205
Greig, Janet, 295
Grey, George, 198
Griffith, C.S., 41
Griffith, Sir Samuel, 43, 48, 55
Grimwade, Sir Russell, 168
Grossbard, A., 374
Gruen, F., 382
Grutzner, John, 185
Guilfoyle, W.R., 251
Gunn, Mrs Aeneas, 202
Gurnett-Smith, A.F., 375
Gurney, T.T., 151
Guthrie, F.B., 171
Guy, J. Allan, 230

Haddon, A.C., 199, 202, 204, 207, 210
Hake, C.N., 171, 172, 175
Haldane, J.B.S., 310

Hale, George Ellery, 60
Hales, Archdeacon, 46
Hall, G. Stanley, 226
Hall, P., 281
Hall, Robert, 375
Hall, T.S., 374
Halley, J.J., 105
Hamilton, Sir Robert G.G., 46, 51, 361
Hamlet, W.M., 46, 49, 171, 173
Hanrahan, Lucy, 137
Harper, W.R., 203, 204
Harradence, Rita, 185
Hart, C.W.M., 209
Hartley, W., 375
Hartnung, E.J., 178, 181
Harvey, J.H., 136
Haswell, W.A., 42, 45, 102, 103, 104, 105, 109
Hawke, R.J., 354, 355
Hayter, Henry H., 42, 46
Healy, T.W., 186, 381
Hector, Sir James, 24, 25, 35, 41, 42, 44, 48, 50, 52, 54, 132, 134, 149, 361
Hedley, Charles, 54, 59
Heffron, R.J., 79
Heilbroner, Robert L., 346, 347
Heiser, V.G., 298
Herbert, D.A., 374
Heymann, Erich, 178, 381
Hill, Dorothy, 379, 380
Hill, James P., 103, 105, 379
Hillary, Sir Edmund, 364
Hills, E.S., 141
Hills, P.D., 345
Hinton, Colin Jack, 375
Hoare, Michael, 3, 23
Hodgkinson, Lorna, 233
Hogbin, Ian, 209
Hogg, E.G., 153
Holman, Mollie, 379
Holmes, Austin, 382
Holmes, J. Macdonald, 255, 256, 259
Holroyd, Arthur, 32
Home, Rod, 3
Hooker, Joseph, 100
Hoskins, Lucy, 137
Howard, Amos, 275, 276
Howchin, Walter, 50, 373, 380
Howitt, Alfred W., 54, 133, 139, 198–9, 201, 203, 216, 361, 379
Hudson, E.R., 278
Hughes, W.M., 58, 177, 298
Hunt, D.M., 336
Hunter, Thomas, 232
Hutton, Frederick W., 44, 45, 50, 53, 102, 105, 136, 361, 375
Huxley, Julian, 310
Huxley, T.H., 25, 30, 33, 47
hygiene, 42, 292–304 *passim*

Inglis, Alexander, 152
Inglis, Ken, 22
Inkster, Ian, 52
Institute of Physics, 156–8, 159, 163
Iredale, Thomas, 178
irrigation, 260–1, 275, 280, 281
Isaacs, J.E., 382

Jack, R.L., 42, 48, 133, 135
Jackman, Lloyd, 185
Jackson, W. Roy, 186
Jamison, Dr, 229
Jennings, J.N., 262
Jennings, Sir Patrick, 31
Jensen, H.I., 139
Jensen, R.B., 374
Johnson, L.A.S., 380
Johnston, R., 334
Johnston, R.M., 41, 131
Johnston, T. Harvey, 380
Jones, Barry, 91, 186, 322, 335, 340, 347, 353, 354, 355
Jones, F. Wood, 210
Joplin, Germaine, 139
Judge, P.J., 375
Julius, Sir George A., 61, 76, 178, 330

Kaberry, Phyllis, 209
Kahn, Herman, 347
Kakulas, B., 374
Karmel, P., 381
Keam, P., 373
Keating, Paul, 354
Kenner, James, 178
Keogh, E.V., 301
Kernot, W.C., 32, 33, 41, 47, 48, 54
Kerr, Clark, 347
Keynes, J.M., 343
Kidson, Edward, 155
Kirk, T., 258
Kirkland, John Booth, 171
Kirkland, John Drummond, 169
Kisokau, Karol, 376
Knibbs, Sir George H., 54, 63, 177, 362
Knox, E.W., 171
Knox, K.W., 363
Krefft, Gerard, 26

Laby, T.H., 61, 65, 152, 154, 155–6, 157, 343–4
Lamb, Horace, 150
La Meslée, E.M., 28
Lane-Poole, C.E., 257
Lang, Andrew, 202
Lang, John, 2
Langer, R., 375
Laurie, Henry, 224, 225, 232
Lavoisier, Antoine, 166
Lefroy, C.E.C., 203

Lehany, F.J., 363, 364
Leibius, Carl, 25, 27, 53
Leighton, A.E., 175, 181
Lewis, Essington, 181
Lindsay, Norman, 295
linguistics, 214, 215
Lions, Francis, 179
literature, 44, 47
Litton, Robert T., 32, 33
Liversidge, Archibald: and AAAS 8, 30–63 *passim*, 133, 134, 361, 373; and British AAS visit, 174; as chemist, 166, 169, 170, 171, 174, 176; early career, 22–30
Liversidge Lectures, 179, 262, 380–1
Lloyd, A., 382
Lodge, Oliver, 57
Longman, Heber, 380
Lord, Clive E., 375
Love, E.F.J., 150, 228, 374
Lovell, Henry Tasman, 224, 225, 232, 237
Low, A.R., 381
Lucas, A.H., 374
Lyle, Sir Thomas Ranken, 152, 154
Lyons, L.E., 179, 381
Lyster, F.J., 284

Macadam, John, 169
Macarthur, John, 273
McAulay, Alexander, 153
McAuley, James, 7
Macbeth, A.K., 178, 181–2
McCarthy, F.D., 213
McCarthy, Joseph, 312
McCaskill, Lance, 258
McCombie, H., 174
McCoy, Frederick, 102, 169
Macdonald, Stuart, 353
McEwen, John, 74
Macfarlane, W.V., 380
McIntyre, A.K., 380
Macintyre, Stuart, 5
Mackellar, C.K., 22, 32, 229
Mackerras, Ian, M., 380
Mackie, Alexander, 237
Maclaurin, J.S., 180
Maclaurin, –, 180
McMahon government, 87
Macmillan, J.R.A., 363, 373, 379
McQueen, Humphrey, 2
Madsen, John, 54, 154, 373
Maiden, Joseph Henry, 50, 54, 58–9, 60, 61, 63, 103–4, 112, 135, 172, 204, 254, 362, 373, 380
Maitland, A. Gibb, 374, 380
Malinowski, Bronislaw, 205, 207, 208, 212
Mander, L.N., 381

Mandeville, Tom, 353
Mann, E.A., 374
Manning, F.N., 227
Manning, N.C., 364
Manning, Sir William, 41
Marcuse, Herbert, 316
Marett, R.R., 204, 205
Marien, M., 348
Marshall, C.E., 286
Marshall, Patrick, 139, 363
Marston, Hedley, R., 313, 380
Martin, A.H., 236
Martin, A.P., 22
Martin, Charles J., 103
Martin, L.H., 160, 313
Martyn, D.F., 83–4, 155, 158
Marx, Karl, 199
Massey, H.S.W., 157
Masson, Sir David Orme: 1, 179; and AAAS, 44, 45, 54, 207, 362, 381; and ANRC, 61, 62, 177–8, 212; and British AAS visit, 56–7, 174, 204; as chemist, 137, 166, 169, 170, 171, 173, 174, 176, 177; war service, 58
Masson, Elsie, 205
Masson, Irvine, 175
Masuda, Y., 348
mathematics, 134, 147–63 *passim*
Mathew, John, 201, 204
Mathews, R.H., 204
Mawson, Sir Douglas, 58, 59, 61, 63, 135, 137, 363, 380
mechanics, 134, 147–8
Mellor, David P., 75–6, 179, 181, 184, 330, 381
Melville, Sir Leslie, 381
mental science, 222, 224
mental testing, 231–4
Menzies, Robert Gordon, 80–3, 84, 86, 260, 261, 313
Messel, Harry, 160
metallurgy, 172, 184
meteorology, 8, 41, 42, 47, 149, 152, 153, 155
Miller, E. Morris, 230, 232, 236
mineralogy, 8, 41, 43, 134, 172, 173
Mingaye, J.C.H., 171
mining, 41, 283–7
Mitchell, S.R., 206
Mitchell, Thomas, 130
Moir, R.J., 380
Monash, Sir John, 362
Moran, Jean, 78
Morgan, A., 364
Morgan, G.T., 174
Morgan, Lewis Henry, 198, 199
Morris, Edward, 44, 47
Morrison, William L., 88, 321
Morton, A., 104, 375

Morton, Mrs Alex, 46
Moseley, H.G.J., 57, 174
Moseley, H.N., 29
Mountford, C.P., 211
Moyal, Ann M., 3, 74, 88, 90, 364
Mueller, Baron Ferdinand von, 27, 43, 44, 48, 50, 117, 251–2, 361
Mueller Memorial Medal, 379–80
Mules, J.H.W., 279
Murphy, R.K., 169
Murray, Sir Hubert, 50, 203, 206, 217, 362
Murray, K.L., 33
Murray, Sir Keith, 82
Murray, R.A.F., 131
Muscio, Bernard, 226
Musso, L.A., 276
Myers, C.S., 202, 226

Nadel, George, 3
Nangle, J., 235
Nanson, E.J., 151
Nash, W.H., 33
National Health and Medical Research Council (NH&MRC), 299, 304
Needham, Joseph, 310
Nevile, J., 331
Newbery, James Cosmo, 169
Newbigin, W.J., 178
Newnham, I.E., 381
New Zealand Association of Scientists (NZAS), 319
Nicholls, K.D., 375
Norman, Sir Henry, 46
Nossal, Sir Gustav, 321, 364, 379
Nyholm, Ronald, 179, 185

O'Connor, D.J., 373
Oeser, Oscar A., 225, 239
Oldham, R.D., 136, 137
Oliphant, Marcus L.E., 10, 77, 81–2, 83, 85, 86, 160, 182–3, 314, 363, 379
Olver, N.H., 374
Onslow, Earl of, 51
Osborn, T.G.B., 109, 110
Oxnam, D.W., 374

Page, Earle, 207
Palmer, Vance, 2
Park, N., 382
Parker, T.J., 375
Parker, A.J., 381
Parker, H.T., 233–4, 236, 237
Parker, Mrs K. Langloh, 202
Parker, T.J., 102
Parkes, Sir Henry, 31, 32
Parkin, J., 381
Parnell, Thomas, 154
Pasco, Commander, 45
Passmore, Alan, 240

Paterson, 'Banjo', 2
Patterson, R.A., 261
Pawsey, J.L., 161
Pearson, J., 375
Penfold, A.R., 381
Pennycuick, S.W., 178
Perkins, J.A., 278
Perlmutter, Patrick, 186
Petterd, W.F., 134
Pettit, Roland, 185
Petty, Bruce, 317
pharmacy, 167, 172, 179, 181–2, 183–4
Phillips, Arthur, 2
Phillips, G.E., 233–4
Phillips, L.F., 381
philosophy, 224, 237, 238
physics, 40, 41, 49, 134, 147–63 *passim*; nuclear, 160, 312–15
Piddington, Ralph, 196, 209, 213
Pigot, E.J., 135
Pitt, F.R., 374
Pockley, –, 90
Pollock, James A., 41, 58, 152, 154
Pond, J.A., 41
Pope, William J., 57, 174
Popper, Karl, 347
Porteus, Stanley, 209, 233
post-industrial society, 345–9, 353–4
Potter, Charles, 284
Potts, Kevin, 185
Powdermaker, H., 209
Prendergast, Kathleen, 137
Prescott, J.A., 255, 380
Price, J.R., 184, 381
progressivism, 300
Pryor, L.D., 375, 380
psychology, 222–38
Pudney, R.L., 274
Pugwash movement, 314
Purnell, W.E., 374
Pye, H., 277

Quilty, P., 375
Quinan, K.B., 175

Radcliffe-Brown, A.R., 204, 205, 207, 208, 209, 210, 211, 215, 217
Rae, Ian, 186
Raggatt, H.G., 285–6, 287
Ratcliffe, Francis N., 110, 114–15, 256
Rattigan, G.A., 381
Ray, S.H., 202
Reddaway, W.B., 381
Rein, Wilhelm, 225
Reingold, Nathan, 6
Rennie, Edward H., 41, 43, 47, 167, 170, 171, 172, 174, 362, 373
research and development, 90–1, 331–3, 337, 339
Richards, H.C., 141

Richardson, A.E.V., 363
Richardson, L.R., 375
Rickards, R.W., 381
Rigg, Sir Theodore, 363, 381
Ringwood, A.E., 380
Rivers, W.H.R., 202, 204
Rivett, A.C. David: and AAAS, 174, 363, 374; and ANRC, 62; and British AAS visit, 57; as chemist, 175, 177, 380; and CSIR, 64, 76–7, 91, 109, 178, 330; and science, 58, 82–3, 311
Roberts, S.H., 65
Robertson, J. Steel, 42
Robertson, R.N., 86–7, 363, 379, 380
Robin, A.F., 105
Robinson, Gertrude, 174
Robinson, Robert, 174
Rockefeller Foundation, 206, 207–9, 210–11, 298
Roe, Michael, 3, 300
Rolleston, Charles, 22
Ross, A.D., 156–7, 158, 160, 374
Ross, I.G., 364
Roszak, Theodore, 316
Roth, W.E., 199, 201, 203, 216, 294
Rowley, Charles, 214
Roxas, Sixto K., 381
Royal Anthropological Society of Australia, 202
Royal Australian Chemical Institute (RACI), 63, 166–7, 176–7, 179, 182–3, 184, 185, 186
Royal Society of New Zealand, 24, 84
Russell, Bertrand, 314
Russell, H.C.: and AAAS, 27, 32, 34, 40–1, 48, 50, 52, 53, 55, 147, 148, 361; as astronomer, 1, 23, 25, 133, 147, 148–9, 153; on environment, 251, 252, 259
Russell, James E., 236–7
Russell, Roger, 321
Rutherford, Ernest, 57, 174, 178

sanitary science, 42, 45, 48, 49, 171, 292–5, 303–4
Sargeson, A.M., 381
Schedvin, Boris, 3, 331
Schonell, Sir Fred, 364
science: and Britain, 7–8, 19–20, 100–3, 105, 108, 225–6; 'colonial', 3–7, 20–3; co-operation in, 8, 23–35, 44–5; education, 55–6, 78–82, 155–6, 169–70, 213–14, 235–6; histories of, 2–3, 5, 11; policies, 73–5, 82–3, 86–93, 177, 310; problems 65; society, 1–2, 20–3, 51–3, 55, 65, 89, 308–23
Scientists Against Nuclear Arms (SANA), 322
Scott, Alan, 374
Scott, Ernest, 1, 5–6, 66, 363

Scragg, R.F.R., 376
Search, 10, 89–90, 142, 263, 281, 317, 322, 335, 345
Searle, G.R., 230, 300
Seligman, C.G., 202
Selinger, B., 281
Serle, Geoffrey, 3, 6
Service, James, 29
Shand, John, 150
Sharp, R.L., 209
Shaw, Mansergh, 374
Shelton, E.M., 48
Shirley, John, 374
Shorland, F.B., 381
Sidgwick, N.V., 178
Skeats, E.W., 380
Skey, William, 180
Skillen, Elizabeth, 233
Skinner, H.D., 196, 197, 203, 207
Slayter, R.O., 364
Slowey, A.I., 172
Smith, Alfred Mica, 1, 171
Smith, Bernard, 3
Smith, F.B., 1
Smith, Sir Grafton Elliott, 204, 205, 206, 207–8, 212
Smith, H.G., 104, 172
Smith, John, 22, 29, 150, 168, 169
Smith, P.W., 375
Smith, Robert, 168
Smith, S. Percy, 41
Smith, W. Ramsay, 203, 204
Smyth, John, 232, 233
Smyth, R. Brough, 132, 198
Snape, R.H., 382
Society for Social Responsibility in Science, 317–18
sociology, 224, 239–40
Soete, Luc, 352
Sollas, W.J., 57
Solomon, D.H., 381
Sorrenson, M.P.K., 198
Southcott, R., 373
Specht, Ray, 116
Spencer, Sir W. Baldwin: AAAS officer, 206, 362, 374; and ANRC, 207; anthropological work 199, 201, 202, 203, 204, 205; biologist, 102–3, 104, 105; and British AAS visit, 204
Sprent, J.F., 380
Springthorpe, J.W., 294, 296–7, 300
Stanley, G.A.V., 376
Stanley, N.F., 374
Stanner, W.E.H., 209
Stanner, W.E.H., 380
Steele, Bertram D., 170, 173, 174, 175
Stefansson, Vilhjamur, 254
Stephen, Sir Alfred, 22
Stephen, H.W.H., 24
Stephens, W.J., 28, 29, 32, 102, 104

Sticht, Robert, 171, 284
Still, J.L., 363
Stirling, E.C., 41, 54, 203
Stokes, R.H., 381
Stokes, S.J., 256
Stoneman, Ethel, 233, 236
Stout, Alan, 240
Stout, G.F., 225
Strahan, Ronald, 364, 373
Stranks, D.R., 179, 381
Strehlow, Carl, 201, 209
Stretton, H., 328, 340
Strickland, Sir Edward, 33, 34, 42, 361
St-Simon, Henri de, 346
Stuart, T.P. Anderson, 42
Sully, James, 226
Sussmilch, C.A., 137
Sutherland, John, 284
Sutherland, K.L., 364, 381
Sutherland, William, 44, 148, 151, 173, 224
Sutton, Harvey, 295, 296, 297, 300
Suttor, F.B., 229
Swan, John M., 186, 381
Swan, T.W., 381
Sweet, Georgina, 59, 374

Taplin, G., 198
Tasman, Abel, 197
Tate, Frank, 223, 237
Tate, Ralph, 44, 49, 50, 53, 102, 104, 105, 131, 136–7, 361
Taylor, A.J., 134
Taylor, Clara M., 172
Taylor, Norman H., 380
Taylor, T. Griffith, 63, 253–5, 260, 264
Teakle, L.J., 256
technology, 89, 321–2, 326–55
Temple, Diana, 364
Thomas, A.P.W., 41, 102, 105, 375
Thompson, J.P., 33
Thomson, D., 209
Thomson, A.M., 22, 168–9
Thomson, George M., 375
Thomson, J. Ashburton, 294, 303
Thomson, J.M., 374
Thorne, P.G., 374
Thornton, H., 262
Thornton, J.B., 363
Threlfall, Richard, 29, 41, 43, 58, 147, 150, 151, 152
Tillyard, R.J., 380
Tindale, N.B., 213
Titchener, G.B., 225, 232
Titterton, E.W., 313
Tizard, Henry, 57
Todd, Alexander, 183, 184
Todd, Sir Charles, 23, 25, 47, 50, 148–9, 153
Todd, Jan, 52

Trikojus, Victor, 179
Trouton, Frederick, 57
Tryon, Henry, 33, 105
Tylor, E.B., 199, 201

Underwood, Eric J., 364, 379
Underwood, Graeme, 185
unemployment, 343–5, 349–53

Veblen, Thorsten, 327
Verney, Noel V., 374
veterinary science, 277, 279, 280–1
Vincent, J.M., 373
vocational guidance, 234–6
Vonwiller, Oscar U., 311, 344

Waddell, S.T., 374
Wadham, Sir Samuel, 282, 363
Walby, B.J., 364
Walker, David, 2
Walker, Sir Ronald, 381
Walkom, A.B., 54, 362, 363, 373, 379
Wallerstein, Immanuel, 353
Walsh, Alan, 184
Ward, L. Keigh, 138, 373
Ward, R.G., 364
Ward, Russel, 2
Waring, J., 380
Wark, Sir Ian W., 83, 178, 181, 182, 379
Warner, W.L., 209
Warren, S., 375
wars and science, 57–8, 60, 75–6, 111–12, 154, 158–9, 174–6, 181–2, 212, 237–8, 312, 329–31
Waterhouse, D.F., 380
Waterhouse, G.A., 362
Watson, Sir Bruce, 364
Webb, Beatrice, 55
Webb, H.R., 375
Webb, L.J., 380
Webb, Martyn J., 374
Weber, Max, 346
Webster, H.C., 84
Wedgwood, Camilla, 209
Wells, H.G., 66, 346
Wentworth, W.C., 312
Westermarck, Edward, 202
Wheelwright, Ted, 354
White, C.T., 374, 380
White, D.E., 374
White, Sir Frederick, 86, 363, 364, 379
White, Isobel, 202
White, M.J.D., 380
White, Richard, 230
Whitlam, E.G., 87, 354
Whitton, W.I., 381
Wild, J.J., 196, 199, 200, 217
Wild, J.P., 379
Wilken, A., 202
Wilkinson, C.S., 25, 32

Willett, R.W., 375
Williams, F.L., 217
Williams, Gordon J., 375
Wilsmore, Norman T.M., 170, 171, 173, 175, 180, 374, 380
Wilson, A.T., 381
Wilson, James T., 42, 103, 105, 109
Wilson, Sir Ronald, 382
Wise, Bernard, 43
Wise, T., 211
Wiskich, J.T., 373
Wissler, Clark, 208
Womersley, H.B.S., 380
Wood, J.G., 109, 110
Woodall, R., 380
Wood-Jones, F., 380
Woolley, Richard, 153, 160, 363
Woolnough, W.G., 141
Wootten, J.H., 364
Worley, F.P., 178
Worsnop, T., 198
Wran, Neville, 92
Wright, F., 373
Wright, J.F.H., 374
Wyndham, H.S., 236

zoology, 41, 103, 113, 114–16, 262